Introduction to BioMEMS

Introduction to BioMEMS

Albert Folch

CRC Press
Taylor & Francis Group
Boca Raton London New York

CRC Press is an imprint of the
Taylor & Francis Group, an **informa** business

CRC Press
Taylor & Francis Group
6000 Broken Sound Parkway NW, Suite 300
Boca Raton, FL 33487-2742

First issued in paperback 2019

ISBN-13: 978-1-4398-1839-8 (hbk)
ISBN-13: 978-0-367-86496-5 (pbk)

Library of Congress Cataloging-in-Publication Data

Folch, Albert.
  Introduction to bioMEMS / Albert Folch.
    p. cm.
  Includes bibliographical references and index.
  ISBN 978-1-4398-1839-8 (hardback)
    1. Nanotechnology. 2. Biotechnology. 3. BioMEMS. I. Title.

TP248.25.N35F65 2012
660.6--dc23   2012010233

Visit the Taylor & Francis Web site at
http://www.taylorandfrancis.com

and the CRC Press Web site at
http://www.crcpress.com

This book is dedicated to Lisa Horowitz, wife, collaborator, critic, and, above all, friend, without whom this book could have never been conceived.

Contents

List of Figures

Preface

This textbook is based on my experience of teaching a BioMEMS course at the University of Washington's Bioengineering Department (Seattle) for 11 consecutive years. Its strongest virtue, I think, is that it closely follows my syllabus: the PowerPoint slides that are available on the book Web site for instructors' use follow the same sequence as the figures of the textbook. I hope that this one-to-one correspondence between textbook and lectures will help instructors and students. My students have used it for the last two offerings in draft form and have given me very valuable feedback. Due to its length, the book is more adequate for a semester format; for quarter-length courses, the instructor should decide what sections should be reduced. However, the inquisitive student will find plenty of additional material with which to expand his or her mind.

The first task in writing this book was the most difficult one: which works should be featured among the vast body of BioMEMS literature? I constantly reminded myself that I was writing a textbook, not a detailed review article. I wanted to give a sense of the state-of-the-art of the field and, importantly, how we got here. *This is for students*. To attempt this task, I imagined that BioMEMS is like a forest that the students wanted to visit for the first time, so they needed me as a guide. The first and the tallest trees of the forest are very easy to find—to find those, they would not need me as a guide. But the forest has a lot of undergrowth, and not all the nice trees are easy to find—for that they do need me (the textbook or an expert instructor). Most of these trees (a.k.a. the dozens of designs of microvalves and gradient generators) were nameless, yet they are very memorable, and we want our students to remember them (don't we?), so it was necessary to give them names—an arbitrary exercise that had to be done sometime by someone. My apologies if they are not to your liking, but I'm hoping they will be useful for the students to move on in our field.

The monumental task of writing the textbook was a distraction from my research but a necessary one to recapitulate what is noise and what is signal in the field. What should we leave behind and what should we take with us from now on as we educate the new generation of engineers and biologists? This task required quite a bit of introspection and humility, as I realized that my own lab was "guilty"—just as most other leading BioMEMS labs—of producing a lot of results that do not stand the test of time; surely, that is what research is about. This is a process in which I have learned a lot myself. I have tried to convey my immense respect for those who paved the way for others: there is much to learn from the mental process that generates an enlightening spark, the novel idea. For reasons of space, I have focused on the work that has had the highest impact and shown the most creativity, encompassing all the length scales of life—from biomolecules to cells, tissues, and organisms.

This book introduces the nonspecialist reader, in particular, students, to a set of problems in biology and medicine that benefit from—and ideally *require*—the miniaturization of a certain device. By "miniaturization," I simply mean the process of making *small things*, including both reduction in size of things that exist on a larger scale and the fabrication of small things whose big counterparts simply do not exist. Small with respect to what? Here, I arbitrarily define "small" as the size of something that is small with respect to humans, in other words, that which cannot be easily handled with our fingers and whose details cannot be easily seen with the naked eye—anything smaller than about a millimeter. Other definitions of big and small are certainly possible. General microfabrication techniques are covered in Chapter 1 (with their applications to bio-patterning covered in Chapter 2); microfabrication techniques that are specific to microfluidics are covered in a section of Chapter 3, the chapter devoted to Microfluidics. I could have condensed all the microfabrication in one chapter, but in my teaching experience, it has worked well to go back (in teaching Chapter 3) to the concepts of Chapter 1.

WHAT IS BioMEMS?

Everyone understands the "Bio" part of the word—it's the "MEMS" that makes people frown. Historically, MEMS stands for *micro-electro-mechanical systems*, the first of which were built in the 1980s by etching silicon and depositing thin films, as an offspring of microelectronic devices built with silicon-processing technology. Because imagination is not bound by materials nor by acronyms, other MEMS were devised in many other materials with functions that were neither "electro" nor "mechanical"—some were optical, some fluidic, some purely biological, some with nanofabricated components, and some were hybrids of all the above. Nowadays, the MEMS acronym is somewhat too restrictive, but most people are OK with it because there is nothing better.

Because the aim of this book is to embrace a wide body of knowledge, here we use a wider definition of BioMEMS as *any* biomedical device that is fabricated (at least in part) using *any* miniaturization technique. By derivation, BioMEMS is also the engineering field that uses or develops such BioMEMS devices.

As you will see, there are numerous biomedical applications that benefit from "going small." Therefore, the book is very wide in scientific scope: I cover devices that record the electrical activity of brain cells, some that measure the diffusion of molecules in small fluidic channels, and others that allow for probing the expression of genes in large numbers—among many, many others. For the teacher or writer, the main challenge in reviewing this colorful field is that the student or reader cannot be expected to know about the brain's structure *and* thermodynamics *and* genetics and so on. As this textbook is geared primarily toward senior Bioengineering undergraduates, the general assumption is that students are familiar with basic biology concepts, calculus, and organic chemistry.

More importantly, I am aware that I cannot satisfy both the engineering instructor who seeks rigorous quantitative treatment of every subject and the biology student who does not want to see a single formula—I had to find a middle ground. As a compromise, the textbook contains only the most essential formulas, and the majority of the exercises that I propose have no calculations. I made a strong effort to reach out to biology-oriented students who otherwise might never enter the field. Indeed, the majority of the text is written somewhat informally, in the style of biology textbooks and also in the spirit of a classroom discussion, avoiding the electrical engineering or physics format replete with definitions—I wanted this textbook to help ignite the kind of classroom discussions that I have had with my students over the years. However, to avoid getting the student lost by the informal style in places where rigor is paramount, I highlighted in bold the fundamental concepts; those are then defined with precision in a glossary of terms as an appendix that almost grew to become a BioMEMS dictionary—please do use it to expand the accuracy of your learning, if those concepts are new to you.

It has been impossible to fit everything in the book. There are several areas that I believe are extremely important in BioMEMS but should be dealt with once the student is familiar with the basics. Given the increasing importance that microfluidic technology is acquiring in the BioMEMS arena, I had to constantly keep in mind that this was not a microfluidics textbook—there are already a number of great microfluidics textbooks for which this one is no substitute. The formulas for capillary filling of a microchannel or the Laplace equation relating the pressure drop across a curved interface has a great predictive power in microfluidics but had to be left out—the whole book has to be taught in one semester! I'm afraid I've committed what amounts to engineering heresy by not showing the derivation of the Navier–Stokes equation, the fundamental pillar of fluid mechanics, from Newton's second law; I only show a special case of the Navier–Stokes equation that is much more mathematically tractable than the general case and that happens to be the most common situation encountered in BioMEMS—that of incompressible, isotropic, viscous, and isothermal fluids. Another large "hole" is the vast body of nanopatterning techniques, still under intense investigation, that will have a great impact on biomolecule detection. Likewise, the family of surface imaging techniques dubbed *scanning probe microscopies* that now use microfabricated sensors (AFM, STM, etc.) is only marginally covered here, but the technology (to which I devoted my whole Ph.D. thesis!) does deserve a more honorable space in a definitive BioMEMS textbook. Luckily for the student, the subject has been well covered in many textbooks already. Similarly, BioMEMS will always walk hand in hand with the fast-moving field of cellular and molecular imaging, which I cannot even attempt to start covering. I purposefully left out of the book the revolutionary work of Aydogan Ozcan, the brilliant inventor of a new miniature lensless camera, because it did not fit in any of my chapters. I wish I had had space to cover elastomeric optics and an incredibly colorful variety of optical and molecular detection techniques that make BioMEMS a very rich field. After much internal struggle, I decided to delete a beautiful section on fabric microfluidics that reviewed a fascinating new field: the fabric devices, which have features on the millimeter scale, don't quite make the miniaturization cutoff that is required to enter the above "BioMEMS" definition. Also, my opinion (which could be wrong!) is that fabric will never outcompete paper in terms of manufacturing cost and versatility when it comes to deploying molecular assays. For the sake of the students, I had to stop *somewhere*.

The greatest lasting value of this book probably rests in the quality of the figures, most of which are original figures (even if they refer to older work). Ever since I read the marvelous photography book *On the Surface of Things,* by George Whitesides and Felice Frankel, I became aware of the collective artistic value of the field. Several hundred researchers contributed more than 400 original figures for this book, and I shall always treasure their thank you e-mails. (You may be one of them!) It really has taken the BioMEMS village to raise this child. I have been approached by publishers about writing a BioMEMS textbook ever since I posted my BioMEMS class notes on the Web, but for the last seven years, none had accepted my full-color proposal, arguing it was too expensive. I come from a family of publishers and know all too well that it's a matter of spreading the costs over enough books, so I waited (and in the meantime, I wrote) and waited. . . . If the field and the market were not ready for a nice textbook, then I did not want to author it. But in 2009 came Luna Han, a young, dynamic editor from CRC Press, who embraced my request of "color on every page" with a refreshing enthusiasm.

There was one problem though. Most figures are copyrighted, obtaining copyright requests can be lengthy (I don't have a secretary), and many journals charge a hefty fee for reprinting. The nascent *Lab on a Chip* journal welcomed my proposal with a big cheer and allowed me to bypass all the copyrights requests—which was a huge time saver. I soon realized that most people are delighted to be featured in a textbook, so I decided instead to invest my time in e-mailing each author requesting an original figure. This process was very rewarding: the average response time was less than 24 hours(!), usually accompanied by a cheerful encouragement to go on with the project. My deep thanks to all.

The book had two important participants at the University of Washington. First, Professor Paul Yager wrote about one-third of Chapter 4, which deals with the application of microfluidics for analyzing biomolecules, a subject in which he is a world leader and I'm not. Without his jump start, I would have started in the dark. Second, the section that deals with the formulas for flow in rectangular channels (from flow speed to shear stress and forces) in Chapter 3 was supervised by Professor Emeritus of Chemical Engineering Bruce Finlayson, a world expert in Fluid Mechanics (author of *Nonlinear Analysis in Chemical Engineering* and *Introduction to Chemical Engineering Computing*), who always corrected my mistakes with professorial patience and a wonderful smile. He used the finite-element simulation software *Comsol* to nonchalantly verify that the legendary book by Frank M. White (*Fluid Mechanics*) contains an incorrect constant 1/6 in front of the flow rate formula; it should be 4/3 as used in this textbook. (Several other literature sources agreed with 4/3; note that 4/3 and 1/6 differ by a factor of $8 = 2^3$, which can be explained by an error in the definition of the microchannel dimensions, that is, using $2h$ instead of h, because flow rate scales with h^3.)

Finally, before anyone (or everyone) starts screaming at the lack of references, remember that *this is not a review article*; it's a *textbook*. You won't find exhaustive reference lists in this book—only the references that are essential to understand and enter the field. In an age when our students' first impulse is to *Google* (my students check *Wikipedia* for more information *while* I'm lecturing in the classroom!), it seemed futile to waste efforts referencing thousands of articles. My criterion has been that if a reference was important enough to be mentioned in the book, it also deserved a figure as an illustration (with a handful of exceptions due to copyright problems or simply lack of space). For the reader's and the student's convenience, I have placed the full references to the figures inside the figure captions—an innovative procedure that I hope becomes the norm—and spelled the reference out in full, including the title and *all* the authors: deleting authors to save paper seems one of the worse deals that we scientists have ever accepted from publishers. To help the student expand the study of the subject, I have also placed a brief list of key references at the end of every chapter. To encourage classroom participation and help students with self-study (e.g., homework assignments), I have added, in an appendix at the end of the book, a collection of exercises, design challenges, and common misconceptions compiled from past midterms and assignments. Several of my own research group members (Anthony Au, Dr. Nirveek Bhattacharjee, Tim Chang, Dr. Samira Moorjani, Chris Sip, and Dr. Adina Scott) volunteered time to compile exercises and literature references that added to the educational value of the textbook.

I have learned all the biology I know in part by reading textbooks but mostly by interacting very closely with the same person every day, so I think she should be acknowledged here. My last acknowledgment therefore goes to my wife, Lisa Horowitz, who can work as a biology postdoc, raise our two kids, cook dinner every night, and not go crazy when I say that I have an early appointment so she has to take both kids to school. If one day I write an autobiography, I will repeat what all our friends already know: so often do I come home saying I got an awesome idea (at least one night a week!) that she invariably rolls her eyes—but, amazingly, she listens. The last 11 years of my career can be summarized as follows: Once a year (at most), one of these crazy ideas makes it past her filter. She helps me refine it, and then I talk to other biology colleagues, who invariably like it. One of them is so impressed that (s)he becomes a participant in a NIH grant. All the NIH-funded projects in my lab have been ideas approved by her, and she helped in the grant writing at some level or another. The overall tone of the book reflects the passion for biology that she inspires in me every day. (If it were for me alone, I'd be a gadget guy who would never have thought of writing this book.) She's worth millions.

I would like to think that this textbook represents the established body of knowledge in the BioMEMS field, even if it only reflects *my* knowledge of it. If that is not sufficient, I am afraid I don't know any better.

Author

Albert Folch was born in Barcelona, Spain, in 1966. He received his B.Sc. degree in physics from the University of Barcelona (U.B.) in 1989. In 1994, he received his Ph.D. degree in surface science and nanotechnology from the U.B.'s Physics Department under Dr. Javier Tejada's supervision. While completing his Ph.D., he was also a visiting scientist (1990–1991) at the Lawrence Berkeley Laboratory (Berkeley, CA) working on atomic force microscopy under Dr. Miquel Salmeron. From 1994 to 1996, he was a postdoc at M.I.T. developing microdevices under the advice of Martin A. Schmidt (EECS Department) and Mark S. Wrighton (Chemistry). In 1997, he joined the laboratory of Dr. Mehmet Toner as a postdoc at Harvard University's Center for Engineering in Medicine to work on BioMEMS and tissue engineering. In 2000, he joined the University of Washington's Department of Bioengineering (Seattle), where he is presently an associate professor. His laboratory works at the interface between cell biology and microfluidics. In 2001, he received a National Science Foundation CAREER Award, and he now serves on the Advisory Board of *Lab on a Chip*, the leading BioMEMS journal. The Folch lab produces microscopy images of microchannels and cells that it then uses to run an acclaimed outreach program called BAIT (short for "Bringing Art Into Technology"), which funds art exhibits.

How Do We Make Small Things?

MICROFABRICATION TECHNOLOGY IS A BROAD TERM that encompasses literally hundreds of techniques that can be fuzzily classified into dozens of overlapping classes. Of these, three classes have been used extensively in BioMEMS: **photolithography**, **micromachining**, and **soft lithography**. We also cover **micromolding** as a mass-fabrication strategy. Before discussing the fundamentals of these four classes, it is important to step back and ponder briefly on what ways microfabrication is beneficial in addressing biomedical problems. Here, we cover the fundamental concepts of these techniques, as well as their advantages and limitations, as an introduction to the next chapter in which we address micropatterning and microstructuring of biologically interesting material in detail.

1.1 Why Bother Making Things Small?

Interestingly and despite the enormous differences between microelectronics and biomedicine, the forces that have driven the trend toward miniaturization are similar in both fields, as illustrated in Figure 1.1. Microfabrication techniques are beneficial to biomedical applications mostly for three reasons.

1.1.1 The Small-Scale Benefit

First, miniaturization allows for addressing a small *unit*, whose unique function is inherently determined by its size. In microelectronics, an example of such a unit is a transistor; the switching performance of the transistor is enhanced by its smallness. For micromechanical parts, such as the accelerometer inside an airbag deployer, or a fluidic valve, the smaller they are in size, the faster they react, the less energy they require to operate, and the less sensitive they are to ambient vibrations. In biology, an example of a small unit is a cell. Cells have evolved to be exquisitely sensitive to their surroundings, so techniques that allow for the spatial and temporal modulation or probing of the cellular microenvironment are of great interest to basic cell biology studies.

Another example of a small unit in biological research is the tiny volume of fluid inside a pipette. The cost of producing certain biomolecules has driven a trend toward using smaller and smaller fluid volumes, so microfluidic devices offer an instant cost-savings benefit. However, cost is not everything. Fluids behave in nonintuitive ways when confined in small cavities; in addition, reactions in small volumes are more efficient because the molecules cannot diffuse away. Thus, clever devices exploit these microscale phenomena to detect biomolecules or sort cells using less than a nanoliter of solution. We shall refer to the benefit derived from using small scales as the **small-scale benefit**.

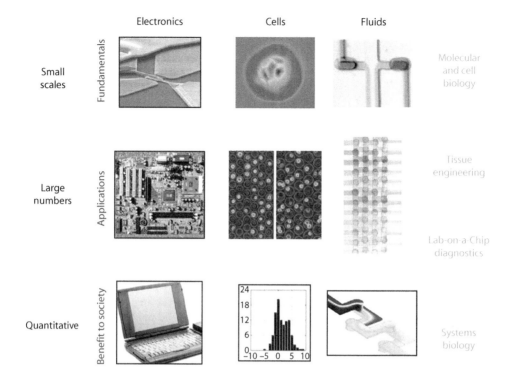

FIGURE 1.1 Benefits of microfabrication.

1.1.2 The High-Throughput Benefit

Second, most miniaturization techniques coincidentally happen to allow for fabricating small units in large numbers, or "batches," for almost the same price as one unit, which lowers the cost per unit. Independently of fabrication costs, large ensembles of small units can often be operated or visualized simultaneously, which lowers the cost of operation. For example, a single transistor cannot be used to calculate 2 + 2, but integrated circuits featuring large arrays of transistors and other microfabricated components can feature a computational power so astounding that it will beat a chess grand master. Biologists have used methods to address small numbers of single cells—such as micropipettes, microscopes, and lasers—for decades. Microfabrication is providing high-throughput approaches to probe and image single cells in large numbers, delivering rich statistics on the variability of cell behavior and gene expression; the ability to control the position of cells with respect to other cells enables cell–cell communication studies that may shed light on how the complex behavior of ensembles of cells emerges from the behavior of each cell. Biochemical assays that had to be run on test tubes one at a time can now be run in parallel in small fluidic channels because fabricating one channel is just as easy as fabricating a thousand. In some cases, automation is almost free in microsystems. We shall refer to the benefit derived from batch fabrication or parallel operation as the **high-throughput benefit**.

1.1.3 The Quantitative Benefit

Last but not least, miniaturized systems are very amenable to quantitative design, in the sense that their performance can be predicted quantitatively even before fabrication starts. Fluids in microchannels flow in a very predictable way, a deterministic behavior that is very different from the turbulence omnipresent in our bathroom's pipes—this quantitative knowledge can be used for a measurement of diffusion in a binding assay, for example. Because of their size, microsystems can be homogeneously treated with chemicals, cleaned, and then coated with thin films of well-characterized thicknesses; inherent in the fabrication process is a detailed physicochemical knowledge of the materials that form each part of the device, which is essential in tailoring biochemical coatings for cell attachment, implants, and enzymatic reactions in a biosensor—very unlike the low-reliability instructions on sanding and priming that one needs to follow to paint the exterior of a house with a durable coating. As a corollary of allowing for quantitative design, microsystems yield quantitative measurements, and this is not only because of the large numbers involved. For instance, in a traditional cell culture, the spacing between cells and the average cell density cannot be controlled independently—we say that the variables are *confounded*—yet they are both important factors in cell behavior; on the other hand, in a micropatterned cell culture in which the cells can be positioned at well-defined spacings, the two variables can be varied independently—we say that they are *decoupled*. We shall refer to the benefit derived from quantitative design and rich-statistics output as the **quantitative benefit**.

What is the result of such benefits? As depicted in Figure 1.1, the small-scale benefit has been most important in *building the fundamentals*, such as increasing the switching speed of transistors and understanding the basic properties of cells and fluids. The high-throughput benefit, on the other hand, has been mainly important for *building applications*, such as integrated circuits, engineered tissue scaffolds, and lab-on-a-chip biosensing systems. These benefits are by no means exclusive: all three benefits added together promise to provide fundamental insights into the variability of single-cell behavior—a paramount goal of systems biology, which aims to yield a global understanding of cell physiology and disease. The high-throughput benefit works together with the quantitative benefit to lower the *cost of fabrication*, and large numbers and small-scale benefits cooperate to reduce the *cost of operation* (e.g., use of small amounts of reagents in many channels in parallel). In addition, cost reduction is a key factor in enabling the dissemination of BioMEMS—e.g., commercializable point-of-care diagnostic microdevices—for the ultimate benefit of society.

It is important to recognize that this analogy between BioMEMS and microelectronics should not be abused. Microelectronic devices deal only with electrons, fundamentally immutable entities, and their economy of scale has thus far obeyed the famous Moore's Law (according to which the number of transistors per unit area increases exponentially, almost doubling every year). In contrast, BioMEMS devices struggle to deliver, analyze, and measure a virtually infinite variety of biomolecules whose function often degrades with time. There is no transistor for BioMEMS, and most likely, there will be no Moore's law for BioMEMS. If a technology choice is based on the previous analogy, it is most likely not founded on solid scientific principles.

1.2 From Art to Chips

Modern microfabrication technology is based on **photolithography**, the patterning of a layer of photosensitive polymer (**photoresist**) by light. The use of photosensitive materials for patterning is not new. A myriad of photographic methods have been used in art-making, reproduction, and printing since the nineteenth century, and they have all been derived from the discovery of light-sensitive substances. Even before photography, in 1827, Nicéphore Niepce coated a copper plate with bitumen (a light-sensitive natural tar), exposed it to sunlight through a waxed paper containing an ink drawing, dissolved the unexposed ink-shadowed areas with benzene, and transferred the pattern to the underlying copper plate using an acid etch. Later known as "photo-engraving," this process soon incorporated photographic film as exposure masks and more convenient, water-soluble substances, such as gelatin or egg whites, which were photosensitized by the addition of potassium dichromate.

WHAT DOES IT TAKE?

DESPITE THE UBIQUITOUS STATEMENT found in BioMEMS papers—"our technique is straightforward and inexpensive"—making small things is not easy, and it is costly. One can become an expert at using the techniques, but that does not make them cheaper. The user fee in a fully staffed clean room of a major research university in the United States is on the order of $100/h per student, although many university clean rooms have caps around $500 to $1000/month per student. The typical time for training a starting graduate student in standard photolithography is several weeks. It involves taking safety training and scheduling a couple of sessions per piece of equipment (one for learning basic operation procedures, the other for the student to demonstrate self-sufficiency) with professional trainers. Minimal training includes the spin coater, the mask aligner, and the developing station. Counting indirect costs, tuition, and benefits, a graduate student costs at least $3000/month in the U.S. So, the total bill for training can average several thousands of dollars. After training, the student is ready to start his or her own microfabrication process, which requires adjusting spin coating parameters, photoresist type, length of exposure, and developing or baking times, depending on the size of the features. This process may easily take another month or two of full dedication to the project. Clearly, microfabrication is not something to pursue unless you *absolutely* have to.

Incidentally, the early printed circuits were etched on copper foil that was masked by photo-patterned gelatin. Photoengraving methods have been further sophisticated by the development of dry photoresist films, which can be rolled onto thin metal plates for exposure and chemical etching. Referred to today by the industry as "photochemical machining" or "photoetching" (misleading names because the etching reaction is not a photochemical one), it is presently used for making stencil masks (i.e., plates containing through-holes), molds (e.g., for injection-molding), and calibration grids (e.g., for microscopy), among other utensils, at low cost.

1.3 Photolithography

1.3.1 Basics: Photoresist and Photomask

Photolithography is the starting point of most microfabrication technologies. In general, as schematized in **Figure 1.2**, it involves the selective exposure to light of a thin coating of **photoresist**. One of the most commonly used types of photoresists contains diazonaphthoquinone (DNQ) as the photosensitive component (absorbing light between 300 and 450 nm), which is mixed with Novolac resin (a phenolic polymer). Ultraviolet (UV) light is shone through a **photomask**

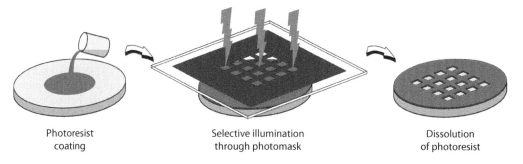

| Photoresist coating | Selective illumination through photomask | Dissolution of photoresist |

FIGURE 1.2 Photolithography.

containing the desired pattern in the form of opaque features on a transparent support. The photoresist is usually spun on a flat substrate from solution to a thin film and dried before exposure. The exposure chemically alters the photoresist by modifying its solubility in a certain **developer** solution. (The developer is specific to the photoresist; in the case of DNQ-Novolac–based photoresists, for example, the exposed areas can be dissolved in a basic solution of tetramethylammonium hydroxide whereas dissolution of the unexposed areas is inhibited by the presence of unexposed DNQ.) By convention, a **positive photoresist** is one that becomes soluble when it is exposed to light (as in **Figure 1.2**) and a **negative photoresist** is one that becomes insoluble.

Unfortunately, because the smallest dust particle distorts the spreading of the photoresist dramatically during spinning, photolithography must be carried out in clean room facilities, which require an expertise uncommon in biological laboratories and are costly to build and maintain. Although there are a number of photolithographic techniques that help in reducing the cost of processing, particularly when the resolution requirements are less stringent, in general, clean room–based photolithography remains inherently expensive. Despite its shortcomings, photolithography is widely used beyond microelectronics processing because it is extremely reliable. It has been optimized to produce high yields over large areas; the fabrication of a typical desktop computer's microprocessor, for example, consists of more than twenty photolithographic steps, so it had better be reliable! It should be noted that photolithography is a *subtractive* patterning process; that is, patterning is carried out by selectively removing material (photoresist in this case).

1.3.2 Black or White versus Gray Scale

Take a look at **Figure 1.2** and you will notice that *all* three steps consist of a modification of the substrate that is homogeneous from edge to edge of the wafer: (1) the photoresist thickness is uniform, (2) the light dose is the same on all the exposed areas (i.e., the photomask has only two levels of opacity: fully transparent or fully opaque), and (3) the whole surface is homogeneously exposed to the developer solution. As a result, the final features are of a single height—it is a "black-and-white" process.

In many cases, features of various (and variable) heights are needed. Ideally, one would like a process in which feature height is not encoded by the spinning speed (which is homogeneous across the wafer), but that could be varied across the wafer, in other words, that could be specified in the photomask in the form of grayscale variations. In practice, it has been hard to develop such translucent materials. It is, however, possible to make three-dimensional (3-D) features by patterning various photoresists (each with its own developer chemistry) in multiple steps (**Figure 1.3**). Suspended bridges with various lengths and suspended plates with micropits and microvilli are successfully formed.

There are a number of techniques that are capable of "**grayscale photolithography**," that is, of patterning multiheight features in one step, but they are generally more specialized, more expensive, or restricted in terms of the patterns that can be produced. A recently developed digital projector formed from an array of micromirrors—used in some movie theaters—promises to become the standard for photolithographic patterning because the fabrication of a photomask is no longer necessary, and is inherently gray scale (see Section 1.3.6.1). As shown in Section 1.6.6.2, a low-cost approach to grayscale photolithography uses fluids as a feature of the photomask (which is itself a microfluidic device).

1.3.3 Resolution

In microfabrication, we refer to the "resolution" of any part of a process as the smallest linewidth that can be achieved—it depends on the materials being patterned, on the substrate on which they are being patterned, on the photomask, on the photoresist and equipment used, and on environmental parameters such as temperature and humidity. Unless special tricks are used, the resolution of a process is always limited by the lowest resolution of each of its subprocesses.

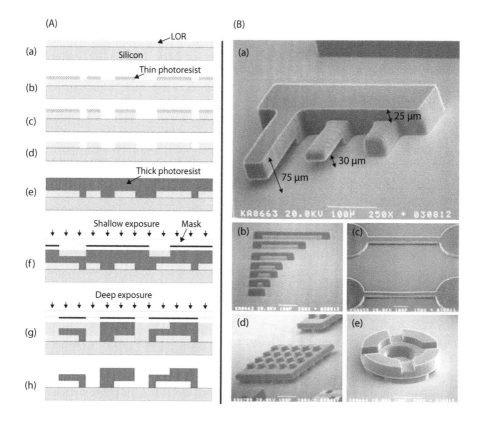

FIGURE 1.3 3-D photoresist structures. (A) Fabrication process for 3-D positive photoresist structures. (a) Spin coating of LOR (a positive photoresist) on Si. (b) Photolithography of thin photoresist using AZ 340 developer. (c) Development of unmasked LOR in AZ 400K developer. (d) UV exposure and dissolution of thin photoresist in AZ 340. (e) Spin coating of thick photoresist. (f) Shallow exposure. (g) Deep exposure. (h) Development of exposed thick photoresist and whole LOR in AZ 400K. (B) SEM pictures of suspended 3-D photoresist structures. The levitation height is approximately 30 μm and recessed depth by shallow exposure is approximately 25 μm. (From Yun, K.S. and E. Yoon, "Fabrication of complex multilevel microchannels in PDMS by using threedimensional photoresist masters," *Lab. Chip.*, 8, 245, 2008. Reproduced with permission from The Royal Society of Chemistry.)

(For example, there is no point in using a very expensive projection system if the photomask has very low resolution.) A high-resolution photomask is typically made of chrome (features) on glass, which is placed in contact with the photoresist film during exposure. For applications requiring lower resolution, a transparency film (e.g., containing features made with toner ink on a plastic support) suffices. Usually, the photomask is drawn in black and white using drawing software and electronically sent to the printing facility. Some printing services, such as those used by publishers and graphic designers, now offer the printing of letter-size transparencies at 65,000 dots per inch at very low cost. Unfortunately, 65,000 dpi does not correspond to the expected 65,000 inch/dot = ~0.4 μm/dot because of mechanical inaccuracies of the printer head. Nevertheless, ~5-μm-diameter circles and ~5-μm-wide lines or gaps with marginal "graininess" are faithfully reproduced. In any case, care must be taken to ensure that the photomask is in close contact with the photoresist layer. In microelectronics and other high-resolution applications in which contact between the photomask and the substrate is not desirable, the mask is usually demagnified by optical projection during exposure; alternatively, exposure of the photoresist may be done serially by scanning a focused electron beam, ion beam, or laser beam. A requirement of high resolution is that the photoresist film should be thin; for example, less than

1 μm thick if better-than-1 μm-resolution is sought. (As the reader can guess, the word "resolution" is application-dependent and thus cannot be described by a simple number: any given degree of graininess may or may not be a problem depending on the application.)

1.3.4 The SU-8 Era: High Aspect Ratios

A prevailing rule of thumb in microfabrication is that thicker photoresist layers (or larger height/width aspect ratios) are harder to micropattern. Because most photoresists have been developed for manufacturing microelectronic circuits, in which the micropatterns are always thin, they are designed to work well for photoresist thicknesses of a micron or less. Typically, layers thicker than ~20 μm are not easily spin-coated or patterned with traditional photoresists. In the past, this limitation slowed the development of many microfluidic applications for two reasons: (a) very shallow channels have high flow resistances, so running fluids through them often involves nonobvious pumping schemes or tricks; and (b) when the choice of materials to make microfluidic devices was limited to glass and silicon, transferring the photoresist pattern to the underlying substrate required inconvenient or expensive etching methods.

A novel type of photoresist, called **SU-8**, was introduced by IBM in the late 1990s for MEMS applications to produce features with high aspect ratios. SU-8 is a negative photoresist, a type of photocurable epoxy. When acid is photochemically released inside the photoresist by exposure to UV light (peak absorption at 365 nm), the epoxy groups react and cross-link (forming an insoluble polyester network). An example of a 53-μm-deep test pattern is shown in **Figure 1.4**.

Given the salient features of SU-8, it is not surprising that the vast majority of microfluidic devices nowadays are fabricated by micromolding from SU-8 molds:

* SU-8 (which can be spun up to ~1 mm in thickness) allows for vertical sidewalls, so microfluidic devices have a reproducible, rectangular cross-section; rectangular geometries are amenable to fluid dynamics modeling of fluid behavior.

* A practical advantage of SU-8 is that, unlike many other photoresists, areas that have been exposed correctly have an extremely low solubility in the developer; hence, timing of the developing step is not as critical as with other photoresists.

* SU-8 allows for high aspect ratios; 10:1 height/width aspect ratios and vertical side walls are easy to achieve over a wide range of heights (as long as well-collimated light is used and a close contact between the mask and the photoresist is ensured).

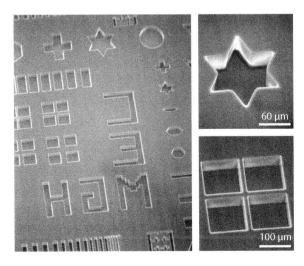

FIGURE 1.4   SU-8 photoresist. (Figure contributed by the author.)

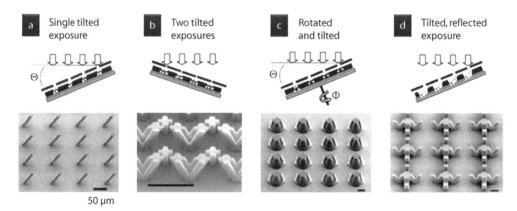

a Single tilted exposure   **b** Two tilted exposures   **c** Rotated and tilted   **d** Tilted, reflected exposure

50 μm

FIGURE 1.5 Tilted microstructures. (From Han, M., W. Lee, S.-K. Lee, and S.S. Lee, "3D microfabrication with inclined/rotated UV lithography," *Sens. Actuators A: Phys.*, 111, 14–20, 2004. Reprinted with permission from Elsevier; see also the excellent review on SU8 photolithography by del Campo, A. and C. Greiner, "SU-8: A photoresist for high-aspect-ratio and 3D submicron lithography," *J. Micromech. Microeng.* 17, R81–R95, 2007.)

- Tilted structures can be fabricated simply by tilting the substrate (again, as long as well-collimated light is used), as shown in **Figure 1.5**.

- Overhanging features (e.g., a cantilever) can be made by depositing a sacrificial shadow mask (e.g., an opaque pad of metal) on unexposed SU-8. Thus, the areas shadowed by the pad are not cross-linked by subsequent exposures to light (of that layer or subsequently spun layers).

1.3.5 Biocompatible Photoresists

Most types of photoresists are not biocompatible—cells die when cultured on top of them and they are not authorized for use in implantable devices. However, there are some notable exceptions: **photosensitive poly(dimethyl siloxane)** (PDMS), **Ormocer**, and **water-soluble photoresists based on poly (N-isopropylacrylamide) (poly-NIPAM)**.

- As we will see in Section 1.6.2, PDMS is an extremely important polymer in BioMEMS, mostly patterned by molding from a microfabricated master because it cures thermally by a very simple procedure. However, it can also be photosensitized by the addition of a photo–cross-linker, which circumvents the molding step and allows for precise alignment of multilayer structures.

- **Ormocer** (short for "organically modified ceramics") was developed by the Fraunhofer Institute in Germany, originally as a dental implant polymer. It is a negative resist requiring ~170°C curing temperatures. Two dental filling composites based on Ormocer (Definite and Admira) are available.

- Recently, Stefan Diez's group at the Max Planck Institute of Molecular Cell Biology and Genetics in Dresden, Germany, has developed a water-based processing of a photoresist based on a modified (photocleavable) copolymer of the thermoresponsive polymer **poly-NIPAM** (**Figure 1.6**). Upon UV illumination, the copolymer's photocleavable groups are deprotected, which results in an increased low critical solution temperature. The nonirradiated and irradiated copolymer can be dissolved sequentially in water at different temperatures, a process that is more biocompatible than previous solvent-based processing.

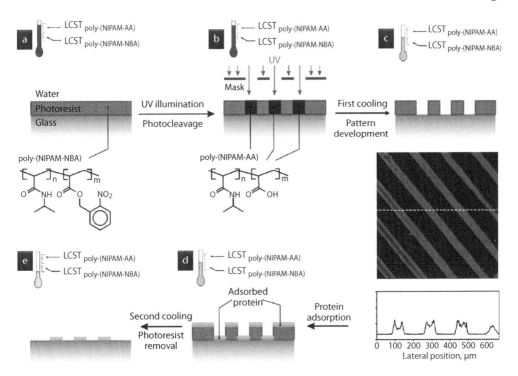

FIGURE 1.6 Water-soluble thermoresponsive photoresist for protein patterning. (From Ionov, L. and S. Diez, "Environment-friendly photolithography using poly(*N*-isopropylacrylamide)-based thermoresponsive photoresists," *J. Am. Chem. Soc.*, 131, 13315–13319, 2009. Reprinted with permission of the American Chemical Society.)

1.3.6 Maskless Photolithography

Photomasks are inconvenient, to say the least. Why can't we fabricate *every* substrate the way photomasks are fabricated? That would be too slow, and that is the very reason why we use photomasks...but for certain applications (mainly the fabrication of microfluidic masters), it is starting to make sense to fabricate the masters simply by raster-scanning a laser over SU-8 photoresist with a photomask generator (essentially, a laser writer). It takes several hours to expose

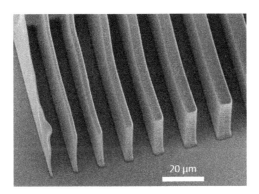

FIGURE 1.7 Laser lithography of 30-μm-thick SU-8 structures. (Courtesy of Wallace, P. and S. Braswell, Nanotechnology User Facility, University of Washington, Seattle, WA.)

a 4-in. wafer master (independently of the feature density), but exposures can be done overnight (**Figure 1.7**). It is straightforward to produce high aspect ratios ranging from 1 to 500 μm in height in SU-8. A closely competing technology is Texas Instruments' **Digital Micromirror Device** (DMD).

1.3.6.1 The Digital Micromirror Device: Affordable Maskless Photolithography, at Last

In 1987, Texas Instruments invented a device, dubbed the DMD, that consists of an array of hundreds of thousands of aluminum-coated silicon micromirrors (made by silicon micromachining techniques), each 16 μm across in size, that can be individually deflected (by ±10–12 degrees) so as to project the light of just about any lamp that is focused on the array (**Figure 1.8a** and **b**). The reflected light is then projected onto a screen, where each pixel corresponds to a micromirror. To produce gray scales, the mirrors are toggled on and off very quickly (1024 shades of gray are possible based on the ratio of on-time to off-time). Originally developed for display applications (the micromirrors flip positions at video rates!), it has recently been incorporated into a commercial projection UV photolithography system. The user simply places a photoresist-coated substrate on the screen and directly projects the mask on the substrate, skipping the mask production step. Xiang Zhang's group at UCLA has incorporated the DMD into a stereolithography system (whereby UV light is used to cure a resin only on the focal spot) to produce complex 3-D shapes (**Figure 1.8c** and **d**). It is foreseeable that the price of DMD-based systems will gradually be reduced and, in less than ten years, photomasks will be a tool of the past.

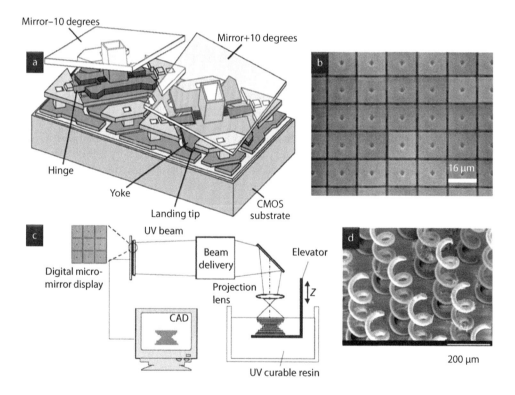

FIGURE 1.8 Digital micromirror device. (From Sun, C., N. Fang, D.M. Wu, and X. Zhang, "Projection micro-stereolithography using digital micro-mirror dynamic mask," *Sens. Actuators A: Phys.*, 121, 113–120, 2005, Reprinted with permission from Elsevier.)

1.4 Micromachining

The photoresist pattern may later be transferred onto the underlying substrate to form nonplanar structures by a variety of processes including chemical etches and deposition of materials; although the concept is almost as old as photolithography, it is now known by the term coined for it in the 1980s—"**micromachining**." The term micromachining also includes other strategies such as laser patterning. Historically, the micromachining field started in the 1980s as a spin-off of microelectronics, using recycled facilities (as many chemicals used for etching are incompatible with electronics) and the same silicon wafers that had been conveniently developed for integrated circuit processing (photoresist spreads well on smooth, circular wafers). Glass and quartz wafers were introduced soon for applications requiring electro-osmotic flow, and provided a transparent substrate. Hence, silicon and glass have always been considered the "traditional substrates" among the micromachining community.

Some micromachined devices are already ubiquitous and have had a huge impact in society, such as the accelerometers in your car's airbag deployers or the nozzle in the printer head of your office's inkjet printer. The airbag deployer is based on a thin slab of silicon that hangs over an etched cavity; because of its minuscule size, the slab reacts extremely fast. The inkjet nozzle is also an etched cavity covered with a thin membrane that has a microfabricated hole; a small heater inside the cavity vaporizes the ink, which forms a bubble that ejects a tiny droplet out of the hole, onto a nearby surface such as a sheet of paper. Because of the size of the cavity, the ink cools down in less than a millisecond and is ready again for the formation of another bubble (i.e., it can print fast), and owing to the size of the hole, the printer has a resolution superior to that of the eye.

Micromachining techniques are not simple for a nonexpert and require a lot of fine-tuning of materials processing and expensive facilities. However, several devices can be made at once on the same wafer, and the price of processing several wafers is almost the same as processing one; so, the cost per unit can be quite low, ensuring fast commercial dissemination for everyday objects.

1.4.1 Etching: Wet versus Dry, Isotropic versus Anisotropic

The etching chemistry must be chosen such that it attacks the substrate but not the photoresist, or with as much selectivity as possible (**Figure 1.9**). Chemical etches can be carried out using solutions (a "**wet etch**") or using a reactive-gas plasma (a "**dry etch**," also termed "**reactive ion etch**"). We distinguish between a chemical etch that proceeds in every direction homogeneously ("**isotropic etch**") and one that proceeds in a preferential direction ("**anisotropic etch**"). Virtually every imaginable combination of chemical etches, substrates, and masking materials has been tried in micromachining. When micromachining is performed by the addition or subtraction of shallow layers on the surface of the wafer, we term it "**surface micromachining**"; on the other hand, when the processing involves deep etches through a good portion of the thickness of the wafer, we speak of "**bulk micromachining**." Nearly all commercial MEMS devices (e.g., accelerometers) are fabricated by surface

Photoresist   Selective etch   Photoresist
pattern   removal

FIGURE 1.9 Etching using a photoresist mask.

micromachining. See Further Reading at the end of this chapter for references on etching and microfabrication.

1.4.2 Deposition and "Lift-Off"

The photoresist pattern is almost never the last step of a microfabrication process. For example, it can be used to mask the deposition or growth of another layer (such as a uniform gold thin film), as shown in **Figure 1.10**; the photoresist is removed with a solvent, a procedure known as "**lift-off**."

As an example, let us consider the creation of a suspended membrane of silicon nitride across a silicon cavity. It would require the following steps, approximately:

- Deposition of a silicon nitride layer on a silicon wafer using a high-temperature furnace.

- Photolithography to create a photoresist pattern atop the silicon nitride layer.

- Dry etch of the silicon nitride (that is not protected by photoresist) until the silicon substrate is reached; typically, a reactive gas SF_6 would be introduced into the plasma etcher chamber.

- Removal of the photoresist with acetone or with an oxygen plasma; at this point, the silicon nitride layer would contain the same pattern as the original photoresist layer.

- Wet etch of the silicon (that is not protected by the silicon nitride) with an aqueous solution containing a strong base such as KOH, which etches the Si(100) crystalline plane around 100 times faster than the Si(111) plane, resulting in a pit with slanted walls; if the pit is wide enough, the etch can reach the other side of the wafer, creating a suspended silicon nitride membrane.

This process has been the starting process of many micromachined devices. As a relatively simple example, consider the fabrication of a silicon tip on a silicon nitride cantilever (**Figure 1.11**). Still, it involves three photolithography (P) steps, two deposition steps (D), three dry-etching steps (E_d), and three wet-etching steps (E_w, not counting the three dissolutions of the photoresist), in this order: $D-P-E_d-E_w-E_w-D-P-E_d-P-E_d-E_d$. All the etches used different chemicals or gases. **Figure 1.11** only shows the last etching step (a dry etch that forms the Si tip by

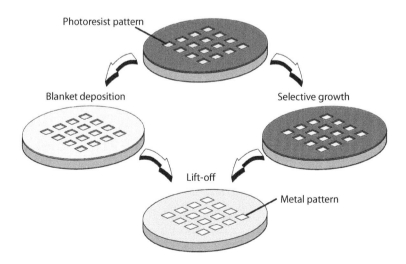

Photoresist pattern

Blanket deposition

Selective growth

Lift-off

Metal pattern

FIGURE 1.10 Metal deposition and lift-off.

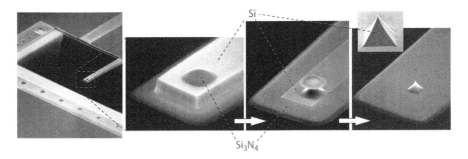

FIGURE 1.11 Micromachining of a cantilevered tip. (From Folch, A., M.S. Wrighton, and M.A. Schmidt, "Microfabrication of ultra-sharp Si tips on Si3N4 cantilevers for atomic force microscopy," *J. Microelectromech. Syst.*, 6, 303, 1997. Figure contributed by the author.)

underetching a silicon nitride square atop a Si layer, which itself is resting on a cantilever made of silicon nitride). And that is a relatively simple device!

Micromachining processes often include many such photolithography/etching steps, but unfortunately, the structural materials that are (traditionally) "allowed" are very few—mostly (crystalline or amorphous) silicon, silicon oxide, and silicon nitride. Yet, it is remarkable how many structures can be manufactured with just a few combinations of materials and chemistries. Using electrochemical deposition of metals and ceramics, and planarization processes borrowed from the integrated circuit chip industry, multilayer devices with arbitrary 3-D geometries that incorporate moving parts (gears, tweezers, etc.) with 2-μm resolution are possible (Figure 1.12). However, the processing cost of the first wafer ranges from $10,000 to $100,000, depending on the complexity of the device (the cost per device can be lowered if there are several devices per wafer and as production proceeds).

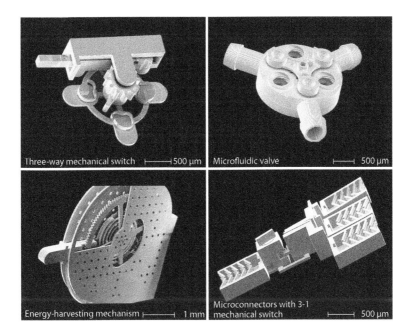

FIGURE 1.12 State-of-the-art of 3-D micromachining. The structures are made of ceramics. Images courtesy of Microfabrica, Inc. (Note: the figure is not meant to endorse Microfabrica's technology over other competing technologies.)

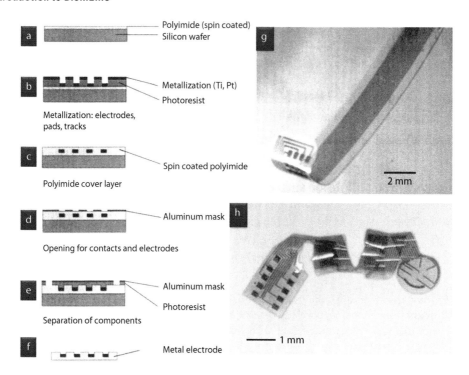

FIGURE 1.13 Metal patterns on a flexible plastic substrate. (From Stieglitz, T., H. Beutel, and J.-U. Meyer, "A flexible, light-weight multichannel sieve electrode with integrated cables for interfacing regenerating peripheral nerves," *Sens. Actuators A: Phys.*, 60, 240, 1997. Reprinted with permission from Elsevier.)

1.4.3 Nontraditional Substrates

The MEMS community has always pushed the development of technologies for patterning substrates other than the traditional silicon and glass wafers, which can be conveniently spin-coated with photoresists. Rectangular glass slides can also be spin-coated, but the photoresist does not spread well at the edges. Strategies to pattern and handle plastics in clean rooms have attracted the most attention in BioMEMS because plastics are cheap and flexible, which makes them ideal candidates as interfaces with delicate body parts in disposable biomedical devices as well as parts of cell culture devices integrated in cell culture petri dishes. One polymer that has had good acceptance in clean rooms because of its good compatibility with silicon processing has been polyimide. It supports metallization well and can be etched with various dry etching steps (typically, a sacrificial aluminum mask is patterned on top of the polyimide to transfer the pattern to the polyimide layer). **Figure 1.13** shows the fabrication process of a set of electrodes on a polyimide ribbon intended for implantation.

Methods for microstructuring other nontraditional materials that support fluid flow, such as gels, paper, and fabric, will be covered in Section 3.5, under the chapter devoted to microfluidics (Chapter 3).

1.4.4 Laser Cutting

An inexpensive CO_2 laser can be used to cut through a thin plastic roll, which conveniently comes prepackaged with a layer of solid glue (warning: the laser actually melts the plastic, which produces toxic fumes, so the whole apparatus must be placed in a fume hood). Each cut can be thought of as a channel when various layers of plastic are stacked together. Melting of plastic causes the formation of a lip at the edge of the cut that ultimately limits the resolution of the technique. This

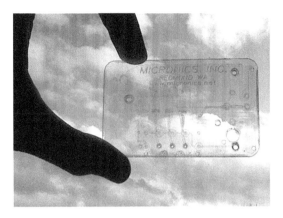

FIGURE 1.14 Laser-cut laminated devices for diagnostics. Image contributed by the author. (Disclaimer: The author has collaborated with Micronics, however, the image is not meant to endorse Micronics' devices. Micronics and the laboratory of its founder Paul Yager, at the University of Washington's Bioengineering Department, are early developers of laser-cut laminate technology.)

lamination approach, in combination with **laser cutting**, has been used successfully by various companies to make low-cost microfluidic devices for medical diagnostics (**Figure 1.14**).

1.4.5 Multiphoton Lithography

Multiphoton lithography, also called **direct laser writing**, allows for the fabrication of arbitrary 3-D structures (**Figure 1.15**) and is based on multiphoton polymerization of an adequate photoresist. The center wavelength of the laser is chosen to coincide with the transparent spectrum of the photoresist, and it takes two or more photons (absorbed simultaneously on the same tightly focused spot) to expose the photoresist. The beam spot is moved relative to the sample by displacing the sample with a programmable piezoelectric XYZ sample stage. Once the exposure is finished, the unexposed photoresist is dissolved away in a developer solution. Unfortunately, the price of these systems is still in the range of hundreds of thousands of dollars.

Jason Shear's group (from the University of Texas at Austin) has extended the concept of direct laser writing to the use of photo–cross-linkable biomolecules in aqueous environments

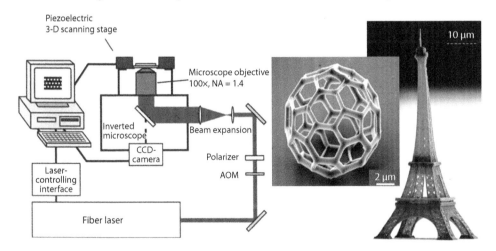

FIGURE 1.15 Direct laser writing. (Courtesy of Nanoscribe GmbH Germany.) (Note: the figure is not meant to endorse Nanoscribe's technology over other direct laser writing systems.)

instead of photoresist. In particular, his group has focused high-powered pulsed lasers to deposit biocompatible structures made of photo–cross-linkable albumin *in the presence* of live cells (**Figure 1.16a**). The non–cross-linked protein is simply rinsed away after the fabrication procedure (so there is no development step). In one experiment, "corrals" (which actually have a roof that is not visible because it is out of focus) were built to trap bacteria (**Figure 1.16b and c**); a microsphere trapped inside a 28-µm-diameter circular microchamber is seen to do a whole orbital turn in 0.3 to 0.4 seconds (corresponding to a linear velocity of ~150 µm/s) because of the clockwise swimming of *Escherichia coli* (**Figure 1.16b**). In another experiment, albumin lines were used to constraint the growth of axons within corrals (**Figure 1.16d**). Other more complex structures such as two-story "apartment buildings" were also possible.

Serial writing with a laser is limiting because writing 3-D structures is a slow process that involves scanning the beam in all three dimensions, and programming the beam movements can be a painstaking procedure. A more efficient way is to reduce arbitrary shapes to slices using imaging software and to project the laser beam with a DMD (see Section 1.3.6.1). The Shear lab has, indeed, coupled multiphoton lithography successfully with a DMD to produce arbitrary microstructures of photo–cross-linked bovine serum albumin (BSA) (**Figure 1.17**). A titanium sapphire

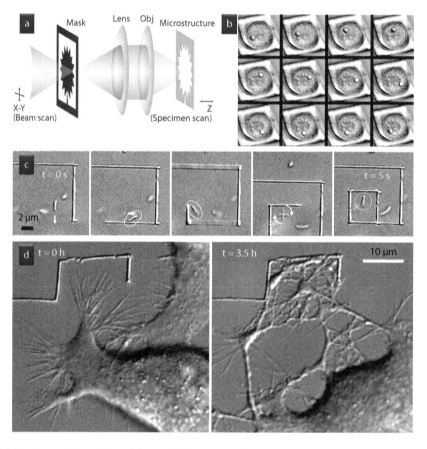

FIGURE 1.16 Laser deposition of "corrals" for containing cell growth and motility. (From Kaehr, B., R. Allen, D.J. Javier, J. Currie, and J.B. Shear, "Guiding neuronal development with in situ microfabrication," *Proc. Natl. Acad. Sci. U. S. A.*, 101, 16104, 2004. Copyright (2004), National Academy of Sciences, U. S. A.; Kaehr, B. and J.B. Shear, "High-throughput design of microfluidics based on directed bacterial motility," *Lab. Chip.*, 9, 2632, 2009. Reproduced with permission from The Royal Society of Chemistry.)

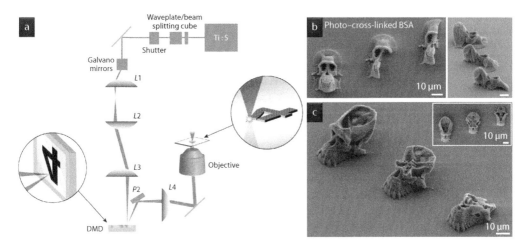

FIGURE 1.17 Multiphoton lithography with the DMD. (From Nielson, R., B. Kaehr, and J.B. Shear, "Microreplication and design of biological architectures using dynamic-mask multiphoton lithography," *Small*, 1, 120–125, 2009. Reproduced with permission of Wiley-VCH Verlag GmbH & Co. KGaA.)

laser is raster-scanned across the DMD chip, which typically projects approximately 2-minute sequences of more than 100 photomasks while the focal plane is vertically stepped in synchrony (step size ~0.3–1 µm). Spiral ramps that guide the motility of *E. coli* were demonstrated.

1.5 Micromolding

Traditional microfabrication as reviewed previously, that is, involving photolithography and micromachining, is very limited in the choices of materials that can be patterned. For many applications, silicon, photoresist, and metals are not appropriate materials, and the methods used to process them cannot be extrapolated to other materials. Polymers, on the other hand, exist in endless varieties. They have become ubiquitous in our daily life as key parts of furniture, toys, home appliances, cars, and scientific instruments, to name a few. For the same reason, polymers are becoming increasingly important in the development of novel microdevices such as micro-optical, micromechanical, and microfluidic systems. Because traditional silicon or glass microdevice fabrication is essentially based on chemical etching and dissolution processes that are not well suited for shaping polymers, most efforts in polymer-based microfabrication methods have focused on **micromolding** strategies, which offer the additional benefit of high-throughput replication once a suitable master mold is developed.

Given a certain mold, not all polymers replicate equally well. In general, polymers are synthesized from their monomers as fluids, solidified, and made available in pellets. To mold a presynthesized polymer, it must be either melted or dissolved in the appropriate solvent, applied and solidified again on the mold, and separated from the mold. The polymer can also be formed (i.e., synthesized) on the mold directly. If the mold is not damaged by the process, the mold can be reused again for high-throughput fabrication, which helps reduce the overall cost.

Polymers can be divided into two large classes depending on the strength of interactions between the monomers: **thermoset polymers** (also called **duroplastic polymers**), which are heavily cross-linked through an irreversible curing reaction and are therefore hard; and **thermoplastic polymers**, which can be molded into shapes above the glass transition temperature and freeze into a glassy state when cooled sufficiently. We also discern **elastomeric polymers** (which can be thermoset or thermoplastic), a class of polymers that can be stretched and compressed by virtue of their weakly cross-linked polymer chains. Accordingly, the three most widely used methods for micromolding polymers are **injection molding** (for thermoset and thermoplastic polymers), **hot embossing** (for thermoplastic polymers), and **soft lithography** (for elastomeric polymers).

1.5.1 Injection Molding

If the polymer is applied onto the mold from the molten phase, it is usually a viscous liquid that does not flow well, so the mold cannot have narrow, deep features that might otherwise trap bubbles. The molten polymer, in addition, degrades quickly in air, so it is best to quickly apply the polymer at high pressure using a pneumatic piston—a technique dubbed **injection molding**. A vast majority of the polymeric objects that surround us—toys, furniture, bottles, etc., as well as the widely used polystyrene petri dishes and flasks for cell culture—are molded by injection molding. Unfortunately, the required pressures are incompatible with the brittle silicon substrates used in photolithography, so the master must be fabricated on a more solid substrate, usually a metal surface, onto which the photoresist pattern is transferred by etching (the photoresist does not survive the injection molding temperatures). The large differences in thermal expansion between the mold and the polymer can make release difficult as the temperature is lowered. In addition, melting of the polymers generates hazardous vapors that must be dealt with, typically by using a fume exhaust system that is exorbitantly expensive to install.

1.5.2 Hot Embossing

When only surface features on a flat polymer substrate are sought, heating the mold above the glass transition temperature (T_g) of a thermoplastic polymer and pressing it onto the polymer substrate results in embossed features—a method termed **hot embossing**. (The technique is also known as **nanoimprinting** or **nanoimprint lithography** in the nanotechnology community, but "hot embossing" is more descriptive.) Compared with injection molding, hot embossing does not require pneumatic injection and does not produce appreciable vapor amounts, so it can be easily implemented on a laboratory benchtop. The success of the replication is determined by the T_g (the lower the better) and by the difference in thermal expansion coefficient with respect to the mold (the smaller the better). Examples of polymers that replicate well with hot embossing are poly(methyl methacrylates) (PMMA; $T_g = 106°C$), polycarbonates ($T_g = 150°C$), polystyrenes ($T_g = 80°C–100°C$), and cyclic olefin copolymers ($T_g = 138°C$). All these materials have a similar linear thermal expansion coefficient in the range of 60 to 90×10^{-6} K^{-1}. Hot embossing is the "poor man's version" of injection molding—its main disadvantage is that the molding of deep features produces large displacements of material that have to be redeposited somewhere, usually at the top of the trenches, forming nonplanar accumulations that are feature-dependent.

1.5.3 Curable Polymers

If the polymer is applied onto the mold as a solvent-based solution, shrinking of the polymer as the solvent is evaporated can be problematic—for example, residual stress, resulting in warping after release, and loss of resolution. Thus, a favored strategy is to *create* (or "cure") the polymer irreversibly *on* the mold by using the monomer solution and a cross-linker or initiator. Epoxies, polyurethanes, and silicones are examples of **thermosetting polymers** (also called **thermosets**) that can be formed this way. (Despite the name, the curing reaction can be done by heat, a chemical reaction, or irradiation.)

A particularly successful micromolding polymer is the elastomer PDMS, a type of silicone rubber which is thermally cured as a two-component mixture. PDMS micromolding is the basis of a family of sister techniques that George Whitesides and colleagues at Harvard ingeniously baptized as "**soft lithography**" (from the softness of PDMS). These techniques have had such an enormous impact on BioMEMS that they deserve a section of their own.

1.6 Soft Lithography

George Whitesides' group has pioneered a family of microfabrication techniques collectively dubbed "**soft lithography**" which have in common the use of a micromolded piece of PDMS—an elastomeric polymer that is best described as transparent rubber, although it does not belong to the carbon-based rubber family—to generate a pattern or form a device. The stamp is generally made by replica-molding from a microfabricated master. Hence, soft lithography typically requires access to some other microfabrication method for the fabrication of the starting mold.

GEORGE WHITESIDES

 GEORGE M. WHITESIDES (BORN AUGUST 3, 1939, Louisville, Kentucky) is an American chemist and professor of chemistry at Harvard University. He has contributed to a number of areas, including NMR spectroscopy, organometallic chemistry, molecular self-assembly, soft lithography, microfabrication, microfluidics, and nanotechnology. As of December 2011, he held the highest Hirsch index rating (169) of all living chemists (i.e., he has 169 publications that have been cited at least 169 times each). His laboratory employs around 40 people. With the invention of soft lithography (the first paper in alkanethiol microstamping was published by *Applied Physics Letters* in July 1993), he has revolutionized BioMEMS and microfluidics. Recently, Whitesides has developed processes for producing analytical microdevices made of paper, which promises to stir a second revolution in microfluidics.

Whitesides received his A.B. degree from Harvard College in 1960 and earned a Ph.D. in chemistry from Caltech (focusing on the use of NMR spectroscopy in organic chemistry) in 1964 under the tutelage of John D. Roberts. Whitesides began his independent career as an assistant professor at the Massachusetts Institute of Technology (MIT) in 1963 and remained there until 1982. While at MIT, he played a pivotal role in the development of the Corey–House–Posner–Whitesides reaction. In 1982, Whitesides moved his laboratory to the Department of Chemistry at Harvard University. He is currently the Woodford L. and Ann A. Flowers University Professor at Harvard, one of only 21 University Professorships at the institution. He is the author of more than 1,000 scientific articles and is listed as an inventor on more than 50 patents. Whitesides has mentored more than 300 graduate students, postdocs, and visiting scholars. He ranked 5th on Thomson ISI's list of the 1000 most cited chemists from 1981 to 1997. Whitesides has cofounded more than 12 companies with a combined market capitalization of more than $20 billion, including Genzyme, GelTex, Theravance, Surface Logix, Nano-Terra, and WMR Biomedical. He serves on the editorial advisory boards of several scientific journals, including *Angewandte Chemie*, *Chemistry & Biology*, and *Small*, and is the Chairman of *Lab on a Chip*.

Among other awards, Whitesides is the recipient of the American Chemical Society's Award in Pure Chemistry (1975), the Arthur C. Cope Award (1995), the National Medal of Science (1998), the Kyoto Prize in Materials Science and Engineering (2003), the Dan David Prize (2005), the Welch Award in Chemistry (2005), the Priestley Medal (2007), the Dreyfus Prize in the Chemical Sciences (2009), and the Benjamin Franklin Medal in Chemistry (2009). Whitesides is a member of the American Academy of Arts and Sciences, the National Academy of Sciences, and the National Academy of Engineering. He is also a fellow of the American Association for the Advancement of Science.

George Whitesides is also a scientist with a deep appreciation for imagery, photography, and writing (as testified by the overall style and clarity of his papers—Whitesides' students have shared online his cleverly simple and very helpful "manual of style"). Together with his long-time collaborator and pioneer scientific photographer, Felice Frankel, he has published two beautiful photography books on his experiments, *On the Surface of Things* (2008) and *No Small Matter* (2009). The first is arguably the most delightful book on microfabrication ever published.
[*Excerpt adapted from Wikipedia.*]

Soft lithography differs fundamentally from photolithography in the way surfaces are patterned, as illustrated in **Figure 1.18**. Two classes of soft lithographic patterning strategies can be distinguished depending on where the material is selectively deposited: in the areas where the stamp contacts the surface ("**microstamping**" or "**microcontact printing**") or where it does not contact the surface ("**microfluidic patterning**"). These techniques will be reviewed in detail later.

Soft lithography has gained widespread popularity among MEMS researchers because the methods are inexpensive and straightforward once the mold is fabricated: the replication procedure requires neither expertise nor equipment—unlike other micromolding methods—and PDMS is commercially available by the gallon at low cost. In addition, PDMS has particularly interesting optical, surface, and mechanical properties that make it an ideal material for many applications in biomedicine for which traditional MEMS materials (silicon and glass) are unsuited. This means that the devices can be disseminated among a wider community than was possible before. Thus, soft lithography represents an *enabling* technology.

1.6.1 Basics of Soft Lithography

The starting procedure in soft lithography is the microfabrication of a master mold. The next step, which is the molding of PDMS, is deceptively simple, as shown in **Figure 1.19**. A PDMS prepolymer is prepared as a ~10:1 ratio by weight mixture of two components, the monomer solution and the cross-linker, and is poured over the master. Because the monomer solution is viscous, both mixing of the two components as well as the pouring can cause air bubbles to be trapped at the corners of small features, so it is standard practice to place the prepolymer-covered mold in a house vacuum until the bubbles rise to the surface of the liquid and pop. The cross-linking reaction—which is slow at room temperature but can be sped up by heating (e.g., ~4 hours at 65°C)—causes the PDMS to solidify to a material that feels and looks very much like a transparent rubber. After curing, the molded block is manually peeled off.

As opposed to other micromolding methods, the fact that the replication starts from a room temperature fluid that has time to conform to the smallest features ensures that the replication is

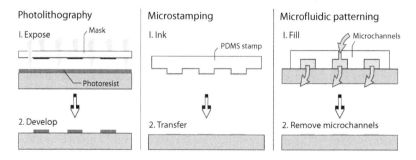

FIGURE 1.18 Photolithographic versus soft lithographic patterning methods.

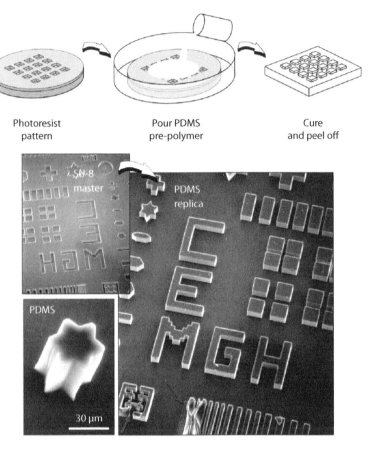

Photoresist
pattern

Pour PDMS
pre-polymer

Cure
and peel off

FIGURE 1.19 PDMS micromolding. The linewidth of the letters "CEMGH" is 50 µm. Red arrows point to PDMS structures that, because of their high aspect ratios (>5:1 height/width), have not replicated well. (Figure contributed by the author.)

exceptionally faithful—features smaller than 100 nm can be replicated; importantly, extremely flat surfaces in the master are also replicated as extremely flat PDMS surfaces. For some features such as long, high aspect ratio trenches, the direction in which the replica is peeled off the master can be important to avoid tearing the PDMS. High aspect ratio (>5 or so) columns or walls have very low integrity and can collapse against the surface (**Figure 1.20**).

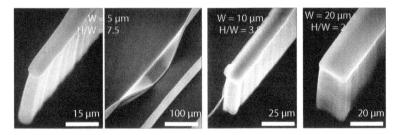

FIGURE 1.20 Structural integrity of PDMS walls. The ridges on the PDMS walls were present on the walls of the trenches from which these structures were replicated. W indicates width and H/W indicates height-to-width "aspect" ratio. (From Folch, A., A. Ayon, O. Hurtado, M.A. Schmidt, and M. Toner, "Molding of deep polydimethylsiloxane microstructures for microfluidics and biological applications," *J. Biomech. Eng.*, 121, 29, 1999. Figure contributed by the author.)

1.6.2 The Magic of PDMS

Most certainly, soft lithography owes its tremendous impact on BioMEMS to the unique physicochemical properties of PDMS. To understand this, it helps to take a glance at the chemical formula in **Figure 1.21**.

This molecule is a polymer whose backbone is –Si–O–, in other words, a "siloxane." Two methyl groups hang from the silicon atoms; hence, its name "poly(dimethyl siloxane)," and occasionally, instead of a methyl group, there is a cross-link to another backbone. The backbone should not be thought of as flat and rigid like the chemical formula cartoon implies; rather, it is a very flexible, 3-D chain because each Si–O bond can be symmetrically rotated. Siloxane polymers are better known as silicones, of which there are thousands of varieties depending on the groups that hang from the silicon atoms and the degree of cross-linking. Qualitatively speaking, more cross-linking yields more hardness and less elasticity.

PDMS has a unique combination of properties that can be very advantageous for BioMEMS devices:

❋ *PDMS is very elastic.* The elasticity of a material is characterized by its Young's modulus of elasticity, usually symbolized as E, which gives a measure of the force you need to apply to elongate a piece of the material. More strictly speaking, E is the ratio between the stress applied to the material (stress = force/cross-sectional area) and the strain resulting from it (strain = % elongation). For PDMS, E varies wildly around ~1–100 N/m^2 (MPa). (For a hard plastic such as PMMA, i.e., Plexiglas, E takes values on the order of a few giga Pascals). There is no consensus of the unique value in the literature because E is very sensitive to the degree of cross-linking (which depends on the exact monomer/cross-linker ratio, the curing temperature, and the curing time; everyone's recipe is slightly different). Also, PDMS is a hyperelastic material, in other words, its Young's modulus is not a constant but it decreases as it is stretched. This imprecise knowledge of the Young's modulus of PDMS poses a challenge to the design of microdevices that incorporate sensors and actuators made of PDMS; typically, thin mechanical structures or membranes of PDMS that can be deflected by large amounts. The Poisson ratio of PDMS is typical of rubber, ~0.5 (which means that if a piece of PDMS is elongated by, say, 10%, it will become 5% thinner). In practice, the high elasticity of PDMS means that the surface features can be stretched by amounts much larger than their dimensions, which makes it possible to release the replica with no apparent tear, replicating the smallest details even on sidewalls (**Figure 1.20**); a less elastic material could remain trapped in deep or undercut features, thereby resulting in a damaged replica or damaged mold.

❋ *PDMS is soft.* As a result, a PDMS stamp is unlikely to damage hard surfaces. Also, the PDMS surface will envelop a particle on the surface, minimizing pattern distortion away from the particle, which makes it possible to assemble and use devices outside of a clean room with acceptable yields. (A somewhat exaggerated claim could be that PDMS enables "benchtop microfabrication.") However, the softness can also impair the structural integrity of PDMS walls and channels (**Figure 1.20**), and make it difficult to align multilayer structures.

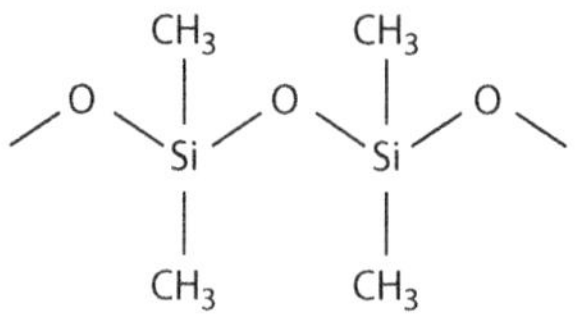

FIGURE 1.21 PDMS chemical formula.

❋ *PDMS is optically transparent to light at wavelengths down to 300 nm.* Even for UV light, in the range 365 to 435 nm (at which the mercury lamp lines used in photolithography are found) the transmittance of PDMS is 92% to 93.3% (water, by comparison, has 99.97%, and many glasses range in the 98%–99%). At 20°C, the index of refraction of PDMS is 1.43, very similar to that of glasses with low index of refraction. The optical resemblance with glass, combined with the possibility of controlled deformation on a microscale and micromolding capabilities, make PDMS a very attractive optical element, which has spawned a wide range of applications collectively termed **elastomeric optics**.

❋ *The surface of PDMS is hydrophobic*, presumably because of the abundant presence of methyl groups ($-CH_3$) at the surface (see **Figure 1.21**). The range of contact angles (a measure of the shape of a water droplet on a surface, which is commonly used as a measure of hydrophobicity—see Section 2.1.2) measured for PDMS is approximately 90 to 120 degrees, a fairly large value. (For 90 degrees, the droplet looks exactly like a semisphere.) An important combined consequence of being soft and hydrophobic is that it forms *spontaneous seals with most smooth dry surfaces* (even in the presence of small asperities). The surface can be rendered hydrophilic by an oxidation treatment or by coating it with a hydrophilic layer (see below).

❋ *PDMS is relatively inert, but its surface can be oxidized, etched, and derivatized.* Most researchers would agree that it requires a chemist's knowledge and skills to attach molecules to PDMS chemically. Because of the stability of methyl groups and the siloxane chain, PDMS has a relatively low reactivity with most chemicals. Wet-etching of PDMS has been reported with a 3:1 mixture of tetrabutylammonium fluoride ($C_{16}H_{36}FN$) and N-methyl-2-pyrrolidinone (C_5H_9NO) (one of the few solvents that do not swell PDMS; see next section). A gas mixture of CF_4 and oxygen can be used to plasma-etch PDMS. A favored method (invented by George Whitesides' group in 1998) is to expose the surface to a highly oxidant gas, such as an oxygen plasma or UV ozone, that presumably substitutes surface methyl groups by hydroxyl groups ($-OH$). Hydroxyl groups are much more reactive, which facilitates a number of chemical reactions for attaching biomolecules to the PDMS surface. Hydroxylated surfaces are also very hydrophilic, so oxygen plasma has become a common approach to facilitate wetting of PDMS microstructures (such as microchannels or deep microwells; other methods for facilitating wetting will be reviewed in Section 3.6.1). Unfortunately, the hydrophilic state "wears off" and the surface reverts to hydrophobic in a few hours (unless it is kept under water). UV grafting of polymers (a method by which monomers are polymerized onto a surface, with the polymerization reaction initiated by a UV exposure) has been used chemically to attach useful polymers such as acrylic acid, acrylamide, poly(ethylene glycol) (PEG), and monomethoxyl acrylate, which have various degrees of hydrophilicity. Unlike plasma-oxidized surfaces, these coatings are stable (even in air), so they can be used in applications in which it is necessary to tailor or preserve the charges on the surface (such as to drive electro-osmotic flow, see Section 3.3.2, or to prevent adsorption of biomolecules, see Section 2.2.2).

❋ *Biocompatibility.* Partially because of its inertness, PDMS is very biocompatible (e.g., silicones are used for breast implants). For cell culture applications requiring direct contact between the PDMS and live cells for long periods of time, non-cross-linked monomers can be toxic, but "autoclaving" (a hospital procedure to sterilize surgical equipment consisting of 2–4 hours of exposure to 120°C–150°C and high pressures) eliminates the problem—and, for the same price, eliminates bacterial and fungal contamination from the PDMS surface in preparation for cell culture.

❋ *Swells when exposed to solvents.* This swelling is a serious obstacle when the PDMS surface is used as a stamp using a solvent-based ink and trying to register the stamp features to an underlying substrate pattern because the pattern of the swelled stamp is difficult to predict and evolves with time as the solvent is allowed to evaporate before stamping. When filling microchannels (not bonded to a surface) with a swelling solution, the microchannels might detach, spilling their contents onto undesired areas; hence, swelling can be a limiting factor for patterning certain polymers from solvent-based precursors using microfluidic channels. Examples of "strong" solvents that swell PDMS are ethanol, acetone, hexane, and tetrahydrofuran, whereas examples of solvents that hold acceptably in microchannels without breaking the conformal seal are N-methylpyrrolidinone, dimethylformamide, or dimethylsulfoxide. Swelling of PDMS can be advantageous as a strategy to open the pores of PDMS and introduce chemicals; benzophenone (a photoactivatable cross-linker) mixed with 10% acetone has been introduced into PDMS (as deep as ~50-μm-thick layers) to chemically immobilize other polymers such as PEG, a protein adhesion deterrent.

High permeability to gases and fluids. Values that can be found in the literature for gas permeability for PDMS are on the order of several hundred Barrers (1 Barrer = 10^{-10} cm^3 [STP] cm/[cm^2 s cmHg]) for air's gases (N_2, CO_2, and O_2), with the permeability to N_2 and CO_2 being similar and the permeability to O_2 being approximately three times higher than N_2 and CO_2. In comparison, the values for a plastic such as PMMA are two to four orders of magnitude smaller (a fraction of a Barrer). These values vary wildly with the degree of cross-linking (which determines the porosity of the elastomer). Anecdotally, common short-chain oils such as citronella can be readily smelled through a PDMS membrane. Water also diffuses readily through PDMS (diffusion coefficient, D ~3–6 × 10^9/ m^2 s^{-1}), compared with D ~2 × 10^6/m^2 s^{-1} for PMMA, three orders of magnitude smaller.

High thermal expansion coefficient. PDMS expands considerably when heated compared with most solids, in particular, compared with the silicon mold. In the common range of curing temperatures from $T_1 = 20°C$ to $T_2 = 70°C$, the thermal expansion coefficients of PDMS and silicon are approximately $C = 3.1 × 10^{-4}$ K^{-1} and 2.6 × 10^{-6} K^{-1}, so during the cooling process, PDMS shrinks about 100 times more than silicon, by a factor of $C(T_2-T_1) = 1.55\%$ and 0.013%, respectively.

1.6.3 Microstamping

This method, originally named **microcontact printing** and popularly referred to as **microstamping**, was first devised to "print" organic molecules. Microstamping is conceptually equivalent to the old rubber stamp-based printing technique called "flexography." It essentially consists of the contact-transfer of the material of interest from a microfabricated rubber stamp onto a surface only on the areas contacted by the stamp (**Figure 1.18**). In the original paper, a PDMS stamp soaked with alkanethiols was used to selectively form an alkanethiol self-assembled monolayer that resisted a gold etchant only on the areas contacted by the stamp. Thus, in contrast to photolithography, microstamping is an *additive* patterning method. Because of its additive nature, microstamping can be applied to nonplanar surfaces. In-registry multilayer patterning can be achieved by backing a thin stamp with a rigid glass support.

The most attractive aspects of microstamping are:

1. Replica-molding the stamp from the master is straightforward, nonhazardous, and inexpensive (in terms of both materials cost and human labor).

2. The molding procedure leaves the master intact; thus, the most expensive part of the process, photolithography—or any other method used to pattern the master—is needed only once.

3. The patterning procedure usually does not damage the sample or the stamp (thus it can be reused).

The most common concerns when using microstamping are:

1. The printed molecules can diffuse laterally, or they can be deposited from the gas phase (or both), into areas not contacted by the stamp, resulting in a broadening of the features. The severity of broadening from the gas phase depends on the molecule's vapor pressure and is not a concern for proteins. However, broadening due to lateral diffusion is mediated by a fluid layer, the presence of which depends on surface hydrophilicity and ambient moisture, and it may affect small protein patterns.

2. Manual handling of the stamp exerts forces on the stamp that cause distortion of the pattern. This is usually not a problem for single-pattern applications, but for applications that require alignment of two different patterns, one must devise some scheme to handle the stamp appropriately, depending on the size of the features to be aligned. The problem is exacerbated by the fact that misalignment cannot be corrected—by the time the user has realized that the stamp is not in the right position, printing has already occurred: contact with the surface is a one-time chance. To overcome pattern misalignment, one may use multilevel stamps, in which the pattern that gets printed/inked depends on the vertical force applied during stamping/inking (see Section 2.4.2).

3. PDMS expands with temperature, so the patterns are sensitive to the ambient temperature.

4. The stamp can buckle between features, especially for patterns in which most of the area is *not* printed or for stamps that have very shallow features relative to the spacing between features (Figure 1.20); there is no easy way to predict buckling of PDMS (the Young's modulus of elasticity changes with strain—"hyperelasticity"—, a behavior that varies widely with the exact curing procedure, as explained earlier), so researchers have found it simpler to optimize the pattern design by trial and error. Several variants of PDMS stamps have been proposed to alleviate this problem, among them (a) printing with flat stamps that have been selectively inked ("microcontact inking"; Figure 1.22); this method may be constrained by the lateral diffusion of the ink molecules on the surface of the stamp (e.g., protein patterns are faithfully printed but alkanethiol patterns are blurred); (b) printing with stamps mounted on a hard backing; and (c) printing with stamps made of harder silicones.

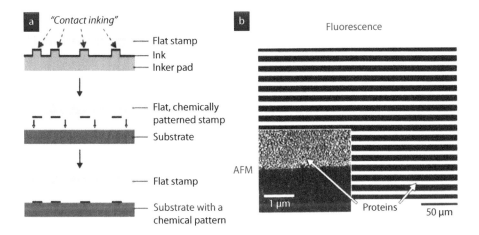

FIGURE 1.22 Selective inking of a flat stamp. (From Geissler, M., A. Bernard, A. Bietsch, H. Schmid, B. Michel, and E. Delamarche, "Microcontact-printing chemical patterns with flat stamps," *J. Am. Chem. Soc.*, 122, 6303, 2000. Reprinted with permission of the American Chemical Society.)

5. For some applications in which surface purity is paramount, there is the concern that, at least under certain conditions, small amounts of dimethylsiloxane monomer seem to remain on the areas where the PDMS stamp has contacted the surface.

6. Fabrication of the stamp is a very low-throughput process because it takes hours to produce a single PDMS replica and peeling it off the master is not easily automated. This limitation has, no doubt, slowed down the adoption of soft lithography by the industry.

In summary, microstamping is inferior to photolithography in pattern fidelity (the stamp deforms—a severe problem because it is pattern-dependent) and in fabrication throughput (molding is time-consuming—a severe problem because it is difficult to automate), but clearly superior in cost-effectiveness and materials versatility.

1.6.4 Microfluidic Patterning

An interesting variation of microstamping takes advantage of the fact that the PDMS stamp can be designed to form a network of microchannels on the areas where the stamp does not contact the surface (see **Figure 1.18**). The microchannels can thus be used to deliver fluids onto selected areas of a substrate, a strategy we refer to as **microfluidic patterning**. The fluid can, depending on its nature, be either cured into a solid itself, used as a vehicle to deposit a material that remains when the microchannels are peeled off, or used to remove (i.e., dissolve or etch) underlying material. Therefore, microfluidic patterning can be an additive as well as a subtractive technique.

As opposed to microstamping, in (additive) microfluidic patterning, the material is added where the PDMS does *not* come into contact with the surface. The fluids are blocked from wetting the substrate in the areas where the PDMS contacts the surface thanks to the unique property of PDMS that it self-seals reversibly against another smooth, dry surface (see properties of PDMS mentioned in Section 1.6.2). This is probably because of a combination of its elastomeric nature allowing for a highly conformal contact and its hydrophobic surface, which impedes wetting of liquids into the PDMS–substrate interface. On the other hand, solvents easily wet PDMS, which can be used to quickly fill microchannels with a liquid precursor to the polymer ("prepolymer") by capillary action—a technique developed by Whitesides' group and dubbed **micromolding in capillaries** or **MIMIC** (**Figure 1.23**). Other materials such as crystals and beds of packed beads have also been patterned by MIMIC.

Microchannels featuring deeper and wider profiles can be filled by pressure-driven flow over larger areas than shallower microchannels. Shown in **Figure 1.24** are computer tomography images of polyurethane microstructures. Two sets of straight microchannels were sealed facing each other at 90 degrees, filled with the polyurethane precursor, then exposed to UV to cure the precursor into polyurethane (**Figure 1.24a**); at these dimensions, manual stacking (and bonding with polyurethane precursor again) produces reproducible 3-D structures (**Figure 1.24b**).

In 1987, Friedrich Bonhoeffer's group, then at the Max Planck Institute in Tübingen (Germany), was the first to use micromolded PDMS microchannels to microfluidically deliver biomolecules such as cell membrane fragments for axon guidance experiments (which came to be known as the "stripe assay" in the axon guidance community; see **Figure 6.41** in Section 6.5.1.1); Bonhoeffer's technique required a porous substrate to capture the biological material. Hans Biebuyck and colleagues at IBM Zurich have been credited for being the first to demonstrate, in 1997, microfluidic patterning of purified proteins on solid substrates (a somewhat incremental claim with respect to Bonhoeffer's, considering the IBM study was published in *Science*; see **Figure 2.18** in Section 2.4.3). The most attractive aspects of microfluidic patterning are:

1. The experimental protocols traditionally used for coating the surface of an open petri dish with any given biomolecule are translatable directly to the microfluidic format: the solution of interest is introduced in a microchannel and allowed to deposit onto the surface under the same conditions as the traditional protocol that has been known, often for decades, to work well for that surface.

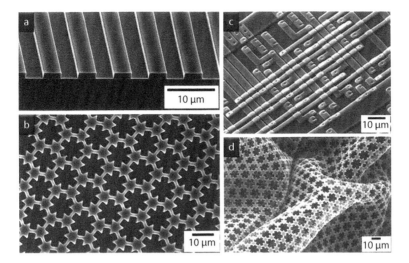

FIGURE 1.23 Micromolding in capillaries. SEMs of polyurethane microstructures on Si/SiO$_2$ formed using a commercially available, UV-curable polyurethane prepolymer used for optical parts. (a) Lines on Si/SiO$_2$. (b) Complex pattern containing interconnected features. (c) Quasi–3-D pattern containing multiple thicknesses molded in a single step. (d) Free-standing film formed by MIMIC followed by dissolution of the support in HF. (From Kim, E., Y. Xia, and G.M. Whitesides, "Micromolding in capillaries: Applications in materials science," *J. Am. Chem. Soc.*, 118, 5722, 1996. Reprinted with permission of the American Chemical Society.)

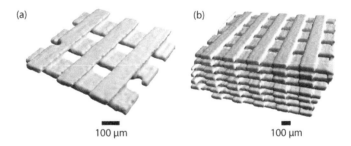

FIGURE 1.24 Microfluidically patterned polyurethane 3-D structures. (From Folch, A., S. Mezzour, M. During, M. Toner, and R. Muller, "Stacks of microfabricated structures as scaffolds for cell culture and tissue engineering," *Biomed. Microdevices*, 2, 207, 2000. Figure contributed by the author.)

2. The material to be patterned does not need to be dried. Thus, the concerns about protein denaturation that apply to microstamping do not apply to microfluidic patterning.

3. Replica-molding the microchannels from the master is straightforward, nonhazardous, and inexpensive (in terms of both materials and human labor).

4. The molding procedure leaves the master intact; thus, the most expensive part of the process, photolithography—or any other method used to pattern the master—is needed only once.

5. The patterning procedure usually does not damage either the sample or the channels (thus, the channels can be reused).

The most common concerns when using microfluidic patterning are:

1. The patterns must be connected; this limitation can be solved by bonding the micro-channels to a PDMS stencil, which limits the solution–substrate contact areas to the stencil holes (see Section 1.6.5).

2. For micron-sized channels in which the surface-to-volume ratio is very high, absorption of material onto the walls of the microchannels can be significant, especially in the case of proteins (see Section 2.4.3). This property can be used to the researcher's advantage to create gradients of, for example, etchants or biological solutions, in the direction of the channel length.

3. Microchannels with high fluidic resistance are difficult to fill and flush, which, in practice, limits the resolution of the technique. More specifically, microchannels of a few microns in depth can be challenging, unless they are very short. Once the channels are filled, the solutions cannot be exchanged easily (e.g., for flushing or further surface treatment) except by diffusion or specialized capillarity pumps. If resolution is not a consideration, wider/deeper channels can be perfused continuously (or the solutions exchanged easily) with pressure-driven flow.

4. Forces on the PDMS can detach the microchannels and spill the contents of a micro-channel into adjacent microchannels. Forces can be produced by hasty manual handling of the microchannels, especially during connection and disconnection of the inlets, or by changes in substrate/ambient temperature (PDMS expands when heated).

5. The microchannels can buckle, particularly for patterns in which most of the area is to be coated or for microchannels that are very shallow; the walls between channels cannot be arbitrarily thin or they will detach. Here, the hyperelasticity of PDMS also makes it difficult to model PDMS buckling, so researchers have found it simpler to

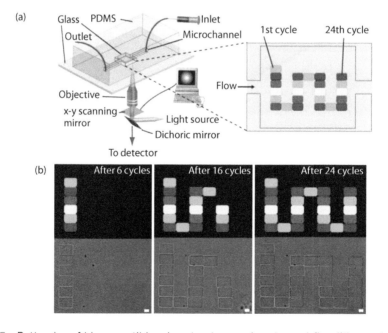

FIGURE 1.25 Patterning of biocompatible microstructures using stopped-flow lithography. All scale bars are 200 μm. (From Cheung, Y.K., B.M. Gillette, M. Zhong, S. Ramcharan, and S.K. Sia, "Direct patterning of composite biocompatible microstructures using microfluidics," *Lab. Chip.*, 7, 574, 2007. Reproduced with permission of The Royal Society of Chemistry.)

optimize pattern design by trial and error. As an empirical rule of thumb, PDMS microchannels (or walls) that have aspect ratios between 1:1 and 1:10 work well for most applications.

6. Solvent-based solutions cause PDMS to swell; hence, PDMS microchannels will detach if filled with solvents for extended periods of time.

7. For some applications in which surface purity is paramount, there is the concern that, under certain conditions, small amounts of dimethylsiloxane monomer may also remain on the areas where the PDMS has contacted the surface.

8. Fabrication of the microchannel is a very low-throughput process because it takes hours to produce a single PDMS replica and peeling it off the master is not easily automated.

9. Introduction of the solutions into the microchannels and starting/stopping flow is not easy to automate either, which limits microfluidic patterning to research applications.

Note that, in microfluidic patterning applications in which the deposition of material is directed by light, another degree of versatility is introduced by patterning the light as well. Patrick Doyle's group at the Massachusetts Institute of Technology (MIT) originally developed this modality, dubbed **stopped-flow lithography**. An example is shown in Figure 1.25. The

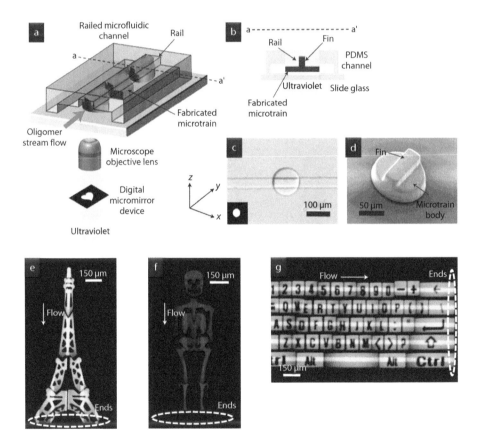

FIGURE 1.26 "Railed" microfluidic channels for guided self-assembly. (From Chung, S.E., W. Park, S. Shin, S.A. Lee, and S. Kwon, "Guided and fluidic self-assembly of microstructures using railed microfluidic channels," *Nat. Mater.*, 7, 581, 2008. Reprinted with permission from Macmillan Publishers Ltd.)

polymer consisted of 40% w/v of poly(ethylene glycol)-diacrylate (PEG-DA), with 2% w/v of Irgacure 2959 as photoinitiator. For clarity of visualization, PEG-DA in each cycle was doped with a combination of green, red, and blue fluorescent dyes.

A common misconception is that microfluidics is simply "assisting" the process because the actual patterning is performed with light. Note, however, that the microchannel is used to define the height of the microstructure, which allows for bypassing the traditional photoresist spin-coating step and, as a result, the whole clean room operation—clearly, the microchannel is not merely assisting but *enabling*. Moreover, it is possible to produce more complex structures and organize them downstream if the microchannel contains a guide (groove) on the roof, so the laser-fabricated structures can be directed downstream with a system of "rails" (Figure 1.26).

Another variation of stopped-flow lithography, called **lock-release lithography** (Figure 1.27), uses a microchannel with a very thin roof that is "inflated" on completion of the chemical reaction to release the microstructures. By sequential filling of the microchannel with different chemistries (Figure 1.28), it is possible to fabricate composite particles.

In summary, microfluidic patterning is inferior to photolithography in pattern fidelity and inferior to microstamping in convenience and resolution. As microstamping, it is pattern-dependent (unlike photolithography, the procedures used to produce one pattern may have to be changed for another pattern, and not all patterns are possible) but it is superior to photolithography in cost-effectiveness and materials versatility. For biological patterning, the absence of a need for further processing (e.g., obviating the protein drying step as needed for microstamping) can be invaluable with respect to traditional protocols.

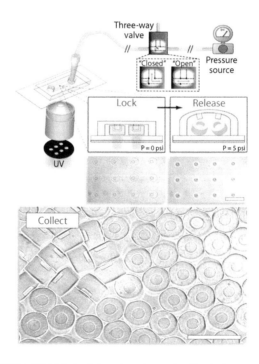

FIGURE 1.27 "Lock–release lithography" in microchannels. Scale bars are 100 μm. (From Bong, K.W., D.C. Pregibon, and P.S. Doyle, "Lock release lithography for 3D and composite micropartices," *Lab. Chip.*, 9, 863, 2009. Reproduced with permission from The Royal Society of Chemistry.)

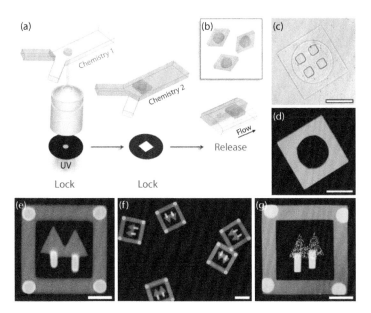

FIGURE 1.28 Synthesis of composite particles using lock-release lithography. Scale bars are 100 μm (c, d, and f) and 50 μm (e and g). (From Bong, K.W., D.C. Pregibon, and P.S. Doyle, "Lock release lithography for 3D and composite microparticles," *Lab. Chip.*, 9, 863, 2009. Reproduced with permission from The Royal Society of Chemistry.)

1.6.5 Stencil Patterning

Patterning with a thin piece of material containing holes ("**stencil patterning**") is an idea that was borrowed from art-making and has been used for decades to shadow the deposition of metal vapors. The first reported application of microfabrication technology to biology, by S. B. Carter in 1967, used a custom-made, foil-metal stencil to shadow the deposition of palladium islands of various shapes on a piece of acetate plastic; cells adhered only to the palladium. However, a metal stencil cannot be used to pattern material from the liquid phase because the stencil does not form a close contact with the surface. A PDMS stencil is an attractive solution because, as we have seen, PDMS seals well. Molding holes, however, is more challenging than molding features on a surface because the molding process must guarantee that PDMS is excluded from the top of the microfabricated features (or else the "holes" are occluded by a thin PDMS membrane). Exclusion is accomplished in essentially three ways:

- Spinning the master after addition of the PDMS prepolymer; the spinning speed must be large enough to exclude PDMS from the top of all the features. This method is probably the easiest one to optimize, but it requires a dedicated spinner and it results in stencils that are not flat at the top surface because of the meniscus formed where the air–PDMS prepolymer interface contacts the vertical walls of the master's features.

- Capping the master with a cover (which effectively creates a microchamber with columns), then filling the microchamber with PDMS prepolymer. This method is sometimes referred to as "microfluidic molding" (**Figure 1.29**). It is very reliable for dense, regular arrays; however, a stencil with only one hole or a few holes cannot be made with this method. Both the top and bottom surfaces of the stencil are flat after release.

- Pressing a cover against a master *after* PDMS prepolymer has been dispensed onto the master (i.e., "exclusion molding"; **Figure 1.30**). This method usually requires fine-tuning of the applied force. Not just *any* surface can be used as the cover—the cover

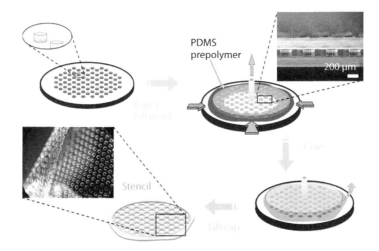

FIGURE 1.29 Fabrication of PDMS stencils by microfluidic molding. (From Tourovskaia, A., T. Barber, B. Wickes, D. Hirdes, B. Grin, D.G. Castner, K.E. Healy, and A. Folch, "Micropatterns of chemisorbed cell adhesion-repellent films using oxygen plasma etching and elastomeric masks," *Langmuir*, 19, 4754, 2002. Reprinted with permission.)

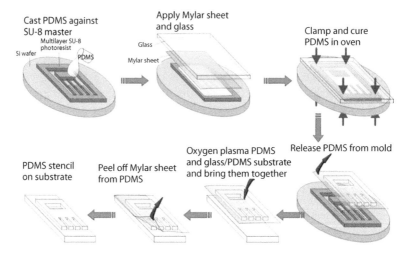

FIGURE 1.30 Fabrication of PDMS stencils by exclusion molding. (From Hsu, C.-H., C. Chen, and A. Folch, "'Microcanals' for micropipette access to single cells in microfluidic environments," *Lab. Chip.*, 4, 420, 2004. Reproduced with permission from The Royal Society of Chemistry.)

must be soft enough to form a good contact and hard enough to bridge features in the master without sagging. A solution to this problem has been the use of multilayer covers; however, the layer that comes into contact with the master is made of a thin sheet of plastic, and pressure is applied to the plastic by a hard surface such as a glass slide. Mylar (acetate film) happens to stick particularly well to cured PDMS; when Mylar is used as the cover for exclusion molding, cured PDMS—surprisingly enough—prefers the flat Mylar surface to the microstructured surface of the master, even though the PDMS is not irreversibly bound to the Mylar (i.e., the PDMS stencil can still be peeled off the Mylar).

PDMS stencils have been used to mask (i.e., "protect the underlying substrate from") many processes, such as protein adsorption, cell attachment, metal deposition, gas plasma etching, or water-based etchants. Obviously, stencils deform a lot when handled so they should not be used in applications in which alignment is critical. The first PDMS stencils were developed as masks for patterning regular arrays (of cells, in which alignment is not critical), but they also find a use in 3-D microfluidic circuits, in which the holes in the stencils can be used to connect fluid channels at two different levels. For microfluidic applications, exclusion molding of PDMS with a Mylar cover is very powerful because the Mylar layer (with the PDMS features on it) can be manipulated with high lateral precision in an aligner, which enables the formation of 3-D networks.

1.6.6 Dynamic Substrates

Soft lithography allows for fabricating thin PDMS membranes that can be regarded as dynamic microelements (e.g., they can be deflected by pressure application or can be used to confine moving fluids). A new generation of "tunable" substrates has been presented that can be used as "inflatable" molds or as grayscale photomasks.

1.6.6.1 Tunable Molds

The use of polymers for the fabrication of microdevices is attractive because polymer parts usually can be mass-produced at low cost by replica-molding and because their physicochemical properties can be tailored by appropriate monomer choice. However, most micromolding processes are limited by the range of features that can be produced in the master—with standard photolithography, features of uniform height and vertical sidewalls—and by the need for a different mold for each device design. The author's lab has devised molds whose features (cavities topped with PDMS membranes) can be individually deformed (or "tuned") to a continuum range of heights and smooth profiles by selective pressure application (**Figure 1.31**). The pressure settings, not just the original master's topography, determine the mold topography, which allows for creating many replicas with dissimilar microstructures from one mold.

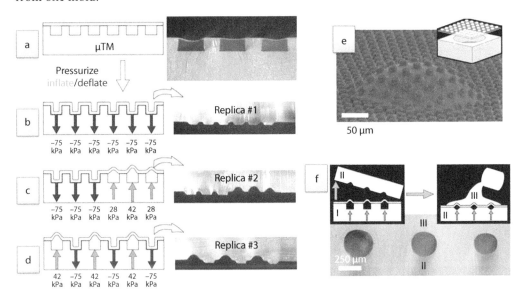

FIGURE 1.31 Tunable micromolding. (From Hoffman, J.M., J. Shao, C.-H. Hsu, and A. Folch, "Elastomeric molds with tunable topography," *Adv. Mater.*, 16, 2201, 2004. Reprinted with permission.)

It is also possible to replica-mold PDMS from liquid hydrogel features, as demonstrated by Bincheng Lin's group (Dalian Institute of Chemical Physics, China). In this process, a chemical pattern of poly(acrylamide) (pAAM) is UV-grafted onto a glass surface and, when dipped in a glycerol solution, the glycerol wets the pAAM gel pattern only. Glycerol replicates well in PDMS, which allows for generating interesting structures containing gradients of heights using a single mask (**Figure 1.32**).

Both methods have been shown previously, are straightforward to implement, and provide a simple route for microfabricating structures that are difficult, expensive, or impossible to produce with existing methods.

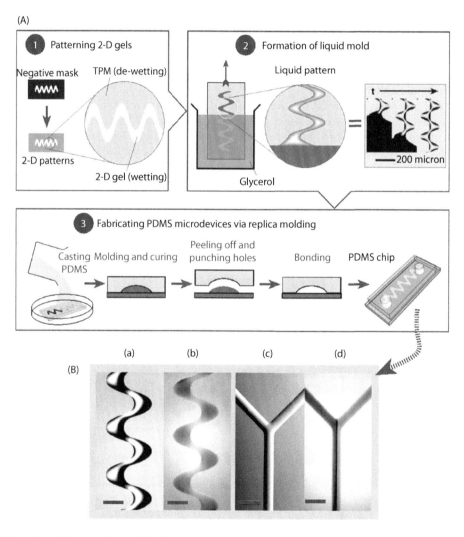

FIGURE 1.32 Micromolding of PDMS from liquid patterns. Scale bar is 1 mm. (From Liu, X., Q. Wang, J. Qin, and B. Lin, "A facile "liquid-molding" method to fabricate PDMS microdevices with 3-dimensional channel topography," *Lab. Chip.*, 9, 1200, 2009. Reproduced with permission from The Royal Society of Chemistry.)

1.6.6.2 Microfluidic Photomasks for Grayscale Photolithography

Although adequate for integrated circuit fabrication, "black-and-white" photolithography is ill-suited for the fabrication of structures such as 3-D micromachines, micro-optical components, or photochemical/biomolecular gradients, which are best produced with grayscale approaches capable of generating a range of opacities. Presently, grayscale photolithography can be realized with scanning lasers, micromirror projection displays, high–energy beam–sensitive glass photomasks, ultrahigh resolution "halftone" photomasks, or with metal-on-glass photomasks in which each grayscale level is determined by a different metal thickness. Unfortunately, all these grayscale approaches only *exacerbate* the costs/turnaround times of standard photolithography. The Folch lab has developed "**microfluidic photomasks**" (µFPMs) which contain liquid features that can be addressed (i.e., altered) with PDMS microchannels. The liquid can be any UV-absorbing water-soluble dye. The µFPMs contain a virtually unlimited number of gray scales and can be manufactured rapidly and inexpensively. The µFPMs are transparent sheets of PDMS that enclose microfluidic channels next to one of the surfaces of the sheet. The method, schematized in **Figure 1.33**, consists of five steps: (a) micropatterning of a silicon master, (b) replica-molding of PDMS from the master, (c) bonding of the PDMS mold against a PDMS thin film (~10 µm thick) to form microchannels and introduction of dye solutions of desired concentrations into the microchannels, (d) placing the µFPMs in contact with a photoresist film and exposure of the photoresist to UV light using standard photolithography equipment, and (e) developing the photoresist. Note that the resulting structure has multiple heights (**Figure 1.33g** and **h**), despite the fact that the starting photoresist height, the microchannel heights, and the exposure were homogeneous. PDMS is inexpensive and the fabrication of a µFPM takes less than 4 hours. The master and the µFPM can both be reused endlessly if operated with care.

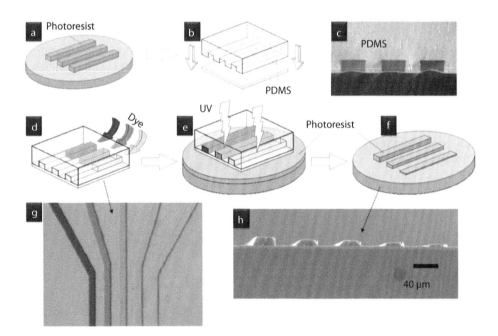

FIGURE 1.33 Microfluidic photomasks for grayscale photolithography. (From Chen, C., D. Hirdes, and A. Folch, "Gray-scale photolithography using microfluidic photomasks," *Proc. Natl. Acad. Sci. U. S. A.*, 100, 1499–1504, 2003. Reprinted with permission from the National Academy of Sciences.)

1.7 Hydrogel Devices

PDMS is not the only soft material "out there." As a matter of fact, another class of material—a **hydrogel**—might be more interesting for certain applications. Consider agarose, for example, most commonly used for electrophoresis in biological laboratories. Its hardness can be tailored (from "yogurt-ish" to "PDMS-ish") simply by changing the percentage weight of water added to a dried agarose powder that is sold by the bucket. It is not as transparent as PDMS, but it is translucent. Two examples are shown in **Figure 1.34**.

Agarose stamps are uniquely useful for micropatterning materials. Compared with a PDMS surface, which can only be "inked" at the surface, the gel has the advantage that the bulk of the gel can also be "loaded." The total amount of molecules that can be loaded depends only on the thickness of the stamp; the time it takes to load a given amount is a strong function of the diffusivity of the molecules in the gel (i.e., of the gel density and the molecule size). To facilitate the handling of the hydrogel stamp, an 8% to 12% w/w solution of agarose is used. The agarose is degassed in vacuum and gelated on the mold (typically made of PDMS). As in any micromolding method, depending on the feature pattern, the stamp at this point should contain a faithful complement of the master's features after the gel is peeled off the master mold. Finally, the stamp is loaded with the solution of interest simply by soaking the stamp in the solution for a while. Intermediate drying of the gel is critical because if it is too wet, the trenches between the features are filled with liquid, and if it is too dry, the gel loses its properties. (This is not unlike PDMS microstamping, in which we shall see that drying of the stamp is also critical, especially when printing proteins.) Still, typical methods for drying are simple: blow-drying with nitrogen, placing the stamp on filter paper, and so on, after which the stamp is left alone for a while to homogenize its water content.

A striking demonstration, coming from the group of Bartosz Grzybowski at Northwestern University in Illinois, is provided by loading (i.e., *soaking*) the stamp with an etchant (**Figure 1.35a** through c). When loaded with a copper etchant, the stamp will carve holes in copper with the shape of the features in the stamp. On hydrophilic surfaces such as glass, lateral spreading of the etchant is avoided by etching under oil. It should be noted that the stamp not only delivers the reactants but also absorbs the products of the chemical reaction—diffusion acts both ways. The gels also make it possible to prepare and microfabricate features on free-standing foils, which are grown by electroless deposition on the surface of the gel from the reactants that are soaked into the gel (**Figure 1.35d** and e).

As we shall see, hydrogel devices have been used successfully as stamps to print biological material—ranging from proteins to cells—as microfluidic gradient generators, and as scaffolds for tissue engineering.

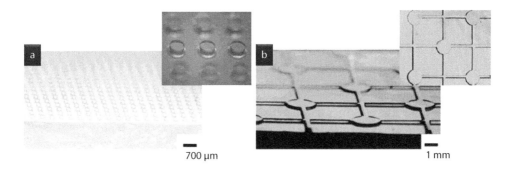

FIGURE 1.34 Agarose stamps. (From Weibel, D.B., A. Lee, M. Mayer, S.F. Brady, D. Bruzewicz, J. Yang, W.R. DiLuzio, J. Clardy, and G.M. Whitesides, "Bacterial printing press that regenerates its ink: Contact-printing bacteria using hydrogel stamps," *Langmuir*, 21, 6436, 2005. Reprinted with permission from the American Chemical Society.)

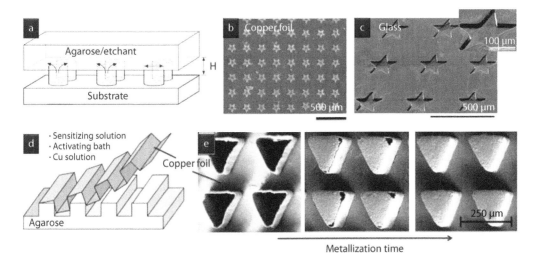

FIGURE 1.35 Depositing and etching of posts and wells using agarose stamps. (a–c: Smoukov, S.K., K.J.M. Bishop, R. Klajn, C.J. Campbell, and B.A. Grzybowski, "Cutting into solids with micro-patterned gels," *Adv. Mater.* 17, 1361, 2005; d and e: Smoukov, S.K., K.J.M. Bishop, R. Klajn, C.J. Campbell, and B.A. Grzybowski, "Freestanding three-dimensional copper foils prepared by electroless deposition on micropatterned gels," *Adv. Mater.* 17, 751, 2005. Reproduced with permission from Wiley-VCH Verlag GmbH & Co. KGaA.)

1.8 Nanofabrication Techniques

In many applications, UV light does not provide enough resolution to define the features necessary to probe a given phenomenon. This textbook is not the place to review the vast array of techniques—many under intense development—that exist for fabricating nanoscale objects, but we will cite two broad families because of their widespread use: **electron beam lithography** and **scanning probe lithography**. Unfortunately, both are *serial patterning* techniques—patterning twice as many features takes twice as long.

1.8.1 Electron Beam Lithography

Electron beam lithography uses the electron gun of a scanning electron microscope (SEM; the same gun used for imaging) to inject a highly focused beam of electrons into the sample (**Figure 1.36**). The e-beam resist is usually a thin layer of PMMA, which is a positive e-beam resist; SU-8 is a negative e-beam resist. Unfortunately, the patterning resolution is dictated not so much by the e-beam spot size (which can be focused down to ~1 nm diameter by accelerating the e-beam to voltages >100 kV) but by the secondary electrons emitted by the sample in a circumference that can vary from 10 to 100 nm (depending on the material and charging conditions).

1.8.2 Scanning Probe Lithography

Scanning probe lithography uses a sharp tip in proximity (not necessarily in contact) with the surface of interest. The tip, an integral part of a **scanning probe microscope** (SPM), is typically used to image the surface. A proximity signal (e.g., a tunneling current, or the deflection of the cantilever where the tip is mounted) is used to adjust the vertical distance between the tip holder and the sample as the tip is scanned across the sample. To perform nanoscale lithography, the tip of the SPM is used to modify the surface (**Figure 1.36**)—either indentation, (electro)chemical

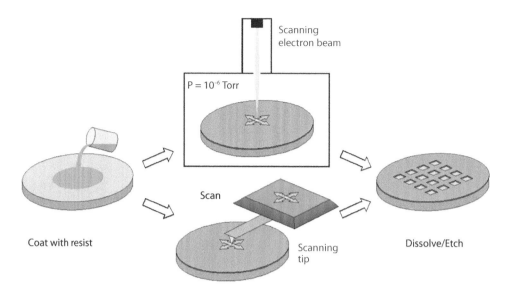

FIGURE 1.36 Nanoscale lithography.

growth, charge deposition, etc.—by temporarily disabling the proximity signal feedback. If the tip has been used to modify a resist layer, the resist must be developed with a developer solution to transfer the pattern to the substrate.

1.9 Fabrication Based on Self-Assembly: A "Bottom-Up" Approach

In his often-cited visionary lecture "*There is plenty of room at the bottom*" in 1959 at the American Physical Society meeting at Caltech, the physicist Richard Feynman explained very graphically the vast amount of information storage density available on surfaces down to nanometer scales. In it, he explained how one could envision manipulating atoms if it were possible to build nanoscale machines. This talk historically marked the birth of the field of nanotechnology. The present view, however, is radically different. Although it is true that there exist nanoscale machines that manipulate molecules atom by atom, it is also true that they existed well before humans: Nature has been inventing them for millions of years in the shape of proteins and RNA enzymes. Nevertheless, Feynman's talk inspired a whole generation of scientists. Yet (undoubtedly for lack of time) the talk did not address a crucial aspect of nanoscale machines and surfaces: at short distances, surfaces exert enormous attractive forces that will keep most molecular entities stuck to them like glue.

Did someone say glue? Maybe *that* can be put to some good use. Perhaps at very small scales, we can make use of those natural attractive forces to cause objects to stick to each other spontaneously in ordered arrays. Actually, this "bottom-up" approach already has a technical name: it has been termed **self-assembly**. Self-assembly is actually based on an old observation: that if you take thousands of spheres, such as balls or oranges, and put them in a container, they *always* pack to minimize the spacing left between them—because that minimizes their gravitational energy. This can be generalized to beads on a sticky surface or in thin liquid films, as was known since the 1980s: place a drop of water containing latex or silica beads on a surface, let it dry, and the surface tension as the water recedes creates a monolayer of beads in a near-perfect hexagonal array arrangement—which minimizes the capillary forces between the spheres.

All systems have a tendency to settle into a minimal energy state. It is straightforward to realize that, as long as the beads form a monolayer (i.e., as long as they do not stack up), the spaces between the beads can now be used to shadow-deposit several materials; hence, many research groups have jumped at the opportunity to use such bead monolayers to generate ordered arrays of nanopatterns having resolutions that rival those achievable with expensive electron beam lithography systems—but for pennies. This technique is now referred to as "**particle lithography**," of which we will see an application to protein patterning in Section 2.4.5.

Before particle lithography was born, George Whitesides and coworkers at Harvard University had already applied the concept of self-assembly to single heterobifunctional molecules of the right length. Certain classes of molecules form what are now known as **self-assembled monolayers** or **SAMs** (see Section 2.2.1), which assemble via electrostatic interactions, and Whitesides' group extended it to the assembly of millimeter-sized objects via capillary interactions (Figure 1.37). To obtain self-assembly on a millimeter scale ("meso-scale"), selected faces of polygonal small objects were metallized and covered with tin solder (which was allowed to cool down and solidify); introduction of large numbers of these small polygons in a heated rotating bath of KBr melted the solder (Figure 1.37a), but the solder faces eventually "found" each other and stayed stuck by capillarity, magically producing large assemblies over time whose shape depended on the shape and configuration of the individual components (Figure 1.37b). With clever design, it was possible to fabricate 3-D electrical circuits (Figure 1.37c), folding structures (Figure 1.37d), millimeter-scale models of protein folding, and aggregates that reconfigured their structures when their environment changed. Such bottom-up thinking is now commonplace in the design of devices with advanced materials

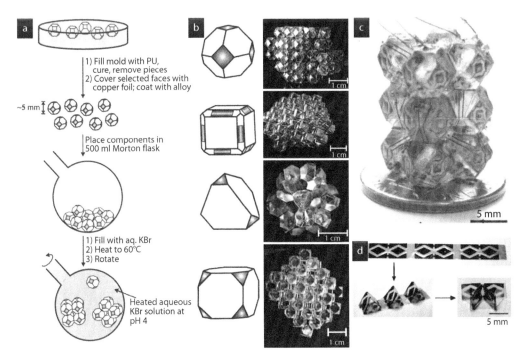

FIGURE 1.37 Mesoscale self-assembly. (a and b: From Breen, T.L., J. Tien, S.R.J. Oliver, T. Hadzic, and G.M. Whitesides, "Design and self-assembly of open, regular, 3D mesostructures," *Science*, 284, 948, 1999; c: From Gracias, D.H., J. Tien, T.L. Breen, C. Hsu, and G.M. Whitesides, "Forming electrical networks in three dimensions by self-assembly," *Science*, 289, 1170, 2000; d: Bruzewicz, D.A., M. Boncheva, A. Winkleman, J.M. St. Clair, G.S. Engel, and G.M. Whitesides, "Biomimetic fabrication of 3D structures by spontaneous folding of tapes," *J. Am. Chem. Soc.*, 128, 9314–9315, 2006. Reprinted with permission of the American Chemical Society.)

and complicated geometries in which traditional fabrication technologies cannot reach, such as in the assembly of components on curvilinear surfaces (e.g., an electronic contact lens, see Section 8.2.6), in tissue engineering, in microrobotics, etc.

1.10 Summary

Photolithography, the miniaturization of components with light as used routinely for the fabrication of microelectronics, constitutes the starting point for miniaturization in BioMEMS. Because of its materials and costs constraints, however, most of the current work on BioMEMS relies on soft lithography, based on the patterning of materials with the elastomer PDMS, a material that has many useful properties for biopatterning and for device fabrication. The first step of soft lithography is the fabrication of a master mold using photolithography. PDMS replicas of this mold can be used to make stamps and microfluidic devices, which are then used to selectively deposit the material of interest in selected locations. Approaches based on hydrogels (e.g., agarose), biocompatible photoresists, and self-assembly are emerging as powerful alternatives or complements to PDMS for biopatterning.

Further Reading

Abgrall, P., Conedera, V., Camon, H., Gue, A.M., and Nguyen, N.T. "SU-8 as a structural material for labs-on-chips and microelectromechanical systems," *Electrophoresis* **28**, 4539–4551 (2007).

Ariga, K., Hill, J.P., Lee, M.V., Vinu, A., Charvet, R., and Acharya, S. "Challenges and breakthroughs in recent research on self-assembly," *Science and Technology of Advanced Materials* **9**, 1–96 (2008).

Banta, S., Wheeldon, I.R., and Blenner, M. "Protein engineering in the development of functional hydrogels," *Annual Review of Biomedical Engineering* **12**, 167–186 (2010).

Bishop, K.J.M., Wilmer, C.E., Soh, S., and Grzybowski, B.A. "Nanoscale forces and their uses in self-assembly," *Small* **5**, 1600–1630 (2009).

del Campo, A., and Greiner, C. "SU-8: a photoresist for high-aspect-ratio and 3D submicron lithography," *Journal of Micromechanics and Microengineering* **17**, R81–R95 (2007).

Madou, M. *Fundamentals of Microfabrication*, 752 pp. CRC Press (2002).

Maruo, S., and Fourkas, J.T. "Recent progress in multiphoton microfabrication," *Laser & Photonics Reviews* **2**, 100–111 (2008).

Qin, D., Xia, Y., and Whitesides, G.M. "Soft lithography for micro- and nanoscale patterning," *Nature Protocols* **5**, 491–502 (2010).

Wang, W., and Soper, S.A. (editors), *Bio-MEMS: Technologies and Applications*, 477 pp. CRC Press (2007).

Weibel, D.B., DiLuzio, W.R., and Whitesides, G.M. "Microfabrication meets microbiology," *Nature Reviews Microbiology* **5**, 209–218 (2007).

Wikipedia article on etching, "Etching (microfabrication)," see http://en.wikipedia.org/wiki/Etching_(microfabrication).

Xia, Y., and Whitesides, G. "Soft lithography," *Angewandte Chemie (Int. Ed.)* **37**, 550–575 (1998).

2

Micropatterning of Substrates and Cells

"Traditional" materials used in microelectronics fabrication—namely, silicon, metals, photoresist, and others—are not interesting for the study of biomolecules (proteins, DNA, etc.) and cells. Because many efforts in BioMEMS have been focused on micropatterning materials such as proteins, cells, and biomedically interesting polymers, there is a wide array of "biopatterning" techniques now available. Here, we only discuss techniques that involve some type of miniaturization, which necessarily excludes the potentially powerful use of focused lasers. Indeed, lasers have been used to "pick and drop" single cells ("laser traps"), or to laser-ablate tiny portions of cell culture substrates (with the cells on them) so as to "shoot" them against a "receiver" substrate; however, these techniques have had little success in the applications covered here, likely because of their serial nature and concerns over the deleterious effect of intense radiation on cell physiology.

Before we proceed to a description of those techniques, we should ask ourselves the same question that BioMEMS researchers considered back in the 1980s: Can we use photolithography to pattern proteins using the traditional liftoff procedure, the same way one uses it to micropattern, say, a gold layer? The answer can be quite involved and depend on the application (see "A thought experiment" text box for a case example), but a short safe answer is, "if possible, no." Many biopatterning methods use photoresist-based photolithography at some point in the process. Overall, however, the application of photolithography to biopatterning faces two major obstacles: (1) most chemicals used in standard photolithography are toxic to cells and are denaturants to biomolecules, and (2) biological solutions, because of their rich ionic and molecular content, are a menace to the finely tuned conductivity of a semiconductor circuit and are therefore banned from most clean-room facilities originally designed for microelectronics applications. The methods reviewed in this chapter constitute creative solutions to dodge either one or both of these problems. In other words, how can we pattern proteins or cells without making them enter the clean room or come in contact with photoresist?

A THOUGHT EXPERIMENT

The following thought experiment will help us understand some of the limitations imposed by photolithography. Let us try to pattern a protein layer by liftoff: we first create a photoresist micropattern on a substrate (step 1), then we deposit the protein (step 2), and finally lift off the photoresist micropattern (step 3).

Step 1 seems straightforward and can be done in the clean room but, in practice, we encounter two caveats: (a) spin-coating the substrate with a thin layer of photoresist requires a flat substrate (so that the excess photoresist flies off the edge), hence the convenient petri dish (which serves as a container for the cell culture solution and is *itself* the cell culture substrate) cannot be used; and (b) the photoresist solution contains organic solvents that attack polystyrene, the most widely used cell culture substrate (the plastic petri dish). After overcoming this little disappointment, we would probably settle on a more inert but equally attractive and foolproof substrate such as glass.

Step 2 can use one of a number of protein immobilization steps that are detailed in the next section and should be relatively easy because many chemical methods to attach proteins to surfaces were designed to work on glass. Apparently, there is no major problem there—except that the management of the clean room will not like it that our surface was exposed to inorganic ions: other users doing microelectronics consider inorganic ions "contamination" because the ions diffuse readily into the silicon substrate and damage its conductivity. Removing the small ions uses chemical etches that also remove the proteins. So, what can we do? The only hope is that the clean room has equipment dedicated to "contaminating" processes—which is not unthinkable if you are in a university with a MEMS-friendly facility.

If you are able to overcome step 2, which presents a practical challenge, step 3 presents a more fundamental problem: How can we remove the photoresist without affecting the protein layer, trying to leave absolutely no photoresist? (We are concerned that the photoresist might be toxic to cells—after all, on the bottle, it does not say we can eat it.) Unfortunately, the vast majority of photoresists are organic polymers that are soluble in organic solvents only; hence, to remove the photoresist layer, we have to expose the protein-coated substrate to an organic solvent. However, proteins have evolved for millions of years to adopt precise shapes in the presence of water, and it is the water that provides the protons that constitute the hydrogen bonds necessary to stabilize the protein structure. An organic solvent will exclude water around the protein and make drastic changes to its natural shape (a process called *denaturation*), which may result in a loss of activity. The loss of activity is more likely to be a problem for enzymes, specialized proteins that catalyze biochemical reactions. Nevertheless, the early BioMEMS literature is full of reports in which the use of solvents did not seem to hinder significantly the bioactivity of proteins denatured by solvent exposure. This insensitivity to denaturation may be real (e.g., the exact 3-D shape of the protein is not always relevant for the assay because the activity of the protein is localized to a small active site composed of a few peptides) or apparent (e.g., the protein was already denatured by virtue of the fact that it was adhered to a substrate).

Before we proceed to explain how the techniques outlined in Chapter 1 can be applied to micropattern these "new" materials, it is important to consider the *importance of surfaces*. In most BioMEMS devices, the interaction of cells or biomolecules with a surface of the device plays a significant role in the device's function. However, we must not satisfy ourselves with the thought that biomolecules and cells magically stick to surfaces—understanding the physicochemical, biochemical, and biophysical principles that cause or deter adhesion is crucial to designing a successful device. Also, not all surface treatments are compatible or easily combined with all microfabrication techniques, so a minimal knowledge of **surface engineering** can be very helpful in implementing a BioMEMS idea. Here, we cover the principles and techniques for micropatterning **proteins**, **cells**, **nucleic acids**, and certain **synthetic polymers**.

2.1 Interaction between Surfaces and Biomolecules

If you were a biomolecule, you would find that surfaces are very *sticky*, and that once stuck to them, it would be very hard for you to release yourself. Also, the forces you would feel likely would be so strong as to flatten you against the surface, not unlike a hypergravity world. In surface science language, the phenomenon in which a molecule adheres to a surface is called **adsorption**. If the adsorption is mediated purely by physical forces, it is termed **physisorption**. If a chemical bond forms between the molecule and the surface, then we speak of **chemisorption**. These interactions exert forces on the biomolecule, distorting its shape. Extreme shape changes can result in loss of (some or all) bioactivity, a phenomenon known as **denaturation**. (Denaturation can occur through all sorts of causes, not just by the proximity of a surface—for example, the exposure to a solvent causes a protein to be denatured.)

Physisorption is mediated by four different forces or interactions, and in all of them, water plays a crucial role:

* Biomolecules are invariably polar, which means they have a dipole moment that is attracted to the charges and dipoles present in the solid (whose surface we are considering). Dipoles always induce dipoles in other nearby molecules or atoms. The added action of all the fluctuating, dipole-induced dipole interactions is known as **van der Waals force**, which is always attractive. The range of this force for two simple molecules in a vacuum is only a few nanometers. The situation becomes extremely complex in the presence of a solvent such as water, which must be considered a third player in its own right.

* Biomolecules are also invariably charged, and so are most surfaces in physiological fluids. The **electrostatic force** between a biomolecule and a surface depends on the ionic concentrations present in the solution and on the surface functional groups. Unlike the well-known Coulomb force between two charges in a vacuum, the electrostatic forces in a physiological fluid are hard to predict because they are extremely sensitive to very small concentrations of ionic species (often undetermined) in the solution, which partially can shield the electrostatic field felt by more distant ions. The forces associated with the formation of complexes of ions and water are termed **hydration forces**.

* A hydrogen atom can dissociate partially from a donor (electronegative) atom and be shared with another electronegative atom, forming a **hydrogen bond**. When a highly polar group such as –OH (abundant on glass surfaces and many cell culture polymer substrates) is immersed in water, the hydrogen atom is so loosely bound to the polar group that it often prefers to altogether lose its electron (becoming a **proton, H^+**) and leave the group –O^- on the surface. The proton can then associate with either water (forming the **hydronium ion H_3O^+**) or another polar group on the surface of a biomolecule (through a hydrogen bond). This competition highlights the important role of water in the interaction between biomolecules and surfaces. Needless to say, this process is highly dynamic: protons actually hop from molecule to molecule millions of times a second.

* As a result of the strong associations that water forms with polar groups, nonpolar groups and water tend to repel each other (a net, loosely defined interaction referred to as **hydrophobic force**); thus, nonpolar groups are left with no other choice but to stick to each other. It is not so much that nonpolar groups "like" each other but that they are "bullied" into "hugging" each other by the surrounding water network of protons and hydronium ions.

As a corollary of the above, it is hard to design (or find) a surface to which proteins and nucleic acids (both large biopolymers with many charged groups and dipoles) will *not* physisorb. However, there are also molecular-scale random events that can counteract this propensity to adhere to surfaces:

❄ Thermodynamic fluctuations (parts of the biomolecule bending to an adherence-unfavorable position, collisions with other biomolecules in the solution, etc.) can cause the biomolecule to leave the surface, a phenomenon known as **elution**. Certain surfaces favor elution more than others, and surfaces (or surface coatings) that altogether deter protein physisorption have attracted a lot of interest in BioMEMS, as we will see. Elution can also be provoked by changing the solvent (e.g., water for ethanol), as this disrupts all the hydrogen bonds and electrostatic interactions that were keeping the molecules in place.

❄ A biomolecule physisorbed onto a surface can be substituted or displaced by another molecule that is attracted to the surface with a larger force.

As a second corollary, it follows that physisorbed coatings are unstable. Hence, for applications such as biosensors and implants, historically, many efforts have been directed toward chemically anchoring biomolecules on surfaces to create surface coatings that are stable. But is it worth the trouble?

2.1.1 Physisorption versus Chemisorption

Before we review the most common chemical immobilization strategies used in BioMEMS, we briefly point to frequent misconceptions about the stability of chemisorbed layers in biological fluids. The question that concerns us is the choice that the BioMEMS researcher must make when a biomolecular coating for a surface is sought: Should the biomolecule under consideration be physically adsorbed or chemically linked to the surface?

The advantages of physisorption are simple. As opposed to chemisorption:

❄ Physisorption procedures require no expertise or special equipment.

❄ Physisorption can be used in combination with most surfaces, in particular with the ubiquitous, inexpensive glass slides or polystyrene petri dishes used as cell culture substrates.

❄ Physisorption does not *require* surface treatment (on most surfaces); however, in some cases, it may be advantageous to treat the surface to enhance physisorption (e.g., tissue-culture polystyrene petri dishes are treated with a plasma discharge that is believed to introduce charged groups on the surface of polystyrene, which thus becomes more adhesive to proteins and cells).

Reciprocally, those advantages can be listed as disadvantages of a chemisorption method:

❄ Chemisorption requires a specialized substrate that allows for a chemical linkage between the biomolecule and the surface; because the functional groups at the surface of organic polymers are not always well characterized, these materials, otherwise extremely appealing in BioMEMS for many other reasons, are not compatible with a number of the chemistries commonly used to immobilize biomolecules (as we will see).

❄ Chemisorption often requires the expertise of a chemist (not always available in biological laboratories). In modern laboratories with stringent safety standards, it also mandates the presence of a chemical hood, proper exhaust in the building, and waste disposal procedures. Implementation of safety training and precautions, even when it is commonsensical, is bound to generate tedious paperwork and inspections (believe me). So, even a chemist will tell you that physisorption is simpler than chemisorption.

Still, the notion that, in principle, chemisorption allows for a more stable anchorage between the biomolecule and the surface has historically appealed to many groups in BioMEMS. However, the degree and usefulness of this "stability" can be challenged:

❄ Many biomolecules can also physisorb atop each other, thus, ensuring a chemical linkage between the bottom layer and the surface does not *necessarily* ensure a stable coverage either; in applications involving direct contact between the substrate and a cell, the space between the cell membrane and the surface can be filled by proteins secreted by the cell; severe versions of this secretion process are the basis of the *encapsulation* of implants by fibroblasts or the *fouling* of surfaces by bacteria.

❄ Because biomolecules do not react with the most widely used artificial substrates, it is necessary to use one or more linkage groups. The chemical groups involved in the chemisorption reaction might cross-link unwanted molecular species, which can confound the results; this problem can be more pronounced for multistep chemical reactions because the efficiency of a chemical reaction never reaches 100%. We will see several examples in these pages.

❄ In some cases, the biomolecule itself may be degraded by natural processes (e.g., cells can secrete proteases, enzymes that degrade proteins); in that case, whether the biomolecule is chemisorbed or physisorbed is irrelevant.

2.1.2 Hydrophilic versus Hydrophobic

What surfaces, then, favor physisorption? A common misconception in the BioMEMS literature is that protein physisorption can be explained or predicted by some measure of the affinity of the surface to *water*—a property known as **hydrophilicity**. Hydrophilicity is not really a magnitude (only a property) but is commonly quantitatively described by the **contact angle** between a droplet and a surface, which is the angle between the tangent of the fluid–air interface and the surface at the point of contact. (Note that hydrophilicity is identified at the interface of three materials: water, surface, and gas.) For a given surface, the droplet of a "wetting liquid" would form a contact angle smaller than 90 degrees, that is, a droplet would look like less than half a sphere; a "nonwetting" droplet would form a contact angle larger than 90 degrees and would look like bigger than half a sphere. A **hydrophilic surface** is one that *water* can wet, and a **hydrophobic surface** is one that water cannot wet, although the distinction can be fuzzy (especially for contact angles around 90 degrees) because there is a continuum of hydrophilicities. As it turns out, maximum physisorption is observed for surfaces that are moderately hydrophobic (as are most organic polymers), for which increased protein physisorption is *correlated* with increased hydrophobicity (larger contact angles). However, hydrophilicity—or, conversely, **hydrophobicity**—alone does not *really* explain the affinity of the protein with the surface because neither very hydrophobic (such as Teflon) nor very hydrophilic surfaces (such as gels) support protein physisorption. Clearly, a satisfactory mechanistic explanation of physisorption must account for the forces that attract the biomolecule toward that surface; the question is extremely complex because the precise structure of water at the fluid–surface interface and at the biomolecule's surface is still, after decades of intense research, not fully understood.

2.1.3 Cell Attachment to Substrates

The molecular mechanisms by which cells recognize certain substrates as suitable for attachment have largely been elucidated. Cell adhesion is mediated by cell membrane–bound receptors. In particular, **integrins**, a family of heterodimeric transmembrane proteins that are linked to the cytoskeleton on the cytoplasmic side of the membrane, recognize specific peptide sequences present in the cell-secreted fibrillar meshwork of proteins and polysaccharides known as **extracellular matrix** (**ECM**). Thus, integrins establish a mechanical link not only between the membrane and the ECM substrate but also between the ECM and the cytoskeleton. Moreover, integrins aggregate in organized structures termed **focal adhesions**. More importantly, in most cell types, certain biochemical signals essential for cell growth, function, and survival are triggered by integrins on attachment: without attachment, the cell undergoes apoptosis or programmed cell death.

In the 1980s, Erkki Ruoslahti and Michael Pierschbacher at the La Jolla Cancer Research Foundation identified and synthesized the short peptide sequences in ECM that are responsible for cell attachment and spreading when they are recognized by integrins. A notorious one, present in fibronectin and collagen, is the tripeptide arginine–glycine–aspartic acid (**RGD**) sequence. Numerous studies have shown that surfaces to which cells do not adhere normally, when grafted with RGD-type sequences on their surface, become cell-adherent. The recognition of these peptides by integrins is so specific that, if a surface is prepared with a mutant of RGD that differs by a single peptide, the cells do not attach to it anymore. Note that other ECM proteins differ in the peptide sequences that are recognized by cells for cell attachment—for example, laminin's is the pentapeptide tyrosine–isoleucine–glycine–serine–arginine (**YIGSR**), commonly used in neuronal cell culture.

2.2 Surface Engineering

Surface chemistry methods have been used to tailor the adhesiveness of biomolecules and cells on artificial surfaces and implantable biomaterials for a long time. Howard Weetall, working for Corning (New York), was the first to show in 1969 that enzymes may be immobilized on glass via an organosilane linkage and yet retain their activity. Usually, these methods seek the formation of an intermediate monolayer of organic molecules bearing specific functional groups by reaction of their reactive end group with a clean surface of the appropriate reactivity. Glass has traditionally been the substrate of choice for surface chemical immobilization of biomolecules because, unlike most organic polymers, it has a well-characterized surface composition. In principle, in the absence of contaminants, the heterobifunctional nature of these molecules should ensure that the thin film is monomolecular. Upon formation of a complete monolayer, the chemical nature of the surface is no longer defined by the composition of the underlying substrate but by the exposed functional groups of the monolayer. In the case of implantable materials, this strategy is often used to mask the chemical identity of the implant (which would otherwise cause an immune reaction) or to control the mechanical integration of the implant with the surrounding tissue.

2.2.1 Self-Assembled Monolayers

As depicted in **Figure 2.1**, some organic molecules form a close-packed "self-assembled" monolayer (**SAM**). As the SAM molecules react (i.e., chemisorb) with a surface from solution, they cooperate to minimize the free energy of the surface by spontaneously forming a densely packed, crystalline-like layer—hence the term "self-assembly"—of single-molecule thickness. In the case of alkanethiols on gold, the fully extended alkyl chains spontaneously tilt from the surface normal by 20 to 30 degrees to optimize the packing. The best packing occurs for intermediate lengths: for alkyl chains, if the chains are too long (more than ~16 carbons), random molecular motions overcome the attractive forces between the chains, and if the chains are too short (less than ~10 carbons), the attractive forces overcome the thermal noise.

For reasons of ease of synthesis, the most widely used SAM molecules are the long-chain **alkanethiols** ($CH_3–(CH_2)_{n-1}–SH$, $n > 9$) on metals such as Au, Ag or Cu—via bonding of the sulfur end to the metal surface—and the **alkanesilanes** (or alkanesiloxanes) on silicon or various oxides—via reaction of the silane/siloxane end group with a surface hydroxyl group (see **Figure 2.1**). Alkanephosphonates on zirconium oxide, alkaneisonitriles on platinum, and alkanecarboxylic acid or alkanehydroxamic acid on aluminum oxide have also been shown to form SAMs.

SAMs of different end-functionalities have been used to tailor surface properties such as, for example, adhesion, lubrication, wettability, corrosion, or protein adsorption. Moreover,

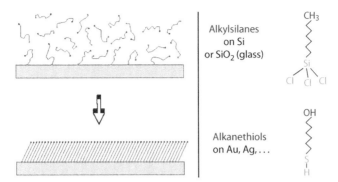

FIGURE 2.1 Self-assembled monolayers.

SAMs may be functionalized with reactive groups to which biological material is later attached. Proteins may then be immobilized either directly or by attaching an intermediate "tether" or cross-linker onto the monolayer.

SAMs whose end-functionalities are recognized directly by specific biomolecules are also possible. In the early 1990s, Massia and Hubbell created SAMs functionalized with the specific peptide sequences (such as RGD) recognized by cell adhesion receptors during cell attachment and spreading. These peptide layers are more stable against heat treatment and pH variation than ECM protein coatings. The advantage of promoting cell attachment via covalently linked biospecific peptide sequences is that surface coverage and composition is easily controlled and temporally stable, whereas physisorbed protein layers may, in time, elute from the surface as a function of time or be digested by cell-secreted proteases. Other SAMs with biospecific end-functionalities have been designed to recognize histidine-tagged proteins or biotinylated biomolecules.

SAMs may be formed on materials other than glass, silicon, or metal, such as organic polymers. By comparison to glass, however, organic polymers have a poorly characterized surface composition and nanoscale topography, which makes it difficult to design reactive chemistries. The technology of SAM formation onto organic polymers dates only to the early 1990s and still faces many challenges. Unlike glass, which has a high density of surface hydroxyl groups, the surfaces of organic polymers need to be "activated" with reactive groups such as amines, alcohols, carboxylic acid, or thiols before covalent immobilization. A favorite method of activation is the exposure to a plasma (a highly ionized gas); the gas can be air (the plasma is created with a simple handheld glow discharge gun) or pure oxygen (the plasma is created inside a microwave oven that has been fitted with an oxygen inlet and a vacuum pump). Hydrophobic polymers may be rendered hydrophilic (presumably by introduction of a number of undetermined reactive groups such as alcohols, aldehydes, carboxylic acids, ketones, etc.) by plasma-activating the surface so it can be silanized. Aminated polystyrene petri dishes are now commercially available, although the process used for amination is usually not specified. Despite the surface chemistry challenges, (transparent) polymers are an attractive cell culture substrate for their low cost, transparency, and molding capabilities. Still, alkanethiolate SAMs are favored by chemists because the synthesis of alkanethiols is highly versatile (by the standards of SAM chemists). For SAM-based cell culture studies, thin (<12 nm) films of Au on glass are popular because they are almost transparent and still allow for alkanethiol derivatization.

It is also possible to design monolayers of synthetic polypeptides that selectively bind to solid supports such as particular metals, semiconductors, and plastics. The polypeptide is heterobifunctional, with a metal binding peptide sequence on one end and a biofunctional (e.g., cell-binding, enzymatic, etc.) sequence on the other end. These so-called **solid-binding peptides** have received great attention lately because their synthesis is straightforward and aqueous-based (compared with organic chemistry methods, which require toxic solvents), can be genetically engineered using *Escherichia coli* cell cultures, and peptide libraries can be easily screened in a multiwell format. The solid-binding peptide end can be selected to bind specifically to, say, a particular crystal face of gallium arsenide. In such monolayers, the order or the degree of completion of the monolayer is typically not a strong requirement for their proper function, whereas many alkanethiol-based SAMs do not function properly unless the alkanethiol chains self-align.

HOW WELL ASSEMBLED IS SELF-ASSEMBLED?

IN THE BioMEMS FIELD, the term "SAM" has been used somewhat loosely. Many immobilization methods are based on short "tether" molecules, resulting in disordered or poorly packed (i.e., lacking self-assembly) monolayers. Poor order seems to have an adverse effect on the reactivity of the SAM as well as on its surface free energy and wetting properties, but the influence of surface order or monolayer packing on cell or protein attachment is still unclear. Some cell types have been shown to be sensitive to substrate surface functionalities even if the substrate is coated with protein before cell seeding. This suggests that the rich (yet poorly defined) chemical complexity of tissue culture polystyrene is advantageous for cell culture, and forewarns that cell function data obtained on monofunctional SAM substrates should be interpreted with caution.

2.2.2 Deterring Protein Physisorption

Because proteins readily physisorb from solution onto most materials, any strategy for creating protein micropatterns on a surface must face the propensity of that surface to physisorb proteins from solution—or the arduous task of removing or inactivating them. However, the design of surfaces that inhibit cell attachment and protein physisorption is an ongoing challenge.

This problem is not new. Biologists have long been interested in deterring protein physisorption for studies that require immunostaining (i.e., detection of biomolecules with antibodies that specifically recognize them)—if the antibody sticks to everything, then it will be no good for detecting the target antigen against the background. Researchers have found their best ally in their fight against nonspecific protein physisorption in…a protein! Luckily, it is a very abundant (thus inexpensive) one: albumin, a major component of the blood. For practical reasons, most albumin used in biological research is extracted from the blood of cows—the oft-cited **BSA** (for "**bovine serum albumin**"). BSA readily sticks to surfaces and to other biomolecules (including itself, forming aggregates), so it effectively "blocks" binding sites. Antibodies, with a strong affinity to a small fragment of the biomolecules they recognize, presumably displace BSA for binding. Luckily also, BSA does not contain any known integrin-binding sequences so it is suited for deterring cell attachment in cell culture studies (e.g., to form cellular micropatterns, see **Figure 2.2**). However, it should be noted that proteins present in the medium or secreted by the cells after attachment may displace an underlying protein layer, and cells secrete proteinases that digest proteins in their migration tracks, so the selectivity does not last permanently (**Figure 2.2**). As we shall see below, more sophisticated surface engineering methods are available for questions that require the cells to be confined to their regions for periods lasting days or even weeks.

Unfortunately, the physicochemical interactions underlying protein adhesion to surfaces have not been thoroughly elucidated. As pointed out previously, the common belief that protein adhesiveness correlates with the hydrophobicity of the surface is not solidly founded because neither very hydrophilic surfaces (e.g., hydrogels) nor very hydrophobic surfaces (e.g., fluoropolymers) support protein adhesion. Contributions from van der Waals interactions, electrostatic forces, and electron donor/acceptor (also called "hydration" or "Lewis acid/basic") interactions

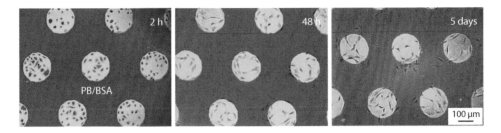

FIGURE 2.2 Degradation of cell attachment selectivity at a BSA/fibronectin interface on a polystyrene (PS) surface. (From A. Tourovskaia, T. Barber, B. Wickes, D. Hirdes, B. Grin, D. G. Castner, K. E. Healy, and A. Folch, "Micropatterns of chemisorbed cell adhesion-repellent films using oxygen plasma etching and elastomeric masks," *Langmuir* 19, 4754, 2002. Figure contributed by Anna Tourovskaia.)

play intertwined roles in the attraction between a surface and the polar as well as apolar amino acid side chains on a protein. Exposing the surface to a surfactant (a compound that lowers the surface tension of a liquid) during protein adsorption can substantially decrease physisorption on hydrophobic surfaces.

Colossal research efforts have focused on the development of materials that do not support platelet or microbial adhesion for implantation and surgery. Examples of nonadhesive surfaces often encountered in the BioMEMS literature are cellulose acetate, agarose, **poly(ethylene glycol) (PEG)**, fluorocarbon polymers, and poly(vinyl alcohol). Surfaces may also be engineered to present functionalities that repel protein and cell adhesion; following are a few notable examples:

> When a glass surface derivatized with (3-mercaptopropyl)trimethoxysilane (i.e., a silane molecule terminated in a thiol group) is exposed to UV light in air, the thiol terminal oxidizes to sulfonate (presumably via reaction with UV-generated ozone from air), which can resist protein adsorption by as much as 75% to 90%, depending on the protein.

> By functionalizing PEG with $SiCl_3$, PEG can be reacted with hydroxyl groups on glass or silicon surfaces as well as on PEG itself, thus forming a multilayer of PEG film the thickness of which is controlled by the duration of the reaction (**Figure 2.3**). Functionalized PEGs have not always been commercially available. (For example, a company called Shearwater Polymers sold them for many years but went out of business and it was not until recently that another one, LaySan Bio from Alabama, took their place in the market.) PEGs with other biofunctionalities such as biotin and cross-linkers are also available.

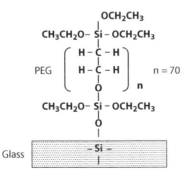

FIGURE 2.3 Direct immobilization of PEG on silicon or glass. (From D. Irimia and J. O. M. Karlsson, "Development of a cell patterning technique using poly(ethlylene glycol)disilane," *Biomed. Microdevices* 5, 185, 2003. Figure contributed by Daniel Irimia.)

❋ Thin films of a co-polymer of PEG and acrylic acid support no cell adhesion even when challenged with high cell seeding densities over weeks. However, if RGD-like peptides are grafted to exposed carboxylic acid moieties of the copolymer, cells attach and spread onto the film although in itself the surface does not support protein adhesion. In this system, cell surface attachment specificity studies are not confounded by the presence of proteins in the culture medium or the secretion of ECM by the cells.

❋ Thin films of "interpenetrating" polymer networks (IPN) of acrylamide and PEG (**Figure 2.4**) can be photopolymerized sequentially onto glass (which is derivatized with allyltrichlorosilane SAM to initiate the polymerization reaction). These coatings, which were developed by Kevin Healy and colleagues (then at Northwestern University in Chicago), are very stable and have shown negligible levels of protein physisorption and cell invasion for weeks. **PEG-IPN** can be micropatterned by photolithography or by using stencils (see Section 2.6.1.2).

❋ To mimic the low protein adsorption characteristics of PEG-coated surfaces, protein physisorption on a gold surface can be blocked by derivatizing the surface with an alkanethiol SAM presenting oligomers of ethylene glycol [$(-CH_2CH_2O-)_nR$, $n = 3–7$, R = H or CH_3], commonly referred to as a "PEGylated thiol" SAM (**Figure 2.5**). Similar results can be achieved with alkanethiol SAMs presenting tri(propylene sulfoxide) groups, which are structurally unrelated to the oligo(ethylene glycol)

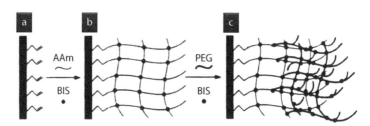

FIGURE 2.4 Interpenetrating networks of PEG and acrylamide resist cell and protein adhesion. (From J. P. Bearinger, D. G. Castner, S. L. Golledge, A. Rezania, S. Hubchak, and K. E. Healy, "P(AAM-co-EG) interpenetrating polymer networks grafted to oxide surfaces: surface characterization, protein adsorption, and cell detachment studies," *Langmuir* 13, 5175, 1997. Reprinted with permission of the American Chemical Society. Figure contributed by Kevin Healy.)

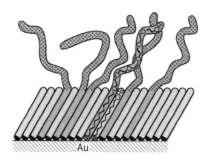

FIGURE 2.5 PEG-thiol SAM. (From K. L. Prime and G. M. Whitesides, "Self-assembled organic monolayers: Model systems for studying adsorption of proteins at surfaces," *Science* 252, 1164, 1991. Figure contributed by George Whitesides.)

group but share with it the potential for accepting hydrogen bonds as well as the solvation and chain flexibility characteristics; this functional similarity between tri(propylene sulfoxide) and oligo(ethylene glycol) groups suggests a possible general principle for protein physisorption. By linking RGD peptides to the end-group of a PEGylated SAM, physisorption of ECM proteins can be blocked and at the same time the precise surface concentration of adhesive ligands needed for cell adhesion can be tailored. PEGylated SAMs, introduced by George Whitesides' lab, have become very popular in BioMEMS because, being alkanethiols, they can be easily patterned by microstamping (see Section 1.6.3). However, thiols desorb with time under water, which raises concerns for long-term studies, and require the deposition of a gold layer, which adds cost and attenuates light for microscopic inspection.

2.2.3 Cross-Linkers

The majority of immobilization methods used in biopatterning require not only a functionalized SAM but also a "**cross-linker**"—a molecule that covalently bridges two molecules, in this case, the biomolecule of interest to the end-group of the SAM. Generally used for protein conjugation, most of these cross-linkers are now commercially available from a variety of sources. Cross-linkers may have two different reactive ends ("**heterobifunctional cross-linkers**") or the same functional group at both ends ("**homobifunctional cross-linkers**"). For our purposes, here, we only mention the few that appear most often in the text. **Glutaraldehyde** is a common, inexpensive homobifunctional cross-linker that links amino groups by reaction with its aldehyde groups. For example, an aminosilane SAM that is exposed to a glutaraldehyde aqueous solution readily immobilizes glutaraldehyde, thus presenting a second aldehyde group for subsequent immobilization of a molecule which contains amino groups—such as a protein. Other groups that are highly specific toward amines and present in many heterobifunctional cross-linkers are **succinimidyl esters** (e.g., *N*-hydroxysuccinimide ester—always abbreviated as **NHS ester**), isothiocyanates, and sulfonyl chlorides. These amine-reactive groups conjugate with aliphatic nonprotonated amines, hence the reaction is faster at slightly basic pH. **Maleimido** functionalities, on the other hand, react specifically with thiol groups—not with amino groups. Maleimidoacetic acid NHS ester is thus a cross-linker that links amino groups to thiol groups. A cross-linker may also contain a photoreactive group that reacts nonspecifically upon exposure to light of a certain wavelength. Albeit expensive, photoreactive cross-linkers have great potential for biopatterning because the reaction can be light-addressed through a standard chrome mask or with a focused laser. **Benzophenones**, **azides,** or **diazirines** are examples of photosensitive groups often present in commercially available cross-linkers. For instance, the amine-reactive cross-linker 4-azidosalicylic acid NHS ester contains an azide group; the thiol-reactive cross-linker 4-(*N*-maleimido)benzophenone contains a benzophenone group. Benzophenones have the advantage, over azides and diazirines, of being chemically more stable, insensitive to ambient light or moisture, reactive with unreactive C–H bonds even in the presence of solvent water, and activatable at wavelengths (350–360 nm) that are not damaging to biomolecules.

2.3 Micropatterns of SAMs

The use of SAMs in micropatterning biomolecules is appealing because the adhesiveness of the surface is engineered at a molecular level and, at the same time, patterning is reduced to patterning the SAM. **Figure 2.6** depicts the main four strategies used to pattern SAMs.

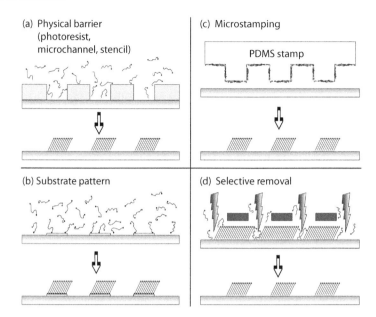

FIGURE 2.6 SAM micropatterning.

2.3.1 Selective Blocking of SAM Formation

Philip Hockgerber's group, then at AT&T Bell labs, is credited with creating the first micropattern of a SAM (which they used to micropattern neurons, as we see later). In their approach, a photoresist micropattern acts as a physical barrier that blocks the formation of the SAM on the photoresist-covered areas (**Figure 2.6a**). After removal of the photoresist with a solvent such as acetone, the bare areas can be (and often are) derivatized with a second SAM. Selective blocking of the substrate during derivatization also may be achieved by applying PDMS barriers such as microchannels or stencils.

The approach used by Kleinfeld and Hockberger has been so influential that it deserves careful analysis. The diagram shown in **Figure 2.7** illustrates the steps to achieve the desired outcome. The first half of the process involves straightforward photolithography to create a photoresist pattern. After developing the pattern (with the photoresist developer), we must ensure that there are no organic residues on the surface that might impede the silanization reaction. How can we tell whether there are organic residues? The "trick" most often used in the literature has been to follow the wettability of the surface before and after silanization: because the methyl-terminated silane is hydrophobic, water dewets those areas—which is interpreted as a sign that surface derivatization occurred. Strictly speaking, this interpretation is incorrect because the dewetting pattern would have occurred on just about *any* surface exposed to the methyl-terminated silane—the silane can also physisorb and react with itself (silanes typically have three terminal reactive ends) next to the surface, whereupon drying (the silanization step is always followed by a drying step to terminate the reaction) will form a hydrophobic film on the surface that is difficult to remove, even with thorough rinsing. In any case, to ensure that a chemical reaction occurs with the surface, the best practice is to use a harsh cleaning method (such as exposure to an oxygen plasma or "ashing") to start with. However, it is impossible to guarantee that all three terminal reactive groups of all the silane molecules will react with the glass surface; in other words, most certainly some of the silanes will have reacted with other silanes, and the "SAM" will not look as neatly ordered as in **Figure 2.7**. This concern does not apply to alkanethiol SAMs because thiols only have one terminal reactive group and will not condense with each other.

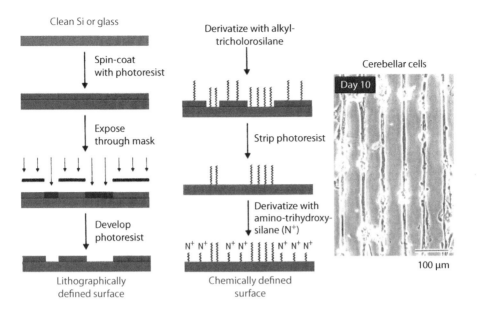

FIGURE 2.7 First SAM micropattern. (From D. Kleinfeld, K. H. Kahler, and P. E. Hockberger, "Controlled outgrowth of dissociated neurons on patterned substrates," *J. Neurosci.* 8, 4098, 1988. Figure contributed by David Kleinfeld.)

2.3.2 Selective Formation of SAM on Prepatterned Surfaces

This approach requires a previously micropatterned surface of the appropriate materials; the SAM molecules react with only one of the materials (**Figure 2.6b**). A gold microelectrode pattern on a glass substrate, for example, may be derivatized with an alkanethiol SAM leaving the glass intact. *Both* regions of a micropattern may be derivatized with different organic precursors if the chemical reactions are "orthogonal," that is, if each type of SAM molecule reacts with only one region. It is also possible to block SAM formation by applying a strong reducing potential to the electrode(s), so that different electrodes can be derivatized with different SAMs by switching the voltages and the solutions.

2.3.3 Microcontact Printing of SAMs

A microfabricated elastomeric stamp may be used to contact-transfer SAM precursor molecules onto a surface only on the areas contacted by the stamp (**Figure 2.6c**). In the original work by Kumar and Whitesides, the stamp was used to "print" hexadecanethiol molecules (CH_3–$(CH_2)_{15}$–SH in ethanol) onto gold. The procedure has been adapted to microstamp a long list of self-assembling molecules and surfaces, such as alkylsiloxanes onto Si or SiO_2 surfaces, and alkanephosphonic acids on aluminum oxide, among many others. In general, the degree of order in the printed SAM is similar to that of SAMs formed directly from solution. Inking with alkanethiols in ethanol works particularly well because the ethanol quickly evaporates and is absorbed into the PDMS, leaving a homogeneously thin oily film on the stamp that is very efficiently transferred on contact; transferring other SAM molecules such as siloxanes that are sensitive to moisture can be trickier because of the mandatory stamp-drying step (see Section 1.6.3).

2.3.4 Selective Modification/Removal of SAM

This approach starts with the formation of the SAM over the whole substrate surface, then proceeds with local modification or removal of the SAM in selected areas (**Figure 2.6d**). Methods used for SAM removal can range from ablation (with laser or an ion beam), degradation (with an electron beam), electrochemical desorption (with the tip of a scanning tunneling microscope), mechanical shear (with the tip of an atomic force microscope), or an oxygen plasma etch (the latter masked with a photoresist pattern or a stencil).

2.4 Micropatterns of Proteins

Proteins can be used as the active sensor element in cell-free devices (e.g., in immunoassays) and can play a cell-signaling role in cell-containing scaffolds or surfaces. Hence, driven by biotechnology, cell biology, and tissue engineering applications, there has been great interest in techniques that allow for micropatterning proteins. Here, we review only the most widely used methods, although the repertoire is much more colorful than presented here.

2.4.1 By Light

Photolithography is not really suited to directly micropattern proteins—as one would use it to micropattern, say, a gold film; the solvents present in the photoresist solution and in the developer severely denature proteins because the solvents repel the water that the proteins need to stabilize their structure (see Section 2.1). Nevertheless, before the advent of soft lithography, photolithography was the only method available to micropattern proteins, and several technological variations on the same theme were attempted. In one highly cited scheme, photolithography was used to micropattern an aminosilane SAM on a background of alkylsilane SAM (**Figure 2.8**). Note that the first steps were exactly as in Kleinfeld's scheme (**Figure 2.7**)—so the same general concerns regarding the completion of silane reactions apply here. Next, the pattern was bathed in glutaraldehyde, a cross-linker that binds to amines on both ends; one of the ends of the cross-linker could presumably be free to bind amine groups on proteins. Alternatively, the glutaraldehyde simply cross-linked amino groups on the aminosilane SAM. The glutaraldehyde pattern was exposed to fluorescently-labeled horseradish peroxidase (HRP) to demonstrate that there was more protein attached on the aminosilane areas than on the background. (It is

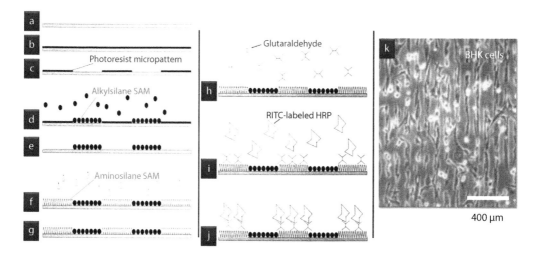

FIGURE 2.8 Micropatterning of proteins and cells via aminosilane and alkylsilane SAMs using photolithography. (From Stephen Britland, Enrique Perez-Arnaud, Peter Clark, Brian McGinn, Patricia Connolly, and Geoffrey Moores, "Micropatterning proteins and synthetic peptides on solid supports: A novel application for microelectronics fabrication technology," *Biotechnol. Prog.* 8, 155, 1992. Figure contributed by Stephen Britland.)

impossible to rule out protein adsorption with fluorescence without a calibration standard.) Fibroblast-like baby hamster kidney (BHK) cells were also shown to attach selectively to the aminosilane patterns in the presence of 10% calf serum.

What if the proteins themselves contained photoreactive groups, so as to be amenable to selective photochemical immobilization directly? In principle, there are two ways one could do this: either by cross-linking a photoreactive group in an existing protein, or by genetically engineering the protein. David Tirrell and colleagues at Caltech were able to synthesize the non-natural photosensitive amino acid *para*-azido-phenylalanine (*p*N$_3$Phe in **Figure 2.9**) in the bacterium *E. coli* and incorporate it into extracellular proteins such as elastin and fibronectin. Upon exposure to UV light (365 nm), the azidophenyl groups form a highly reactive nitrene group that cross-links the protein to almost anything in proximity (so reactive that they undergo carbon–hydrogen bond insertion, enabling patterning on virtually any surface). Essentially, Tirrell's group has demonstrated a genetically engineered "protein photoresist" that can be used for producing biosensors, cellular micropatterns (**Figure 2.9**), and others.

Tirrell's photoreactive proteins may find their most useful application in the fabrication of protein gradients and 3-D protein structures for tissue engineering and cell biology. A collaborative group led by Xiang Zhang from the University of California at Berkeley and Sarah Heilshorn (who studied with Tirrell) at Stanford University has used a Digital Micromirror Device (see Section 1.3.6.1) to immobilize Tirrell's photoreactive proteins in gradients and 3-D structures (**Figure 2.10**). A 380 nm UV-LED light source was used, resulting in a light power density on the sample plane of 0.2 to 1 mW/cm²; approximately 3 to 5 minutes of exposure time were required for a spun (dried) film of protein and 5 to 20 minutes were required for immobilization from the liquid phase, depending on the optical reduction used. After irradiation, samples were washed by agitation in 0.1% sodium dodecyl sulfate (a denaturant, to remove protein that is not covalently immobilized) for at least 4 hours and thoroughly rinsed with water. The projections of neuronal-like PC12 cells were observed to align with the topographical troughs and ridges of a linear 3-D protein pattern.

However, photoreactive proteins are not easy to produce and the one that you are interested in may not be commercially available for a while. In 2003, a clever chemist at Texas A&M

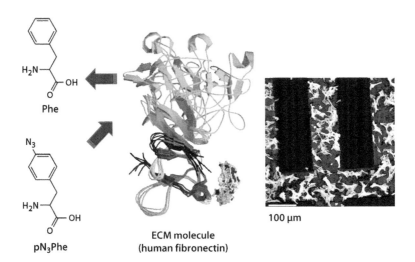

FIGURE 2.9 Micropatterning of genetically engineered photoreactive proteins. (Adapted from Isaac S. Carrico, Stacey A. Maskarinec, Sarah C. Heilshorn, Marissa L. Mock, Julie C. Liu, Paul J. Nowatzki, Christian Franck, Guruswami Ravichandran, and David A. Tirrell, "Lithographic patterning of photoreactive cell-adhesive proteins," *J. Am. Chem. Soc.* 129, 4874–4875, 2007. The fibronectin structure is from the Protein Data bank archive (www.wwpdb.org). Reprinted with permission of the American Chemical Society.)

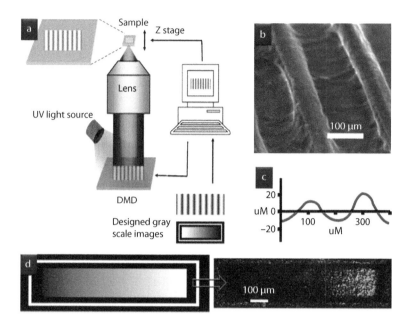

FIGURE 2.10 Micropatterning of gradients of photoreactive proteins using the digital micromirror device. (From Sheng Wang, Cheryl Wong Po Foo, Ajithkumar Warrier, Mu-ming Poo, Sarah C. Heilshorn, and Xiang Zhang, "Gradient lithography of engineered proteins to fabricate 2D and 3D cell culture microenvironments," *Biomed. Microdev.* 11, 1127, 2009. Reprinted with permission from Springer Science+Business Media.)

University and Paul Cramer presented a general aqueous-based strategy for patterning proteins on surfaces that exploits the principle that photobleaching creates photogenerated radicals. These radicals, he hypothesized, can be used for attaching organic molecules to surfaces. The patterning process starts by coating the surface with BSA, which acts as a "sticky surface" for the photobleached molecules. As photobleachable molecules, Cramer's group demonstrated the deposition of biotin-4-fluorescein (0.025 mg/mL, photobleachable with blue light for 30 minutes) and of Alexa 594–labeled anti-dinitrophenyl IgG (0.25 mg/mL, photobleachable with yellow/green light for 2 hours). The IgG pattern was visible as deposited (each IgG, which contains three to four fluorescent labels, was only partially photobleached); to visualize the biotin pattern, a solution of Alexa 488-streptavidin was added. This photochemistry has been recently implemented by Santiago Costantino's group (University of Montreal, Canada) both in a laser-based setup and in a wide-field illumination setup (**Figure 2.11**). By placing a spatial filter (such as a liquid crystal display) in the image plane of where a camera's CCD chip would normally be placed and using the camera port as an illumination port, it is possible to project arbitrary computer-generated patterns onto the sample with any light source (**Figure 2.11a**) and deposit proteins with high dynamic range fidelity (**Figure 2.11b**). With this setup, Cramer's photobleaching scheme (**Figure 2.11c**) has achieved submicron resolution (**Figure 2.11d**) and multiprotein patterning capability (using lasers of different wavelengths to immobilize proteins tagged with different fluorophores).

The high-energy femtosecond pulsed laser of a multiphoton microscope, equipped with a computerized scanner and selectable frequencies, is a powerful tool for biopatterning—for both adding and deleting material. As a notable example, Jason Shear and colleagues from the University of Texas at Austin have used multiphoton imaging lasers to photo-cross-link albumin, thus constructing biocompatible nonadherent microstructures in the presence of the cells (**Figure 2.12**) that can be used to guide axons and trap bacterial cells (see **Figure 1.16**). Others have used the laser to desorb proteins selectively, thus deleting patterns in the presence of the cells.

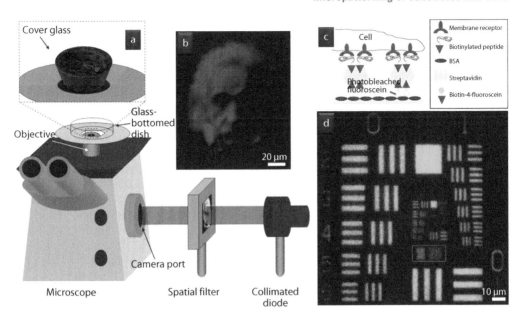

FIGURE 2.11 Photopatterning of proteins using photobleaching-induced adsorption. (From Jonathan M. Bélisle, Dario Kunik, and Santiago Costantino, "Rapid multicomponent optical protein patterning," *Lab Chip* 9, 3580–3585, 2009. Reproduced with permission from The Royal Society of Chemistry.)

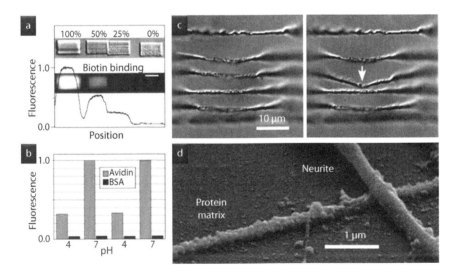

FIGURE 2.12 Selective in situ photo-cross-linking of albumin. (From Bryan Kaehr, Richard Allen, David J. Javier, John Currie, and Jason B. Shear, "Guiding neuronal development with in situ microfabrication," *Proc. Natl. Acad. Sci. U. S. A.* 101, 16104, 2004. Copyright (2004) National Academy of Sciences, U. S. A. Figure contributed by Jason Shear.)

2.4.2 By Microstamping

From all we know about proteins, microstamping them seems like a dangerous proposition: to be stamped, they need to be dried—and dried proteins form crystals, irreversibly losing their 3-D structure and with it some or all of their bioactivity. Many researchers, however, thought that it was worth the try—and, surprisingly, it worked (at least for the applications intended).

Why does it work? First, let us see how it is done. Hans Biebuyck's group at IBM Zurich is credited with stamping one of the first protein micropatterns back in 1998, achieving remarkable resolution (**Figure 2.13**). (Bruce Wheeler's group, then at the University of Illinois at Urbana-Champaign, submitted a similar, lower resolution study a few months earlier.) A solution of protein (20–200 µg/mL) is deposited on an untreated stamp for 10 to 60 minutes. Next, the stamp is blow-dried with a nitrogen stream. After a 1-second contact between the stamp (inked with fluorescently labeled IgG proteins) and a glass surface, the transferred pattern shows homogeneous coverage and no smearing of the master. The efficiency of the transfer is higher than 99%, with virtually no ink left on the stamp. The patterns are recognized specifically by antibodies, demonstrating some level of bioactivity. So far so good.

But it is less simple than it seems—try it yourself. The procedure is vitally sensitive to the choice of substrate, ambient humidity, and surface conditions. It turns out that when printing proteins, the hydrophobicity of both the stamp and the stamped surfaces are critical, likely for two reasons: (a) the water adsorbed on the surface hydrates the proteins, acting as a vehicle for the transfer of the proteins; and (b) the charges on the sample surface (which determine hydrophobicity) are strong determinants of the attraction of the proteins by the surface (see Section 2.1). When a flat stamp (i.e., without features) inked with proteins is applied onto a surface that has a SAM micropattern with a hydrophobicity contrast, such as regions with a methyl ($-CH_3$, hydrophobic) termination surrounded by regions with a carboxylic acid termination ($-COOH$, hydrophilic), then more protein is transferred onto the CH_3-terminated regions (**Figure 2.14**).

The key parameter appears to be the *differential* of wettability between the stamp and the surface, which has to be above a certain threshold to ensure the transfer of a full protein monolayer;

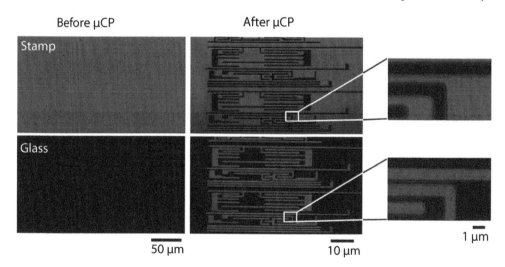

FIGURE 2.13 Microstamping of proteins. A fluorescently tagged antibody (red) on a PDMS stamp (containing 300-nm-deep features) is transferred to glass (black) only on the areas where the stamp contacts the surface. Note how the stamp becomes darker (black) in the areas where it has contacted the glass, showing that the stamp has lost more than 99% of the protein to the glass. (From André Bernard, Emmanuel Delamarche, Heinz Schmid, Bruno Michel, Hans Rudolf Bosshard, and Hans Biebuyck, "Printing patterns of proteins," *Langmuir* 14, 2225, 1998. Figure contributed by Emmanuel Delamarche.)

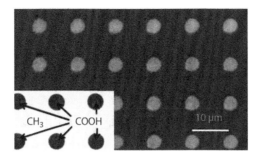

FIGURE 2.14 Substrate wettability determines efficiency of protein transfer in microstamp. (From John L. Tan, Joe Tien, and Christopher S. Chen, "Microcontact printing of proteins on mixed self-assembled monolayers," *Langmuir* 18, 519–523, 2002. Reprinted with permission of the American Chemical Society. Figure contributed by Chris Chen.)

for example, if the PDMS is derivatized with a very hydrophobic functionality ($-CF_3$), proteins are transferred well to moderately hydrophilic surfaces, but if the PDMS is derivatized with a hydrophilic functionality ($-NH_2$), then the surface must be very hydrophilic to ensure efficient transfer (**Figure 2.15**). Unfortunately, PDMS derivatization is still a nonquantitative science because laboratory-specific, small changes in the procedures involved (PDMS curing, oxygen plasma oxidation, and silanization) can have dramatic effects in the wettability of PDMS, which itself can change with time and curing conditions.

Microstamping proteins with a PDMS poses a secondary problem: the PDMS stamp deforms, so the features are difficult to align on top of each other when several layers are being patterned. The most obvious solution is to make the PDMS stamp more rigid—either by cross-linking it more or by mounting a thin slab of it on a thick glass support. Chris Chen's group has proposed a more creative solution based on multilayer stamps (**Figure 2.16**). Thus, the stamp has several (discrete) levels, and the surface (whether for inking or for printing) can be brought into contact with any of those levels by adjusting the vertical pressure applied with the stamp.

It may well be that PDMS is not that well suited for printing proteins: after all, the PDMS surface cannot act as reservoir for the ink (the protein). We need something *moist*. So why not directly use a

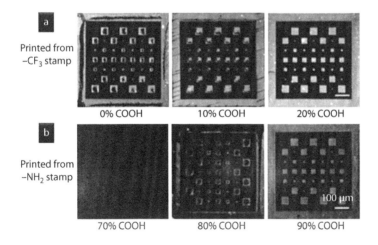

FIGURE 2.15 Hydrophobicity differential influences protein microstamping. (From John L. Tan, Joe Tien, and Christopher S. Chen, "Microcontact printing of proteins on mixed self-assembled monolayers," *Langmuir* 18, 519–523, 2002. Reprinted with permission of the American Chemical Society. Figure contributed by Chris Chen.)

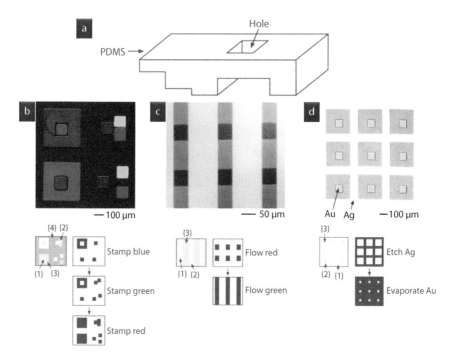

FIGURE 2.16 Fabrication of aligned microstructures with a single elastomeric stamp. The stamp has three or four levels, as indicated, separated by distances $d_2 = 1$ μm, $d_3 = 2$ μm, and $d_4 = 50$ μm. Each stage of compression prints onto the substrate blue-, green-, and red-labeled proteins that had been adsorbed onto (or filled into) the areas labeled {1}, {2}, and {3}/{4}, respectively. (From J. Tien, C. M. Nelson, and C. S. Chen, "Fabrication of aligned microstructures with a single elastomeric stamp," *Proc. Natl. Acad. Sci. U. S. A.* 99, 1758, 2002. Copyright (2002) National Academy of Sciences U. S. A. Figure contributed by Chris Chen.)

hydrogel that, for the same price, will keep the proteins in their hydrated state? That was exactly the rationale behind using agarose stamps by a team led by George Whitesides from Harvard University (Figure 2.17). Repetitive stamping without re-inking (up to 100 times!) at resolutions down to 2 μm was demonstrated. Furthermore, it is possible to print gradients of (two) proteins by first forming the gradients on the stamp by diffusion, something that is not possible with a PDMS stamp.

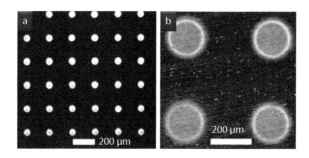

FIGURE 2.17 Selective deposition of proteins using agarose stamps. (a) Pattern of 50 μm spots of TRITC-BSA printed onto PDMS using a stamp with 50 μm diameter posts. (b) Pattern of FITC-collagen spots printed onto glass using a stamp with 200 μm diameter posts. (From Michael Mayer, Jerry Yang, Irina Gitlin, David H. Gracias, and George M. Whitesides, "Micropatterned agarose gels for stamping arrays of proteins and gradients of proteins," *Proteomics* 4, 2366–2376, 2004. Copyright Wiley-VCH Verlag GmbH & Co. KGaA. Reproduced with permission. Figure contributed by George Whitesides.)

Interestingly, it is possible to "stamp-erase" a protein pattern, at least under certain conditions. Katharina Maniura-Weber and colleagues from the Swiss Federal Laboratories for Materials Testing and Research have covalently linked glutaraldehyde to a PDMS stamp and have shown that, when applied in air to a protein-coated surface, it "picks up" physisorbed protein. A glutaraldehyde-coated flat PDMS stamp was used to remove fibronectin from plateaus of a fibronectin-coated microwell substrate; after exposing the microwells to Pluronic (a cell-repellent polymer, see Section 2.6.1.4), which adsorbed on the plateaus, cells were observed to attach on the microwells but not on the plateaus. In some cases, this strategy may be more cost-effective than creating the positive pattern, although the procedure needs to be optimized for each biomolecule and surface type to minimize the amount of protein left. The approach may be applicable to create surface gradients of biomolecules.

Despite the convenience, direct microstamping of proteins is less intuitive than it seems (adjusting the wettability and ambient conditions can be particularly challenging) and thus has been gradually displaced by methods that deposit the proteins directly from the fluid phase, in which traditional surface immobilization protocols apply. Not surprisingly, microfluidic methods have enjoyed immediate acceptance by the biological community and are rapidly gaining popularity.

2.4.3 By Microfluidic Patterning

The group of Hans Biebuyck (a former postdoctoral researcher from George Whitesides' lab who established his own lab at IBM Zurich) has been traditionally credited with the first micropattern of proteins using microfluidics. (This record is technically incorrect; it is often overlooked that, in 1987, the laboratory of Friedrich Bonhoeffer used PDMS microchannels molded from a silicon master to microfluidically deliver membrane fragments for axon guidance experiments; see Figure 6.41 in Section 6.5.1.1.) The landmark *Science* 1997 article by Biebuyck's group reported the delivery of just two different immunoglobulin solutions through two adjacent PDMS microchannels filled by capillarity (Figure 2.18). The microchannels were then removed before the application of antibodies against the proteins to show that the proteins had been delivered and immobilized successfully at the intended locations. The technique's merit is in its sheer simplicity: it only requires a stamp and a solution—no expertise, no drying, no other strange protocol!

There are several difficulties associated with patterning proteins using microfluidic devices. First, as explained in Section 1.6.4, protein loss to the walls can be a significant issue in micron-sized channels in which the surface-to-volume ratio is very high (Figure 2.19). However, one can exploit this phenomenon to one's advantage to generate soluble or insoluble gradients within the channels.

Second, the resolution of microfluidic patterning is limited mostly by the ability to fill the channels with the protein solution of interest, and to flush them after adsorption is complete. Capillarity filling works well for micron-sized channels that have been made hydrophilic (e.g.,

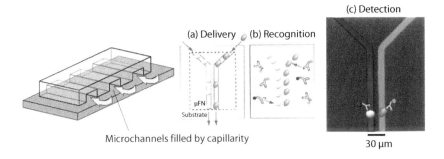

FIGURE 2.18 First microfluidic pattern of proteins. (From E. Delamarche, A. Bernard, H. Schmid, B. Michel, and H. Biebuyck, "Patterned delivery of immunoglobulins to surfaces using microfluidic networks," *Science* 276, 779, 1997. Figure contributed by Emmanuel Delamarche.)

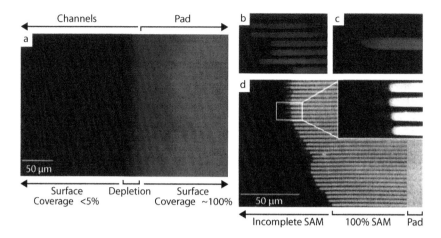

FIGURE 2.19 Depletion of reactants in small microchannels. Reactants are lost to the walls of the PDMS channels during microfluidic patterning for channels with high surface-to-volume ratio. (a) Fluorescent antibody solution fills the microfluidic channels completely and recognizes the surface-immobilized antigen but loses all the antibody to the walls in the first 100 μm for a 3-mm-long channel 1.5 μm × 3 μm in cross section. The pattern of depletion can show plug (b) or tapered profiles (c) depending on whether filling is intermittent or continuous, respectively. (d) Hexadecanethiol solution (1 mM) fills the channels and reacts with gold (protecting it from a subsequent etch) but loses all the alkanethiol molecules to the PDMS in the first portion of the channels; the light-gray regions are alkanethiol-protected gold; the unprotected gold has been etched away. (From Emmanuel Delamarche, André Bernard, Heinz Schmid, Alexander Bietsch, Bruno Michel, and Hans Biebuyck, "Microfluidic networks for chemical patterning of substrates: Design and application to bioassays," *J. Am. Chem. Soc.* 120, 500, 1998. Figure contributed by Emmanuel Delamarche.)

by exposure to an oxygen plasma etch). However, once the channels are filled, the solutions cannot be exchanged easily (e.g., for flushing or further surface treatment) except by diffusion or specialized capillarity pumps. If resolution and reagent cost are not a consideration, wider/deeper channels can be perfused continuously (and the solutions exchanged easily) with pressure-driven flow, making the loss of protein to the channel walls irrelevant (Figure 2.20).

Third, microfluidic channels (which are essentially one-dimensional arrays) often are not easily configured to deliver two-dimensional arrays of reagents. To circumvent this limitation, Emmanuel Delamarche and colleagues at IBM Zurich devised a clever orthogonal binding approach, which they term "mosaic immunoassay" (Figure 2.21). The first step of the mosaic immunoassay is to deliver the antigens (different antigens or different dilutions of the same antigen) with the microfluidic device which is then removed, followed by blocking of the unpatterned surface with BSA (Figure 2.21b); the second step is to add the different antibodies through different microchannels. This time, the microchannels are rotated 90 degrees with respect to the original orientation (Figure 2.21c), so that each row of antigens gets probed by each column of antibody (Figure 2.21d through f). In this way, it becomes possible to achieve large number of repeat readouts (Figure 2.21e) or to obtain dose–response curves for a large number of antibodies in parallel (Figure 2.21f).

These methods have inspired the development of sophisticated microfluidic devices that locally deposit proteins and cells in selected locations of a substrate. We have already seen some of these devices in Section 1.6.4 when we covered stopped-flow lithography. More microfluidic patterning devices are covered later as we delve deeper into the subject of microfluidics.

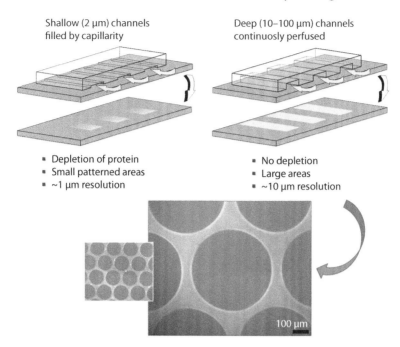

Shallow (2 µm) channels filled by capillarity

Deep (10–100 µm) channels continuosly perfused

- Depletion of protein
- Small patterned areas
- ~1 µm resolution

- No depletion
- Large areas
- ~10 µm resolution

100 µm

FIGURE 2.20 Capillary versus pressure-driven flow in microfluidic patterning. (From Albert Folch and Mehmet Toner, "Cellular micropatterns on biocompatible materials," *Biotechnol. Prog.* 14, 388–392, 1998. Figure contributed by the author.)

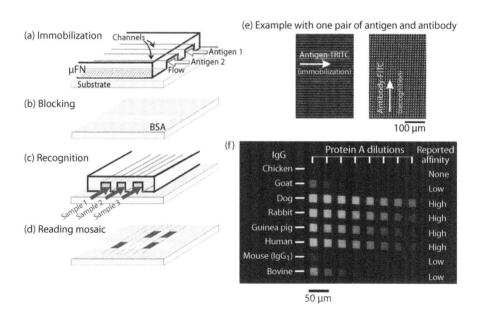

(a) Immobilization — Channels — Antigen 1 — Antigen 2
µFN — Flow — Substrate

(b) Blocking — BSA

(c) Recognition — Sample 1, Sample 2, Sample 3

(d) Reading mosaic

(e) Example with one pair of antigen and antibody

Antigen-TRITC (immobilization) — Antibody-FITC (recognition) — 100 µm

(f)

IgG	Protein A dilutions	Reported affinity
Chicken		None
Goat		Low
Dog		High
Rabbit		High
Guinea pig		High
Human		High
Mouse (IgG₁)		Low
Bovine		Low

50 µm

FIGURE 2.21 "Mosaic" immunoassays fabricated by microfluidic patterning. (From A. Bernard, B. Michel, and E. Delamarche, "Micromosaic immunoassays," *Anal. Chem.* 73, 8, 2001. Reprinted with permission of the American Chemical Society. Figure contributed by Emmanuel Delamarche.)

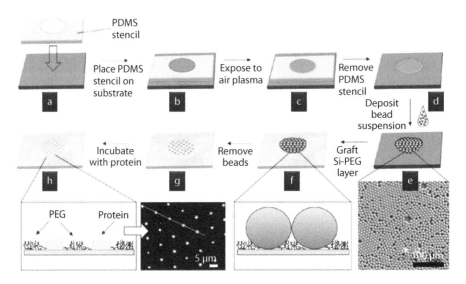

FIGURE 2.22 Protein array patterns by particle lithography. (From Zachary R. Taylor, Krupa Patel, Travis G. Spain, Joel C. Keay, Jeremy D. Jernigen, Ernest S. Sanchez, Brian P. Grady, Matthew B. Johnson, and David W. Schmidtke, "Fabrication of protein dot arrays via particle lithography," *Langmuir* 25, 10932–10938, 2009. Figure contributed by David Schmidtke.)

2.4.4 By Self-Assembly

As we have seen, patterning generally requires the lengthy production of a photomask, a master, and some sort of template (e.g., a PDMS channel or stencil); the other alternative (serial patterning of every substrate with a laser) is not pretty. From an outsider's perspective, patterning is (to put it plainly) painful.

But let us reconsider why we need photomasks: they are intended for applications in which we need to generate *arbitrary* patterns. How about all the applications that simply require microarrays of protein dots—diagnostics, many cellular micropatterns, and so forth? For those applications, there *must* be a simpler way to generate regular patterns of dots. One such way, introduced in Section 1.9, is based on particle lithography, a fabrication technique in which beads self-assemble in packed arrays because of short-range attractive forces; the beads are used simply as a sacrificial contact mask to deposit another material, in this case, proteins. The first report on the use of particle lithography to pattern proteins dates from 2002 by Gang-Yu Liu's group at Wayne State University. However, it was not until 2009 when David Schmitke and colleagues from the University of Oklahoma were able to address the eternal problem of background adsorption, thereby making particle lithography useful for biosensor and cellular micropattern applications. In their study, particle lithography was used to mask the grafting of PEG; upon removal of the beads, the beads left behind small "shadows" where protein could be immobilized (**Figure 2.22**).

2.5 Micropatterns of Cells on Nonbiomolecular Templates

The ability to produce microstructures made of cells is important for building microengineered tissue constructs, for making biosensors, and for the study of cell biology in general. Historically, cell biologists resorted to clever approaches to recreate different degrees of cellular organization in the laboratory. Harrison, for example, studied cell migration on spider webs as early as 1912. In 1945, Paul Weiss at the Rockefeller Institute in New York studied how spinal ganglion cells aligned their axons along grooves milled into mica. Adam Curtis and Malini Varde, then at the University College (London), followed in 1964 with substrates made from polystyrene as replicas of diffraction gratings, which also contain very fine aligned features. In 1971, Vassiliev and colleagues of the Academy of Medical Sciences in Moscow used blank polyvinylchloride music records to study how fibroblasts reacted to microtopography. The development of "micro-cultures" of cells on dried protein spots (i.e., made by depositing small droplets of collagen on a nonadhesive surface such as agar or polystyrene and letting them dry) was a major advance pioneered by David Potter at Harvard Medical School in 1976 and still used in many laboratories—it is possible to spray hundreds of spots at once with a spray bottle that costs 50 cents! Fine lines that guide cell attachment can similarly be created by scratching agar, phospholipid films, or ECM protein. Albeit ingenious, the technology used in these studies could not address the structural dimensions, chemical heterogeneity, or precise repeatability over large areas found in live tissue. Microfabrication technology offers the potential to control cell–surface interactions, cell–cell interactions, and cell–medium interactions on a micrometer scale.

In this section, we review methods that use the preference of cells to attach to certain non-biological materials, as determined by empirical observation rather than by rational design or biological principles. Although most of these methods have been abandoned, an understanding of their weaknesses and strengths served as the basis for later work.

2.5.1 Cells on Micropatterns of Metals or Oxides

S. B. Carter from Imperial Chemical Industries, Ltd., in England is credited with the first recorded BioMEMS publication. He was the first to report, in October of 1967, surfaces with patterned cellular adhesiveness using microtechnology (Figure 2.23). These were not his first cellular patterns because he had already published in *Nature* in December of 1965 an ingenious method to create patterns of mouse fibroblasts onto palladium (which had been evaporated on acetate, the nonadhesive portion of the patterns); as a mask, Carter had used a wire in contact with the acetate substrate, which had the advantage that it created a smooth metallic gradient on the acetate, resulting in substrate-directed cell motion. This cell behavior is known as "**hap-totaxis**," a term that was coined by Carter in 1965.

For his October 1967 article, he repeated the shadow-evaporation technique but he substituted the wire for a "perforated" 15-μm-thick stencil mask made of nickel. Although the method used to make the "perforations" is not described, the article mentions that "optical reduction"

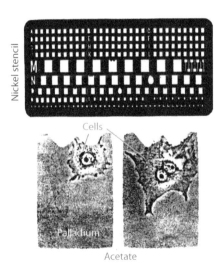

FIGURE 2.23 Microengineering cell shape, back in 1967. (From S. B. Carter, "Haptotactic islands: A method of confining single cells to study individual cell reactions and clone formation," *Exp. Cell Res.* 48, 189, 1967. Adapted with permission of Elsevier.)

has been used and shows a stencil mask containing ~100 × 150 μm squares and 10 to 20 μm lines (**Figure 2.23**), suggesting that the features were made by photochemical machining. Foreseeing the feasibility of massively parallel single-cell assays, he studied the spreading of single cells confined to adhesive islands.

Carter's idea of patterning with a removable (and reusable) stencil mask is attractive for its conceptual simplicity, compatibility with a variety of surfaces, low toxicity, and low cost. However, metallic and oxide substrates, by virtue of being chemically foreign to in vivo cellular environments, might arguably trigger unexpected, nonphysiological cell responses. Also, metal stencils do not seal with the substrate, thus fluids and cells might not obey the pattern as with other stencil materials.

Carter's technique was so ahead of his times that it was used essentially unaltered by others for more than twenty years. Here are some highlights:

- 1973: Albert Harris from Yale University used the same shadow-evaporation concept in combination with finer stencil masks—transmission electron microscope (TEM) grids (33–50 μm lines/square). He applied it to a variety of cell lines and substrates and noticed that there is a materials hierarchy for cell adhesion, for example, cells attached to palladium (Pd) on Pd/acetate patterns but attached to the polystyrene background on Pd/polystyrene patterns.

- 1975: In his investigation of axon guidance in chicken embryo neuron cultures, Paul Letourneau, then at Stanford University, studied the hierarchical preference of growth cones for a set of surfaces and found that growth cones showed a poor preference for the tissue culture plastic background over the Pd patterns but a strong preference for the background when the background had been coated with polyornithine or collagen before shadow-evaporation of Pd; this article marked an inflection in the field, for it became clear that natural proteins could be used as adhesive substrates (see **Figure 2.26** in Section 2.6). Interestingly, in glial cell cocultures, the growth cone-to-polyornithine adhesion was roughly as strong as adhesion to glia.

- 1976: Guenter Albrecht-Buehler from the Cold Spring Harbor Laboratory in New York used gold islands shadow-evaporated onto glass (through TEM copper grids) to study the exploratory function of filopodia during spreading of 3T3 cells. He observed that cells did not extend lamellipodia unless some filopodia, in a rapid scanning motion, had contacted a gold area; 1 to 2 hours after seeding, most of the cells had migrated to gold-coated areas. This suggested that the directional preference of 3T3 cells during migration is a response to the substrate adhesiveness as sensed by the filopodia. He also noticed that, 24 hours after plating, the cells did not display any preference for gold versus glass, which can be attributed to secretion of adhesive factors.

- 1978: Following Harris' studies, Bengt Westermark at the University of Uppsala in Sweden further improved the attachment selectivity by shadow-evaporating Pd onto a thin layer of air-dried agarose. When seeded on such surfaces, the attachment and spreading of glial cells was entirely confined to the Pd islands, and their growth stopped at confluence even though they had a free edge bordering the agarose "sea."

- 1980: Jan Pontén and Lars Stolt, also at the University of Uppsala, adopted the same Pd/agarose system as Westermark, but used a 14-μm-thick titanium mask with arbitrary patterns microfabricated by photolithography and chemical etch for the shadow evaporation step. They confirmed Westermark's findings on a variety of cell lines down to single-cell, 26-μm-diameter islands, suggesting that cell contacts are not necessary for the production of a quiescent growth state.

- 1986: Grenham Ireland and colleagues at the University of Manchester in England created adhesive Pd islands onto substrates previously coated with poly-2-hydroxyethyl methacrylate (poly-HEMA) by metal evaporation through a 5-μm-thick copper stencil mask containing feature linewidths of approximately 20 μm. The shape of single cells confined to spread on an island was constrained dramatically to the shape of the island. By fluorescently staining for actin and vinculin 3T3 cells spread on circle-, triangle-, or line-shaped islands, they proved in a follow-up article in 1990 that cytoskeletal organization depends on island shape, for example, the focal points tended to accumulate to the periphery of the islands, whereas the total focal adhesion area was independent of the shape of the cell (see Section 6.2.1).

2.5.2 Cells on SAM Micropatterns

In a seminal article first-authored by a young David Kleinfeld, Philip Hockberger's group (then at AT&T Bell Labs in New Jersey) created the first micropatterns of SAMs, and for the purpose of patterning the outgrowth of neurons. As explained in Section 2.3.1, they created patterns of alkyl-trichlorosilanes on an aminoalkyl-trimethoxysilane background [with either aminopropyl (A-P), ethylenediaminepropyl (EDA-P), diethylenetriaminepropyl (DETA-P), or acetylated EDA-P functionalities] using photolithography (Figure 2.7). Here, we discuss in detail their experimental rationale, now somewhat outdated, as it provides many valuable lessons for the BioMEMS student.

Kleinfeld et al.'s choice of aminosilanes as a cell-adhesive surface was stated as an attempt to mimic poly-D-lysine (PDL), a poly–amino acid containing amino side groups, which is a widely used adhesive surface coating in neuron culture as an inexpensive mimic of ECM. Surfaces derivatized with diamines (EDA-P) and triamines (DETA-P)—but not monoamines (A-P)—did, indeed, promote adhesion of embryonic mouse spinal cells and perinatal rat cerebellar cells, and morphology was assessed to be very similar to that of cultures on PDL.

The goal of using methyl-terminated alkylsilanes was "to recreate the hydrophobicity of certain surfaces which inhibit cell adhesion." This is also arguable: as we will see, other researchers have created cellular micropatterns in which cells attach precisely to methyl-terminated areas. This discrepancy can probably be explained by the fact that cellular adhesiveness is highly dependent on a number of parameters, including cell type and medium composition. Indeed, even in their own system, Kleinfeld et al. observed that the alkylsilane-derivatized areas inhibited neuron adhesion only if serum was present in the seeding medium. Patterns of the cerebellar cells were preserved for at least 12 days, which allowed for the development and observation of electrical excitability. In addition, the different behavior of monoamines with respect to diamines and triamines cannot be evaluated without taking into account the completeness of the monolayer and that a covalently bound layer of hexamethyldisilazane, a photoresist adhesion-promoter used in their study, was not removed from the surface before aminosilane modification.

Using Kleinfeld et al.'s method, Geoffrey Moores' group at the University of Glasgow showed in 1992 that fibroblast-like BHK cells attached and aligned to aminosilane patterns on a methylsilane background (see Figure 2.8). In view of the fact that cell adhesion and spreading were greater on the aminosilane and untreated surfaces, they hypothesized that "adhesive factors such as fibronectin" present in the medium (containing 10% calf serum) physisorbed preferentially to those regions. They also quantified the evolution of BHK fibroblast adhesiveness contrast as a function of exposure to serum-containing medium, and found that it was reduced by approximately 60% in 24 hours. In addition, it was observed that cells tended to align to and elongate along parallel tracks of dimethyldichlorosilane on a bare glass background when the grating period was large (>4 μm).

Hockberger's group later found that neuroblastoma cells, osteosarcoma cells, or fibroblasts attached preferentially to the aminosilane areas (independently of the type of alkylsilane in the background) using patterned SAMs of alkylsilanes (either N-octadecyldimethylchlorosilane or dimethyldichlorosilane) and N-(2-aminoethyl)-3-aminopropyl-trimethoxysilane. The attachment selectivity was first interpreted as a result of the association between cell surface proteoglycans and the positively charged amine. Bone-derived cells did not comply with the pattern unless they were seeded in the presence of serum or unless serum proteins were prephysisorbed onto the aminosilane/alkylsilane pattern. Later, the group of Kevin Healy, spawned from that of Hockberger's, determined that the requirement for regioselective attachment and spreading was the presence of vitronectin (an ECM protein) in the serum. This strongly favors the alternative hypothesis that preferential attachment is mediated by integrin binding to physisorbed ECM protein. Indeed, by using a radial flow apparatus, it was found that, within 20 minutes of seeding, the strength of adhesion in the presence of serum was significantly larger on aminosilane-derivatized areas as compared with alkylsilane-derivatized or bare areas, whereas it was similar on both areas after longer (2-hour) culture periods, presumably because of physisorption of endogenously secreted ECM. Preferential attachment was time-dependent: cells first attached randomly across the SAM-patterned surface, but organized onto the aminosilane areas within 30 minutes. Cells spread both on the aminosilane and into the alkylsilane areas in the presence of serum. Interestingly, when cultures were extended for 15 and 25 days, the matrix synthesized by the cells was preferentially mineralized on the aminosilane regions.

At the Naval Research Laboratory in Washington, DC, a group led by Jeff Calvert and Jacques Georger (in the 1990s) developed techniques for patterning SAMs using deep-UV laser-assisted photolysis of aminosilane SAMs and formation of perfluoroalkylsilane SAM on the irradiated areas. Neuroblastoma cells, fetal rat hippocampal neurons, and porcine aortic or human umbilical vein endothelial cells were observed to attach and grow selectively onto the aminosilane patterns whereas the perfluoroalkylsilane background resisted adhesion. The attachment selectivity was attributed to electrostatic attraction between the positive charges present on the amine-terminal SAM (absent on the perfluoroalkyl SAM background) and the negative charges present at the cell surface. However, proteins adsorbed from the serum-containing medium could have played an important role as well. Indeed, a subsequent study showed that prephysisorption of human fibronectin onto aminosilane SAMs on glass did not affect adhesion, spreading, and proliferation of human umbilical vein endothelial cells when these were seeded in 2% fetal bovine serum. Importantly, human umbilical vein endothelial cells were shown to differentiate into oriented neovascular cords after the addition of basic fibroblast growth factor in long-term (7–10 days) culture. With this system, David Stenger at the Naval Research Laboratory, together with Carl Cotman and colleagues at the University of California (Irvine), were able to create in 1998 circuit-like micropatterns of hippocampal neurons, which were shown to form functional axo-dendritic synapses and could be patterned in defined axonal/dendritic orientations (see Section 6.5.2).

Many groups have reported numerous variations on these previous strategies with variable success. The variations use different SAM patterning strategies (**Figure 2.6**) and different SAM terminal groups, but are often difficult to compare because they use different cell types or different medium formulations (in particular, the use of serum in the medium is a confounding factor). This past research has not been in vain, though—it has helped us to understand that, because cell attachment is mediated by integrin signaling, *the SAM itself must be designed to present a bioactivity (promoting or inhibitory)*, by either directly presenting the integrin ligand, providing an adhesive surface for the integrin ligand (e.g., an ECM protein), or blocking protein physisorption.

2.5.3 Cells on Polymer Micropatterns

This section reviews biopatterning strategies that rely on the fabrication of a polymeric template, but excludes work in which a polymer background was molecularly designed (rather than observed) specifically to *repel* protein adsorption or cell attachment (or both).

Several cell patterning strategies have used thermally sensitive or photosensitive polymers. In 1973, on the observation that mouse and hamster fibroblasts do not adhere to paraffin, N. G.

Maroudas, then at the Imperial Cancer Research Fund in London, devised an original method to create patterns of paraffin on tissue culture substrates. A TEM grid wetted with a solution of paraffin was placed onto a tissue culture dish and heated to 50°C to 60°C; because of thermal shrinkage, paraffin retracted from the tissue culture surface and coalesced under the TEM grid. The grid could be removed mechanically, and cells were observed to attach only to the resulting 100-μm-square "islands" of tissue culture substrate surrounded by the (nonadhesive) paraffin background. Although simple, this method has the limitation that paraffin cannot easily be patterned by photolithographic methods.

Poly(N-isopropylacrylamide) (poly-NIPAM) is a thermoresponsive polymer which is insoluble in aqueous solutions above a certain lower critical solution temperature (LCST = 32°C in pure water and 25°C in pH 7.4 saline). It is possible to culture cells on poly-NIPAM and then induce cell detachment by lowering the incubation temperature below the LCST point. Detachment by temperature manipulation is appealing because it circumvents treatment with supplements such as trypsin, which are damaging to cells. Several groups, many of them in Japan, have focused on this strategy since the 1990s.

In 1998, Yukio Imanishi and Yoshihiro Ito's groups at Nara Institute of Science and Technology in Ikoma, Japan, synthesized a random copolymer of NIPAM and acrylic acid (20:1) and photosensitized it by coupling the carboxylic acids with azidoaniline (a molecule which contains a photosensitive phenylazide group). Thus, thin layers of this photosensitized copolymer can be attached to (and patterned onto) tissue culture polymers by UV irradiation. They selectively exposed a copolymer film on polystyrene to UV light through a chrome mask and washed away the unexposed copolymer with cold water. Fibroblasts attached onto both the copolymer patterns and the polystyrene background, but could be detached selectively from the copolymer areas on switching below the LCST (21.5°C). The concept is ingenious because it is based on selective *detachment* (rather than selective *attachment*) of cells, presumably mediated by the high hydration of the substrate. However, selective detachment was observed only when the fibroblasts were cultured in serum-free medium—it failed when the patterned surfaces were exposed to a fibronectin solution or serum-containing medium before seeding because fibronectin was able to physisorb onto both the unprotected polystyrene background and the copolymer pattern. Furthermore, the detachment temperature was cell type–dependent (10°C for rat hepatocytes and 20°C for bovine endothelial cells), which raises reliability questions. To optimize the "attachability" and "detachability," Yuichi Mori's group at the Japan Research Center in Kanagawa cast-dried and exposed a mixture of type I collagen and poly-NIPAM solutions and found that approximately 100% attachability and detachability of human dermal fibroblasts seeded in 10% fetal bovine serum could be obtained if the collagen content was 4% to 5% and if the UV exposure was 2000 J/m^2; it was hypothesized that cross-linking of collagen to poly-NIPAM occurs on UV irradiation. Sheets of fibroblasts were selectively detached after 3 days of culture, resulting in the formation of cellular spheroids, the size of which could be controlled by changing the pattern dimensions. This work was later extended to a total of 23 cell types, but data on long-term cultures are still unavailable.

The thickness of the poly-NIPAM layer may be a critical factor. A study in 1996 led by Françoise Winnik at RIKEN in Saitama, Japan, reported reduced levels (60%–70%) of fibrinogen and ribonuclease A physisorption on ultrathin (<100 Å) poly-NIPAM layers grafted onto aminosilane- or methacrylsilane-derivatized silicon oxide. Because the poly-NIPAM layer could be patterned by laser ablation through a stainless steel stencil mask, and featured reduced levels (70%–90%) of cellular attachment compared with bare glass, micropatterns of neuroblastoma-glioma cells could be demonstrated. However, no temperature effect on protein physisorption or cell attachment was observed within the range of 20°C to 40°C, which suggests that poly-NIPAM does not undergo LCST phase change when grafted to such thin layers.

In 1996, Takehisa Matsuda and Takashi Sugawara at the National Cardiovascular Center Research Institute in Osaka, Japan, synthesized photoreactive copolymers containing phenylazide-conjugated monomers (Figure 2.24). The phenylazide groups form highly reactive nitrene

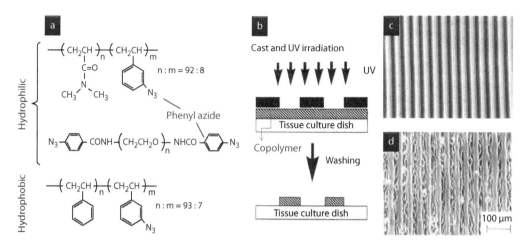

FIGURE 2.24 Micropatterning of photoreactive copolymers. (From Takehisa Matsuda and Takashi Sugawara, "Control of cell adhesion, migration, and orientation on photochemically microprocessed surfaces," *J. Biomed. Mater. Res.* 32, 165, 1996. Reprinted with permission of John Wiley and Sons. Figure contributed by Takehisa Matsuda.)

groups on UV irradiation. The hydrophilicity of the unirradiated copolymer could be designed by choosing different monomers. When a thin film of copolymer was cast on a polymeric surface such as polystyrene or polyvinyl alcohol, and exposed to UV irradiation through a chrome mask, only the exposed parts of the copolymer film bonded to the underlying surface; hence, creating a polymer-on-polymer pattern with high hydrophilicity contrast. They were able to create micropatterns of photoreactive (hydrophobic) styrene copolymer cast onto (hydrophilic) polyvinyl alcohol and micropatterns of photoreactive (hydrophilic) dimethyl acrylamide copolymer cast onto (hydrophobic) polystyrene. Endothelial cells were observed to attach with high selectivity only on the hydrophobic parts of the substrate. Unfortunately, no details were given on possible protein exposures and the selective attachment was interpreted only as a result of differential surface wettability—it was claimed that the photosensitized copolymers prevented the adsorption of ECM protein onto the irradiated, highly hydrophilic areas. With this system, nevertheless, selective attachment and growth of neuroblastoma cells for more than 1 month was demonstrated, and endothelial cells were observed to migrate only along adhesive tracks. Because little experimental details and no protein physisorption data were provided, the question remains whether cell attachment was indeed mediated by physisorption of proteins. The same strategy was later extended to other copolymers.

Of all photosensitive polymers, the most widely available are the photoresists used in microelectronics processing, and their photochemistry is extremely well characterized. So why not give it a try? In the early 1990s, André Kléber and colleagues at the University of Bern (Switzerland) studied the behavior of neonatal rat heart cells cultured on photoresist patterns (on glass). They found that the photoresist resisted the attachment of cells and the shape and orientation of the individual myocytes were a function of photoresist channel width. Surprisingly, it did not show signs of cytotoxicity as assessed by morphology and electrophysiology measurements for up to 17 days of culture. The cytotoxicity and attachment selectivity of photoresist for other cell types remains to be assessed (or, more likely, many laboratories keep testing them and do not report their negative results). The Yoshikawa group at the Osaka National Research Institute published a study in which a range of photoresists and surface modification schemes routinely used in microelectronics was tested for neuron and glial cells seeded in serum-containing medium. The cells attached preferentially onto the 0.6-µm-thick photoresist patterns on a glass background only for certain photoresists; the most suitable surface functionality was obtained using low exposed diazo-naphtoquinone/novolak imidazole–doped (a commercially available, positive tone) photoresist. They found that the attachment selectivity was determined by the change in surface functional groups rather than by the hydrophobicity contrast.

In sum, the development of more sophisticated surface chemistry techniques and microfluidic delivery methods that deposit biologically relevant ligands without long-term toxicity concerns has displaced the use of nonbiological materials in cell patterning.

2.6 Micropatterns of Cells on Biomolecular Templates

All the methods described in this section were designed to take advantage of the existence of natural biorecognition interactions between molecules present in the cell membrane and molecules present on the adhesive substrate. Note that, given multiple observations, attachment selectivity is remarkably dependent on the medium's serum content, some of the more advanced work reviewed in the preceding section already incorporated the notion that cells attach preferentially to ECM proteins. Here, we refer to the interactions that allow cell attachment and trigger cell spreading as "biological adhesiveness." The substrate molecules may play a passive role, such as ECM proteins or specific peptide sequences recognized by integrins in the cell membrane, or they may play an active role, such as antibodies recognizing molecules on the cell membrane. Protein physisorption to the pattern background may be impeded by interposing a "barrier" between the protein solution and the surface—either a physical barrier (e.g., a stencil) or a chemical barrier (e.g., a SAM which repels protein physisorption). The physisorption barrier may or may not be removable, or may be removed by chemical or physical means. Physisorption onto the background may be circumvented altogether by patterning the proteins in dry conditions. The methods reviewed in the next section encompass most of the combinations of the different imaginable strategies (chemisorption/physisorption, permanent/removable barrier, and ECM/peptides).

In 1975, Paul Letourneau, then at Stanford University, used Carter's nonbiological shadow-evaporation method to create palladium islands over polyornithine or collagen coatings—and observed that the palladium was now repulsive relative to the polyornithine or collagen (Figure 2.25). Importantly, this article shifted the thinking on cellular micropatterning because it incorporated the notion that cells could adhere to a micropattern by their specificity to specific biomolecules.

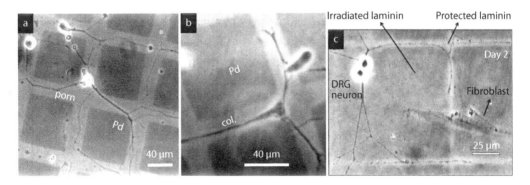

FIGURE 2.25 Shadow-patterning of metals and laminin to pattern neuronal growth. (a and b, from P. C. Letourneau, "Cell-to-substratum adhesion and guidance of axonal elongation," *Dev. Biol.* 44, 92–101, 1975. Adapted with the permission of Elsevier; c, from J. A. Hammarback, S. L. Palm, L. T. Furcht, and P. C. Letourneau, "Guidance of neurite outgrowth by pathways of substratum-adsorbed laminin," *J. Neurosci. Res.* 13, 213–220, 1985. Figure contributed by Paul Letourneau.)

ADULT HUMOR IN BIOMEMS

PAUL LETOURNEAU MAY HAVE MANAGED TO CONTRIBUTE the only bit of adult entertainment to the otherwise sexually-neutral BioMEMS literature. This was back in 1975, when Letourneau was a graduate student at Stanford University, after he conducted a study with 8-day-old chick embryo neurons cultured on micropatterned substrates. Intentionally or not, he decided to label the polyornithine tracks as "porn" (see **Figure 2.25a**), and neither the reviewers nor the journal editor seem to have challenged his choice of words. Perhaps they were amused by it, as well as by the inevitable conclusion, recorded forever in history: that chicks prefer porn to palladium. The pun takes such a hilarious turn if one reads the abbreviation of palladium "Pd" in French slang that it is hard not to conclude that Paul Letourneau is, in fact, a comical genius.

In 1985, reasoning that metals are foreign to in vivo microenvironments, a team led by Letourneau, now at the University of Minnesota, uniformly coated a surface with laminin and placed a microfabricated TEM grid atop the surface. Next, they UV-irradiated the laminin to "inactivate" it. Because of some unknown mechanism, the cells (both the somas and their axons) were repelled by the "inactivated laminin" areas (**Figure 2.25c**). The "laminin inactivation" method by Letourneau—which has not been confirmed to work for other cell types or even other neuronal types, or other media formulations—is not used nowadays (yet it is still being cited), but it is reproduced in this textbook because it represents the first effort to use only natural biomolecules to microengineer cell attachment.

2.6.1 Micropatterns of Physisorption-Repellent Background

Once researchers realized that the right approach to biopatterning was to use biomolecules for the cell-adhesive part of the pattern, the bulk of the research efforts shifted to finding approaches and materials that would repel cells for the nonadhesive part of the pattern. This challenge is twofold: (a) there are cell-adhesive proteins in the cell culture medium that will adhere in the background areas; and (b) cell-dependent or cell-independent processes might cause the degradation of nonadherent coatings and promote their cell invasion. We shall see that not all approaches are equally effective at keeping the cells at bay, and many are cell type-dependent. Typically, cell types that can secrete large amounts of ECM (such as fibroblasts) and cells that can project processes over small nonadhesive regions (such as neurons sprouting neurites) are more difficult to contain.

2.6.1.1 HEG-Thiol SAM

In an article that has been cited more than 700 times since 1994, a team led by George Whitesides at Harvard University and Donald Ingber at Children's Hospital (Boston) microstamped patterns of methyl-terminated alkanethiol SAM on gold and filled the background with alkanethiols terminated in a **hexa(ethylene glycol)** (**HEG**) group (**Figure 2.26**). When such a surface (colloquially referred to as "**PEG thiol**") is exposed to ECM protein solution, protein physisorbed onto the methyl-terminated areas only, and selective cell attachment onto the protein templates could be demonstrated. SAMs of tri(propylene sulfoxide)-terminated alkanethiols were also shown to resist protein adsorption and the attachment of endothelial cells for at least 24 hours. Most strikingly, the HEG-derivatized areas prevent not only cellular attachment but spreading as well. As shown in **Figure 2.26b**, the spreading and shape of individual cells can be constrained within adhesive "islands" surrounded by HEG-terminated areas. Because the function and growth of a cell depends on its degree of attachment and spreading onto the substrate, engineering a large population of cells with a particular cell shape could be used to elicit, identify, or simply discern certain cellular responses dictated by cell shape. In 1994, this team observed that a reduction in DNA synthesis and an increase in apoptosis rates in hepatocyte micropatterned cultures correlated with a reduction of

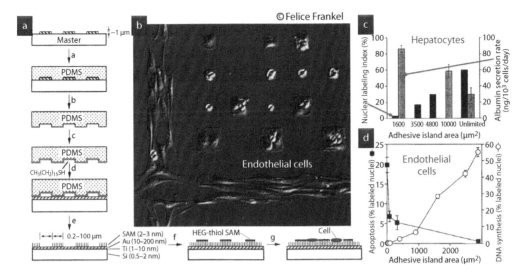

©Felice Frankel

FIGURE 2.26 Microengineering cell shape and function using microstamped SAMs. (From Rahul Singhvi, Amit Kumar, Gabriel P. Lopez, Gregory N. Stephanopoulos, Daniel I. C. Wang, George M. Whitesides, and Donald E. Ingber, "Engineering cell shape and function," *Science* 264, 696, 1994; and Christopher S. Chen, Milan Mrksich, Sui Huang, George M. Whitesides, and Donald E. Ingber, "Geometric control of cell life and death," *Science* 276, 1425, 1997. Figure contributed by George Whitesides.)

island size (see **Figure 2.26c**). In addition, albumin secretion rates, which are known to decrease with increasing culture time as part of a hepatocyte dedifferentiation process, decreased at a slower pace for smaller islands. This indicates that, at least to some degree, cell function can be tailored by modifying its shape. In a follow-up, highly-cited article, the same laboratories presented in 1997 a study on how endothelial cells proliferate and go into apoptosis when constrained into single-cell islands; as it turns out, both behaviors correlate with island size (see **Figure 2.26d**).

2.6.1.2 PEG Interpenetrated Networks

There is a wealth of evidence that, regardless of the chemistry used, micropatterning of PEG is a successful long-term strategy for confining cell spreading. Kevin Healy and coworkers have also demonstrated precise control of the shape of individual cells by photopatterning thin layers of PEG-IPN of poly(acrylamide) and PEG onto glass (see Section 2.2.2 and **Figure 2.4**). Vitronectin preferentially physisorbed onto aminosilane-derivatized glass areas surrounded by a PEG copolymer background and directed the selective attachment and spreading of osteoblasts, as shown in **Figure 2.27**. Cells constrained to attach to small (<900 μm²) islands were not able to organize

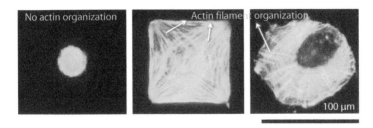

FIGURE 2.27 Cell shape and cytoskeleton organization constrained by PEG-IPN micropatterns. The images show actin filament immunostaining. (From C. H. Thomas, J.-B. Lhoest, D. G. Castner, C. D. McFarland, and K. E. Healy, "Surfaces designed to control the projected area and shape of individual cells," *J. Biomech. Eng.* 121, 40, 1999. Figure contributed by Kevin Healy.)

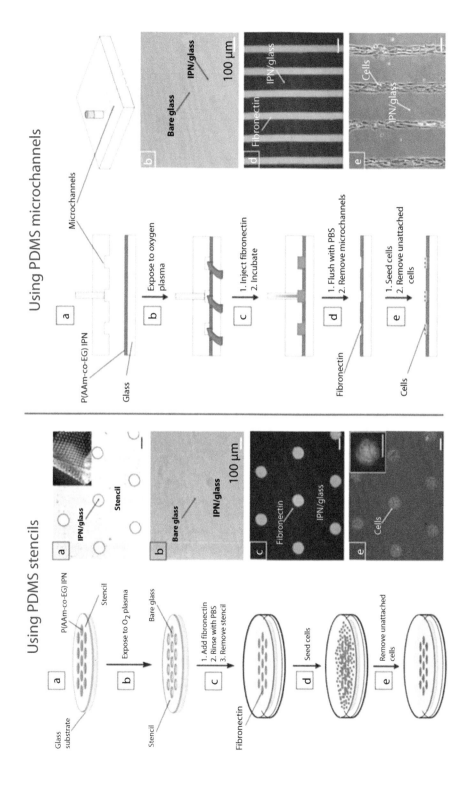

FIGURE 2.28 Micropatterns of cells on PEG-IPN using elastomeric masks. (From Anna Tourovskaia, Thomas Barber, Bronwyn T. Wickes, Danny Hirdes, Boris Grin, David G. Castner, Kevin E. Healy, and Albert Folch, "Micropatterns of chemisorbed cell adhesion-repellent films using oxygen plasma etching and elastomeric masks," *Langmuir* 19, 4754–4764, 2003. Figure contributed by Anna Tourovskaia.)

their cytoskeleton (Figure 2.27, left), whereas for larger islands that still constrained cell spreading, long-term (21 days) cytoskeleton organization was dictated by the shape of the adhesive area.

The stability and nonfouling durability of the PEG-IPNs is hard to match, and it is certainly superior to the thiol-based SAMs, at least for cell patterning applications. However, they cannot be patterned in as convenient a way as a simple microstamping step as SAMs can. Nevertheless, in a collaboration with the Healy lab, the author's lab at the University of Washington in Seattle micropatterned PEG-IPNs using elastomeric stencils and microchannels (Figure 2.28). The stencil or the microchannels serve as an etching mask to protect the PEG-IPN from being etched by an oxygen plasma and also to define the areas in which adhesive protein will be deposited on bare glass. Fibroblasts and muscle cells were observed to attach and spread for weeks only on the etched, PEG-free areas.

2.6.1.3 Direct Photolithography of PEG-Silane

Wouldn't the simplest approach be to buy PEG that is derivatized so that it can be immobilized on a surface of choice? For a while, the company Shearwater (now out of business) offered PEG-silane and, indeed, Jens Karlsson's group (then at the Center for Engineering in Medicine of the Harvard Medical School) seized the opportunity to pattern cells (on glass) simply by direct photolithographic masking (Figure 2.29). The approach worked extremely well, except for the "small" problem that Shearwater's supply was cut off for a long time (until LaySan Bio from Alabama recently replaced them), so building a laboratory based on these strategies can be risky. Next on the BioMEMS researcher's wish list would be that the group be photoactivatable, so that one could skip the photoresist step and patterning could be extended to polystyrene petri dishes and PDMS.

2.6.1.4 Pluronic

In 1996, Patrick Bertrand and colleagues from the Université Catholique de Louvain in Belgium devised a novel concept to selectively block protein physisorption based on forcing the proteins to compete for physisorption sites with a surfactant. First, a photoresist pattern was photolithographically defined on injection-molded polystyrene sheets. To avoid the attack of polystyrene by strong solvents, the photoresist solution was dissolved (50%) in water and an ethanol-soluble developer was used. The photoresist pattern served as a mask to protect the underlying polystyrene areas from becoming hydrophilic during a subsequent oxygen plasma oxidation. Photoresist was then stripped with ethanol. After dipping the substrate in a phosphate-buffer solution of the surfactant Pluronic F68 (a triblock copolymer of polyethylene oxide, polypropylene oxide, and polyethylene oxide, ~150 µg/mL) and ECM protein (either collagen or fibronectin, ~33 µg/mL) for 3 hours, it was observed that adsorption of Pluronic predominated onto the hydrophobic areas, effectively blocking protein physisorption. Rat adrenal pheochromocytoma cells were shown to adhere preferentially onto the hydrophilic areas in serum-containing medium.

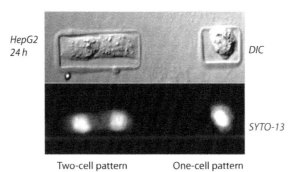

FIGURE 2.29   Cells on photolithographically defined PEG-silane micropatterns. (From D. Irimia and J. O. M. Karlsson, "Development of a cell patterning technique using poly(ethlylene glycol)disilane," *Biomed. Microdevices* 5, 185, 2003. Figure contributed by Daniel Irimia.)

2.6.1.5 PLL-g-PEG Copolymers

In 2003, Gaudenz Danuser (then at ETH Zürich) and colleagues pioneered the spontaneous molecular assembly of a polycationic PEG-grafted copolymer, poly-L-lysine(20 kDa)-g{3.5}-PEG(2 kDa) (PLL-g-PEG), which physisorbs onto surfaces and can be readily loaded onto stamps from aqueous solutions. Here, g expresses the *grafting ratio* (g = 3.5), which is the average number of PLL monomer units per PEG side chain. Peptide(RGD)-functionalized PLL-g-PEG is also possible. Not surprisingly, the approach has become very popular as a cell patterning strategy, and PLL-g-PEG is now commercially available from Surface Solutions (based in Zürich). A group led by Matthieu Piel and Michel Bornens at the Institut Curie in Paris have shown that PLL-g-PEG can be microstamped, leading to high-resolution ECM patterns (that are used to modulate the shape of single cells) when the samples are exposed to a fibronectin solution (**Figure 2.30**). This group has been able to micropattern PLL-g-PEG by selectively degrading it with deep UV light (using a low-pressure mercury lamp, at 185 and 254 nm wavelengths); after deep UV light exposure, the PEG carbons (C-O-C) turn to carboxyl groups and are observed to support protein and cell attachment.

2.6.1.6 Micropatterns of Cell-Repellent Hydrogels

PEG is not the only cell-repellent material out there—most nonprotein hydrogels lack cell attachment motifs and their soft, highly hydrated surface structure is too mobile to support protein or cell adhesion. Mehmet Toner's group at Harvard Medical School demonstrated a microfluidic approach to create micropatterns of hydrogels such as agarose, which has been known to deter protein physisorption and cell attachment for decades. The trenches of deep PDMS microchannel networks were filled with hot (~80°C) molten agarose and the agarose was allowed to solidify; next, the PDMS/agarose textured surface was incubated in a fibronectin solution, which presumably physisorbed only onto the bare PDMS areas. Hepatocytes were observed to attach only onto the fibronectin-coated PDMS "islands" and to limit their spreading precisely to the

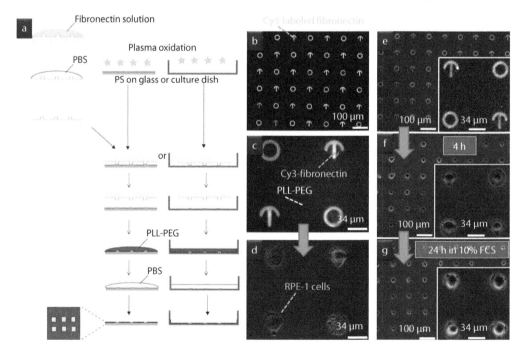

FIGURE 2.30 Deterring cell attachment and spreading with PLL-PEG copolymers. (From Ammar Azioune, Marko Storch, Michel Bornens, Manuel Théry, and Matthieu Piel, "Simple and rapid process for single cell micro-patterning," *Lab Chip* 9, 1640–1642, 2009. Figure contributed by Manuel Théry and Matthieu Piel.)

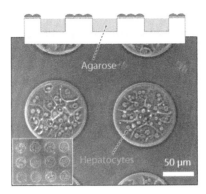

FIGURE 2.31 Confinement of cells using micromolded agarose hydrogel micropatterns. (From Albert Folch and Mehmet Toner, "Cellular micropatterns on biocompatible materials," *Biotechnol. Prog.* 14, 388–392, 1998. Figure contributed by the author.)

border of the island for several days (Figure 2.31). The same technique was later applied by others to produce islands of single cells.

In 1998, Takehisa Matsuda and coworkers at the National Cardiovascular Center Research Institute in Osaka, Japan were able to synthesize a thiolated poly(vinyl alcohol) (SH-PVA) hydrogel which can be straightforwardly immobilized on gold in aqueous solution (Figure 2.32). They

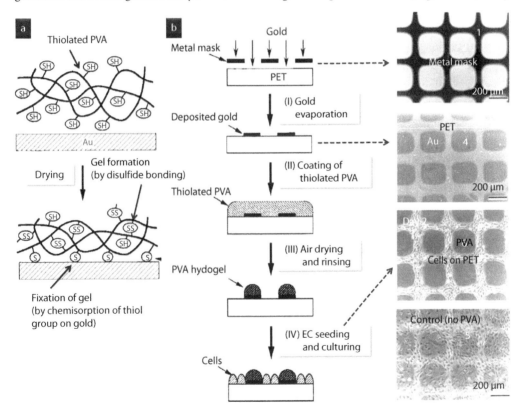

FIGURE 2.32 Simultaneous immobilization and formation of a hydrogel thin film on gold surfaces for cellular micropatterning. (From Y. Nakayama, K. Nakamata, Y. Hirano, K. Goto, and T. Matsuda, "Surface hydrogelation of thiolated water-soluble copolymers on gold," *Langmuir* 14, 3909–3915, 1998. Reprinted with permission of the American Chemical Society. Figure contributed by Takehisa Matsuda.)

created a gold micropattern by shadow-evaporation onto a poly(ethylene terephthalate) (PET) sheet, and a SH-PVA micropattern formed only on the gold areas. After physisorbing ECM protein onto the PET background, bovine endothelial cells seeded in 15% serum were observed to attach and spread for at least 2 days on the PET areas only.

2.6.2 Micropatterning Cell–Substrate Adhesiveness

We have seen the previous methods to immobilize proteins, polymers, and SAMs, including techniques to prevent the physisorption of proteins and to deter cell attachment. Here, we cover the most prominent methods for designing which areas of a substrate the cells will adhere to. We note again that glass and most polymers are covered immediately by a physisorbed protein layer when placed inside a protein-containing biological fluid. Combining the fact that it is surprisingly difficult to completely remove this layer and that cell anchorage is exquisitely sensitive to trace amounts of ECM protein, several researchers have been able to produce cellular patterns on microfabricated templates of physisorbed (or dried) protein or by focal delivery of the cell suspensions (using microfluidic devices), avoiding chemical immobilization methods altogether.

There are now hundreds of published cell patterning techniques, of which we cover the most successful and historically relevant. Some of these methods rely on sophisticated chemical surface modification, some require microfluidic devices, yet others only involve a simple stencil—but not all deliver the same results, resolution, or throughput. The researcher's needs should determine which technique is best suitable for his or her application.

2.6.2.1 Physical Masking of Background with Photoresist

In 1998, Mehmet Toner's group at Harvard Medical School introduced a method to micropattern cocultures of two cell types that is now out of vogue (thanks to soft lithography) but it is explained in detail here because it still has a lot of pedagogical value. The process, outlined in **Figure 2.33a**, takes advantage of differences in adhesiveness between each cell type and consists of chemisorbing a collagen pattern and then filling the background of the pattern by physisorbing albumin (which does not support cell attachment)—the hepatocytes' integrin receptors recognize RGD-like sequences on the collagen (and not on the albumin areas), so selective attachment of the hepatocytes on the collagen occurs.

What is surprising is that it works because the collagen is patterned using a traditional lift-off process whereby the photoresist is removed with acetone, a strong solvent that is a strong denaturant. In other words, the cells do not really care, for the purposes of cell *attachment*, that collagen has been denatured. It makes sense: the cells do not really attach to the whole collagen molecule—only to a small sequence of peptides that has not been altered. We note that the chemisorption reaction chosen was an aminosilane SAM linkage followed by a glutaraldehyde cross-linker, which presumably chemically immobilizes the protein. As it turns out, if one skips the aminosilane and glutaraldehyde steps, collagen patterns are also formed (collagen physisorbs so strongly to the surface that acetone denaturation is not enough to remove it from there) and, most important, the hepatocytes seem to be equally functional. This *also* makes sense—the cells only see the protein, not the surface under it. This whole process teaches us that there are more practical strategies; namely, ones based on physisorption and direct delivery of proteins, to achieve selective cell attachment.

How about the fibroblasts? They attach to the albumin areas because they are seeded in a serum-containing medium that contains a lot of other proteins that promote cell adhesion, even though albumin does not. Nobody knows if these proteins are adhering on top of the albumin, or the albumin desorbs, or the fibroblasts help degrade the albumin first, or all of the above. (Interestingly, the fibroblasts that settle on top of hepatocytes display a remarkable capacity to "find" the substrate even when it seems that the hepatocytes are occupying all of it—either the fibroblasts win some fierce battle or the hepatocytes were not really occupying it: in a few hours, if properly labeled, fibroblasts can also be seen tightly packed in between the hepatocytes, although in low numbers.)

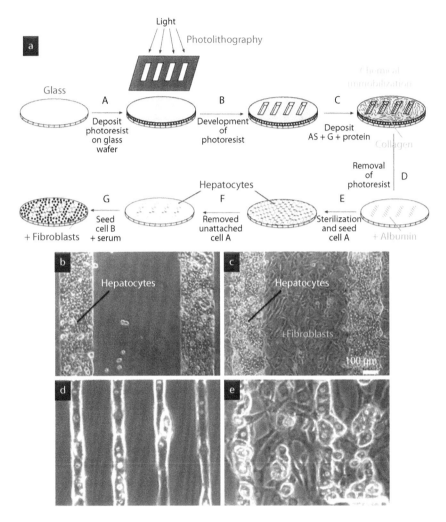

FIGURE 2.33 Microfabricated cocultures. (From S. N. Bhatia, U. J. Balis, M. L. Yarmush, and M. Toner, "Probing heterotypic cell interactions: Hepatocyte function in microfabricated co-cultures," *J. Biomater. Sci. Polym. Ed.* 9, 1137, 1998. Reprinted with permission of John Wiley and Sons. Figure contributed by Sangeeta Bhatia.)

WHY MICROFABRICATED COCULTURES?

COCULTURES ARE USUALLY CREATED by mixing two cell types at random at a certain cell–cell ratio and have been used extensively as an in vitro system to recreate the cell–cell interactions between different cell types ("heterotypic interactions") as found in vivo. However, because of the random nature of the seeding process, many different interactions occur which confound data analysis and which do not necessarily correspond to the interactions so exquisitely structured in vivo. Microfabricated cocultures, as compared with traditional random cocultures, represent the next step in the attempt to simulate in vivo interactions by allowing the researcher to specify cell density and the total length of contact (or "heterotypic interface") between the two cell populations independently of cell–cell ratio.

Using this system, Toner and colleagues showed that primary hepatocytes from rat liver in serum-free medium attach selectively to the collagen areas and do not spread onto the albumin areas for the first 12 to 24 hours (**Figure 2.33b** and **d**). On the other hand, 3T3-J2 fibroblasts did not respect the collagen/albumin template and attached nonpreferentially to all collagen and albumin-coated areas (**Figure 2.33c** and **e**), presumably because fibroblasts can attach to their own, abundant ECM protein secretions. Hence, a micropatterned coculture of hepatocytes and fibroblasts can be achieved by seeding fibroblasts *after* full attachment and spreading of the hepatocytes.

Obviously, the method is not restricted to patterning cocultures only—the nonadhesive background may remain bare of cells. Using a liftoff process similar to that of Toner's group's, Helen Buettner and colleagues at Rutgers University made micropatterned stripes of laminin (chemisorbed to an aminosilane SAM via a glutaraldehyde linkage) on a physisorbed albumin background to study nerve growth cone dynamics. The mean outgrowth length along 20- or 30-µm-wide laminin stripes was observed to be smaller than that on uniform laminin surfaces, whereas outgrowth direction was strongly biased in the direction parallel to the stripes.

The liftoff, solvent-based patterning strategy outlined in this section deserves further commentary. In essence, it is based on the removal of photoresist with a strong organic solvent while exposing the protein pattern to the solvent. This raises two important concerns. First, many polymeric materials used in tissue culture (e.g., polystyrene) are attacked by organic solvents and must therefore be ruled out as the substrate. Second, proteins undergo partial denaturation when exposed to most solvents. Notwithstanding, the work reviewed previously demonstrates that cells attach, spread, grow, and function on denatured ECM proteins, presumably because the degree of denaturation does not mar integrin binding to the ECM peptide fragments (see Section 2.1.3). This is not surprising in view of the fact that such denatured micropatterns, which are routinely visualized by secondary immunofluorescence, feature affinity binding with their natural antibody. However, other long-term functions, more sensitive to the state of denaturation of the underlying ECM protein or to the presence of residuals from the acetone-stripped material, might be impaired by this culture technique. Also, the method cannot be combined with other biomolecules such as antibodies that are irreversibly damaged by the solvent. In 1997, A. H. Bates and coworkers from the Western Regional Research Center reported a creative solution to this problem: the biomolecular layer was first immobilized over the whole substrate, covered with a thin layer of sucrose (a well-known stabilizer of the tertiary structure of proteins on drying) to protect it from subsequent processing, and dried. Thus, a layer of sucrose-covered immobilized antibodies could be later coated with photoresist without damage; the photoresist was then patterned by UV exposure and dissolved on the exposed areas, leaving the sucrose-protected layer exposed. After the sucrose was dissolved in water, the exposed antibody could be removed by a brief oxygen plasma etch. The photoresist/sucrose-protected antibody could be nondestructively uncovered by dissolving the photoresist in acetone and the sucrose in water. Finally, the antibody micropattern was shown to retain its full immunoreactivity. No cellular applications were reported.

Most interestingly, it appears that the chemisorption procedure may be skipped altogether, at least for certain conditions and cell types. What happens if one performs the protein immobilization procedure devised by Toner's group but, against conventional wisdom, skips the silane derivatization and the glutaraldehyde steps and exposes the (now physisorbed) protein to solvent to lift off the photoresist? This is exactly what Bruce Wheeler and colleagues. Then at the University of Illinois at Urbana-Champaign did, with remarkable success. They created PDL patterns on glass by blocking designated areas with a photoresist pattern and physisorbing PDL on the exposed areas; interestingly, removal of the photoresist by sonication in acetone did not result in complete removal of the physisorbed PDL nor compromised its cellular attachment function (**Figure 2.34**). They seeded B104 neuroblastoma cells, a cell line shown to possess neuron-like properties such as electrical excitability, bipolarity, and neurotransmitter production, and to be induced into differentiation by dibutyryl-cyclicAMP, an agent which also stops their proliferation. The cytophobicity of glass was found to be superior than that of a phenyltrichlorosilane (PCTS) SAM; whereas compliance of somata attachment and neurite growth to the PDL pattern was overall poorer than that

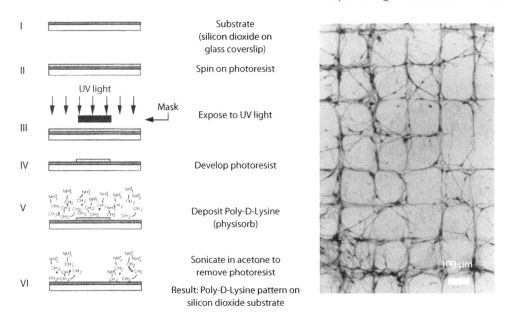

I — Substrate (silicon dioxide on glass coverslip)

II — Spin on photoresist

III — UV light / Mask / Expose to UV light

IV — Develop photoresist

V — Deposit Poly-D-Lysine (physisorb)

VI — Sonicate in acetone to remove photoresist

Result: Poly-D-Lysine pattern on silicon dioxide substrate

100 µm

FIGURE 2.34 Micropatterns of neuroblastoma cells using liftoff and physisorbed PDL. (From Joseph M. Corey, Anna L. Brunette, Michael S. Chen, James A. Weyhenmeyer, Gregory J. Brewer, and Bruce C. Wheeler, "Differentiated B104 neuroblastoma cells are a high-resolution assay for micropatterned substrates," *J. Neurosci. Methods* 75, 91–97, 1997. Reprinted with permission of Elsevier. Figure contributed by Bruce Wheeler.)

achieved for hippocampal primary neurons on PCTS SAMs, it was improved substantially by the addition of dibutyryl-cyclicAMP and represented a quick assay to evaluate the attachment selectivity on candidate materials and to predict the substrate's suitability for primary neuron patterns.

2.6.2.2 Selective Photochemical Immobilization of Proteins

In 1996, Yukio Imanishi's group from Kyoto University pioneered the use of photoreactive growth factors. As an example, photoreactive insulin was synthesized by coupling with azidobenzoic acid. Insulin photoimmobilized onto tissue culture polystyrene enhanced the growth of anchorage-dependent cells such as Chinese hamster ovary cells and mouse fibroblasts, with greater mitogenic activity than free insulin. Micropatterns of insulin were created on PET by exposing the substrate to UV light through a mask in the presence of a photoreactive insulin solution. The immobilized insulin did not enhance cell attachment but transduced a growth signal to the cells. When the medium was depleted of serum, cell growth was observed only for the cells on immobilized insulin. The researchers also synthesized photoreactive polyallylamine by coupling with N-[4-(azidobenzoyl)oxy]succinimide and grafted it onto polystyrene by UV irradiation. Next, the azidophenyl-derivatized polyallylamine was conjugated with mouse epidermal growth factor (EGF). Thus, photoreactive EGF was micropatterned onto polyallylamine-derivatized polystyrene by UV illumination through a mask. Although cells attached everywhere across the substrate, seeding at low densities resulted in segregated cellular stripes when the stripes were far apart (~100 µm) because cell growth was only observed in the EGF-immobilized areas. When pattern widths smaller than the cell were used (~2 µm), patterned cell growth did not occur because all cells proliferated. Regardless of the mechanism used to immobilize the proteins and the cells, Imanishi's work introduced the novel concept of micropatterning a growth factor.

2.6.2.3 Removable Microfabricated Stencils

The abovementioned masking strategies for blocking protein chemisorption onto the background require *chemical* dissolution of the mask (e.g., the photoresist pattern) after the protein immobilization step, thus exposing the protein pattern to strong solvents. One may also design stencil masks that can be *physically* removed (e.g., peeled off). In fact, Carter's work (see **Figure 2.23**) constitutes an early example of stencil masks applied to micropatterning materials with different cellular adhesiveness. A collaborative team led by Mehmet Toner (Harvard Medical School) and David Beebe (then at University of Illinois at Urbana-Champaign) micromolded approximately 50- to 100-μm-thick PDMS stencils (see Section 1.6.5) which could be applied nondestructively onto a substrate to mask any aqueous surface chemistry. Because the stencil forms a seal with the substrate, surface modification occurs only on areas that are exposed through the holes. After rinsing, the PDMS stencil may be peeled off, leaving a protein micropattern. Note

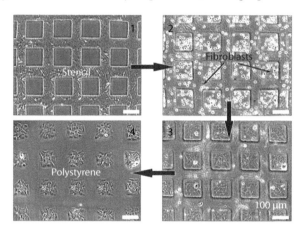

FIGURE 2.35 Elastomeric stencils for cell patterning. (From A. Folch, B.-H. Jo, O. Hurtado, D. J. Beebe, and M. Toner, "Microfabricated elastomeric stencils for micropatterning cell cultures," *J. Biomed. Mater. Res.* 52, 346, 2000. Figure contributed by the author.)

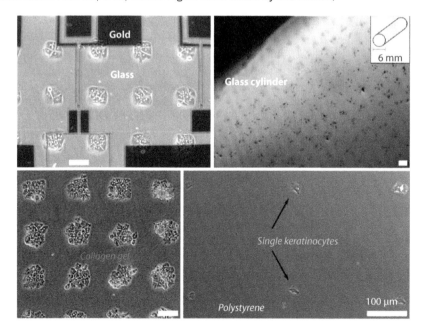

FIGURE 2.36 Cellular microstructures created with PDMS stencils. (From A. Folch, B.-H. Jo, O. Hurtado, D. J. Beebe, and M. Toner, "Microfabricated elastomeric stencils for micropatterning cell cultures," *J. Biomed. Mater. Res.* 52, 346, 2000. Figure contributed by the author.)

that with this method, now used extremely, the micropattern background is never exposed to protein solution because it is covered by PDMS during derivatization. After the cells are seeded and allowed to attach (both onto the substrate and onto the stencil), the stencil is removed, and cellular "islands" with the same shape as the hole remain (Figure 2.35).

A big advantage of stencil technology over other cell patterning strategies is that it can be used straightforwardly in a biological laboratory. It can be applied by hand or with tweezers onto the everyday cell culture surface (which can be heterogeneous, rounded, or even a gel, as shown in Figure 2.36), peeled off after cell attachment without harming the cells, combined with virtually all cell types (down to single-cell resolution), and reused after a simple ethanol wash. Unfortunately, the stencil is limited in that it requires some manual dexterity to fabricate it, works best with regular arrays, and cannot easily produce subcellular spacings.

2.6.2.4 Microstamping of Protein Patterns

As discussed in Section 2.4.2, the generalization of alkanethiol microstamping methods to proteins is not straightforward but it has been worked out to an acceptable degree. In 1998, Harold Craighead and coworkers at Cornell University successfully microstamped (onto plain glass) patterns of PLL, which had been allowed to physisorb onto PDMS. A 2×2 cm PDMS stamp was applied for 15 minutes under a weight of 50 g. Atomic force microscopy of the PLL micropatterns revealed that the coverage was on the order of 0.5 nm and that it was not uniform. Before stamping, the stamp surface was rendered hydrophilic "by treatment in a plasma cleaner/sterilizer" (presumably in oxygen ambient), which the authors claim it caused an "increase in protein adsorption of the stamps" (data not shown). After plasma treatment, the stamp was wet with protein solution for 15 minutes and blow-dried with nitrogen. Dissociated hippocampal neurons selectively attached to 1 μm-resolution, microstamped PLL patterns after 2 to 4 hours in 10% serum. The cells complied to the pattern for at least 3 days when cultured upside down approximately 500 μm above a glial monolayer. Almost at the same time, that same year, two other groups (Hans Biebuyck's from IBM Zurich and Bruce Wheeler's, then at University of Illinois at Urbana-Champaign) also reported having successfully microstamped protein micropatterns using somewhat different protocols—and several other groups have added successful variants. (Unfortunately, more than a decade later, it is not clear yet which protocol works best.) However, an attractive aspect of microstamping is that it is possible to fabricate aligned protein patterns with a single stamp using variable pressure application (see Section 2.4.2), which has been used to create cocultures with a single stamp (Figure 2.37).

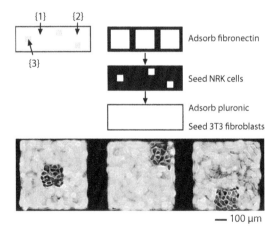

— 100 μm

FIGURE 2.37 Fabrication of cocultures of NRK cells and fibroblasts using a multilevel stamp. (From J. Tien, C. M. Nelson, and C. S. Chen, "Fabrication of aligned microstructures with a single elastomeric stamp," *Proc. Natl. Acad. Sci. U. S. A.* 99, 1758, 2002. Copyright (2002) National Academy of Sciences, U. S. A. Figure contributed by Chris Chen.)

2.6.2.5 Selective Microfluidic Delivery of Proteins

As seen in Section 1.6.4, ECM templates may be deposited on the surface directly from the soluble phase using a microfluidic device that is sealed on the surface of interest, an approach that may be considered the "negative" of microstamping (the protein is deposited where the stamp does not contact the surface). The technique can be used to create micropatterns of peptides, DNA, organic polymers, ceramics, proteins, etc.

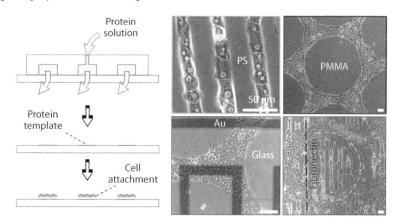

FIGURE 2.38 Cellular micropatterns using microfluidic networks. (From Albert Folch and Mehmet Toner, "Cellular micropatterns on biocompatible materials," *Biotechnol. Prog.* 14, 388–392, 1998. Figure contributed by the author.)

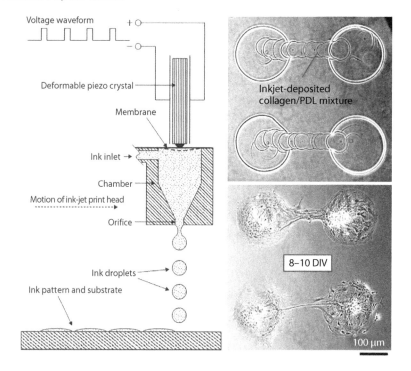

FIGURE 2.39 Cellular micropatterns using inkjet printing. (From Neville E. Sanjana and Sawyer B. Fuller, "A fast flexible ink-jet printing method for patterning dissociated neurons in culture," *J. Neurosci. Methods* 136, 151–163, 2004. Reprinted with permission of Elsevier. Figure contributed by Neville Sanjana and Sawyer Fuller.)

In 1998, Mehmet Toner's group at Harvard Medical School pioneered the use of microfluidics to create ECM protein templates for selective cell attachment; the templates were physisorbed on a variety of biocompatible materials, such as polystyrene, PDMS itself, polycarbonate, and PMMA (or heterogeneous surfaces containing metal circuits or more than one protein), to produce patterns of hepatocytes and keratinocytes on collagen or fibronectin. After the channels were flushed and the elastomer was removed, cells attached only on the protein template (Figure 2.38).

Scientists have fancied for a long time the use of readily available devices that deliver tiny amounts of solutions of proteins and cells to selected locations of a substrate—it is a simple dream: apply the device, and the cells selectively attach to the desired locations. That is what microfluidic devices are about, but the technology is not readily available to everyone. As it turns out, there is a microfluidic device that is commercially available: the inkjet printer! Because of its convenience, several groups have explored the use of printers to deposit scaffolds for cell culture for a long time. Among them, Sawyer Fuller, then at MIT, constructed a custom-built inkjet printer capable of depositing protein droplets (gap resolution, 8 ± 2 μm; smallest islands, 65 ± 5 μm). Dissociated rat hippocampal neurons recognized patterns of collagen/PDL mixture on a PEG background up to 10 days in culture and were electrophysiologically and immunocytochemically normal compared with control cultures (Figure 2.39).

The device developed by Fuller's group is simple but requires the protein to be deposited from the soluble phase onto a dry surface. Can this concept be generalized to work under fluids, to keep the protein in its natural (nondenatured) state? Emmanuel Delamarche's group at IBM Zurich has precisely developed a microfluidic probe that can be used as a noncontact "fountain pen" to write or deposit materials on surfaces—including onto live cells (see Section 3.9.4).

2.6.2.6 Selective Microfluidic Delivery of Cell Suspensions

The concept of microfluidic protein patterning pioneered by Friedrich Bonhoeffer and Hans Biebuyck (Section 2.4.3) can be generalized to cell suspensions (Figure 2.40). In this case, the microchannels must feature deep (~100 μm) channel dimensions to enable the direct injection of a cell suspension. With shallower (<50 μm) microchannels, cell accumulation occurs at the channel entrance, which effectively reduces the cell suspension density in the channel. After cell attachment (~1 hour) under non-flow conditions, the microchannels are removed and a cellular micropattern remains. The technique can be used straightforwardly to micropattern several other cell types simultaneously; because cell attachment is a highly metabolic, oxygen-dependent process, cell types such as hepatocytes or neurons that have high oxygen uptake rates might require a specialized oxygenation scheme. The method uniquely allows for micropatterning homogeneous surfaces at very low cost. Importantly, it can be used in combination with cell types such as fibroblasts that secrete large amounts of ECM and thus do not attach selectively to templates of ECM protein.

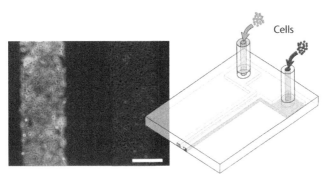

FIGURE 2.40 Selective microfluidic delivery of cells. Scale bar is 100 μm. (From A. Folch, A. Ayon, O. Hurtado, M. A. Schmidt, and M. Toner, "Molding of deep polydimethylsiloxane microstructures for microfluidics and biological applications," *J. Biomech. Eng.* 121, 28, 1999. Figure contributed by the author.)

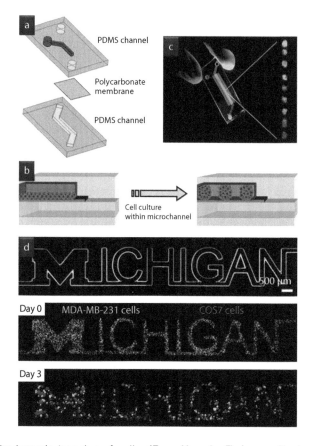

FIGURE 2.41 Hydrodynamic trapping of cells. (From Yu-suke Torisawa, Bor-han Chueh, Dongeun Huh, Poornapriya Ramamurthy, Therese M. Roth, Kate F. Barald, and Shuichi Takayama, "Efficient formation of uniform-sized embryoid bodies using a compartmentalized microchannel device," *Lab Chip* 7, 770, 2007. Reproduced with permission from The Royal Society of Chemistry. Figure contributed by Shu Takayama.)

With the proper design, flow patterns may be established within the channel to deposit the cells at particular locations. Shuichi Takayama and colleagues from the University of Michigan at Ann Arbor have sandwiched porous membranes between two microchannels to micropattern the attachment to cells within microchannels (**Figure 2.41**). The flow is forced to go from the upper microchannel into the lower microchannel through the membrane, but the cells (which do not fit through the pores) get hydrodynamically trapped at the membrane surface, which serves as a long-term cell culture surface. When MDA-MB-231 cells and COS7 cells were simultaneously patterned in a coculture, the cells self-aggregated and formed a spheroid while maintaining the shape of the letters (**Figure 2.42d**). A very similar idea had been used back in 1987 by Friedrich Bonhoeffer's group at the Max Planck Institute (Tübingen Germany) to immobilize cell membrane fragments in stripes, the basis for the famous "stripe assay" for axon guidance studies (see **Figure 6.41** in Section 6.5.1.1).

2.6.2.7 Selective Physisorption on a Microtextured Surface

Alan Rudolph and coworkers at the Naval Research Laboratory in Washington, DC fabricated in 1996 microtextured surfaces with deep trenches in a variety of biomedical polymers and selectively physisorbed proteins onto the mesas by carefully dipping the microstructures in a protein solution. As a result, only the mesas, not the trenches, were coated with protein solution. With this method, they demonstrated the selective attachment of NCTC-929 cells onto

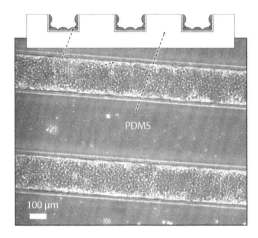

FIGURE 2.42 Cells selectively attached on microtextured surfaces. (From Albert Folch and Mehmet Toner, "Cellular micropatterns on biocompatible materials," *Biotechnol. Prog.* 14, 388–s392, 1998. Figure contributed by the author.)

fibronectin-coated PDMS mesas. The inverse is also possible: by perfusing a PDMS microchannel with fibronectin and using the PDMS surface as the cell culture surface to seed the cells in trenches, Mehmet Toner's group at Harvard Medical School succeeded in creating "quasi–3-D" cell cultures that wrapped around the walls of the trenches (**Figure 2.42**). An extension of this method is the trapping of cells in microwells (see Section 5.3.1).

2.6.3 Cells on Chemisorbed Patterns of Specific Peptide Sequences

Cell attachment and spreading may be promoted on surfaces derivatized with RGD and similar peptide sequences identified by cell adhesion receptors on the cell membrane (see Section 2.1.3). In light of these findings, several groups have developed efforts to micropattern cell attachment peptides on a range of materials.

2.6.3.1 Selective Attachment of Peptides on Heterogeneous Surfaces

In 1995, Patrick Aebischer and coworkers, then at Lausanne University in Switzerland, hydroxylated selected areas of a fluoropolymeric surface by radiofrequency glow discharge through a stencil mask. Oligopeptides containing the laminin fragments YIGSR and IKVAV could be immobilized by their C-terminus via a nucleophilic substitution reaction (in the presence of potassium carbonate) or, alternatively, by their N-terminus using carbonyldiimidazole as a coupling agent. Neuroblastoma cells were shown to attach preferentially to the peptide-derivatized surface in 10% fetal calf serum.

Similarly, Wolfgang Knoll's group from the Max Planck Institute for Polymer Research in Mainz, Germany, created in 1996 micropatterns of a synthetic peptide derived from the neurite outgrowth-promoting domain of the B2 chain of laminin (**Figure 2.43**). First, a glass surface was derivatized with an aminosilane SAM. Then, the SAM was photoablated on certain areas by exposure to a high-radiance (~10 J/cm^2) deuterium lamp through a chrome mask. Once the aminosilane pattern was created, the cysteine-labeled peptide was coupled to the aminosilane areas via a cross-linker, *N*-(γ-maleimidobutyryloxy) sulfosuccinimide ester, which reacts with amines and thiol groups (in this case, the amine groups on the SAM and the thiol group on the cysteine). The irradiated areas were observed to be cell repellent. The peptide-modified stripes guided the growth of rat hippocampal neurons with a morphology resembling that of neurons cultured on laminin-coated surfaces. The same chemistry has been used to make neurons adhere onto a field-effect transistor array.

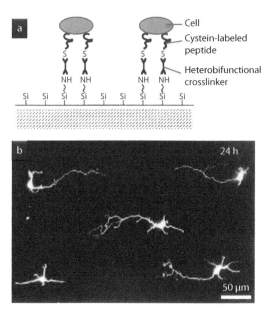

FIGURE 2.43 Neuronal micropatterns using cell-adhesion peptides. (From Mieko Matsuzawa, Paivi Liesi, and Wolfgang Knoll, "Chemically modifying glass surfaces to study substratum-guided neurite outgrowth in culture," *J. Neurosci. Methods* 69, 189–196, 1996. Reprinted with permission of Elsevier.)

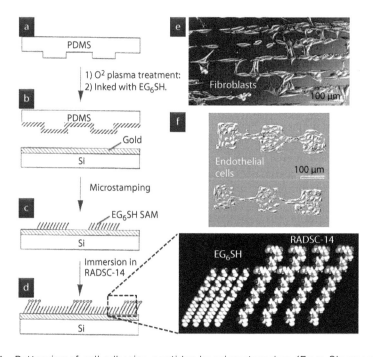

FIGURE 2.44 Patterning of cell-adhesion peptides by microstamping. (From Shuguang Zhang, Lin Yan, Michael Altman, Michael Lässle, Helen Nugent, Felice Frankel, Douglas A. Lauffenburger, George M. Whitesides, and Alexander Rich, "Biological surface engineering: A simple system for cell pattern formation," *Biomaterials* 20, 1213–1220, 1999. Reprinted with permission of Elsevier. Figure contributed by George Whitesides.)

Micropatterning peptides on surfaces that support protein physisorption immediately raises the question of whether cells will, sooner or later, attach to ECM proteins present in the medium or secreted by the cells. Increased control over cell-substrate interactions may be achieved by immobilizing cell-adhesion peptides on PEG copolymers or PEG-functionalized SAMS. In 1998, Molly Shoichet's group at the University of Toronto, in Canada, devised a simple method to micropattern cell attachment peptide sequences on a PEG background. First, PEG-aldehyde was immobilized on an aminosilane SAM on glass. A microfabricated grid was then placed in contact with the PEG surface and used as a stencil mask for gold sputtering, resulting in gold islands on a PEG background. Finally, commercially available peptide sequences labeled with cysteine (which contains a gold-reactive thiol group) were immobilized on the gold areas. Hippocampal neurons seeded under serum-free conditions were observed to attach selectively to the peptide-modified areas only and displayed neurite extension statistics similar to or greater than those found in neurons seeded on laminin. Similarly, a team led by George Whitesides at Harvard University and Alexander Rich at MIT microstamped hexa(ethylene glycol)-terminated alka-nethiol SAMs on gold; next, they chemisorbed (on the nonstamped areas) a cysteine-ended oligopeptide containing a cell-adhesion peptide sequence to demonstrate the selective attachment of cells (human epidermal carcinoma cells, 3T3 mouse embryo fibroblasts, and bovine aortic endothelial cells) onto the peptide areas in medium containing 10% serum (Figure 2.44).

2.6.3.2 Selective Photochemical Immobilization of Peptides

In 1995, Patrick Aebischer's group, then at Lausanne University, coupled the laminin fragment CDPGYIGSR to a photosensitive benzophenone or diazirin group. Thus, the peptide could be immobilized onto various materials, such as hydroxylated fluorinated ethylene propylene, poly(vinyl alcohol), and glycophase glass, by exposure to UV light from a high-pressure mercury lamp. Benzophenone is a particularly convenient photosensitive group because, unlike azide-based compounds, it is reversibly excitable and stable in ambient light. Peptide micropatterns were created by selective exposure through a mask, and selective neuroblastoma cell attachment onto the peptide areas was demonstrated. Later, this photochemistry was applied by laser to an agarose gel to form 3-D patterns of the laminin fragment in agarose.

Similarly, in 1995, Takashi Sugawara and Takehisa Matsuda from the National Cardiovascular Center Research Institute in Osaka, Japan, reacted the amine terminus of synthetic peptide sequences with 4-azidobenzoyloxysuccinimide, a photoreactive and amino-reactive cross-linker; on UV exposure, the peptide (now linked to a photosensitive group) could be immobilized onto poly(vinyl alcohol) in micropatterns by selective exposure through a chrome mask. Bovine endothelial cells attached selectively onto the peptide patterns in medium containing 15% fetal calf serum.

In 1997, a large collaborative team led by Wei-Shou Hu from the University of Minnesota also created micropatterns of synthetic cell-adhesive peptides that had been photosensitized by coupling to benzophenone. As a nonadherent background surface, they used a SAM of hexaethyleneglycol-undecanethiolate on gold, a surface designed to repel cell and protein adhesion. When this surface was UV-illuminated under a solution of photosensitive oligopeptides, it became cell-adherent. By scanning a UV laser beam or by UV-illuminating the surface through a chrome mask, they were able to create oligopeptide (adhesive) micropatterns on a nonadhesive background. Because the density of immobilized RGD groups (i.e., the adhesiveness of the surface) can be fine-tuned by adjusting the exposure, they were able to create *gradients* of adhesiveness by varying the duration of exposure to the UV laser beam. Fibroblasts attached to these surfaces at a density that increased with increasing oligopeptide density.

2.6.3.3 Selective Microfluidic Delivery of Peptides

In 1998, a collaborative team led by Kevin Shakesheff from the University of Nottingham in the United Kingdom and by Robert Langer from MIT used microfluidic patterning to create peptide templates for cell attachment (Figure 2.45). First, amine-terminated PEG was coupled to NHS-biotin, and lactide was polymerized from the hydroxy end of PEG. The obtained

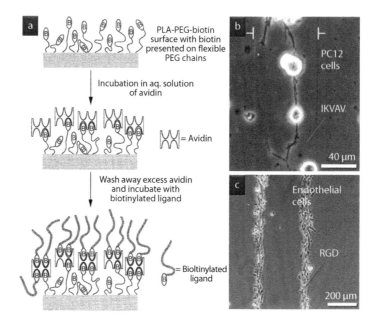

FIGURE 2.45 Microfluidic patterning of cell-adhesion peptides on PEG backgrounds. (From Nikin Patel, Robert Padera, Giles H. W. Sanders, Scott M. Cannizzaro, Martyn C. Davies, Robert Langer, Clive J. Roberts, Saul J. B. Tendler, Philip M. Williams, and Kevin M. Shakesheff, "Spatially controlled cell engineering on biodegradable polymer surfaces," *FASEB J.* 12, 1447, 1998. Figure contributed by Kevin Shakesheff.)

polylactide–poly(ethylene glycol)–biotin repels cell adhesion and is biodegradable. A film of the biotin-conjugated copolymer was droplet-cast and dried onto polystyrene. PDMS microchannels were sealed against the coated surface and an avidin solution was allowed to fill the microchannels by capillary action; avidin binding to the copolymer-biotin groups thus generated an avidin pattern onto the copolymer background. Subsequently, the channels were flushed and removed. When the avidin micropattern was exposed to a solution of biotinylated peptides, the peptides were shown to attach only to the avidin-derivatized areas. IKVAV and RGD peptide sequences were used to demonstrate the selective attachment of PC12 nerve cells and bovine aortic endothelial cells, respectively, in serum-free medium. This surface engineering approach has the merit that it is generalizable to any biotinylated ligand and any surface to which avidin physisorbs.

2.6.4 Other Cell Micropatterning Strategies

The list of cell patterning techniques does not end here. We have not even considered techniques to pattern cells in 3-D, which will be reviewed in Chapter 7 when we cover Tissue Microengineering. In the next section, we present techniques that do not fit in any of the previous categories, yet they have had a notable impact in the field of BioMEMS.

2.6.4.1 Selective Biorecognition by the Substrate

In 1998, Harold Craighead's group from Cornell University implemented the microstamped version of "panning," a method commonly used to sort cells. When a glass or plastic plate is coated with antibodies against proteins present on the membrane of a certain cell type, the plate will *immunocapture* that cell type preferentially, so that the rest can be rinsed away—it's a method used for cell sorting. The approach is ingenious: rather than relying on the cell's ability to recognize and, thus, preferentially attach to a biomolecular micropattern, these researchers demonstrated that a template of an antibody against *E. coli*, made by microstamping onto an unmodified silicon surface, selectively "captures" the bacteria in the antibody-covered areas. The concept potentially could be

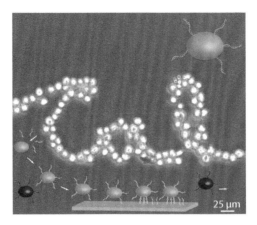

FIGURE 2.46 Cell patterning using DNA barcodes. (From Erik S. Douglas, Ravi A. Chandra, Carolyn R. Bertozzi, Richard A. Mathies, and Matthew B. Francis, "Self-assembled cellular microarrays patterned using DNA barcodes," *Lab Chip* 7, 1442–1448, 2007. Reproduced with permission from The Royal Society of Chemistry.)

used for patterning several cell types simultaneously if different antibodies to selected membrane proteins of each cell type featuring small cross-reactivities can be produced in a practical manner.

Why limit ourselves to immunocapture? Indeed, Richard Mathies, Matthew Francis, and coworkers at the University of California at Berkeley have decorated cells with DNA strands (which they termed "DNA barcodes") and used patterns of complementary DNA immobilized on the surface to capture cells in selected locations (**Figure 2.46**).

2.6.4.2 In Situ Microfabrication of Protein Structures

In 2004, a group led by Matsuhiko Nishizawa from Tohoku University in Sendai, Japan, demonstrated that physisorbed albumin can be electrochemically removed from the substrate on

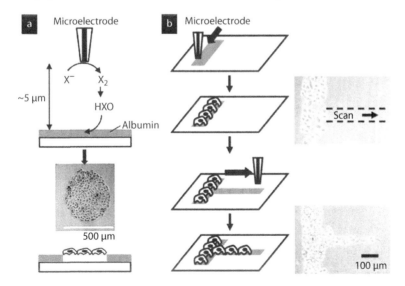

FIGURE 2.47 Selective electrochemical inactivation of albumin. (From Hirokazu Kaji, Masamitsu Kanada, Daisuke Oyamatsu, Tomokazu Matsue, and Matsuhiko Nishizawa, "Microelectrochemical approach to induce local cell adhesion and growth on substrates," *Langmuir* 20, 16–19, 2004. Figure contributed by Matsuhiko Nishizawa.)

exposure to an oxidizing agent such as HBrO; the agent is produced locally with a microelectrode "pen" (a Ag/AgCl electrode positioned 5 μm above the surface biased with 1.7 V for 30 seconds) in the presence of a Br$^-$–containing solution (25 mM KBr), which is compatible with cells (Figure 2.47a). Thus, Nishizawa's group was able to "erase" patterns of BSA in the presence of HeLa cells and observe the growth and migration of new cells in the albumin-free areas (Figure 2.47b).

Nishizawa's technique is highly specialized in that, so far, it has only been demonstrated to work for albumin. But how about using light, instead of electrical current, to pattern proteins in situ? We have already seen how the high-energy femtosecond pulsed laser of a multiphoton microscope can be used for writing protein structures (Figure 2.12). Jason Shear and colleagues at the University of Texas in Austin have photo-cross-linked nonadherent albumin "corrals" in the presence of cells (neurons and bacteria; see Figure 1.16).

2.6.4.3 Microstamping of Cells

It is possible to locally deposit cells from an agarose stamp, both with bacterial and mammalian cell types. The hydrogel stamp, instead of being loaded with an etchant (see Section 1.7), can be loaded with a bacterial suspension using the same simple protocol. The bacteria get "printed"

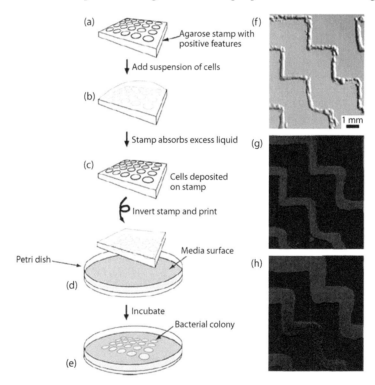

FIGURE 2.48 Microstamped bacterial colonies. (a) Schematic of an agarose stamp. (b) A bacterial suspension is added to the agarose stamp (3% agarose). (c) The cells deposit on the surface of the stamp after excess liquid is removed. (d) Stamp is applied to an agar plate (a common bacterial culture substrate). (e) Bacterial colonies form on the areas contacted by the stamp. (f–h) Images depicting the growth of the patterns of bacteria over time: (f) bright-field image after 10 hours of growth (no photoluminescence was detected); (g) 20 hours; (h) 40 hours. (From Douglas B. Weibel, Andrew Lee, Michael Mayer, Sean F. Brady, Derek Bruzewicz, Jerry Yang, Willow R. DiLuzio, Jon Clardy, and George M. Whitesides, "Bacterial printing press that regenerates its ink: Contact-printing bacteria using hydrogel stamps," *Langmuir* 21, 6436, 2005. Reprinted with permission of the American Chemical Society. Figure contributed by George Whitesides.)

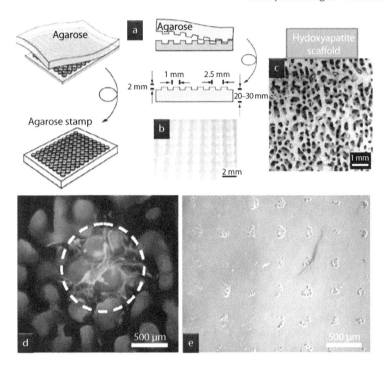

FIGURE 2.49 Direct transfer of mammalian cells with agarose stamps. (From Molly M. Stevens, Michael Mayer, Daniel G. Anderson, Douglas B. Weibel, George M. Whitesides, and Robert Langer, "Direct patterning of mammalian cells onto porous tissue engineering substrates using agarose stamps," *Biomaterials* 26, 7636–7641, 2005. Reprinted with permission from Elsevier.)

and form colonies only on the areas contacted by the stamp (Figure 2.48), and they can even replicate on the stamp—a sort of "ink" that regenerates itself.

This idea has also been demonstrated for patterning osteoblasts (mammalian cells) onto hydroxyapatite (Figure 2.49), a highly porous substrate, so it can presumably be extended to other mammalian cell types and a variety of porous substrates used in biotechnology (e.g., filters) and tissue engineering (e.g., biodegradable polymers).

Importantly, the gel can be loaded with a gradient of biomolecules—for example, by contacting it only partially or with an uneven volume of the loading solution—which has implications in studies of chemotaxis.

2.6.4.4 Electroactive Substrates

George Whitesides at Harvard University and his former postdoctoral researcher Milan Mrksich, now a professor at the University of Chicago, developed a powerful method that allows for *changing* the adhesiveness of the cell-adherent substrate using electroactive SAMs. The method is based on the observation that cell-repellent triethylene-glycol-terminated thiol SAMs (EG_3-C_{11}) desorb when a voltage pulse of approximately 1.2 V is applied to the gold substrate with respect to the cell culture medium. Before application of the voltage pulse, cells were confined to micropatterns in normal growth media for 24 hours, but after the voltage pulse, they started to migrate across the "bare" gold (presumably already covered with proteins secreted by the cells or deposited from the medium) and underwent normal growth and cytokinesis (Figure 2.50).

This principle is very versatile. Stripes (or any pattern) of ECM protein can be fabricated alternating with initially inert electroactive monolayers, so that a population of cells can be seeded to attach to the protein stripes but not the monolayer; next, on electrical activation of the monolayer, the spaces between stripes become adhesive and a second cell population can be

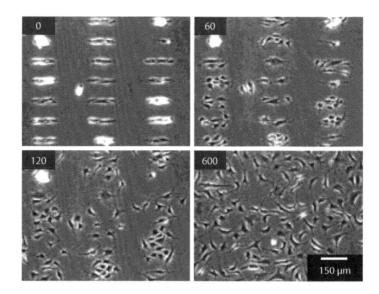

FIGURE 2.50 Electroactive release of cells from micropatterns. Numbers indicate minutes elapsed after application of the voltage pulse. (From Xingyu Jiang, Rosaria Ferrigno, Milan Mrksich, and George M. Whitesides, "Electrochemical desorption of self-assembled monolayers noninvasively releases patterned cells from geometrical confinements," *J. Am. Chem. Soc.* 125, 2366–2367, 2003. Reprinted with permission of the American Chemical Society. Figure contributed by George Whitesides.)

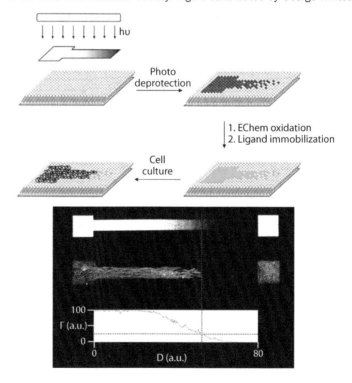

FIGURE 2.51 Dynamic surface gradients using electroactive SAMs. (From E. W. L. Chan, S. Park, and M. N. Yousaf, "An electroactive catalytic dynamic substrate that immobilizes and releases patterned ligands, proteins, and cells," *Angew. Chem. Int. Ed.* 47, 6267–6271, 2008. Figure contributed by Muhammad Yousaf.)

seeded, producing a micropatterned coculture. Another variant is to form electroactive SAMs of cell-adhesion peptides (instead of SAMs of nonadhesive EG_3-C_{11}), and cells detach when the cell-adhesion peptides are electrically desorbed. This work represents probably the most exquisite spatiotemporal and molecular control over cell–substrate interactions.

In 2007, Muhammad Yousaf (formerly a student of the Mrksich laboratory) and colleagues from the University of North Carolina at Chapel Hill were able to bring photochemistry into play, to generate…gradients of electroactive SAMs (Figure 2.51)! In addition, the cells can be released from the substrate after attachment. Such dynamic surface gradients will no doubt become invaluable tools in studies of cell polarization and cell migration.

2.7 Summary

This chapter has reviewed the application of the techniques covered in Chapter 1 for patterning biological material such as proteins and cells, as well as other materials that are used for cell and protein patterning (such as SAMs and polymers). Cells are exquisitely sensitive to the physicochemical properties of the substrate to which they attach. Chemists have spent great efforts to engineer the surface of materials with SAMs and polymers to modulate (prevent as well as enhance) cell adhesion with nanometer precision and, occasionally, with temporal control. Using simple microfluidic devices, it is now possible to create micropatterns of cells on templates made of physisorbed ECM proteins (on a variety of biocompatible polymers); or to pattern cells on electroactive cell-adhesion RGD peptides simply by applying a small voltage pulse—the spectrum of existing techniques is very wide and the dilemma of choosing the best one is application-dependent (and expertise-dependent).

Further Reading

Baneyx, F., and Schwartz, D.T. "Selection and analysis of solid-binding peptides," *Current Opinion in Biotechnology* **18**, 312–317 (2007).

Falconnet, D., Csucs, G., Michelle Grandin, H., and Textor, M. "Surface engineering approaches to micropattern surfaces for cell-based assays," *Biomaterials* **27**, 3044–3063 (2006).

Folch, A., and Toner, M. "Microengineering of Cellular Interactions," *Annual Review of Biomedical Engineering* **2**, 227–256 (2000).

Lim, J.Y., and Donahue, H.J. "Cell sensing and response to micro- and nanostructured surfaces produced by chemical and topographic patterning," *Tissue Engineering* **13**, 1879–1891 (2007).

Madou, M. *Fundamentals of Microfabrication*. CRC Press (2002).

Mrksich, M. "Using self-assembled monolayers to model the extracellular matrix," *Acta Biomaterialia* **5**, 832–841 (2009).

Nie, Z., and Kumacheva, E. "Patterning surfaces with functional polymers," *Nature Materials* **7**, 277–290 (2008).

Smith, R.K., Lewis, P.A., and Weiss, P.S. "Patterning self-assembled monolayers," *Progress in Surface Science* **75**, 1–68 (2004).

Stevens, M.M., and George, J.H. "Exploring and engineering the cell surface interface," *Science* **310**, 1135–1138 (2005).

3

Microfluidics

MICROFLUIDICS IS THE FIELD that studies and exploits the behavior of fluids confined to small volumes, such as microchannels, droplets, jets, thin water films, and so on. A good definition of "small" is one that has at least one dimension less than 1 mm. At this scale, most fluids behave in nonintuitive ways because capillary forces and viscous forces that are usually negligible on a larger scale become the predominant forces. If the fluid volume under study has a dimension of less than 1 μm, it generally belongs to the realm of *nano*fluidics (in which continuum mechanics do not apply) and is outside the scope of this textbook.

WHAT IS MICROFLUIDICS?

HISTORICALLY SPEAKING, MICROFLUIDICS is the daughter of electrophoresis, the technique for separating mixtures of molecules into its components using electrical fields. The first microfluidic devices ever invented were all conceived to improve electrophoretic separations. Note that a thin sheet of fluid (which can be produced with two glass plates and a spacer, for example) fits in our definition of microfluidic, and that's exactly how the first microfluidic devices for electrophoresis were built back in the 1940s. In July 1939, Oxford University biochemist J. St. L. Philpot sent a manuscript to the Transactions of the Faraday Society describing the use of laminar flow and orthogonal voltage to separate proteins. In 1950, Kurt Hannig from the Max Planck Institute (Munich) designed a paper electrophoresis device in which gravity-assisted flow would trickle down a filter paper as proteins were subjected to an orthogonal field—the first paper microfluidics device. Hannig would also design a "free-flow" version (only fluid, no paper) of the same device in 1961, which would become the basis of the McDonnell Douglas free-flow electrophoresis design that was launched and successfully tested in space in 1982. Note, however, that all of these systems confined the fluids to sheets or, at most, wide channels (made by milling, for example); to confine the fluids to smaller channels, or chambers, it would be necessary to use the silicon microfabrication techniques that were developed in the 1980s.

At the beginning of the 1990s, it became clear to many analytical chemists that miniaturizing fluidic systems could provide several major advantages, and that microfabrication provided a means to make such miniaturized systems. Some of these advantages were obvious, such as reducing the volume of reagents and waste, and allowing the analysis of very small samples. The potential for achieving the holy grail of making analytical systems better, faster, and cheaper drew many of us into this field. The discovery that there were unexpected features of the behavior of fluids in microchannels was surprising, at least in the engineering community. This field became known as "**microfluidics**." In this chapter, we will introduce the highlights of this new field.

Biology began exploiting the principles of microfluidics a few billion years before the 1990s. For example, bacteria are so small that when they swim through low-viscosity fluids like water, they stop within microseconds when their flagellar motors reverse—they have so little momentum that, to them, water appears as viscous as honey appears to us. Blood capillaries in higher animals exploit the fact that diffusion of small molecules over short distances is very efficient, so turbulent flow is not necessary to provide rapid transport of oxygen to tissues. It is worth the effort for the BioMEMS student and researcher to ponder for a few minutes why one should spend time and money to build a microfluidic device—after all, it may not be as advantageous as one might think, and it is often less straightforward than the all-too-abbreviated experimental sections of scientific articles would lead you to believe.

3.1 Why Go Small?

Microfluidic devices generally confer six large classes of advantages, as graphically depicted in **Figure 3.1**:

(A) Flow in microchannels is **laminar** (nonturbulent) and thus the flow patterns and concentrations can be mathematically modeled, making quantitative predictions of

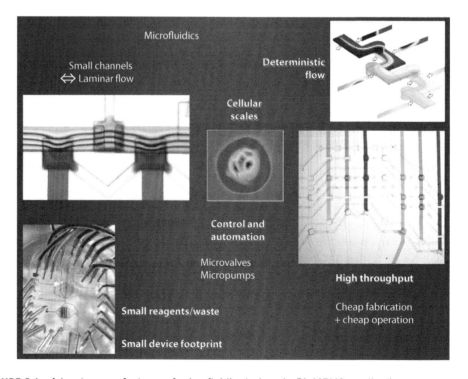

FIGURE 3.1 Advantageous features of microfluidic devices in BioMEMS applications.

the biological environment of cells and other biochemical reactions possible. This **deterministic behavior** of flow in microchannels is a unique feature that does not have a parallel in the macroscopic world and thus often requires clever engineering to exploit it.

(B) Microchannels, as their name indicates, can be fabricated easily on a scale similar to cellular scales (or smaller), so they can be used to probe (sub)cellular phenomena, to seed and sort (single) cells, to build systems that imitate physiological parameters, and so on.

(C) For a relatively small incremental cost, one can integrate **microvalves** and **micropumps** into the same device, which enables automated control of fluids that reduces human error and operator costs.

(D) Microfluidics as a technology has been predicted to provide an economical advantage, especially when the cost is normalized to a per unit basis, in terms of both fabrication (**batch fabrication**) and operation (especially when combined with microvalves and micropumps). Thus, it should enable **high-throughput** experiments. The technology is, however, young and critics caution that this vision is not (and may never be) as fully realized as the exponential progression of integrated circuit improvement known as Moore's law, simply because there is no such thing as a "microfluidic transistor" on which to sustain the same economy of scale as there was for the electronic transistor.

(E) Typically, microfluidic devices consume much smaller amounts of reagents and produce much smaller amounts of waste than their macrofluidic analogues.

(F) Microfluidic devices occupy much smaller footprints than their macrofluidic counterparts and (provided the inputs and the reagents are packaged appropriately) are potentially portable to resource-poor settings such as underdeveloped countries.

3.2 Microscale Behavior of Fluids

In this section, we discuss the physicochemical properties of fluids and their unique behavior in microchannels.

THINK HONEY

WE ARE ALL FAMILIAR with the behavior of fluids from our experience—from what it feels like to swim in a lake to the small but non-zero effort required to mix milk into our morning cup of coffee. We know about the importance of momentum and viscosity in fluidics; if you push on a toy boat in the bathtub, it will drift for a few seconds before slowing to a halt, but we know from the movies that if you try to stop the *Titanic*, it will take you longer than it will take to reach the iceberg. Most of this understanding we glean from the world where momentum is very important, and viscosity is less important. To mix coffee and milk faster, we stir harder. A good exception is the behavior of very viscous fluids, like honey and peanut butter. We know that if we put the honey on one piece of bread and peanut butter on the other, we can return to that sandwich hours or weeks later and we'll still see the honey and the peanut butter sitting separately on their respective sides of the sandwich. No amount of shaking of the sandwich will make much difference because it is not a very effective method of mixing high-viscosity fluids.

3.2.1 Viscosity

Viscosity is one of those things that we all (roughly) understand, but for which we rarely get a clear definition. Viscosity is the property of a medium that allows transfer of shear forces from one object to another within that medium. For example, we all understand that if we need to stir some raw sugar into our latte, stirring the latte with a spoon will (eventually) move the coffee in circles, but the cup is not going to move much. On the other hand, use the same spoon to stir a cup of pure honey (assume you use the same force you used with the latte), and the cup will move with the spoon, but you will not do much to the honey. Why? Because the viscosity of the honey is a few thousand times higher than that of water, so the motion of the spoon is more efficiently transmitted to the cup.

The conventional definition for the **dynamic viscosity** μ (also called **absolute viscosity**) is the tangential force per unit area required to slide one plane with respect to another a unit distance apart at unit velocity. In the CGS system of units, the unit of dynamic or absolute viscosity is the poise:

$$1 \text{ poise} = \text{dyne s/cm}^2 = \text{g/cm s} = 1/10 \text{ Pa s}$$

The usual units are centipoises, cP, in which units the viscosity of pure water is approximately 1 at room temperature.

The other commonly used measure of viscosity is **kinematic viscosity**, ν, which is the absolute viscosity divided by the density, ρ.

$$\nu = \frac{\mu}{\rho}$$

For water, because the density is approximately 1 g/cm³, the value for ν in centistokes (cSt) is, again, approximately 1.

Note that to differentiate the kinematic viscosity (the Greek letter nu, ν) from the velocity of a fluid in equations, most texts use the Roman letter u for velocity, rather than the Roman letter v, which, in some fonts designed with no scientific insight, is indistinguishable from ν.

3.2.2 Nondimensional Analysis: Reynolds Number and Peclet Number

Engineers are very fond of reducing extremely complex phenomena to simple "dimensionless" expressions that allow one to predict behavior on the basis of one number. For flow, there is a quantitative dimensionless expression that allows one to predict the behavior of a system based on the fluid(s) involved, the dimensions of the container and how fast the fluid is moving—the **Reynolds number** (**Re**). The definition for the Re, which predicts whether the system will be dominated by viscosity or momentum, is as follows:

$$Re = \frac{\rho u L}{\mu} \tag{3.1}$$

where:
- u = average flow speed (for example, the flow speed in a river, in a pipe, or in a blood artery)
- L = characteristic length (for example, the diameter of a cylindrical channel or the thickness of the peanut butter layer)
- μ = viscosity

Another useful dimensionless quantity in microfluidics is the Peclet number. The Peclet number relates the rate of advection (i.e. transport by the fluid) to the rate of diffusion, so it appears in mixing phenomena and is defined by:

$$Pe = \frac{uL}{D} \qquad (3.2)$$

where:
   u and L were previously defined
   D = diffusion coefficient

3.2.3 Laminar Flow

In our honey–peanut butter sandwich, Re is dominated by the extremely high viscosity of the fluids, whereas in our cup of coffee, the large dimensions of the cup combined with the low viscosity of the coffee brings Re to values in which mixing by turbulence is possible. Typically at Re approximately higher than 2000, flow is turbulent (random velocities in time and space), which enhances mixing. At very low Re, flow is regular, *laminar*, and does not occur at all unless the fluid is pushed continuously by a pressure gradient. Of course, one could have fluid flow in a channel the size of a house under low Re conditions, as long as the flow rate was low and viscosity of the fluid very high. Conversely, at very high flow rates and with low viscosity fluids (for example gases), you can have turbulent flow.

We live in a world in which most of our experiences reflect high Re (Re $\gg$ 1) largely because for water (and to a much greater extent, air) the viscosity is low, and we are used to fairly large dimensions. For blood in capillaries or water in microfabricated channels, Re is generally much lower than 1, although one can always increase the flow velocity (by pumping harder) to bring the Re to arbitrarily high values and generate turbulence. However, for our purposes, we will only consider Re substantially lower than 2000. In this low-Re world, flow is laminar, fluids only move when pushed, and transport of matter across flow lines is primarily by diffusional mixing.

TURBULENCE? (ALMOST) NEVER

FLOW AT RE LOWER THAN 2000 is usually "laminar," rather than turbulent. *Lamina* is Latin for a metal blade; laminar flow can be modeled by a stack of infinitesimal layers, each slipping past at a steady velocity. The velocity of the fluid perpendicular to the laminae changes smoothly with position, and, at a given point in a laminar system, is constant with time (as long as the driving force is constant). At the submillimeter scale at which we are concerned, most flow is laminar rather than turbulent. Most liquids of interest are, for all practical purposes, incompressible, so they move in response to differences in pressure; such fluid motion can be perfectly predicted by the Navier–Stokes equation. Today, there are commercial Computational Fluid Dynamics (CFD) packages that do an excellent job of predicting the flows of fluids in arbitrarily shaped channels as long as you input the correct fluid viscosity and the forces moving the fluid.

Note that laminar flow, particularly in cylindrical pipes, is sometimes called Poiseuille (pronounced "pwä-'zɔi") flow, after the French gentleman Jean Louis Marie Poiseuille (1797–1869) who donated the first half of his name for the units of viscosity ("Poise").

Most fluid behavior in microdevices is at low enough velocities that low Reynolds conditions dominate. We generally assume "wall stick" conditions, under which the fluid velocity immediately adjacent to the wall is zero, and increases parabolically to the maximum velocity in the middle of the channel (see next section). When the forces responsible for the motion are constant, flow is laminar, so that the velocity at a point in the device is constant in time, and varies smoothly and predictably in space. This limitation is both the blessing and curse of microfluidics, as we will see in the next chapter.

3.2.4 Parabolic Flow Profile

Most fluid mechanics textbooks start by laying out the general differential equation that governs flow, called the **Navier–Stokes equation**, in the most general possible of cases, which is for any type of fluid in a conduit of any shape and considering that perturbations such as external forces or temperature changes can occur. This general equation, which is simply Newton's second law of mechanics [$F = d(mv)/dt$] applied to fluids (under the assumption of continuum mechanics), can become very complicated and do not belong in this textbook. Fortunately for us, in BioMEMS, we deal almost only with aqueous fluids, which belong to the category of incompressible, uniform, and viscous fluids (the so-called "**Newtonian fluids**")—which simplifies the equation substantially. A notable exception is blood (see box below). The solution to the equation is a function then of the shape of the conduit (whether the microchannel has a circular or a rectangular cross-section, etc.).

An important concept in solving differential equations is that of **boundary condition**. The concept is simple: because the equation contains a derivative, say $x'(t)$, then in order to solve it, we need additional information, otherwise the solution stays undetermined by a constant (because the derivative of a constant is zero): both functions $x_1(t) = 3t^2 + 3$ and $x_2(t) = 3t^2 + 5$ have the same derivative, $x'(t) = 6t$. The term "boundary condition" in mathematics is very appropriate because it usually relates to the actual physical boundaries in an engineering problem. One of the most important boundary conditions in microfluidics, for example, is the **no-slip condition**, which states that the value of the velocity at the walls must be zero. It's a condition that nobody has really observed directly (because nobody can stand there without interfering to watch whether the atoms of the fluid are sticking or slipping over the wall atoms), but there is ample indirect evidence that, on average, fluids in microchannels behave as if they "roll" rather than "dragging their feet." The molecular nature of the fluid is not apparent at the scale of the microchannels.

When one assumes that the fluid is indeed incompressible, Newtonian, and isotropic, and that the flow remains at constant temperature (or that the viscosity η does not depend on temperature) through an arbitrary constant cross-section (i.e., straight channel without leaks, so that the flow becomes purely axial in the direction of x), then the Navier–Stokes equation (the solution of which is the flow velocity profile) gets reduced to the following set of much simpler set of equations:

$$0 = -\frac{\partial p}{\partial x} + \eta \left(\frac{\partial^2 u_x}{\partial y^2} + \frac{\partial^2 u_x}{\partial z^2} \right) \tag{3.3}$$

$$0 = \frac{\partial p}{\partial y} = \frac{\partial p}{\partial z} \tag{3.4}$$

In other words, there are no gradients of pressure in y or z; all flow runs parallel to x, and is driven by the pressure gradient in x. Note that, in general, the x component of the flow velocity $u_x = u_x(y,z)$ will depend on the exact cross-sectional shape of the microchannel. The shape and size of the cross-section of the channel are also boundary conditions necessary for solving the previous equations.

A number of common shapes (and not so common ones), including the circular and the rectangular cross-section, can be solved analytically. The isosceles triangle is also included in the next section because it is a geometry commonly used in more traditional silicon-based microfluidics (resulting from the wet etch of Si(100), see Section 1.4.1 and 1.4.2).

3.2.4.1 Circular Cross-Section

For a channel of circular cross-section and radius r_0, the reader can verify by substitution that the following is a solution of **Equation 3.3**:

$$u_x(r) = -\frac{\frac{dp}{dx}}{4\eta}\left(r_0^2 - r^2\right)$$

(3.5)

The flow profile is parabolic: u_x is maximum at the center of the channel (where $r = 0$) and gradually decreases toward the walls, where it is zero [$u_x(r = r_0) = 0$].

Integrating over the whole area A, one obtains the volumetric flow rate Q:

$$Q = \iint_A u_x(r)\,dA = -\frac{\frac{dp}{dx}}{4\eta}\int_0^{r_0}\left(r_0^2 - r^2\right)2\pi r\,dr = \frac{\pi r_0^4}{8\eta}\left(-\frac{dp}{dx}\right)$$

(3.6)

BLOOD: A NON-NEWTONIAN FLUID

BLOOD IS A VERY IMPORTANT FLUID in BioMEMS, one that eludes the simple mathematical treatment of these pages because it is non-Newtonian. Stated plainly, blood contains a large concentration of red blood cells which tend to elongate under high shear, so its viscosity is a function of the shear rate (unlike a Newtonian fluid, in which it is a constant). In other words, there is a critical shear rate value past which the viscosity decreases with increasing shear rate. This property of blood is critical to life, as it saves the heart enormous amounts of energy when it comes to pumping blood through small capillaries. If red blood cells did not elongate (a property called "shear thinning"), our hearts would have to spend much more energy to make blood reach all the corners of our body.

3.2.4.2 Rectangular Cross-Section

Here, the solutions are more complicated but the method is the same:

$$u_x(y,z) = \frac{16h^2}{\eta\pi^3}\left(-\frac{dp}{dx}\right)\sum_{n=1,3,5\dots}^{\infty}(-1)^{(n-1)/2}\left[1 - \frac{\cosh\left(\frac{n\pi z}{2h}\right)}{\cosh\left(\frac{n\pi w}{2h}\right)}\right]\frac{\cos\left(\frac{n\pi y}{2h}\right)}{n^3}$$

(3.7)

where $2w$ is the width and $2h$ is the height of the microchannel (the origin is (0,0), $-w \leq z \leq +w$, $-h \leq y \leq +h$). This function is plotted in **Figure 3.2a**. Interestingly, the flow profile is also approximately parabolic, but the profile stays constant as one moves away from the sidewalls and becomes smaller near the sidewalls; next to the sidewalls, the no-slip condition prevails because the fluid, because of its own viscosity, "feels" the friction exerted by the walls. The flow profile can be readily visualized by introducing a plug of fluorescent dye (**Figure 3.2b**). An immediate effect of the existence of a nonuniform velocity field across the channel's cross-section is that the dye—*even in the hypothetical absence of molecular diffusion*—will rapidly disperse over a

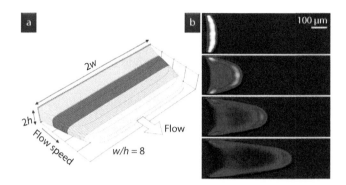

FIGURE 3.2 Flow speed profile for a rectangular channel. (a) Plot of **Equation 3.7** using *w/h* = 8. Figure contributed by Nirveek Bhattacharjee and Albert Folch. (b) Experimental visualization of Taylor dispersion with a fluorescent dye in a 250 μm × 70 μm microchannel under pressure-driven flow. (From Subhra Datta and Sandip Ghosal, "Characterizing dispersion in microfluidic channels," *Lab Chip* 9, 2537–2550, 2009. Reproduced with permission from The Royal Society of Chemistry.)

large area, an effect that has been termed **Taylor dispersion** (named after Sir Geoffrey I. Taylor's observations of flow in tubes in the 1950s). In the case of a non-Newtonian fluid, for the same average flow, we would observe that the fluid flows faster closer to the walls, advancing more like a plug (although there would still be no-slip at the walls), and the pressure needed to drive the flow would be smaller (because of the shear-thinning behavior).

The volumetric flow rate for a rectangular cross-section microchannel can be expressed as:

$$Q = \frac{4}{3\eta} wh^3 \left(-\frac{dp}{dx} \right) \left[1 - \frac{192}{\pi^5} \frac{h}{w} \sum_{n=1,3,5...}^{\infty} \frac{1}{n^5} \tanh \left(\frac{n\pi w}{2h} \right) \right] \tag{3.8}$$

As expected, the flow rate is a sensitive function of the height of the microchannel, its smallest dimension. Incidentally, **Equation 3.8** is also a fascinating function: it can be seen that if the values of *h* and *w* are swapped, the value of *Q*…is preserved! This is not trivial to prove, but it can be verified numerically. This invariance is what we expect from a function that describes the flow rate, a quantity that does not depend on which dimension is named "height" and which one is named "width"; in other words, the flow rate should not change if the channel, or the device, is rotated 90 degrees (the effect of gravity is negligible).

SIR GEOFFREY INGRAM TAYLOR

SIR G. I. TAYLOR (1886–1975) was born in London from a family of mathematicians (his grandfather was George Boole, the famous inventor of Boolean logic). He became a physicist, mathematician, and expert on fluid dynamics and wave theory. He worked on turbulence in the atmosphere which led to the publication of "Turbulent motion in fluids" while teaching at Trinity College. His observations as a meteorologist aboard the ship *Scotia* (International Ice Patrol) in 1913 formed the basis of his later work on a theoretical model of turbulent air mixing. When World War I started, he was sent to apply his knowledge to aircraft design.

After the war, Taylor returned to Trinity and studied turbulent flow applied to oceanography. He also studied a moving body as it passes through a rotating fluid. He did pioneering work on the theory of dislocations in crystals during his appointment as a Royal Society research professor. He also introduced a new approach to turbulent flow through a statistical study of velocity fluctuations.

Even after his official retirement, he made contributions such as developing a method for measuring the second coefficient of viscosity and creating an incompressible liquid with separated gas bubbles suspended in it. Shear viscosity of the liquid resulted in the dissipation of the gas in the liquid during expansion, thus allowing the bulk viscosity to be easily calculated. In 1947, after viewing declassified movies of the first atomic bomb test on 1945 in New Mexico, he was able to correctly estimate the energy released in the explosion using only dimensional analysis. Other late work included the longitudinal dispersion in flow in tubes (in 1953), movement through porous surfaces, and the dynamics of sheets of liquids. In 1969, he published his last research paper at 83 years of age, to describe jets of conducting liquid produced by electrical fields that, to this day, bear the name of **Taylor cone** in his honor. More than a dozen other phenomena, such as the **Taylor–Couette flow** (which occurs when a viscous fluid is confined between two rotating cylinders) and the **Taylor vortex** (an instability occurring in Taylor–Couette flow) are named after him; elsewhere in this book, we describe the **Taylor dispersion** that is ubiquitous in microchannels.

[*Excerpt adapted from Wikipedia. Photograph of G. I. Taylor obtained from* http://old.lms .ac.uk/newsletter/335/335_12.html].

3.2.4.3 Triangular (Isosceles) Cross-Section

The flow velocity profile $u_x(y,z)$ and total volumetric flow rate Q for a microchannel of isosceles triangle cross-section (with base a, height b, and angle subtended by the top vertex ϕ) are:

$$u_x(y,z)=\frac{1}{\eta}\left(-\frac{dp}{dx}\right)\frac{y^2-z^2\tan^2\phi}{1-\tan^2\phi}\left[\left(\frac{z}{2b}\right)^{B-2}-1\right] \tag{3.9}$$

$$Q=\frac{4ab^3}{3\eta}\left(-\frac{dp}{dx}\right)\frac{(B-2)\tan^2\phi}{(B+2)(1-\tan^2\phi)} \tag{3.10}$$

where $B\equiv\sqrt{4+\frac{5}{2}\left(\frac{1}{\tan^2\phi}-1\right)}$ to simplify the previous notations.

3.2.5 Microchannel Resistance

Because of the large surface-to-volume ratio of microchannels, the walls of a microchannel exert a lot of friction when compared with the force required to keep the fluid moving (the fluid's inertia). When the source of energy used to pump the fluid is removed, the fluid stops immediately. In considering pressure-driven flows, an important implication of **Equation 3.8** for the microfluidics designer is the calculation of **microchannel resistance**, which is defined (in analogy with Ohm's law of electricity, $R = V/I$) as the ratio between the applied pressure P (which plays the role of voltage V) and the volumetric flow rate Q (which plays the role of current I).

Note that in all the formulas for Q above, Q is proportional to the pressure gradient $(-dp/dx)$ and does not depend on x, so for any microchannel segment of length L and constant cross-section, we can write

$$R = \frac{\Delta P}{Q} = \frac{\left(-\dfrac{dp}{dx}\right)L}{Q} \tag{3.11}$$

For a rectangular microchannel, we substitute Equation 3.8 into Equation 3.11 to obtain:

$$R = \frac{3\eta L}{4wh^3}\left[1 - \frac{192}{\pi^5}\frac{h}{w}\sum_{n=1,3,5\dots}^{\infty}\frac{1}{n^5}\tanh\left(\frac{n\pi w}{2h}\right)\right]^{-1} \approx \frac{3\eta L}{4wh^3}\frac{1}{1-0.63h/w} \tag{3.12}$$

where the approximation simply states that that the second term of the sum in Equation 3.12 ($n = 3$) is approximately $3^5 = 243$ times smaller than the first term (for "standard" microchannels where $h < w$), so all the terms $n > 1$ can be neglected. If the microchannel has a high aspect ratio, that is, $h \ll w$, then the expression in brackets is approximately 1 and the resistance can be approximated as:

$$R \approx \frac{3\eta L}{4wh^3} \quad (h \ll w) \tag{3.13}$$

Strictly speaking, Equation 3.13 is valid both for $h \ll w$ and $w \ll h$ because the microchannel conserves the same resistance after a 90 degree rotation.

For a circular cross-section microchannel (i.e., a glass or a blood capillary), it is straightforward to see (by substitution of Equation 3.6 into Equation 3.11) that the resistance is:

$$R = \frac{8\eta L}{\pi r_0^4} \tag{3.14}$$

The expression for the pressure drop in a circular pipe (or microchannel) is known as the **Hagen–Poiseuille equation** (Poiseuille derived it experimentally in 1838):

$$\Delta P = R \times Q = \frac{8\eta L Q}{\pi r_0^4} \tag{3.15}$$

It is important to keep in mind that the Hagen–Poiseuille equation only applies to Newtonian fluids.

3.2.6 Shear Stress

Shear stress is mathematically more complex to define because it is a tensor, which means that it is the manifestation of a force that can act in many directions. Like all tensors, it can be manipulated as a matrix of numbers using matrix calculus. The advantage of using matrix calculus is that one can visualize how the x, y, and z components affect each other in a straightforward way. The shear stress tensor (a 3×3 matrix) is usually denoted without a subindex, τ, and each one of its nine components is expressed with a subindex: τ_{xx}, τ_{xy}, τ_{xz}, etc.:

$$\tau = \begin{pmatrix} \tau_{xx} & \tau_{xy} & \tau_{xz} \\ \tau_{yx} & \tau_{yy} & \tau_{yz} \\ \tau_{zx} & \tau_{zy} & \tau_{zz} \end{pmatrix} \tag{3.16}$$

For a Newtonian fluid, these coefficients can be calculated from first principles. Note that the stresses are, in effect, gradients of velocity (with the viscosity μ as a proportionality coefficient):

$$\tau = \begin{pmatrix} 2\mu\dfrac{\partial u_x}{\partial x} & \mu\left(\dfrac{\partial u_x}{\partial y}+\dfrac{\partial u_y}{\partial x}\right) & \mu\left(\dfrac{\partial u_x}{\partial z}+\dfrac{\partial u_z}{\partial x}\right) \\[3mm] \mu\left(\dfrac{\partial u_x}{\partial y}+\dfrac{\partial u_y}{\partial x}\right) & 2\mu\dfrac{\partial u_y}{\partial y} & \mu\left(\dfrac{\partial u_y}{\partial z}+\dfrac{\partial u_z}{\partial y}\right) \\[3mm] \mu\left(\dfrac{\partial u_x}{\partial z}+\dfrac{\partial u_z}{\partial x}\right) & \mu\left(\dfrac{\partial u_y}{\partial z}+\dfrac{\partial u_z}{\partial y}\right) & 2\mu\dfrac{\partial u_z}{\partial z} \end{pmatrix} \qquad (3.17)$$

Fortunately, for a rectilinear flow, $u_y = u_z = 0$ and u_x does not vary downstream (with y or z), so $\tau_{xx} = \tau_{yy} = \tau_{zz} = \tau_{yz} = \tau_{zy} = 0$ and the matrix is much simpler:

$$\tau = \begin{pmatrix} 0 & \mu\left(\dfrac{\partial u_x}{\partial y}\right) & \mu\left(\dfrac{\partial u_x}{\partial z}\right) \\[3mm] \mu\left(\dfrac{\partial u_x}{\partial y}\right) & 0 & 0 \\[3mm] \mu\left(\dfrac{\partial u_x}{\partial z}\right) & 0 & 0 \end{pmatrix} \qquad (3.18)$$

All the information is contained in this matrix, but we are not done yet because, in biology and bioengineering, we are most concerned about the forces applied by the flow onto the cell. The stresses are what causes the forces. To find the (vector) force applied onto cells that are cultured at the floor of the microchannel (here, the plane YZ), we need to multiply our matrix by the vector that represents the surface of our microchannel ("the normal," or Y direction). The result of this operation is a vector:

$$\vec{F} = \vec{n}\cdot\tau = (0\,1\,0)\begin{pmatrix} 0 & \mu\left(\dfrac{\partial u_x}{\partial y}\right) & \mu\left(\dfrac{\partial u_x}{\partial z}\right) \\[3mm] \mu\left(\dfrac{\partial u_x}{\partial y}\right) & 0 & 0 \\[3mm] \mu\left(\dfrac{\partial u_x}{\partial z}\right) & 0 & 0 \end{pmatrix} = \left(\mu\left(\dfrac{\partial u_x}{\partial y}\right),0,0\right) \qquad (3.19)$$

Matrix algebra shows clearly how, although the force is only in the direction of flow (x), it is caused by changes in flow velocity in the y direction. In the case of the rectangular channel (Equation 3.7), we have:

$$F_x = \mu\frac{\partial u_x}{\partial y} = \mu\frac{16h^2}{\eta\pi^3}\left(-\frac{dp}{dx}\right)\sum_{n=1,3,5...}^{\infty}(-1)^{\frac{n-1}{2}}\left(1-\frac{\cosh\left(\dfrac{n\pi z}{2h}\right)}{\cosh\left(\dfrac{n\pi w}{2h}\right)}\right)\frac{\partial}{\partial y}\left[\frac{\cos\left(\dfrac{n\pi y}{2h}\right)}{n^3}\right]_{y=-h} \qquad (3.20)$$

We now take the derivative (the cosine becomes a negative sine), which evaluated at $y = -h$ is ± 1, depending on the value of n, so we obtain:

$$F_x = \mu \frac{16h^2}{\eta\pi^3}\left(-\frac{dp}{dx}\right)\sum_{n=1,3,5\ldots}^{\infty}(-1)^{\frac{n-1}{2}}\left(1-\frac{\cosh\left(\dfrac{n\pi z}{2h}\right)}{\cosh\left(\dfrac{n\pi w}{2h}\right)}\right)\frac{\pi(-1)^{\frac{n-1}{2}}}{2hn^2} \tag{3.21}$$

Rearranging,

$$F_x = \mu \frac{8h}{\eta\pi^2}\left(-\frac{dp}{dx}\right)\sum_{n=1,3,5\ldots}^{\infty}\left(1-\frac{\cosh\left(\dfrac{n\pi z}{2h}\right)}{\cosh\left(\dfrac{n\pi w}{2h}\right)}\right)\frac{1}{n^2} \tag{3.22}$$

This formula cannot be further reduced, even for the special case of the high-aspect ratio microchannel ($h \gg w$ or $w \gg h$), because the hyperbolic cosine is a function that grows extremely fast and any approximation must be valid for all n (but n is inside the cosh). However, the large numbers and the $1/n^2$ term make the summation converge very fast, so a graph can be built with a spreadsheet in a few minutes. The force acting on the cells as a function of position across the channel is plotted in **Figure 3.3** for two different aspect ratios commonly encountered in BioMEMS, $h/w = 10$ and $h/w = 1$. To be fair, we need to specify in which conditions we are performing the comparison: it is not the same to compare a $1:1 = h/w$ microchannel with a $10:1 = h/w$ microchannel at equal flow rates than comparing them at equal (average) flow speeds. Indeed, note that the forces at equal average flow speeds are very similar in both cases (blue and red curves in **Figure 3.3**), although the force in the center of the channel is approximately 4.5% larger for the $h/w = 10$ case. On the other hand, close to the walls, at about a distance corresponding to 5% of the width of the microchannel, the forces are 12% *weaker* for the $h/w = 10$ case (at

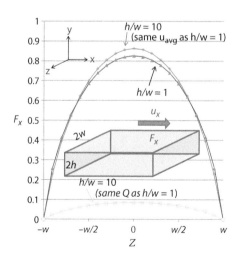

FIGURE 3.3 Force acting on the bottom surface of a rectangular microchannel. For computation purposes, the expression $\mu 8h(-dp/dx)/(\eta\pi^2)$ in **Equation 3.22** is forced to take the value of 1, and as a result F_x is in nonconventional units that depend on the values of h and of dp/dx. (Figure contributed by the author.)

equal average flow speeds) compared with the $h/w = 1$ case, and between 27% (for $h/w = 10$) and 31% (for $h/w = 1$) of the peak value at the center of the channel. This picture changes radically when the comparison is made at equal flow rates (experimentally, the value of dp/dx would need to be adjusted accordingly by almost a factor of 10), because the high aspect ratio microchannel has a much wider cross-sectional area, resulting in much lower flow speeds, and consequently, much lower forces (i.e., the green curve is exactly 10 times smaller than the blue curve in Figure 3.3). In all cases, the force is zero at the walls, as expected from the no-slip condition.

3.2.7 Capillary Flow

For a very small channel on the order of ~1 μm in its smallest dimension, the flow resistance is so large that pressure-driven flow can be impractical (resulting in bursting of the channel, for example). On the other hand, capillary filling scales well as the size of the channel is scaled down. The capillary pressure P_c of a liquid–air meniscus in a rectangular microchannel of width w and height h is:

$$P_c = -\gamma \left(\frac{\cos\theta_b + \cos\theta_t}{h} + \frac{\cos\theta_l + \cos\theta_r}{w} \right) \tag{3.23}$$

where γ is the surface tension of the fluid and θ_b, θ_t, θ_l, and θ_r are the contact angles of the liquid on the bottom, top, left, and right microchannel walls, respectively. This is the pressure that is pulling the flow.

Note that Equations 3.11, 3.12, and 3.13 are still valid here. However, ΔP is not a term produced by, say, a human operator but by the fluid itself. Note that for a capillary open at only one end, we have $\Delta P = P_c + \Delta P_{int}$, where ΔP_{int} is the internal pressure; for channels that are open at both ends, $\Delta P_{int} = 0$ and $\Delta P = P_c$. In this case, we can see that to increase flow speed (which is proportional to P_c), it is to our advantage to (a) decrease the size of the microchannel; and (b) decrease the contact angle (the maximum value of the numerator is 2), for example, by surface functionalization or by adding a surfactant.

3.2.8 Flow through Porous Media

A wealth of applications in microfluidics requires the control of flow through a layer of porous material, for example, a membrane containing nanopores, a slab of gel, or paper. The equation describing the flow rate Q for a fluid through a porous medium under a pressure differential, ΔP, now known as **Darcy's law**, was first established experimentally by Henry Darcy in 1856 and has since been derived from the Navier–Stokes equation:

$$Q = \frac{\kappa A}{\mu L} \Delta P \tag{3.24}$$

where:
    μ = dynamic viscosity (in Pa.s)
    L = length of the porous medium over which the pressure drop ΔP is taking place
    A = cross-sectional area of the flow
    κ = permeability of the medium (in units of area).

3.2.9 Diffusion

The behavior of fluids is not fully predicted by the Navier–Stokes equation, in that it does not consider diffusion. Diffusion is the macroscopic result of the thermally driven microscopic motion of particles (down to the size of atoms). It is inherently irreversible and random in nature, such that the motion of a single particle is not predictable. Fortunately, the motion of *ensembles* of particles is quite predictable, as we experience it in our everyday life when we mix fluids with fluids and fluids with solutes. Diffusion is quantitatively embodied in Fick's three laws, or equations, of which we will touch only on the first two here.

$$J_i = -D \frac{\partial C_i}{\partial x} \quad \text{Fick's first law of diffusion} \tag{3.25}$$

In plain words, the first law states that diffusing species flow down their concentration gradients at a rate that is proportional to the gradient present. The quantity J_i is the flow or flux of the solute i (in units of mass or moles) that crosses 1 cm² of surface in 1 second. C_i is the concentration of solute i in mass units per unit volume. The quantity D, which is one of the most important in microfluidics, is called the **diffusion coefficient**, as defined in the Stokes–Einstein relationship below:

$$D = \frac{kT}{6\pi\eta R_H} \quad \text{Stokes–Einstein relation} \tag{3.26}$$

where:
  k = Boltzmann constant
  T = temperature
  η = viscosity of the solution
  R_H = hydrodynamic radius of the particle.

Note that because the viscosity of a solvent is almost always dependent on temperature, this expression depends doubly on temperature. The hydrodynamic radius is roughly related to the geometric radius of the particle, but depends also on how much solvent moves with the particle, and is difficult to calculate for a nonspherical particle. Overall, the value of D is characteristic of a given particle in a given solvent at a given temperature. Some examples of diffusion coefficients are given in Table 3.1. Note that the diffusion of hydronium ion is anomalously fast, in that the proton can hop from water molecule to water molecule.

Fick's second law of diffusion predicts how diffusion causes the concentration field to change with time:

$$\frac{\partial C_i}{\partial t} = D \frac{\partial^2 C_i}{\partial x^2} \quad \text{Fick's second law of diffusion} \tag{3.27}$$

A simple case of diffusion in one dimension (taken as the x axis) is the introduction of a step concentration $C(0)$ (constant from $-\infty$ to 0) at position $x = 0$ at time 0, which acts as the bound-

Table 3.1 Important Diffusion Coefficients		
Molecule	Molecular Weight	Diffusion Coefficient, D, in Water (μm²/s)
H_3O^+	19	9000
Na^+	23	2000
O_2	32	1000
Glycine	75	1000
Hemoglobin	60,000	70
Tobacco mosaic virus	40,000,000	5

ary condition that allows us to solve Fick's second law. In the case of an input step function, the solution to Equation 3.27 is:

$$C(x,t) = C(0)\frac{2}{\sqrt{\pi}} \int_x^\infty e^{-t^2}\, dt \qquad (3.28)$$

(Other boundary conditions result in different solutions, of course, usually more complicated to derive.) This integral is also called the *complementary error function* ("erfc") and Equation 3.28 is often expressed as:

$$C(x,t) = C(0)\mathrm{erfc}\left(\frac{x}{2\sqrt{Dt}}\right) \qquad (3.29)$$

The length $2\sqrt{Dt}$, which appears in many solutions to the diffusion equation, is called the **diffusion length** and provides a measure of how far the initial concentration has diffused in the x direction in the amount of time t.

A plot of the complementary error function (for which it is assumed that the diffusion length is 1) is shown in Figure 3.4. You can also view this curve as a snapshot of the concentration profile for time $t = 1/(2D)$. An interesting property of this curve is that it keeps its symmetrical sigmoidal shape while it flattens as time goes by, featuring two asymptotes, $\mathrm{erfc}(x) \to 2\ (x \to -\infty)$ and $\mathrm{erfc}(x) \to 0\ (x \to +\infty)$, and the crossover through $\mathrm{erfc}(x) = 1$ at $x = 0$. The physical interpretation of this symmetry is that all the molecules that diffused from the left partition (negative x) are now on the right partition (positive x).

For many purposes, the most important relationship is the following:

$$\bar{x}^2 = 2Dt \qquad (3.30)$$

which relates the mean squared displacement of a molecular species in time by diffusion. It shows that diffusion can move particles short distances very effectively, but over long distances, is less and less a factor in transport. Hence, in microfluidic devices, small molecules can travel short distances very fast, whereas as the distances increase and the diffusion coefficients drop (for larger particles), diffusion becomes a barrier to transport.

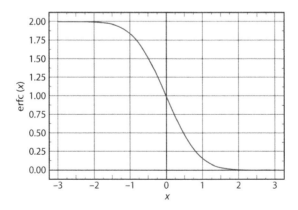

FIGURE 3.4 The complementary error function. This function describes how far a concentration of solute (initially filling from $x = -\infty$ to $x = 0$) has propagated by diffusion in the x direction. The diffusion length $2(Dt)^{1/2}$ is assumed to be 1 for this plot. (Graph adapted from *Wikipedia*.)

3.2.10 Surface Tension, Contact Angles, and Wetting

Surface tension is the property of the surface of a fluid that causes its surface to be attracted to another surface (which can be solid, fluid, or gas)—thus, it is not the property of the fluid alone but a property of its interface with another medium. It has the dimension of force per unit length (e.g., N/m, dyne/cm) or, equivalently, energy per unit area (i.e., J/m^2). When applied to liquids, the concept of **surface energy** (which can be applied to both solids and liquids) is equivalent to surface tension.

Let's consider the case of a liquid wetting the walls of a container (**Figure 3.5**), so there is a liquid–solid interface as well as a liquid–air interface. At the point in which the two interfaces meet, where they form a so-called "**contact angle**" (θ in **Figure 3.5**), all forces must balance. In the horizontal direction, the attractive force f_A cancels the horizontal component of the tension force for the liquid–air interface f_{la}. In the vertical direction, the balance of forces reveals that

$$f_{ls} - f_{sa} = -f_{la} \cos \theta \tag{3.31}$$

The forces are directly proportional to their respective surface tensions, thus:

$$\gamma_{ls} - \gamma_{sa} = -\gamma_{la} \cos \theta \tag{3.32}$$

where:

γ_{ls} = liquid–solid surface tension
γ_{la} = liquid–air surface tension
γ_{sa} = solid–air surface tension

The contact angle is very sensitive to the chemical composition of the surface and of the fluid and is thus a valuable parameter to characterize devices in BioMEMS. Even submonolayer coverages of protein on a surface can dramatically change the character of the surface from hydrophobic to hydrophilic. Although surface energies are difficult to measure independently, contact angles are readily measured with an instrument called a goniometer. **Table 3.2** shows the surface tension of various liquids in contact with air. Note how the surface tension decreases with temperature (i.e., hot fluids tend to wet better than cold fluids). The low surface tension of ethanol explains why it has a contact angle of 0 degrees when it wets glass. Mercury, with the highest surface tension, has a notorious propensity to form droplets (its liquid–solid contact angle on glass is 140 degrees). Biofluids (saliva, blood, urine, etc.), which are very rich in protein and thus have low surface tension (their contact angles on glass can approach 0 degrees, i.e., perfect wettability) are not listed because of their variable compositions.

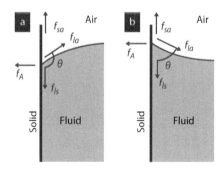

FIGURE 3.5 Forces at contact point. The forces are shown for (a) contact angle θ > 90 degrees (typical of a hydrophobic surface if the fluid is water) and (b) θ < 90 degrees (typical of a hydrophilic surface if the fluid is water). Tension forces are shown for the liquid–air interface (f_{la}), the liquid–solid interface (f_{ls}), and the solid–air interface (f_{sa}). The attractive force is denoted by f_A. (Figure adapted from *Wikipedia*.)

Table 3.2 Surface Tension of Various Liquids in Contact with Air (γ_{la})		
Liquid	Temperature (°C)	Surface Tension, γ_{la} (dyne/cm) (mN/m)
Water	0	75.64
Water	25	71.97
Water	50	67.91
Water	100	58.85
NaCl 6.0 M aqueous solution	20	82.55
Sucrose (55% wt. in water)	20	76.45
Ethanol	20	22.27
Ethanol (11.1% wt. in water)	25	46.03
Ethanol (40% wt. in water)	25	29.63
Isopropanol	20	21.7
Methanol	20	22.6
Glycerol	20	63
Acetone	20	23.7
Acetic acid	20	27.6
HCl 17.7 M aqueous solution	20	65.95
n-Hexane	20	18.4
Mercury	15	487

Source: *Lange's Handbook of Chemistry*, 11th ed., 1973, John A. Jean (Ed.), McGraw-Hill.

Wetting is not a new phenomenon, yet its applications to BioMEMS seem endless. Scientists have observed for centuries how fluids will spontaneously enter thin tubes (a phenomenon termed "**capillary action**") and wick porous media such as paper and sand—entire microfluidic devices are now made integrally in paper (see Section 3.5.1.6). When a drop of water is placed at the inlet of a microchannel (provided its walls are hydrophilic), the water will spontaneously enter the channel, but at a speed that gradually slows down. Emmanuel Delamarche's group at IBM Zurich, in Switzerland, alongside many others, have developed pumps and valves based on capillary action (see Section 3.8.4.1). In the last decade, there have been many demonstrations of devices that exploit selective wetting as a strategy to move fluids or particles and to conduct assays on a microscale. George Whitesides and colleagues at Harvard first used self-assembled monolayers (SAMs) of alkanethiolates (of various terminal groups, with different hydrophilicity) to produce "liquid patterns" on surfaces. A collaborative effort between David Beebe's group and Jeffrey Moore's group at the University of Illinois at Urbana-Champaign produced microfluidic devices in which the fluids were prevented from contacting the sidewalls by stripes of hydrophobic SAM and were guided to stay in the middle of the channel by a stripe of hydrophilic SAM—so the fluid would "magically" flow between the floor and the ceiling of the channel without any walls (the air–fluid interfaces inside the channel would act as "virtual walls"). Similarly, it is possible to use electrical fields to modulate wetting ("**electrowetting**," see Section 3.3.4) for shuttling individual droplets across a flat substrate or in a chamber. The newly demonstrated ability of microfluidic devices to produce a stream of droplets of water in a carrier solution of oil has revolutionized many assays ("**droplet microfluidics**," see Section 3.7).

3.2.11 The Surface-to-Volume Problem

The most serious problem with microfluidic systems is just that they are so small. In shrinking a conventional chemical measurement system, the surface-to-volume ratio increases. This ratio is a serious problem in chemical detection if there is any tendency for an important species to adsorb to the channel walls. In such a circumstance, all pipes between points in the microfluidic device become chromatographic columns; in the worst of cases, the analyte of interest may never reach the point of detection if it adsorbs to the walls and desorbs slowly (or not at all). The smaller the channel, the worse the problem.

The science of this problem is both fascinating and frustrating. There is not just one event that causes loss of analytes from solution, but several. First, there is a close approach by the analyte to the surface by diffusion. If the electrostatics are favorable, the analyte near the surface may become adsorbed loosely to the surface. If it stays in that state long enough, large analytes may make a transition to a different (denatured) configuration that sticks very well to the surface. Each process is vulnerable to being blocked.

Much effort has gone into the creation of nonfouling surfaces for such systems but, regrettably, no perfect material has been found. Some surfaces work very well to prevent adsorption of some compounds. The best of the nonfouling coatings has been polyethylene glycol (PEG or PEO), which has been applied in many different forms. PEG coatings drastically reduce adsorption by most proteins, but they are not particularly stable in the presence of biological samples, and have a few annoying traits, such as triggering platelet activation. The next-best solution has been used for half a century or more, which is adsorbing an "inert" protein to the surface before use by the interesting samples. The most commonly used proteins are albumin (from serum) and casein (from milk); they are both inexpensive and have been shown to stick well to both hydrophobic and some hydrophilic surfaces, and to greatly reduce further adsorption by other proteins. The most common method of using them is to flood the device with a concentrated solution of the protein and to keep the channels filled long enough (minutes to hours, unfortunately), for the protein to adsorb and then partially denature on the surface to make the coating stick better. It is even also sometimes necessary to "bake" on the coatings by subsequent drying at elevated temperatures before use.

An alternate way to prevent adsorption by analytes of interest is to mix your sample with an antifouling compound that does not stick permanently, but simply outcompetes the analytes for the surface sites. Such a "dynamic coating method" can be used with proteins, synthetic antifouling polymers like Pluronic (see Section 2.6.1.4) and even conventional surfactants. Simply adding a high concentration of albumin to a sample can often solve the adsorption problem, but only if adding albumin to the sample does not interfere with the operation of the microfluidic device (e.g., albumin could have a confounding/deleterious effect on the analytical measurement of interest or on the cells cultured in the device, if any).

3.3 Fluids in Electrical Fields

Here, we shall see what happens when an electric field is introduced in a microchannel, or more generally, in a fluid. As it turns out, electric fields can be used to move fluids around (i.e., used as pumps or valves) and to selectively separate certain chemical species from others—not without difficulty. In this section, we review briefly the three most important phenomena that arise when fluids are subjected to electric fields: **electrophoresis**, **electro-osmosis**, and **dielectrophoresis**. Examples of devices that exploit these phenomena will be reviewed in Chapter 4.

3.3.1 Electrophoresis

Electrophoresis—the movement of charged species in a liquid under the influence of an applied electric field—is a well-known phenomenon at the macro-scale. It is used to great advantage in research and industry to fractionate small ions, charged organic molecules, and even macromolecules like proteins and DNA. The field is applied by placing two electrodes at either end of a channel within the fluid; some current flow through the fluid is inevitable. Charged species will move in response to the field. If the medium were a vacuum, they would accelerate continuously until they crashed into the electrode with the opposite charge. In a solution, the charged species (small molecules or microparticles) rapidly reach their terminal velocity because of the frictional forces imposed by the viscous solution. That velocity can be predicted by the following formula:

$$v_i = \frac{z_i e}{6 \pi r_i \mu} E \tag{3.33}$$

where:
   z_i = charge of the ion
   e = charge of the electron
   E = electrical field
   r_i = radius of the ion

The value of μ, the electrophoretic mobility, is characteristic for a molecular species under particular solvent conditions. For small ions, the electrophoretic mobility is generally measured on the basis of conductance of solutions containing the ions. For small monovalent cations, values of μ are on the order of 5 (μm/s)/(V/cm).

For particles much larger than proteins, the electrophoretic mobility (μ) is, according to Smoluchowski, linearly related to its **zeta potential** (ζ), which is, in turn, inversely proportional to the ionic strength of the surrounding environment. The mobility is also dependent on the diffusion coefficient and inversely dependent on the viscosity of the solvent:

$$\mu = \frac{v}{E} = \frac{\zeta D}{4 \pi \eta} \tag{3.34}$$

For proteins, shape matters, as does the specific location of charges and their interaction with the counterflowing counterions in the solution around the proteins. Ironically, for DNA, which exists as a double helix with a constant value of charge per unit length, the electrophoretic mobility is not a function of molecular weight because the charge and frictional forces increase in parallel. As a consequence, it is not possible to separate DNA by "free-flow electrophoresis" (see Section 4.4.2) at all! It is necessary to introduce something into the solution that differentially interacts with different lengths of DNA; hydrogels (and even nongelled solutions) of high molecular weight polymers (such as polyacrylamide) become entangled with the nucleic acid molecules. Because the longer nucleic acid molecules have a greater chance of being caught on the polymer chains, the polymer matrix slows down long DNA molecules more than short ones, providing a very efficient means of fractionation of DNA by chain length.

Electrophoresis has been used to great advantage in microchannels. In fact, electrophoresis was the motivation for producing the first microfluidic devices in history, back in the 1940s and 1950s (see Section 4.4.2 on **continuous-flow electrophoresis**). It turns out that electrophoresis should go faster (and be more efficient) if the accelerating voltage could be increased indefinitely. However, the obligatory current flow in the channel that increases with the voltage increases the solution temperature as well. As temperature rises, the properties of the solution change, the molecules change shape, and, ultimately, the solution can even boil. This imposes practical limits on voltage in real devices. In microchannels, it is possible to use higher voltages because the removal of the heat produced by current flow is faster in narrower channels. The limiting material problem is that if the channel is made of a conductor, there must be a very good insulator on the channel surface to keep the current confined to the fluid channel itself. Silicon is rarely used, but glass and plastics are excellent materials.

Because the velocity of the species in a given electric field depends strongly on the charge of the particle, it stands to reason that any change in the charge of the particle can either slow down or speed up the particle. The strongest influence on the charge of biomolecules (which contain multiple copies of carboxyl, amine, and other titratable groups as their charged species) is the pH. There is always a pH for such macromolecules in which the *net* charge on the molecule as a whole is zero—the isoelectric point. As a consequence, if one can establish a pH gradient, biomolecules exposed to an electrostatic field along the same direction as that gradient can be made to migrate to their isoelectric points, where they will no longer be affected by the field. Molecules can thereby be concentrated and sorted by their isoelectric points in a technique known as **isoelectric focusing** (IEF, see Section 4.4.3). (Note that the positive electrode must be at the low-pH side of the gradient to prevent all the molecules from simply flying off the gradient in one direction.) The IEF technique has gained wide acceptance in the macro-world as one of the two dimensions in so-called two-dimensional (2-D) gel electrophoresis, which allows the identification of thousands of different proteins from one sample.

It is also possible to use IEF in microdevices, with some trade-offs. The unique advantage of microdevices is that one can create the pH gradient using the microdevice itself. Because at sufficiently high voltages to perform electrophoresis, one inevitably produces acid at the anode and base at the cathode by electrolysis of water, the two electrodes always create their own pH gradient. This gradient reaches a steady state in a short time by interdiffusion of the two sets of extreme pH regions, which allows IEF to occur. If the gap between the accelerating voltages is on the order of a few millimeters or less, and the solution is not heavily buffered, an IEF system is automatically created in the microdevice. The downside is that very high accelerating voltages are needed to allow significant separations in small distances, which are hard to establish in a microchannel without the generation of bubbles, which disrupt the orderly alignment of the separated proteins.

3.3.2 Electro-Osmosis

Electrophoresis has a lesser-known cousin—**electro-osmosis**—that is very important in microfluidics. It occurs because of electrophoresis of the charges that reside near channel surfaces. The chemistry of the walls of microchannels is usually such that the wetted channel is charged. For example, the surface of silica contains silanol ($-Si-OH$) groups, whose hydroxyl group will, at neutral pH, lose a proton to become $-Si-O^-$. As the pH is lowered, the walls of a silica channel can be neutral or even positively charged but, for the moment, we can consider that at reasonable pH values in aqueous solvents, the surfaces of silica and most partially oxidized polymer surfaces are negatively charged in water.

This observation becomes interesting because of what a "charged surface" really means in water. The nature of water is such that there is no total charge imbalance within the channel–solution system. However, the *distribution* of charges can be very interesting (see **Figure 3.6**). There are immobilized charges covalently attached to the surface, and a balanced set of other co- and counterions that are distributed near the surface at positions determined by the surface charge, the dielectric constant of the medium, and the temperature. The temperature is crucial because at zero temperature, all the counterions would collapse onto the charged surface and the surface charge would be neutral. Because there is energy available (according to the **Boltzmann distribution**), those ions not covalently attached try to escape to infinity, only to be pulled back by the electrostatic attraction of the surface. This charged sheath of counterions, which resides within a few nanometers or tens of nanometers from the fixed charged surface of the channel (the so-called **Debye layer**, or **electrical double layer**, whose characteristic thickness is the

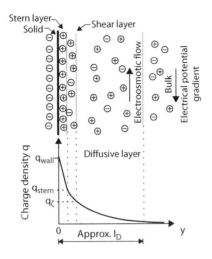

FIGURE 3.6 The role of surface charges in electro-osmosis. (From Dominik P. J. Barz and Peter Ehrhard, "Model and verification of electrokinetic flow and transport in a micro-electrophoresis device," *Lab Chip* 5, 949–958, 2005. Reproduced with permission from The Royal Society of Chemistry.)

Debye length) is free to move by electrophoresis parallel to the channel surface. If a pair of electrodes imposes an external field on the channel, the counterion layer that sheaths the core of the fluid moves together. Because the interaction of the ions with their immediate surrounding solvent molecules is strong, the ions drag the solution with them, so that the electrophoresis pulls the entire column of water with it. The core of the channel is viscously entrained by the moving solvent at the channel walls, so the fluid moves along at the same speed except very close to the walls where the speed is zero (a "**plug flow profile**"), as opposed to a pressure-driven parabolic flow profile. This principle—**electro-osmotic flow (EOF)**–based pumping—is ubiquitously used in capillary electrophoresis systems (see Section 4.4.1).

EOF is found wherever electrophoresis is underway, and can be a significant interfering effect in an electrophoretic separation. You can imagine the problem if you are trying to separate a charged analyte and detect that band by running it past a detector, only to find that the solution in which the analyte is running is pumping itself rapidly in the opposite direction, in which case your analyte band can be made to go in the opposite direction than that intended. EOF can be suppressed either by (a) working with a channel material that has a very low surface charge at the operating buffer temperature; (b) working at very high salt concentrations, which reduces the surface ζ potential; or (c) adding a coating to the surface that suppresses the movement of the first few tens of nanometers of solution relative to the channel walls, like a covalently bound polymer layer.

EOF-based pumping in channels that are much wider than the electric double layer produces good pumping rates, but little ability to pump against backpressure because flow can return along the midline of the channel in which the electric double layer does not penetrate. As the channels become narrower, and enter the regime of nanofluidics, the effective pumping pressures can increase very rapidly, at the expense of lower pumped volumes. In general, scaling down the dimensions of a microchannel favors electrokinetic phenomena because, in narrower channels, the Debye layer occupies a larger proportion of the channel and thus a larger proportion of its volume participates in transport. Conversely, electrokinetic phenomena do not work well on wide channels (i.e., scaling *up* does not work well).

EOF pumping (intentional or otherwise) is only as reliable as the surface potential. Anything that changes the surface charge of the channel wall alters the local pumping driving force, so the overall pump speed can change radically as something adsorbs to or desorbs from the channel walls. This is a particular problem with proteins and "raw" biological samples that contain unknown proteins and other polymers. As a consequence, electrophoresis itself is not a particularly reliable method when using samples that have not been heavily preconditioned. On the other hand, EOF pumping—having a nearly flat velocity profile (except just at the walls, where the no-slip condition still applies)—is not as sensitive to Taylor dispersion as pressure-driven flow, so it is a preferred method for separation techniques in which dispersion adds noise.

3.3.3 Dielectrophoresis

There is an entirely different electrokinetic effect that can be used to great advantage in microfluidics. It is called **dielectrophoresis**, a phenomenon discovered by Herbert Pohl in 1951. As its name suggests, it relies on the difference between the dielectric constant of a particle and the fluid in which it is surrounded for its motive force.

Dielectrophoresis is the motion of uncharged particles as a result of polarization induced by nonuniform electric fields. At a microscopic level, it is caused by induced dipole moments at the interface between the particle and its medium. A particle that is more polarizable than the surrounding medium is attracted toward a region of increasing field strength, a scenario known as **positive dielectrophoresis** (**Figure 3.7a**). Conversely, when the solvent is of higher polarizability than the particle, then the particle is repelled from the high-field regions because the dipoles of the fluid are the ones that move the particle toward the electrical field minima—this situation is known as **negative dielectrophoresis** (**Figure 3.7b**). The dielectrophoretic (DEP) force depends on the particle size, shape and internal structure, and on the magnitude and degree of nonuniformity of the applied electric field. If the lossy dielectric particle is much smaller than the field nonuniformities, the dipole approximation to the DEP force for a spherical particle is as shown in **Equation 3.35**:

$$\mathbf{F} = 2\pi\varepsilon_m R^3 \, \mathrm{Re}\left[\frac{\underline{\varepsilon}_p - \underline{\varepsilon}_m}{\underline{\varepsilon}_p + 2\underline{\varepsilon}_m}(\omega) \times \nabla \underline{\mathbf{E}}^2(\mathbf{r}) \right] \tag{3.35}$$

(from J. Voldman et al., "Holding forces for single-particle dielectrophoretic traps," *Biophys. J.* **80**, 531–541, (2001).)

where:

ε_m = permittivity of the medium surrounding the sphere
R = radius of the particle
ω = radian frequency of the applied field
$\mathbf{r}$ = radial vector describing the spatial coordinates
$\mathbf{E}$ = complex applied electrical field
$\underline{\varepsilon}_m$ and $\underline{\varepsilon}_p$ = complex permittivities of the medium and the particle, respectively.

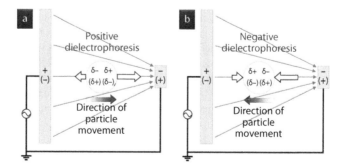

FIGURE 3.7 Positive and negative dielectrophoresis. (From Nicole Pamme, "Continuous flow separations in microfluidic devices," *Lab Chip* 7, 1644–1659, 2007. Reproduced with permission from The Royal Society of Chemistry.)

A positive DEP force propels particles to the field maxima; a negative DEP force propels particles to the field minima. The fields used can be either AC or DC, but in general, AC fields are used because insulated electrodes can be used, allowing them to be used indefinitely in the aqueous medium. At high frequencies on the >1 MHz range, electrolysis-generated H^+ or OH^- ions recombine before diffusing away, which prevents cell-adverse changes in the local pH. Also, AC fields induce reduced transmembrane voltages compared with DC fields: at 20 MHz, a 3 V AC field will only induce a 12 mV transmembrane voltage on HL-60 cells, approximately 20 times smaller than it would at DC.

This force can be used to differentially levitate particles including cells (see Section 5.3.4 on **dielectrophoretic traps**), and to do so in such a way that particles are sorted in time as they progress down a channel—those that float higher than others move at a higher velocity down the channel by residing in a faster-moving fluid stream. It can also be used to concentrate specific particles at specific points on surfaces, and to divert particular particles from one flow stream to another, to cause dynamic concentration.

WHEN SMALLER IS (MUCH) BETTER

Much of the initial microfluidics research impulse was about making things smaller for its own sake, or "just" to save reagents, but overall it was thought that smaller devices would operate in the same way as their macro-counterparts. As it turns out, microfabricating analytical systems is not simply about following a Moore's type law down to a smaller and denser "fluidic microprocessor" to save costs and improve performance. The physics of fluid behavior makes interesting transitions as things reach the microscale, particularly with respect to the aforementioned Reynolds number, and the importance of diffusion for mass transport. Not all things get worse when things get smaller, as a consequence; there are entirely new and extremely useful devices possible at the microscale that have no counterpart at the macroscale. Discovering these new operating principles has been one of the most exciting features of research in this area.

3.3.4 Electrowetting

Electrowetting is the phenomenon whereby an electric field influences the wettability of an electrolyte droplet on a surface (which is typically micropatterned with electrodes). The first observations were made with mercury (a metal fluid) in the 19th century but it has found most applications in the manipulation of small water droplets in the last 25 years. The electrowetting effect can be defined as *the change in solid electrolyte contact angle due to an applied potential difference between the solid and the electrolyte*. The fringing field at the corners of the electrolyte droplet tends to pull the droplet down onto the electrode, lowering the macroscopic contact angle and increasing the droplet contact area. Looking at electrowetting from a thermodynamics perspective, the surface tension of an interface (between the liquid electrolyte and the solid conductor) γ_{SL} is equal to the Gibbs free energy required to create a certain area of the droplet's surface, which contains a chemical and an electrical (charge) term. The electrical component of the Gibbs free energy is the energy stored in the capacitor formed between the conductor and the electrolyte and can thus be modulated by changing the voltage of the electrodes. The chemical component is simply the surface tension of the solid–electrolyte interface in the absence of an electric field (γ_{SL}^0).

Defining C, as the capacitance of the interface; $\varepsilon_r \varepsilon_0 / t$, for a uniform dielectric of thickness t and permittivity ε_r; V, the effective applied voltage, integral of the electric field from the electrolyte to the conductor, then we have the total surface tension between the conductor and the electrolyte as:

$$\gamma_L = \gamma_{SL}^0 - \frac{CV^2}{2} \tag{3.36}$$

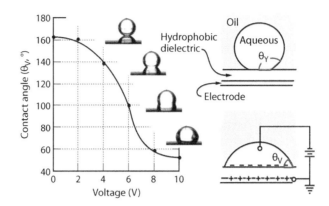

FIGURE 3.8 Electrowetting. (From J. Heikenfeld and M. Dhindsa, "Electrowetting on super-hydrophobic surfaces: Present status and prospects," *J. Adhes. Sci. Technol.* 22, 319–334, 2008. Figure contributed by Jason Heikenfeld.)

The surface tension can be related to more measurable quantities such as the contact angle θ using the **Young equation**:

$$\gamma_{SL} = \gamma_S + \gamma_L \cos \theta \tag{3.37}$$

where:

γ_L = surface tension between the liquid electrolyte and the ambient
γ_S = surface tension between the solid conductor and the ambient

Thus, we can obtain an expression for the contact angle as a function of applied potential, *V*:

$$\theta = \cos^{-1}\left(\frac{\gamma_{SL}^0 - \gamma_L - \dfrac{CV^2}{2}}{\gamma_L} \right) \tag{3.38}$$

An experimental demonstration is shown in **Figure 3.8**. The applied voltages are typically on the range of tens of volts or less.

3.4 Fluids in Acoustic Fields

The manipulation of cells and particles using sound waves has a long history. German physicists August Kundt and Otto Lehmann first reported in 1874 the effect of acoustic forces on dust particles levitating in organ pipes. However, it was not until 1971 that Pond, Woodward, and Dyson discovered that red blood cells could be collected in bands produced by standing acoustic waves in blood vessels in vivo. It has not been realized until fairly recently that the pressure waves induced by (ultra)sounds can be used in microfluidics for a variety of very useful applications. Its main attractive aspect is, undoubtedly, its cost: ultrasound transducers can be obtained for only a few dollars in hardware stores. The two main phenomena are **acoustophoresis** (the separation of particles by acoustic fields) and **acoustic streaming** (used in micromixers).

3.4.1 Acoustophoresis

In 2004, Thomas Laurell from Lund University in Sweden invented a clever microfluidic trick which works very similar to how leaves in a pond form patterns along the troughs of standing waves on the surface of the water: as a standing wave (in the low MHz range) is induced in a microchannel using a piezo transducer, suspended particles entering the channel are positioned by acoustic (pressure wave) forces in the pressure nodal planes (or the antinodal planes, depending on the densities and compressibilities of the particles and the medium, respectively) of the standing wave (see cross-section in **Figure 3.9a** and top view in **Figure 3.9b**). The

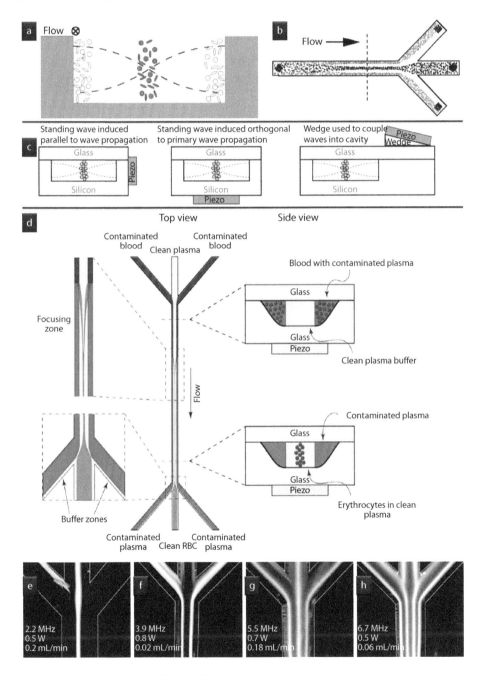

FIGURE 3.9 Acoustophoresis. (From Filip Petersson, Andreas Nilsson, Cecilia Holm, Henrik Jönsson, and Thomas Laurell, "Separation of lipids from blood utilizing ultrasonic standing waves in microfluidic channels," *Analyst* 129, 938–943, 2004. Figure contributed by Thomas Laurell.)

piezotransducer can be positioned perpendicular, parallel, or at an angle to the standing wave (**Figure 3.9c**). A schematic of a separation is shown in **Figure 3.9d**. Four examples of separations for different parameters (forming nodes and antinodes) are shown in **Figure 3.9e** through **h**. Similarly, if a piezotransducer is applied to the external wall of a glass capillary, the cells are "magically" focused to the center of the capillary—which is the basis for Invitrogen's Attune Acoustic Focusing Cytometer.

3.4.2 Acoustic Streaming

A remarkable phenomenon occurs when a vibrating (sound) transducer is placed in a fluid medium: despite the fact that the membrane of the transducer is vibrating *back and forth* (recall that in the classical description of the sound pressure wave, the air molecules transmit all their momentum to their neighbors without net displacement and all the sound's energy is carried by this harmonic oscillator), a net stream of fluid particles does occur in reality—it's just nonaudible! (We can see it in our shirts if we stand very close to a high-power amplifier, however, and you can now buy "loudspeaker guns" in toy stores that shoot harmless air-puffs.) This phenomenon, termed **acoustic streaming**, is a nonlinear effect that is well understood from the Navier–Stokes equation. It is readily visualizable by immersing a piezoelectric transducer in a fluid bath (it works best with glycerol, as depicted in **Figure 3.10a**, because the stream develops more slowly and we have more time to take pictures).

Acoustic streaming can also occur near a boundary or when a boundary is vibrating in a still medium. In 2003, Sascha Hilgenfeldt of the University of Twente in the Netherlands placed an air bubble at the bottom of a cuvette filled with liquid and set it to vibrate at ultrasound frequencies (at these frequencies, the bubble vibrated at resonance, so it vibrated with larger amplitudes) to demonstrate that the fluid around the bubble can vibrate with it at length scales that far exceed the dimensions of the bubble. In other words, the vibrating bubble (in conjunction with other microstructures such as posts and dimples) generates **microstreaming** (**Figure 3.10b**), a principle that is being exploited in microfluidics to create micropumps and micromixers (that can be turned on/off simply by turning on the inexpensive nearby transducer).

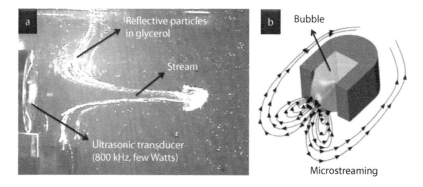

FIGURE 3.10 The phenomenon of acoustic streaming. (a) Acoustic streaming visualized in glycerol. The image was taken a few seconds after the transducer was turned on (the stream is slowly accelerating; the full video can be viewed at http://www.lmfa.ec-lyon.fr/perso/Valery.Botton/acoustic_streaming_bis.html). Figure contributed by Valéry Botton and Jean-Michel Lenoir from LMFA, CNRS, University of Lyon/ECL/UCBL/INSA de Lyon, France. (b) Acoustic streaming near a boundary (e.g., a trapped bubble). (From Daniel Ahmed, Xiaole Mao, Jinjie Shi, Bala Krishna Juluri, and Tony Jun Huang, "A millisecond micromixer via single-bubble-based acoustic streaming," *Lab Chip* 9, 2738–2741, 2009. Reproduced with permission from The Royal Society of Chemistry.)

3.5 Fabrication of Microfluidic Channels

Although nature may have had billions of years to optimize its microfluidic systems, engineers have had only a few decades. Lithographic methods were first used for microfluidics in the 1980s, but have been largely supplanted by methods that make less expensive final devices. Although some microfluidic devices can be very simple, there has been increasing emphasis on creating fluidic systems that allow multistep processes or highly parallel processing of multiple samples. As a consequence, there is a premium on fabrication methods that allow complexity of function through multiplexing, crossing of fluid paths, valving, and other sophisticated architectures typically found in macroscopic fluidic systems like petroleum refining plants.

3.5.1 The Building Materials

Dozens of different materials have been used for microfluidics. Here, we will just mention a few of the most common ones along with their advantages and disadvantages.

3.5.1.1 The "Historical" Materials: Silicon and Glass

Most of the initial microfluidic devices were made using silicon because the original microfluidics researchers were refugees from microelectronics research. Silicon channels have generally been made using conventional photolithography, so pits and channels can be formed in either one or both sides of a Si wafer. The process usually proceeds through the design of one or more photomasks, and photoresist spinning, patterning, and chemical etching (either wet or dry). To make a complete fluidic system, these etched parts must be at least partially sealed. Although gluing is possible with very large crude devices, the most successful devices have used either Si–Si bonding or anodic bonding to a glass with a very similar coefficient of thermal expansion such as Pyrex (because the bonding is done at very high temperatures, a thermal coefficient mismatch would crack the wafer on cooling). The oxide of Si is silica, which has a well-defined pH-dependent surface charge, making it very useful for a variety of microfluidic processes like electro-osmotic pumping.

Etched Si-based devices can be made with very fine features, but are expensive, fragile, and do not hold up to strongly alkaline solutions, which can etch the Si and its oxide coating. Silicon is nearly opaque in the visible and UV regions of the spectrum, so optical methods can only be used by imaging through glass layers bonded to it, generally eliminating transmission monitoring, but allowing fluorescence imaging. The fact that Si is also a semiconductor, and therefore not a good electrical insulator, is also a handicap in one very common microfluidic application—electrophoresis, in which large voltage drops are often required. For capillary electrophoresis on a chip, all-glass devices are the norm. The isotropically etched glass parts can be sealed by apposing two or more parts and increasing the temperature of the entire device to nearly the melting temperature of the glass. The etching processes for both Si and glasses requires very reactive and dangerous chemicals, such as hydrofluoric acid, best done in a professionally staffed microfabrication laboratory; selective glass etching poses additional challenges because very few materials resist a deep glass etch. Consequently, the final devices tend to be expensive. This is fine for devices to be used for days to years (for example, an implanted microfluidic device or an environmental monitor), but not so good for single-use applications (like a point-of-care diagnostics device for use with human blood). Also, the high-temperature sealing processes (upwards of 300°C) destroy any organic molecules in or on the devices, so any biomolecular or antifouling coatings for the devices must be applied *after* assembly.

3.5.1.2 The Advent of Plastics

For low-cost fluidic devices (micro or macro), the selection of choice in the industry has long been injection molding. Many thermoplastics (including inexpensive ones like polystyrene and polypropylene) can be molded into devices with extremely small features, as long as the topology allows removal by opening a mold. Although the end product parts can have a cost of pennies each, the cost of making the first mold is in the tens of thousands of dollars, so the first

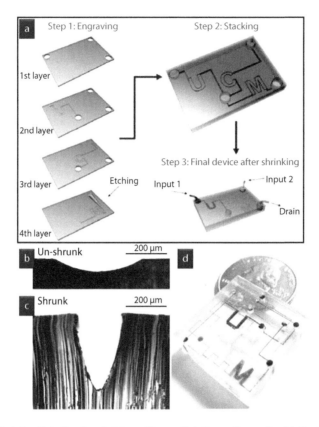

FIGURE 3.11 "Shrinky-dinks" microfluidics. (From Chi-Shuo Chen, David N. Breslauer, Jesus I. Luna, Anthony Grimes, Wei-chun Chin, Luke P. Lee, and Michelle Khine, "Shrinky-dink microfluidics: 3D polystyrene chips," *Lab Chip* 8, 622, 2008. Reproduced with permission from The Royal Society of Chemistry.)

device is very expensive, indeed. As a consequence, researchers have generally avoided injection molding, but manufacturers of disposable devices often must convert a design to one that can be injection-molded to keep the cost-per-device as low as possible.

The inherent shrinkage property of biaxally prestressed thermoplastic sheets (e.g., the toy "Shrinky-Dinks") has also been leveraged for miniaturization: on heating, engraved grooves (using syringe tips or digital cutters) will shrink only in one plane, becoming narrower and deeper (Figure 3.11).

3.5.1.3 A New Kid on the Block: PDMS

As we have seen already in Section 1.6, PDMS has many excellent qualities as a polymer for microfluidics, so it is widely used today. The development of inexpensive methods for patterning PDMS molds, the fidelity with which PDMS can reproduce even nanometer-scale small features in those molds, its ability to be sealed to itself and other oxide-covered surfaces, and its optical transparency have made it the material of choice for researchers. The fact that PDMS is permeable to gases and even to water vapor has proven very useful in some biological applications of microfluidics, as well, although in some cases it could be seen as a liability. The fact that PDMS is a very mobile polymer allows the surface of PDMS to readily change its composition from hydrophobic to hydrophilic and back again. This is not necessarily an advantage. Also, as of this writing, PDMS has not been very amenable to inexpensive mass production (although it is possible, especially if one redefines "mass production"), but has been used in a few commercialized microfluidic devices.

THINKING OUTSIDE OF THE SILICON BOX

BACK IN THE EARLY 1990S, the field of BioMEMS was at a crossroads. BioMEMS had inherited, from the microelectronics industry, a technology that was optimized for silicon, a material that was suboptimal for biomedical applications. On the other hand, it was tempting to use all the second-hand photolithography equipment left behind by the industry, which had moved on to much higher-resolution linewidths. Unfortunately, performing photolithography still meant being stuck with silicon (or glass) wafers and solvent-based photoresist, which was precluding the use of biological material. It was a vicious circle, one that required someone to think outside of the box in coming up with an exit strategy.

The brilliant mind that provided the solution to this problem would have to come from outside the MEMS community, in particular from the chemistry field: Harvard chemist, George Whitesides, who rid the BioMEMS field of its silicon dependence by introducing inexpensive PDMS-based micromolding in 1993. PDMS allows for selectively depositing biological materials and it can be the structural part of optical or biocompatible microfluidic devices.

It is interesting to note that it was not the first time that PDMS had been used to microfabricate a device. The first time it had also been a "MEMS outsider"—a biologist. Friedrich Bonhoeffer at the Max Planck Institute in Tübingen, Germany had used PDMS 6 years earlier to fabricate a microfluidic device for immobilizing membrane fragments for axon guidance experiments in 1987 (see Figure 6.41 in Section 6.5.1.1). Bonhoeffer did not, like Whitesides, exploit the various properties of PDMS to engineer different classes of devices—he concentrated on axon guidance and one design, and thus we now attribute the conception of soft lithography to George Whitesides. In any case, the lesson is clear: MEMS engineers were trapped in a silicon box and needed help to get out of it.

3.5.1.4 Other Polymers

Sheet-formed polymers such as Mylar have proven useful in both research and manufacture of microfluidic devices. Sheets can be modified by ablation using a variety of lasers (from CO_2- to excimer-based lasers), and many commercial instruments allow such ablation of flat sheets under computer control, allowing rapid patterning of sheets into complex structures. As with PDMS fabrication, this device fabrication method allows very rapid transition from computer-based design to finished product; such rapid prototyping is very advantageous in research laboratories. The availability of thin flat layers of adhesives, and even polymer sheets covered with one or two layers of pressure-sensitive adhesive, allows rapid lamination of complex microfluidic devices consisting of from three to dozens of layers. Many of the polymers that can be used are relatively "biocompatible," although almost all must be passivated to prevent adsorption of proteins to them. The glue layers are a remaining liability, as they may both adsorb and leach small molecules; glueless methods using heat-sealable polymers are under development as of this writing. Laser-based methods are not the best for mass production because they are inherently relatively slow, but fortunately, there are other mass production methods that allow the scale-up of the polymer cutting and lamination processes.

Biodegradable polymers, now commonly used in resorbable surgical sutures, are being increasingly used in tissue-engineered implants and in drug delivery, so why not use them to design implantable microfluidic devices that support cell growth? That is the idea behind the effort led by Joseph Vacanti at Harvard Medical School and Jeffrey Borenstein at Draper Laboratory in Cambridge, Massachusetts. In 2004, this team was able to mold and bond simple devices (Figure 3.12) made of **poly(DL-lactic-coglycolide) (PLGA)**, a biodegradable thermoplastic that

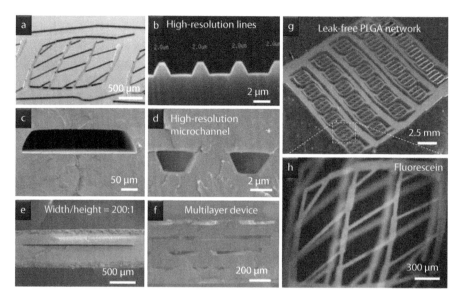

FIGURE 3.12 Biodegradable microfluidics. (From Kevin R. King, Chiao Chun J. Wang, Mohammad R. Kaazempur-Mofrad, Joseph P. Vacanti, and Jeffrey T. Borenstein, "Biodegradable microfluidics," *Adv. Mater.* 16, 2007–2012, 2004. Reprinted with permission from John Wiley and Sons. Figure contributed by Jeffrey Borenstein.)

is approved by the FDA for therapeutics. This copolymer is highly biocompatible because it slowly degrades into harmless acids (the monomers) normally present in the human body by a reaction with water that breaks the ester bonds. The Vacanti/Borenstein fabrication process is entirely based on melt-processing and thermal fusion bonding (avoiding solvents). Scanning electron micrographs of the devices are shown in **Figure 3.12a** through **f**, and filled devices are shown in **Figure 3.12g** and **h**.

3.5.1.5 Hydrogel Devices

Hydrogels are highly porous polymeric matrices closely associated with water. Examples are collagen gel, **Matrigel** (a basement membrane extract), agarose, PEG, and many biodegradable polymers. Hydrogels are extremely appealing for several reasons: (a) most of them can easily be molded from a PDMS mold with a short temperature transition, requiring no further chemistry (others simply require photo–cross-linking, which is also convenient for photolithography); (b) these polymers are highly biocompatible because they are either native ECM proteins (collagen, Matrigel—ideally suited as three-dimensional cell attachment substrates), or very bioinert (agarose, PEG), or biodegradable (polylactic acid); and (c) their high porosity can be used to form devices that create gradients in a time-release manner. Thus, several groups have developed procedures to fabricate devices in hydrogels. The practical challenge is that the hydrogels are often fragile (e.g., if formed as a thin film) and they do not seal well against a dry surface. The very insertion of fluids can be a major problem (because the tubing tends to leak at the contact point). Joe Tien's group at Boston University has shown a tubeless approach to introduce fluids into hydrogel microchannels and to selectively etch away Matrigel microstructures embedded in collagen (using dispase, an enzyme that digests Matrigel but not collagen). The key to their approach is to take advantage of the porosity of the gel, which does allow a certain amount of flow (and can be powered simply by gravity, tilting the channels). Both iron powder and cells embedded in Matrigel were seen to remain in micromolded cavities after the Matrigel was digested (**Figure 3.13**).

Another interesting naturally occurring gel, **alginate**, has been exploited in microfluidics to produce droplets (see Section 3.7) as well as alginate hydrogel microdevices. Alginate is a linear

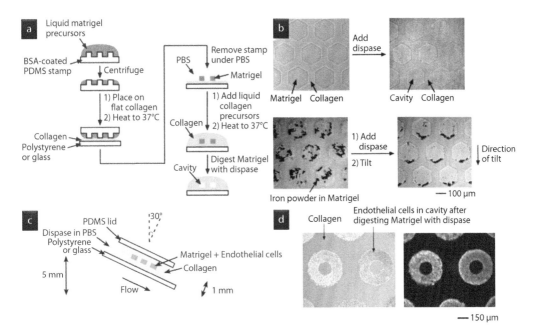

FIGURE 3.13 Cell-seeded ECM-hydrogel microchannels. (From M. D. Tang, A. P. Golden, and J. Tien, "Fabrication of collagen gels that contain patterned, micrometer-scale cavities," *Adv. Mater.* 16, 1345–1348, 2004. Reprinted with permission from John Wiley and Sons.)

copolymer that is extracted from seaweed and that is highly biocompatible: it is edible and is already being used for cell encapsulation by the biomedical industry; thus, it is an optimal candidate for cell-containing microdevices. The gelation process is very simple, requiring the addition of divalent cations such as Ca^{2+}. In 2007, Abraham Stroock's group at Cornell University (Ithaca, New York) presented an injection molding strategy for producing alginate microfluidic devices that can be used as scaffolds for tissue engineering applications (**Figure 3.14**).

Certain hydrogels can be photopolymerized by adding a photoinitiator, so the pattern is defined by the photomask projected onto the prepolymer solution. Samuel Sia's group at Columbia University (New York) has used stopped-flow lithography of PEG-diacrylate (see Section 1.6.4) to fabricate channels and at the same time encapsulate fibroblasts (**Figure 3.15**).

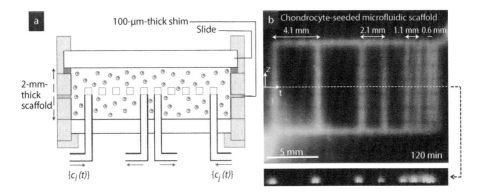

FIGURE 3.14 Alginate microfluidics. (From Nak Won Choi, Mario Cabodi, Brittany Held, Jason P. Gleghorn, Lawrence J. Bonassar, and Abraham D. Stroock, "Microfluidic scaffolds for tissue engineering," *Adv. Mater.* 6, 908–915, 2007. Reproduced with permission from the Nature Publishing Group.)

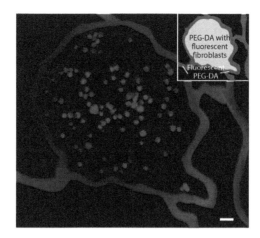

FIGURE 3.15 Fabrication of biocompatible PEG hydrogel microchannels. Scale bar is 50 μm. (From Yuk Kee Cheung, Brian M. Gillette, Ming Zhong, Sharmilee Ramcharan, and Samuel K. Sia, "Direct patterning of composite biocompatible microstructures using microfluidics," *Lab Chip* 7, 574–579, 2007. Reproduced with permission from The Royal Society of Chemistry.)

3.5.1.6 Paper

The use of paper to produce microfluidic devices is rather old, as has been exploited in more than one field. Paper has been used as a medium for electrophoresis of small molecules for a long time: in 1948, Gotfred Haugaard and Thomas D. Kroner of the United Shoe Machinery Co., reported in the *Journal of the American Chemical Society* the separation of amino acids with applied voltage (~100 V) using paper as the medium. It has been used extensively by the electrophoresis community since 1958, when Kurt Hannig invented a device in which gravity-assisted flow was carried by a filter paper layer (see Section 4.4.2); **paper electrophoresis** was later abandoned in favor of other techniques such as capillary electrophoresis, but it may be coming back now that paper can be easily patterned (see below). Paper has also been recognized as a powerful platform to run microfluidic immunoassays for more than a decade now: the commercially available "digital" (yes/no) pregnancy test containing two strips of antibodies that change color based on the human chorionic gonadotropin (hCG) concentration in women's urine is, essentially, a paper microfluidic device (see Section 4.5.1).

Paper is inexpensive, is biodegradable, and most importantly, its natural wicking action acts as a capillarity pump (so no further equipment is required to drive the fluids). Fluid transport in paper is well understood, as it obeys Darcy's law of flow through porous media (see Section 3.2.8). However, the pregnancy test and paper electrophoresis devices are based on a single rectangular strip, which does not offer much in terms of routing fluids to different compartments like most microfluidic devices do. To bring **paper microfluidics** to a new level, technologies for micropatterning paper were recently introduced.

In 2006, a group at Cornell University led by Matthew DeLisa patterned three channels (using laser-cutting) in a nitrocellulose paper membrane to produce gradients (see Section 3.9.5.1). In 2007, George Whitesides and colleagues at Harvard University presented a photolithography process that allows for patterning fluidic channels in paper (Figure 3.16). A piece of paper is impregnated with photoresist, which is then processed by photolithography like a normal wafer. The photoresist features that remain embedded in the paper after development act as hydrophobic walls for the fluids. The cost of the paper devices is much inferior to silicon or PDMS devices, so these devices are targeted for deploying assays in resource-poor settings.

Paul Yager's group at the University of Washington has shown that, interestingly, the paper supports flow at speeds that are fast enough that mixing between adjacent streams is negligible. An example (in which the features are laser-cut) is shown in Figure 3.17. This feature eventually allows

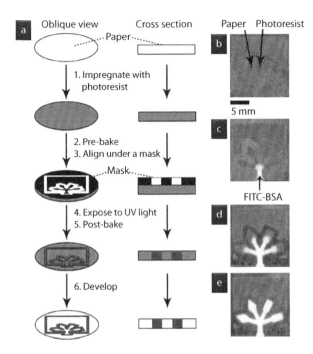

FIGURE 3.16 Paper microfluidics. (a) Schematic of the method for patterning paper with hydrophobic photoresist. A 10 μM solution of FITC-BSA dissolved in 40 mM phosphate buffer (pH 7.4) was wicked into a paper-based device and imaged under UV light (254 nm). The FITC-BSA distributes evenly throughout the channels and test zones. (b–e) Images of a device before (b) and after (c–e) wicking 1 μL (c), 2 μL (d) and 5 μL (e) of the FITC-BSA solution. (From Andres W. Martinez, Scott T. Phillips, Emanuel Carrilho, Samuel W. Thomas III, Hayat Sindi, and George M. Whitesides, "Simple telemedicine for developing regions: Camera phones and paper-based microfluidic devices for real-time, off-site diagnosis," *Anal. Chem.* 80, 3699–3707, 2008. Reprinted with permission of the American Chemical Society. Figure contributed by George Whitesides.)

for a renewed vision in which any heterogeneous-flow microfluidic assay will be implementable in paper. Furthermore, unlike traditional microfluidic channels, in the paper format, the microchannel is (a) readily accessible for sample introduction (so it should be ideal for analyzing "forensic-style" samples—like a "microfluidic swap"—that are usually obtained from soil, skin, sweat, semen, vaginal fluid, or other secretory glands in minute quantities); and (b) it de facto incorporates a filter, which should be invaluable for filtering cell debris, dirt, and other impurities in blood samples.

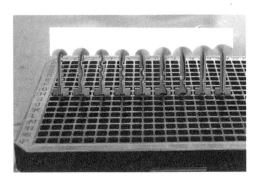

FIGURE 3.17 Laminar flow patterns created in laser-cut paper microchannels. (Figure contributed by the author.)

Last but not least, paper should prove invaluable to bring microfluidics to the classroom (and even K-12 science fairs) because of its (virtually zero) low cost and design-to-prototype time of a few minutes, which allows students to "play" with microfluidics in a rare combination of technology and art class (see Section 9.10 on microfluidics outreach). The use of paper has triggered the imagination of other researchers into exploring the use of other wicking materials such as fabric, which can be weaved into patterns of various hydrophilicities or interwoven in networks using knots—a concept that has spawned a new field dubbed **fabric microfluidics**. It is unlikely that fabric can outperform paper as a material for the deployment of low-cost microfluidic assays because of paper's lower cost and amenability to various micropatterning techniques.

3.5.2 3-D Stacking and Bonding

Once you have made a small untidy stack of individual microfabricated parts of a complex three-dimensional microfluidic device, the next step is to assemble the stack into an integrated system; the first challenge is alignment. If fluids must move from one layer to the next (e.g., see **Figure 3.18**), there must be some method for aligning features on one device with those on the next. In 1998, George Whitesides' group at Harvard University invented a procedure to bond two PDMS surfaces, or a PDMS surface to glass, based on exposing the surfaces to an oxygen plasma. The procedure is short and inexpensive (not accounting for the cost of the oxygen plasma oven) and, most critically, does not add any thickness of "glue" to the final thickness of the device—it only adds hydroxyl reactive groups which, when in contact, produce water molecules (which diffuse away) and leave a highly stable oxygen bridge; if the bonded interface is sliced orthogonally with a razor blade, the bond is so clean that the interface cannot be found. Historically, most soft lithography laboratories adopted this procedure for a good length of time. However, the procedure has two major limitations: (a) the oxygen plasma activation dies off with time (keeping the surfaces under water helps, but complicates the procedure); and (b) it is irreversible, thus if an alignment mistake is made, the device is ruined. Hence, many laboratories have sought other alternatives.

Weijia Wen's group from the Hong Kong University of Science and Technology has been able to produce impressive 3-D PDMS laminated devices by using a PDMS slab derivatized with a fluorosilane SAM ("PFOCTS" in **Figure 3.18**). The slab readily "picks up" cured PDMS features from any SU-8 master (but the features do not bind to the PDMS slab because of the PFOCTS layer) and transfers them to plasma-activated glass or PDMS device, which can be thus can be built layer by layer. This procedure is conceptually similar to the exclusion molding process outlined in **Figure 1.30**. The bonding step may require something like the aligners used in the Si microfabrication industry, but often it is just a matter of steady fingers under a stereomicroscope.

Some lamination tasks require a different approach. A popular way to fix layers to each other without leaks is to incorporate preformed glue layers. The most familiar form of these is a pressure-sensitive adhesive (like that used in "scotch" tape); at least every other layer must be made sticky. Such layers cannot slide past each other, so they must be aligned nearly perfectly on contact. This alignment is often performed by using a jig with two or more precisely machined posts. The posts align with features that are precut into the polymeric layers, allowing them to be lowered onto each other in near-perfect alignment. Once the layers are stacked, they are removed from the jig and can be compressed for final alignment. A similar method can be used for other types of stacking methods not requiring pressure-sensitive adhesives.

A team led by Karen Gleason at MIT has developed a "nano-adhesive" that can be deposited by plasma polymerization conformally onto a wide variety of polymeric and inorganic materials, including silicon, glass, PDMS, polystyrene, polycarbonate, and poly(tetrafluoro ethylene) (**Figure 3.19**). The parts treated with nano-adhesive can be aged for at least 3 months. The nano-adhesive consists of two complementary adhesives (one on each opposing surface), poly(glycidyl methacrylate) (PGMA) and polyallylamine (PAAm), which bond on contact after curing at 70°C. Both polymers are formed by chemical vapor deposition (CVD) processes, and their CVD

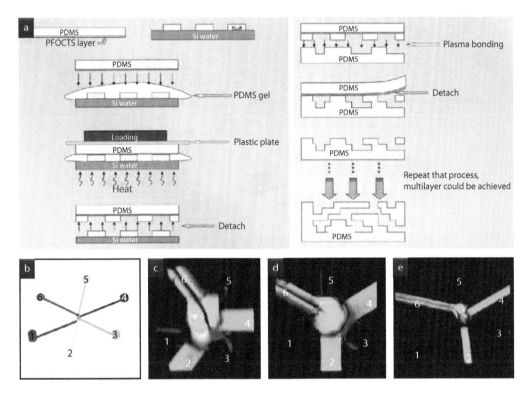

FIGURE 3.18 Fabrication of complex multilevel microfluidics. (From Mengying Zhang, Jinbo Wu, Limu Wang, Kang Xiao, and Weijia Wen, "A simple method for fabricating multi-layer PDMS structures for 3D microfluidic chips," *Lab Chip* 10, 1199–1203, 2010. Reproduced with permission from The Royal Society of Chemistry.)

monomers are commercially available. The thickness of adhesive required to obtain bonding is so small that it is compatible with the fabrication/bonding of 200-nm-wide channels.

Fortunately, CVD is not the only route to achieve good bonding. Nae Yoon Lee's group at Kuyngwon University in Korea has developed a simple, silane-based room temperature process to bond PDMS to plastics (PMMA, polycarbonate, polyimide, and PET; Figure 3.20). The PDMS and plastic surfaces are derivatized with aminosilane SAM (APTES) and epoxysilane SAM (GPTES), respectively, and allowed to react at 25°C for 1 hour—simple! The fact that it takes 1 whole hour for the bonding to complete is advantageous, as it allows the experimenter to bring the two surfaces into contact and, if there is a misalignment, there is plenty of time for a retry (by comparison, oxygen plasma activation of PDMS only allows for a few minutes, at most). Noting that SU-8 has abundant unreacted epoxy groups on the surface even after exposure, Gaozhi Xiao's group from the Institute for Microstructural Science in Ontario, Canada has demonstrated an elegant simplification of the previous process based on the use of a nitrogen plasma (because oxygen plasma was being used for PDMS activation before APTES derivatization anyways) to introduce surface nitrogen groups on the PDMS: up to 1.9% nitrogen can be introduced to the PDMS surface in the form of $C-NH_2$, $C-NRH-$, $N-C-N$, and $C=N-$, as confirmed by X-ray Photoelectron Spectroscopy (XPS) analysis. These groups then slowly react at room temperature with the unreacted SU-8 epoxy groups, allowing for the fabrication of hybrid SU-8/PDMS devices. The Mathies laboratory from the University of California at Berkeley has fabricated PMMA/PDMS hybrid microvalves (PDMS membrane sandwiched/bonded between two PMMA slabs) using a very simple PDMS-to-PMMA bonding procedure: exposure of PDMS and PMMA to a UV ozone cleaner.

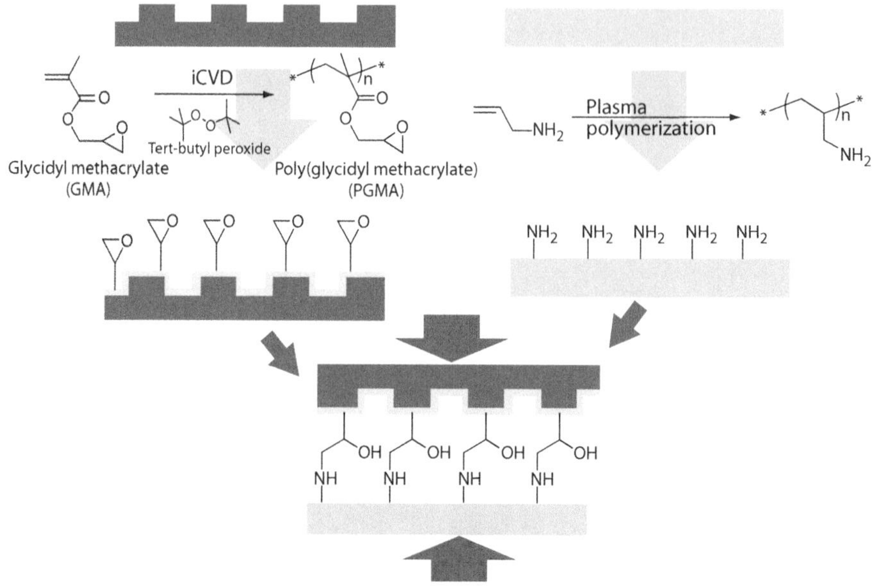

FIGURE 3.19 Nanoadhesive plasma-deposited coatings. (From Sung Gap Im, Ki Wan Bong, Chia-Hua Lee, Patrick S. Doyle, and Karen K. Gleason, "A conformal nano-adhesive via initiated chemical vapor deposition for microfluidic devices," *Lab Chip* 9, 411–416, 2009. Reproduced with permission from The Royal Society of Chemistry. Figure contributed by Karen Gleason.)

Importantly, Jeffrey Zahn's laboratory at Rutgers University has developed a procedure to irreversibly bond polycarbonate, polyethersulfone, and polyester terephthalate membranes to PDMS and glass using 3-aminopropyltriethoxysilane (APTES). The membranes only need to be briefly activated in oxygen plasma, dipped for 20 minutes in a diluted (5% in water) solution of APTES for 20 minutes, dried, and placed on a 80°C hot plate in contact with the PDMS surface (also oxygen plasma–activated).

Every microfluidic researcher has experienced the frustration of a device that fails because of a small design defect, a trapped bubble, or a valve that does not open. In these cases, to facilitate repair, it would be nice if the part of the device that does not work could be isolated—in other words, it would be nice if microfluidic devices could be built modularly in a standardized plug-and-play, LEGO-like format. Nobody has agreed yet on such a standard, but there have been several efforts in building devices from discrete modules, most notably that of Mark Burns and colleagues at the University of Michigan. They demonstrated simple mixers built from 6 × 6 mm blocks, which could connect and establish good seals through the block sidewalls (**Figure 3.21**).

UNDERWATER ADHESIVES FOR MICROFLUIDICS?

In microfluidics, a lot of times, we would like to bond our microchannels when they are already seeded with cells, but then it is too late because there is no available glue that works for wet surfaces. Ten years ago, I shared with my former colleague Ulysses Balis, a very creative M.D., my frustration that water seems to ruin the chemistry of virtually all existing glues, such as epoxy, super glue, etc. His one-word solution: "Barnacles!"—he was thinking of the marine organisms that stick to boat hulls forming layers that are very hard to remove. Actually, *dozens* of underwater organisms, such as acorn barnacles, blue mussels, sandcastle worms, starfish, conus snails, and freshwater caddisfly larva, to name

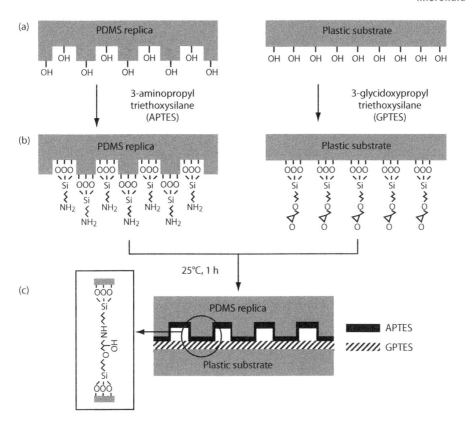

FIGURE 3.20 Room temperature bonding of PDMS to plastics using silanes. (From Linzhi Tang and Nae Yoon Lee, "A facile route for irreversible bonding of plastic-PDMS hybrid microdevices at room temperature," *Lab Chip* 10, 1274–1280, 2010. Reproduced with permission from The Royal Society of Chemistry.)

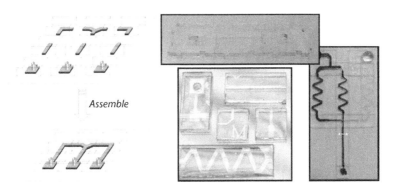

FIGURE 3.21 Modular microfluidics. (From Minsoung Rhee and Mark A. Burns, "Microfluidic assembly blocks," *Lab Chip* 8, 1365–1373, 2008. Reproduced with permission from The Royal Society of Chemistry.)

a few, are capable of secreting proteinaceous fluids that become strong adhesives when in contact with water (which is at a higher pH and ionic strength than their internal environment; to establish a strong bond with the surface, the adhesive proteins seem to displace adsorbed species such as water, ions, and other proteins). These organisms use the secreted adhesive to form structures or communities by gluing themselves to rocks, reefs, boats, or any suitable surface they find. Ulysses' suggestion is still valid—it is surprising that the microfluidics and the materials science communities have not explored this engineering achievement of nature to simplify some of the bonding processes for biomedical microdevices, as it is likely that these bioadhesives are highly compatible with cells and tissues.

3.5.3 Inlets: the "Macro-to-Micro Interface" Problem

Ironically, one of the most problematic aspects of developing microfluidic devices in the laboratory has been figuring out how to connect them to the macroscopic world. For example, you may have developed the world's best chromatographic separation system on a chip, but to push fluids into and out of that device, you still need to connect it to macroscopic pumping systems that have conventional connectors (and conventional internal channel diameters) and that are bulky! Every conceivable method has been tried (and often rejected) for this purpose, from gluing tubing directly to Si chips using epoxy (a method that usually results in filling the holes of the Si chip with extra epoxy, ruining the device) to designing custom connectors for each particular chip design. To date, this remains one of the pesky details that is probably best solved on a case-by-case basis, but two or three particularly creative solutions deserve to be highlighted here for the benefit of the reader.

Brian Cunningham's group at the University of Illinois at Urbana-Champaign has produced a hybrid of the 96-well plate technology and microfluidic channels: the channels run under the wells (which are laid out in the usual 8 × 12 format), and chemically communicate with the wells through an aperture (**Figure 3.22**). A central "common" well in each row serves as an access point for the introduction or withdrawal of reagents for the flow channels. Pressure or suction applied at the common well drives fluid from the common well into the flow channels, or pulls fluid from the 11 analyte wells at the same rate. This "**tubeless microfluidics**" approach allows the user to introduce compounds into the microfluidic network using traditional multipipetters designed for 96-well plates.

Javier Atencia's group at NIST has developed a vacuum manifold that is sealed onto the inlet region of the device, connecting all the inlets at once. A chip featuring 51 inlets and hundreds of microvalves was able to hold 100 kPa (15 psi) on all inlets (**Figure 3.23**).

For some applications, especially for transferring samples *from one chip to another*, it may be simpler to fabricate a dispenser with a razor blade from the end of an array of microchannels, as demonstrated by the groups of Yanyi Huang and of Jianzhong Xi at Peking University in Beijing, China. A version of the device featuring microvalves allowed for filling microwells in arbitrary patterns (**Figure 3.24**).

3.5.4 Microchannel Wall Coatings

We have already seen in detail two phenomena in which surface composition dramatically alters the behavior of a much larger entity that contacts the surface: cell attachment and electro-osmosis. Proper molecular-scale design of cell-adhesive (or cell-repellent) microchannels is important in fields as diverse as cell biology, tissue engineering, and biosensors. Likewise, controlling the charges on the walls of microchannels for EOF is a crucial goal for portable medicine, global health, microsensors, and implantable devices.

As many of the polymers used in microfluidic fabrication (e.g., PDMS, plastics) are hydrophobic, we are usually interested in chemical processes that make them hydrophilic (which facilitates

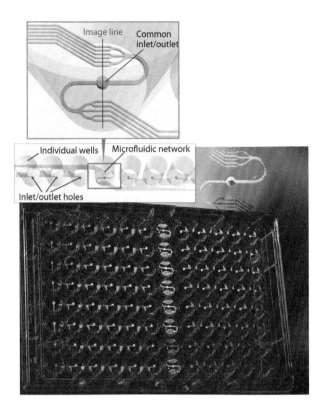

FIGURE 3.22 "Tubeless" microfluidics: a 96-well plate incorporating embedded microchannels. Serpentine flow channel patterns are used to ensure an equal length flow path from each analyte well to the common well. (From Charles J. Choi and Brian T. Cunningham, "A 96-well microplate incorporating a replica molded microfluidic network integrated with photonic crystal biosensors for high throughput kinetic biomolecular interaction analysis," *Lab Chip* 7, 550, 2007. Reproduced with permission from The Royal Society of Chemistry.)

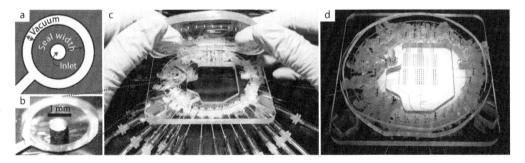

FIGURE 3.23 Vacuum manifold for world-to-chip interface. (From Gregory A. Cooksey, Anne L. Plant, and Javier Atencia, "A vacuum manifold for rapid world-to-chip connectivity of complex PDMS micro-devices," *Lab Chip* 9, 1298, 2009. Reproduced with permission from The Royal Society of Chemistry.)

wetting and filling). We have already seen one way to make a hydrophobic channel hydrophilic which is to chemically modify the surface using a form of oxidation, such as exposure to an oxygen plasma (dry) or a solution containing an oxidizing agent (wet). Alternatively, an intrinsically hydrophobic surface can be coated with a polymer (synthetic or natural) that adsorbs to it and, in turn, exposes a more hydrophilic surface: Pluronics [which are triblock copolymers known as

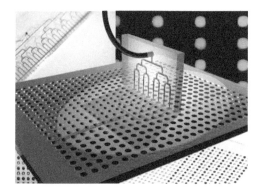

FIGURE 3.24 A chip-to-chip nanoliter dispenser. (From Jianbin Wang, Ying Zhou, Haiwei Qiu, Huang Huang, Changhong Sun, Jianzhong Xi, and Yanyi Huang, "A chip-to-chip nanoliter microfluidic dispenser," *Lab Chip* 9, 1831–1835 (2009). Reproduced with permission from The Royal Society of Chemistry.)

polyoxamers of the form: poly(ethylene oxide)*x* - poly(propylene oxide)*y* - poly(ethylene oxide)*x*] and proteins are two possible alternatives.

Recently, Dong Pyo Kim's group at Chungnam National University in South Korea has developed a several hundred–nanometer-thick silicate glass coating for PDMS. A preceramic polymer—allylhydridopolycarbosilane (AHPCS)—is coated onto the channel and hydrolyzed to form a hydrophilic coating via phase conversion (**Figure 3.25**). The coating confers PDMS an electrophoretic mobility, optical clarity, and solvent resistance that is indistinguishable from that of glass over at least 90 days.

Not just the surface of the devices can be modified. In 2004, Nancy Allbritton and colleagues at the University of California at Irvine presented a novel way to graft polymers onto PDMS,

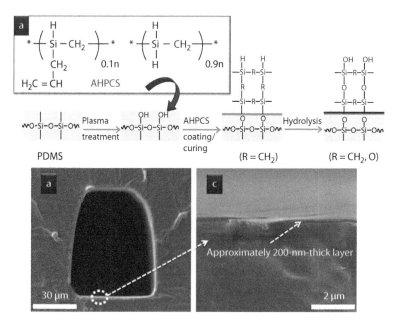

FIGURE 3.25 Glass-like coatings for PDMS. (From Ming Li and Dong Pyo Kim, "Silicate glass coated microchannels through a phase conversion process for glass-like electrokinetic performance," *Lab Chip* 11, 1126–1131, 2011. Reproduced with permission from The Royal Society of Chemistry.)

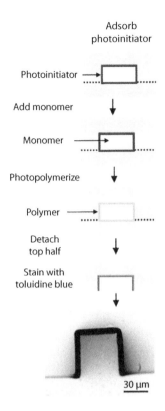

FIGURE 3.26 Grafting of PDMS channels. (From Shuwen Hu, Xueqin Ren, Mark Bachman, Christopher E. Sims, G. P. Li, and Nancy L. Allbritton, "Surface-directed, graft polymerization within microfluidic channels," *Anal. Chem.* 76, 1865, 2004. Reprinted with permission of the American Chemical Society. Figure contributed by Nancy Allbritton.)

making clever use of the fact that PDMS behaves like a "sponge" when it is exposed to strong organic solvents (**Figure 3.26**). When exposed to a solution of the photoinitiator benzophenone in acetone, the benzophenone penetrated into the swollen PDMS and remained there (inside or adsorbed on the surface) after abundant rinsing with water. Next, a monomer solution was introduced into the channel and exposed to UV light for photopolymerization, which started at the benzophenone-covered surface but extended into the bulk of the material. This process is applicable to a variety of monomers and produces very stable layers, which could be useful in fields such as capillary electrophoresis (enhanced electrophoretic mobility was demonstrated), tissue engineering, and cell biology.

3.6 Operation of Microfluidic Channels: Practical Concerns

Filling and running flow through a microchannel is not as trivial as filling a pipe and opening a faucet. Measuring flow velocity is even more difficult. These can be tedious, but it must be done if we are to do *something* with the microchannels. Here, we review practical methods to operate microfluidic channels.

3.6.1 Filling a Microchannel: The "Bubble Curse" and Methods to Jinx It

For a few reasons, filling a microfluidic channel can be harder than filling a garden hose. Even if the viscosity of the fluid is low as is the case of most biomedically interesting aqueous buffers, the importance of surface forces in small channels (another consequence of the high surface-to-volume ratio) can present unique problems. To fill the channel completely, the most important rule is to make sure that the channel "wants" to fill, so the channel walls should be hydrophilic if at all possible. A narrow hydrophilic channel that is open at both ends will fill spontaneously to a certain extent (i.e., by capillary forces, with no external force required), whereas a hydrophobic channel will require pumping of the fluid into the channel. It is important to note that, *once the channel is wet, the hydrophilicity of the walls plays no role in the flow resistance of the channel.*

Few things are more frustrating in researching microfluidic devices, and in developing reliably working practical systems using them, than bubbles. It is possible to fill a dead-end channel with a liquid if the gas that it replaces can be removed through the channel walls; this has been done using either specific gas-permeable membranes (porex, nomex, etc.) at the end of otherwise gas-impermeable channels, or by utilizing the gas permeability of PDMS to eliminate the bubble of gas at the dead end of the channel—just pressurize the fluid slightly, and any bubble will "magically" disappear (something that does not happen with microchannels made of glass or other hard polymers).

Filling a dead-end channel that is not gas-permeable, and removing bubbles that either become trapped in a microchannel during fill, or form during operation, can be one of the most difficult problems with using microfluidic devices—so much so that some researchers base their choice of the material used for building the device on its wettability (Table 3.3). Design of the devices so that they do not form bubbles on filling is an art that requires the proper use of flow rates, gravity, and surface chemistry—even inlet fittings and manual dexterity can be crucial. Here are some tips of the trade for avoiding bubbles on first-wetting:

- Often, the formation of bubbles can be suppressed by adding some surfactant to the wetting solution, but that is not always practical if the solution's composition cannot be compromised (i.e., the surfactant could affect a biochemical reaction or cell behavior in unpredictable ways).

- Sometimes, a better alternative is to prefill the dry device with a gas that is very water soluble, generally CO_2. Because CO_2 readily dissolves in water (forming carbonic acid), any bubbles that are left in the device after filling shrink to zero volume (over seconds to minutes) as the gas dissolves. As long as some acidification of the residing solution is acceptable (otherwise it can be flushed), this is a surefire way to fill almost any device (with water) completely.

- For devices made of PDMS, the surface of PDMS can be oxidized by exposure to an oxygen plasma; this procedure results in very hydrophilic walls that wick water in easily. Conveniently, the plasma oxidation step is the same procedure used to bond the devices; however, the oxidized state seems to revert to its native state after less than approximately 30 minutes in air, so the devices should be filled right after bonding. (The plasma penetrates into the device through the inlet ports, but the penetration length is too short to be practical for most devices.)

- If possible, microchannels with sharp corners should be avoided unless necessary. Sharp corners will almost always trap bubbles.

Table 3.3	Commonly Used Materials for Microfluidic Devices and Their Wettability							
Material	Native SiO_2	Glass	Teflon	HDPE	PMMA	Native PDMS	Oxidized PDMS	Mylar
Difficulty in wetting	Low	Low	High	High	Medium	Medium	Low	Medium

Note: Teflons are hydrophobic fluorocarbons, HDPE is high-density polyethylene, PMMA is poly(methyl methacrylate), also known as Plexiglas, PDMS is poly(dimethyl siloxane), and Mylar is a polyester, also known in fiber form as Dacron.

❋ As mentioned previously, for PDMS devices with complex architectures (e.g., 3-D, several inlets), the most practical way to completely fill a device is to cap all inlets and outlets and steadily introduce the fluid through one inlet pressurizing at a few psi. The liquid will fill the whole device, even if it takes a few hours.

❋ If possible, preconditioning of the device with a charged species that adsorbs strongly to the surface will ensure that the walls are hydrophilic the next time the device is wetted. Albumin (BSA) is widely used for this purpose, and it has the advantage of also blocking the adsorption of other proteins (it saturates protein adsorption sites).

If the bubbles form during the use of the device, which is a particular problem if the device is warmer than stored solutions, or the solution flowing through the device always contains some gas (e.g., blood), there are further problems. Even the most hydrophilic device is prone to forming and trapping a few bubbles. Bubbles change the available volume of the device and change the flow pattern and channel dimensions, so they are a serious problem for maintaining device function over time. In the worst case, it is possible to build an on-device degasser using a gas-permeable membrane (such as a thin PDMS membrane or an expanded hydrophobic polymer membrane such as Goretex), behind which is a vacuum. The addition of a bit of mechanical complexity (such as a vacuum pump and a device with a built-in vacuum manifold) may be the only solution.

3.6.2 Driving the Flow

There are various ways to drive flow, some obvious and some less obvious, some facing obvious hurdles and some facing unexpected ones when implemented on the micron scale. Among them, the most widely implemented ones are:

❋ **Syringe pumps** are extremely precise; however, they are relatively expensive and require power (i.e., are not suitable for resource-poor settings).

❋ **Gravity pumps**, which consist of reservoirs set at a particular height above the inlet, are extremely practical. Their main pitfall is that their pumping rate decreases exponentially as the gravitational energy is gradually depleted from the pump, but this pitfall can be mitigated by making the pump larger. A clever design exists (based on two reservoirs, one higher than the other) that guarantees constant flow until the higher reservoir is empty.

❋ **Peristaltic pumps,** which can be macrofabricated or microfabricated (see Section 3.8.4.3) but in almost all cases require a source of pressure (present in all modern laboratories); these pumps typically require the successive "pinching" of a tube or channel at three locations (imitating peristalsis—the radially symmetric, wave-like propagation of muscle contraction in the digestive tract).

❋ **Capillarity pumps** typically consist of a meniscus inside an incompletely filled microchannel; the surface tension of the meniscus drives the fluid forward so as to fill the empty part of the channel (see Section 3.2.10). The main advantage of capillarity pumps is that they are small and, like gravity pumps, consume no external energy. However, their pumping rate decreases with time as the radius of the meniscus decreases and they are vulnerable to evaporation effects.

❋ **Surface tension–driven pumps** are based on placing droplets at the exit/entrance of a microchannel, and using the surface tension produced by the curvature of the droplet to drive the flow. As with capillarity pumps, the pumping rate is a function of time and is affected by evaporation.

❋ **Evaporation pumps**, based on small, unequal-sized droplets placed above the inlet and outlet of the device, draw their force from differences in water evaporation rates between the inlet and the outlet. Their main pitfall is that water evaporation (in itself difficult to control) changes the local osmolarity of the solution and raises the local reagent concentration, which may be unsuitable for some assays.

- **Electro-osmotic pumps**, which are devices that produce electro-osmotic flow (see Section 3.3.2).

- **Electrowetting** has been used to control the displacement of droplets on dielectrics (see Section 3.3.4).

3.6.3 Flow Visualization

Computation fluid dynamics is a very powerful tool for planning the performance of a microfluidic device, but reality often offers us little surprises; nothing can substitute for observing flow behavior in real devices. Fortunately, there are methods for visualizing flow in microdevices, most of which, logically enough, involve *microscopes*. Most of the materials used for microfluidics are already transparent in the visible range, with the obvious exception of silicon, and even that can be made transparent. (Silicon transmits adequately in the near-infrared.) In general, glass or polymeric materials allow easy imaging. Bubbles are one thing that must be found, as they affect flow so strongly, and fortunately they are easy to see with many imaging techniques.

It is possible to use imaging to monitor (steady and unsteady) flow rates through the channel in several ways. The most common is **particle imaging velocimetry** (**PIV**), a method introduced in 1998 by Juan Santiago and colleagues, in which small imageable particles are introduced into the flowing fluid. The rate of their movement is then measured using one or more images, from which the rate of movement of the fluid that surrounds them can be inferred. This method works very well as long as the particles are not close to walls, which can collide with the particles causing them to move slower than the surrounding fluid. Of course, conventional microscopies force the imaging of multiple depths at the same time, making it difficult to resolve flow rates along the imaging axis. The depth of field of the microscope objective lens also has a large effect on the precision with which the measurement can be made. To avoid this problem, it is possible to illuminate specific portions of the channel with a laser, or to use confocal microscopy to image particles at specific three-dimensional locations.

Imaging, particularly fluorescence imaging, has been used to great advantage to monitor mixing between flowing streams. As one component diffuses into another stream, the fluorescence spreads to fill the volume, so the two- or three-dimensional pattern can be used to determine the diffusion coefficient of the imaged compound, or, if it is already known, the efficiency of the mixing in the particular device.

Flow gauges can be implemented within the microchannels to directly obtain a measure of the flow velocity at one point in the channel. In 2007, Frank Vollmer's group from Harvard University integrated an optical fiber cantilever inside (across) a microchannel such that its bending by the action of the flow was a function of the flow rate (**Figure 3.27**). Later, we will see examples of *microfabricated* flow gauges (Section 3.8.5).

3.7 Droplet Microfluidics

Droplets are extremely interesting containers: they allow for the confinement and manipulation of (bio)chemical and (bio)physical processes in well-defined volumes ranging from nanoliters to femtoliters and, *for the same price*, in a stream format (i.e., making one droplet costs virtually the same as making millions of them). These two powerful characteristics of droplets has attracted many scientists to this young field, now dubbed **droplet microfluidics**, which is now a mature area of research covered by several good review articles.

There are three basic platforms for shuttling water droplets around, one based on electrowetting, one based on oil-filled microfluidic channels, and one based on air-filled microfluidic

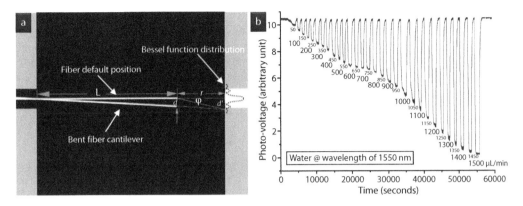

FIGURE 3.27 Measuring flow rate with a bending fiber. (From V. Lien and F. Vollmer, "Microfluidic flow rate detection based on integrated optical fiber cantilever," *Lab Chip* 7, 1352–1356, 2007. Reproduced with permission from The Royal Society of Chemistry.)

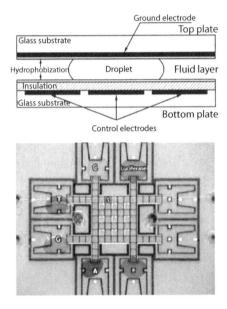

FIGURE 3.28 Common setup for manipulation of droplets by electrowetting. (From H. Ren, R. B. Fair, and M. G. Pollack, "Automated on-chip droplet dispensing with volume control by electrowetting actuation and capacitance metering," *Sens Actuators B Chem* 98, 319–327, 2004. Figure contributed by Richard Fair.)

channels. Other platforms exist (for example, magnetic droplets can be moved around with magnets) but do not have the versatility of these three, so we will not cover them here.

3.7.1 Electrowetting Platform: "Digital Microfluidics"

In the electrowetting platform, the droplets are surrounded by air or oil and requires a layout of electrodes in a parallel-plate glass chamber (Figure 3.28). Pioneering research on electrowetting by Richard Fair and colleagues at Duke University, and many others, has demonstrated the use of electrowetting to manipulate droplets with automation, a concept now dubbed **digital microfluidics** (see Figure 3.28). It is now possible to produce chips that generate small droplets from large reservoirs and take them along paths defined by electrodes to mixing points where biochemical reactions take place. This strategy for discrete fluid manipulation is powerful because the protocols are readily automated (since fluid motion is electrically powered), but has two important shortcomings: (a) the chips have thus far been expensive to produce, compared with all-polymeric devices; and (b) the voltages and the enclosed environment are not friendly to cells, so the chips are mostly limited to noncellular applications such as biochemical reactions. (Recently, Aaron Wheeler's laboratory at the University of Toronto has demonstrated cell culture and selective transfection of HeLa cells within a digital microfluidic platform.)

3.7.2 "Oil Carrier" Microdroplet Platform

The second platform relies on confining the water droplets within the walls of microchannels; the water droplets are surrounded by an oil fluid that acts as the carrier. The shuttling speeds here are orders of magnitude larger than with the electrowetting platform (typical droplet generation rates are on the order of 10,000 droplets/second!), so the "oil carrier" microdroplet platform has gradually displaced the electrowetting platform for throughput-intensive, biorelated applications.

Excluding digital microfluidics, we may adopt a working definition of droplet microfluidics as a family of microfluidic techniques and devices based on the generation of a continuous train of droplets on one fluid suspended in a stream of a different, immiscible fluids. Typically, a surfactant needs to be added to prevent colliding droplets from fusing. Although nonmicrofluidic formats had existed for decades, the first droplet-generating microfluidic device is credited to Stephen Quake's group (then at Caltech) and was presented in January 2001 (**Figure 3.29a**). The generation of the droplet was based on the lateral shear of a jet of water by an impinging flow of oil. Most present droplet generators, however, are based on nozzle designs, the first of which was presented in 2003 by Howard Stone's group at Harvard University (**Figure 3.29b**). Minoru Seki's group at the University of Tokyo also pioneered a microfluidic droplet generator design (presented in March 2001) that has since been abandoned. In all cases, the droplet size is determined simply by the geometry of the junction and the ratio of the water flow rate to the oil flow rate, and the generation frequency depends on the combined flow rates. The monodispersity in

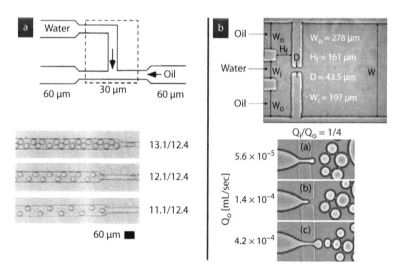

FIGURE 3.29 Generation of droplets in microchannels. (a, from Todd Thorsen, Richard W. Roberts, Frances H. Arnold, and Stephen R. Quake, "Dynamic pattern formation in a vesicle-generating microfluidic device," *Phys. Rev. Lett.* 86, 4163, 2001. Figure contributed by Stephen Quake; b, from Shelley L. Anna, Nathalie Bontoux, and Howard A. Stone, "Formation of dispersions using "flow focusing" in microchannels," *Appl. Phys. Lett.* 82, 364, 2003. Reprinted with permission of the American Institute of Physics. Figure contributed by Shelley Anna, Nathalie Bontoux, and Howard A. Stone.)

droplet sizes can be as low as 1% with accurate control of the flow rates. Silicone oils, perfluorocarbons, oleic acid, Fluid 200 from Dow Corning, and mineral oil, are among the commonly used oils (viscosities on the order of 5 to 50 cst are recommended as a starting point).

Droplet microfluidics, in particular in its "oily microfluidic" platform, has four powerful characteristics:

- The reaction vessel—the droplet—has no surfaces, hence, there is no contaminant adsorption nor any loss of reactants to the walls, so it is ideal for low molecule-count reactions such as single-copy PCR amplification, single-molecule enzymology, etc.

- The reaction vessel is very small, so the reactants diffuse quickly to react with one another (Figure 3.30), and there is less cost in reagents and waste.

- The format provides for multiple (on the order of thousands) repeats, naturally delivering rich statistics.

- The repetition of the assay is inexpensive because it is provided in a "train" format.

A rich variety of biochemical assays have already been implemented in droplet format, and the list only grows by the day: PCR (of single copies of cDNA), protein crystallization screening (1300 trials in 20 minutes, see Figure 4.27), cell encapsulation (see Section 5.3.9), and encapsulation of *C. Elegans*. We note that not all the droplets have to be an aqueous suspension—the generation of alginate gel droplets of various shapes (plugs, disks, spheres, rods, and threads) by the fusion of alginate droplets with $CaCl_2$ droplets, as recently pioneered by Xing-Zhong Zhao's group at Wuhan University in China, is a promising advancement in cellular engineering. Importantly, Bincheng Lin's group from the Dalian Institute of Chemical Physics has presented a microvalve-actuated droplet generator that can produce droplets on demand (Figure 3.31)—all previous droplet generators only function continuously. The size of the droplets is controlled by the amount of time that the microvalve is open. Each microvalve gates an inlet with a different solution (a total of four inlets are featured), and by alternating which inlets are "on" with computer control it is possible to establish sequences and mixtures of droplets. Generation of droplet sizes between 1.3 and 13.3 nL (with volume variations of 7.2% and 1.6%, respectively) at 100 droplets/s was demonstrated. Pei-Yu Chiou's group at UCLA has shown that it is possible to generate droplets

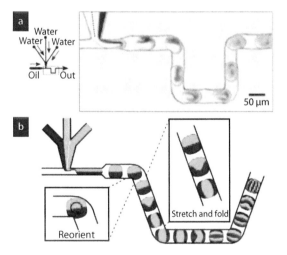

FIGURE 3.30 High-throughput droplet microfluidics. (From Helen Song, Michelle R. Bringer, Joshua D. Tice, Cory J. Gerdts, and Rustem F. Ismagilov, "Experimental test of scaling of mixing by chaotic advection in droplets moving through microfluidic channels," *Appl. Phys. Lett.* 83, 4664, 2003. Reprinted with permission of the American Institute of Physics. Figure contributed by Rustem Ismagilov.)

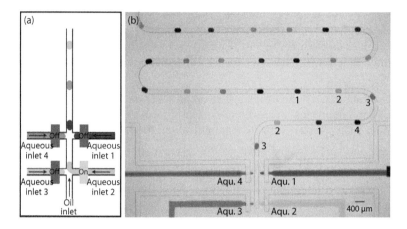

FIGURE 3.31 Microvalve-actuated control of individual droplets. (From Shaojiang Zeng, Bowei Li, Xiao'ou Su, Jianhua Qin, and Bingcheng Lin, "Microvalve-actuated precise control of individual droplets in microfluidic devices," *Lab Chip* 9, 1340, 2009. Reproduced with permission from The Royal Society of Chemistry.)

on demand at rates up to 10,000 droplets/s using a pulse laser-induced cavitation; the system is capable of delivering droplet volumes between 1 and 150 pL with less than 1% volume variation.

3.7.3 "Air Carrier" Microfluidic Platform

The third platform, shown in **Figure 3.32**, was recently invented by Hiroaki Suzuki and coworkers at the University of Tsukuba, Japan. It essentially replaces the oil of the oil carrier

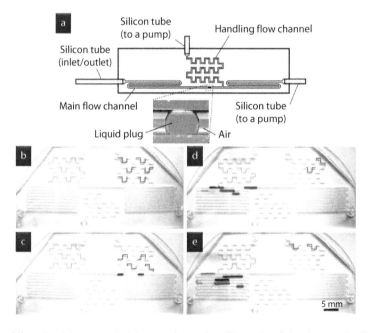

FIGURE 3.32 Microdroplets separated by an air carrier. (From Fumihiro Sassa, Junji Fukuda, and Hiroaki Suzuki, "Microprocessing of liquid plugs for bio/chemical analyses," *Anal. Chem.* 80, 6206–6213, 2008. Reprinted with permission of the American Chemical Society. Figure contributed by Hiroaki Suzuki.)

microdroplet platform by air. Although the concept is simple, the technical challenge is to prevent wetting of the walls (otherwise the droplets would continuously lose material to the walls), which is achieved by a proprietary hydrophobic coating ("Fluorosurf"); droplet speeds of 13 mm/s were demonstrated without breaking the integrity of the droplets. To prevent cross-contamination, a flushing plug can be pushed down the channel. This platform could be ideally suited for cell encapsulation applications because the extracellular medium would be able to rapidly equilibrate its gas concentrations and thus might not suffer the hypoxic conditions that can be encountered in oil carrier microdroplets.

3.8 Active Flow Control

Here we review the miniaturization of various active elements or "actuators"—such as valves, pumps, and others—some of which are inspired in design by their macro-world counterparts, whereas others are based on unique micron-scale phenomena that do not work when scaled up to larger scales. The microvalve and micropump world–expert Nam-Trung Nguyen has undertaken an extensive categorization of all the existing designs based on their operating physical and chemical principles, and ranked them based on their ranges of operation of pressure and response times (Figure 3.33). For example, pneumatic pumps are capable of producing large pressures but the pneumatic pressure is typically delivered through compliant tubing, which introduces capacitance in the lines (so pumping and valving on the order of only several hertz can be achieved in practice). Pumps based on electrostatic actuators, on the other hand, can be extremely fast but cannot deliver large pressures. In practice, some clever designs push the pressures and response times by one order of magnitude, but the graph reveals very valuable rule-of-thumb trends. These values can be used as a starting point for the BioMEMS designer who is considering incorporating microvalves or micropumps, since some operating principles may be fundamentally incompatible with the pressures and response times required by a given application.

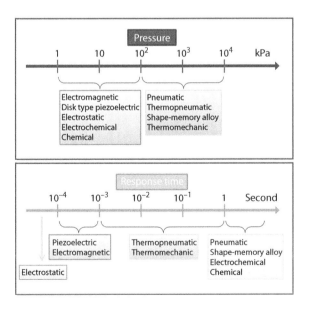

FIGURE 3.33 Pressure and response times typical of actuators used in microvalves and micropumps. (Figure contributed by Nam-Trung Nguyen.)

3.8.1 Microvalves

A large number of **microvalve** designs exist with great variability in performance. To compare all the designs, it would be most valuable to produce a table with a set of benchmark measurements obtained with each design. Unfortunately, there is no standardized performance test against which all microvalve designs are routinely tested in the literature, and the performance of a given design often varies greatly depending on the particular assembling procedure, so here we will review the designs with a somewhat historical perspective and give some insight into their advantages and limitations.

3.8.1.1 Electrokinetic Valving

The earliest implementation of valving in a microfluidic channel is credited to a group led by Andreas Manz, then working at Ciba-Geigy in Basel, Switzerland; in 1993, they reported in *Science* how electro-osmotic flow can be used to quickly switch fluids from one channel to another in a capillary electrophoresis system. **Electrokinetic valving** is best described as a "router," a valving scheme that only works with continuous flow. Many others rapidly followed, such as Michael Ramsey's group at Oak Ridge National Laboratory. In the case of **Figure 3.34**, the sample fluid and the buffer are both attracted by the anode (550 V), so at an intersection, they each turn 90 degrees; when the buffer is left floating, however, the buffer fluid does not flow and the sample fluid is free to proceed straight (as well as turn); as soon as the buffer is connected to the cathode again, the buffer overtakes the channel, resulting in a small plug whose size is a function of the duration of the voltage pulse and the flow rate.

Electrokinetic valving is now only used in a very specific set of applications (mainly, capillary electrophoresis, for which electro-osmotic flow is well suited) because of serious drawbacks: (1) as with all electrokinetic transport (see Section 3.3), it is strongly determined by the surface properties of the channel (in practice, it only works most reliably with glassy surfaces, which are difficult and expensive to micromachine); (2) it is also strongly influenced by the ionic composition of the buffer: at pH > 3, the hydroxyl groups on the walls dissociate, creating negatively charged walls (with pH-dependent charge)—it is the loosely bound positive ions next to the wall that the voltages move; (3) it requires high-voltage sources and switches that are cumbersome, expensive, and unsafe; and (4) as noted previously, this type of valving requires continuous flow to work—what "closes" the "valve" is the flow of buffer—which continuously wastes sample while the valve is closed.

Note that, under neutral pH typical of biological solutions, in the previous device it is the positive ions closest to the (negatively charged) wall which contribute the most to the mass

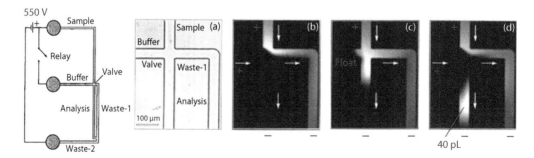

FIGURE 3.34 Electrokinetic valving. (From Stephen C. Jacobson, Sergey V. Ermakov, and J. Michael Ramsey, "Minimizing the number of voltage sources and fluid reservoirs for electrokinetic valving and microfluidic devices," *Anal. Chem.* 71, 3273–3276, 1999. Reprinted with permission of the American Chemical Society. Figure contributed by Michael Ramsey.)

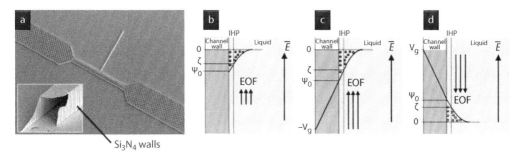

FIGURE 3.35 Flow "FET." (From Richard B. M. Schasfoort, Stefan Schlautmann, Jan Hendrikse, and Albert van den Berg, "Field-effect flow control for microfabricated fluidic networks," *Science* 286, 942, 1999. Figure contributed by Albert van den Berg.)

transport. How about if we artificially charge the wall positively? This is the idea, developed by Albert van den Berg and colleagues at the University of Twente in the Netherlands, which evolved into the "flow **field-effect transistor**" (FET; Figure 3.35), so-called because its operation is reminiscent of the operation of a **FET**. In a traditional (dry) FET, an electrode called the *gate* is used to modulate the current between the other two electrodes, called the *source* and the *drain*: below a certain voltage applied to the gate, no current is allowed through the source–drain circuit. In flow-FET, the gate is an area of the microchannel wall (made of Si_3N_4 instead of glass, for ease of micromachining) that is very thin and that is coated with a metallic strip from the outside. The voltage (positive or negative) applied to the strip determines the sign of the surface-immobilized charges and, thus, the direction of the EOF (Figure 3.35) without having changed the driving electric field ($\bar{E}$ in Figure 3.35). Although automated switching of the driving field requires expensive high-voltage electronics, automated switching of the surface potential can be done with a standard input/output computer card.

3.8.1.2 Centrifugal Microvalves ("Lab-CD")

To circumvent the high-voltage requirement of electrokinetic valving, Gregory Kellogg's group of Gamera Bioscience in Boston introduced in 1998 an original idea: to make the channels rotate so that the fluid is forced through channel constrictions by centrifugal forces; the constrictions thus essentially act as **centrifugal microvalves** (Figure 3.36). This idea was further developed by one of Kellogg's students, Marc Madou, at Ohio State University (now at University of California, Irvine). The device layout is constrained by its own operation requirement (rotation), but by clever design it is possible to sequentially fill microchambers while the disk-shaped ("CD") device is set to spin and stop and spin again (in both directions). The "**Lab-CD**" is usually made of polycarbonate or PMMA onto which holes and trenches can be milled with high precision, although PDMS prototypes also exist. The biggest hurdle for the generalization of this design seems to be its poor compatibility with standard analytical and microscopy equipment although, lately, the technique has received a renewed thrust from a number of groups. Examples of applications demonstrated on this platform include enzyme-linked immunosorbent assays (ELISA), cell lysis, high-throughput screening, PCR, alcohol assay, protein crystallization, and DNA extraction from whole blood, among others.

3.8.1.3 Check Microvalves

Check valves are mechanical devices that allow fluid to flow in only one direction; the mechanical flap that impedes reverse flow forms a seal with the valve seat—and the seal only gets better

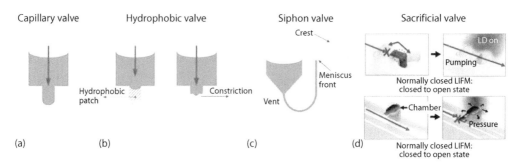

Capillary valve   Hydrophobic valve   Siphon valve   Sacrificial valve

(a)   (b)   (c)   (d)

FIGURE 3.36 Microvalving strategies used in centrifugal microfluidics. (From Robert Gorkin, Jiwoon Park, Jonathan Siegrist, Mary Amasia, Beom Seok Lee, Jong-Myeon Park, Jintae Kim, Hanshin Kim, Marc Madou, and Yoon-Kyoung Cho, "Centrifugal microfluidics for biomedical applications," *Lab Chip* 10, 1758–1773, 2010. Reproduced with permission from The Royal Society of Chemistry.)

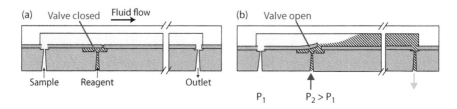

FIGURE 3.37 Flap microvalve. (From Joel Voldman, Martha L. Gray, and Martin A. Schmidt, "An integrated liquid mixer/valve," *J. Microelectromech. Syst.* 9, 295, 2000. Figure contributed by Joel Voldman.)

if reversed pressure is increased. The only way to reopen the valve is to make the downstream pressure lower than the upstream pressure (plus whatever force, e.g., a spring or gravity, is pressing the valve against the seat). Check valves can be implemented on the micron scale using traditional silicon micromachining techniques. A team at MIT led by Martin Schmidt developed a micromechanical flap integrated at the end of a silicon nozzle (see **Figure 3.37**) to control biochemical reactions of two compounds. The nozzle led to a reaction chamber, which had another input nozzle (for the second reactant) and one output. When the first reagent was pressurized at a pressure higher than the chamber pressure (determined by the pressures at the outlet and at the other inlet), the flap would deflect upward (**Figure 3.37b**), opening the valve and producing the chemical reaction in the chamber.

Micromachined **check microvalves** are expensive to develop but extremely inexpensive to commercialize (i.e., batch manufacturing results in low cost per device). Issues of concern are cost of disposal (must be disposed of with sharp objects such as needles) and sensitivity to dust (the valve seat being a hard material, the seal is unforgiving to contamination, so the fluids must be carefully filtered). This device highlights a common problem in the miniaturization of microfluidic assays: the difficulty of metering micron-scale volumes. In macro-scale assays, this problem does not exist because solution volumes are straightforwardly measured with a ruler (by eye) or with a balance—pressure and flow imbalances worry no one in the macro-world.

3.8.1.4 PDMS Microvalve: The Pinch Valve Design or "Quake Microvalve"

Pinch valves use mechanical force to reduce the diameter of a section of the elastomeric tubing through which the fluid is flowing. In 2000, Stephen Quake's group, then at Caltech,

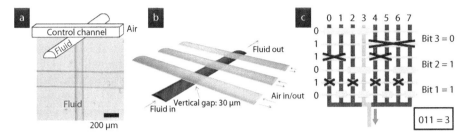

FIGURE 3.38 PDMS pinch microvalves by Quake. (a and b; from Marc A. Unger, Hou-Pu Chou, Todd Thorsen, Axel Scherer, and Stephen R. Quake, "Monolithic microfabricated valves and pumps by multilayer soft lithography," *Science* 288, 113, 2000; c, from Todd Thorsen, Sebastian J. Maerkl, and Stephen R. Quake, "Microfluidic large-scale integration," *Science* 298, 580, 2002. Figure contributed by Stephen Quake.)

devised a way to "pinch" a PDMS microchannel using a second air-filled microchannel on top of the fluid-carrying one and separated by a thin (~30 μm) PDMS gap (Figure 3.38a). The air-filled microchannel is often termed the "control channel." Application of pressure to the control channel pinches the fluid-carrying microchannel only at the location in which the two channels overlap. (For applications in which the high permeability of PDMS to air is a concern in the sense that air could be injected from the control channel into the fluid-carrying channel and nucleate a bubble, filling the control channel with water works just as well.) Large numbers of valves can be operated simultaneously by pressurizing their control channels in parallel or in sequence (or both), opening the possibility of integration of micropumps (Figure 3.38b).

The fabrication of these microvalves, popularly known as "**Quake microvalves**," involves two clever microfabrication tricks that, paradoxically, are also their Achilles heel in cellular applications. First, the microchannel roof is fabricated to have a rounded cross-section—the fluid-carrying microchannel *must have* a rounded cross-section, otherwise the valve does not seal well when the channel is pinched. To make rounded photoresist features, the master is exposed to high temperatures to melt the photoresist, a trick known as "photoresist reflow": the surface tension of the photoresist maintains the width of the photoresist features but evenly spreads their volume (like one sees in a droplet). Second, PDMS is dispensed *while the master is spinning* so as to create a thin layer of PDMS, barely covering the photoresist pattern with a membrane. On top, a block of PDMS featuring one or more PDMS channels is bonded with the PDMS channels facing the membrane side. The areas in which the air-filled and the liquid-filled microchannels overlap significantly constitute a valve (Figure 3.38a). The valves can be operated (depending on their size) at frequencies up to hundreds of hertz if the air pressures are switched via miniature computer-controlled solenoid valves (cylindrical ~1 cm high, ~0.5 cm diameter, commercially available at a low cost for a wide range of applications).

Note the need for this "significant overlap" between the control channel and the fluid-carrying channel. This means that, for any given pressure, a wide control channel may be able to pinch the microchannel below whereas a narrow control channel might not—an advantageous feature for "wiring" a control channel over inactive microchannels. Based on this concept, the Quake group presented a **microfluidic multiplexer** which allows for selecting one of many inlets using a number of control channels that is inferior to the number of inlets. Instead of dedicating one control channel to opening and closing each fluid-carrying microchannel, here, each control channel is used to valve many (to be precise, half) of the fluid-carrying microchannels (Figure 3.38c). A more detailed mathematical treatment of various multiplexer designs is shown in Section 3.8.3.

The Quake microvalves have the following limitations:

Pattern design versatility: The microchannel height and width cannot be designed as separate parameters because the photoresist reflow process melts the whole volume of the photoresist. This means that lines of different width (but originally designed to be the same height) end up having different heights, and lines designed to be of different heights (but the same width) end up having very similar heights. Although devices with microchannels of various heights are possible (e.g., optimizing the photoresist reflow), valving of the shallowest channels is difficult because it involves bending a wider gap of PDMS (because the master is evenly covered by spin-coating with PDMS).

Microscopy: The rounded roof produces a rounded water–PDMS interface that acts as a lens, which severely distorts images in transmitted-light microscopy modes such as phase-contrast microscopy—the most prevailing mode of observation of live cells in culture to date. In practice, the only viable cell imaging option is fluorescence microscopy (in which light is emitted by the sample, so the light is not required to go through the sample if using an inverted microscope). Unfortunately, fluorescence microscopy comes at a cost—in terms of supplies (dyes) and equipment (UV lamp and filters) and because it is limited by the amount of fluorescence signal it can collect, as the signal fades with time and with light exposure (photobleaching).

Fluid dynamics modeling: Unlike for rectangular cross–sections, the equations governing flow in a rounded roof channel do not have an analytical solution, and thus finite-element modeling is the only option to predict shear stress, flow resistance, and others. The cross-section of the microchannel needs to be characterized for each pattern design if finite-element modeling is to be undertaken. Even then, the curved geometry of the channel is not straightforward to input into most modeling packages, which are standardized for rectangular walls and circular pipes. Thus, modeling of flow in microchannels made by photoresist reflow is, at the very least, cumbersome and highly computation-intensive for complex channel architectures. For microfluidic cellular studies, in which knowledge of shear stress is paramount, not performing such flow simulations can amount to walking barefoot on gravel in the dark—it can get painful.

Metering: The volume of a chamber created between two valves is difficult to determine (although it *can* be characterized for each device by fluorescence microscopy). This volumetric uncertainty can be important when using the chambers as chemical reactors, although it is possible to implement a metering scheme.

Portability: The valves are open at rest, requiring energy to close. This implies that the device cannot be "unplugged" and transported to another location (for example, a microscopy facility) if the separation between volumes is to be maintained (e.g., in chemical reactions).

In sum, the Quake microvalves are not optimal in terms of microscopy, compatibility with cells, modeling, metering, and portability. Nevertheless, despite their limitations, without any doubt, they have revolutionized the field of microfluidics.

3.8.1.5 PDMS Microvalve: The "Doormat" Design

In 2000, just three months after Stephen Quake submitted his manuscript to *Science* and while it was still in press, Kazuo Hosokawa and Ruytaro Maeda of Japan's AIST (National Institute of Advanced Industrial Science and Technology) in Tsukuba submitted an entirely different PDMS microvalve design (**Figure 3.39**). Here, the PDMS membrane is contact-transferred as a film, "sandwiched" between the control channel and the fluidic channel. The valve is then formed by a small pad of PDMS membrane suspended over a control channel and it is positioned *under* the

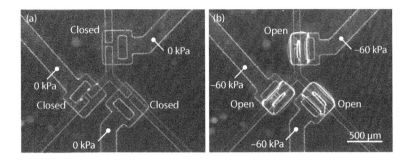

FIGURE 3.39 PDMS "doormat" microvalves. (From Kazuo Hosokawa and Ryutaro Maeda, "A pneu-matically-actuated three-way microvalve fabricated with polydimethylsiloxane using the membrane transfer technique," *J. Micromech. Microeng.* 10, 415, 2000. Reproduced with permission from the Institute of Physics.)

wall separating two microchannels (hence the name "doormat"). As negative pressure is applied to the control channel, the membrane pad deflects downward and the two sides of the micro-channel communicate under the wall (Figure 3.40). This design can be fabricated entirely with SU-8 photolithography, bypassing the need for photoresist reflow and allowing for microchan-nels with rectangular cross-sections.

An example of its operation is shown in Figure 3.40. When the membrane is not deflected (oval pads in Figure 3.40), the valve is closed and no mass transport of red dye occurs (i.e., the valve is closed at rest). Application of vacuum to the control line deflects the membrane down-wards, allowing mass transport (be it convective transport, i.e., flow, or diffusive transport, or both).

The doormat design has none of the aforementioned limitations of the Quake microvalve. Because the valve is actuated from the channel floor, it can actuate both deep and shallow chan-nels. The valve design benefits from all the advantages of SU-8 microfluidics, namely: (a) the channel height is specified independently of its width; (b) the channels have rectangular cross-sections (greatly facilitating fluid dynamics modeling—both analytical and numerical); and (c) the roofs are flat (allowing for phase-contrast microscopy of cells within the channels). Microchambers enclosed by doormat microvalves have volumes that are defined only by pho-tolithography (not by a deformed PDMS wall), so metering is limited simply by the resolution of the photolithographic process and, because the SU-8 sidewalls are straight, calculation of the volumes is trivial (Figure 3.40). Finally, because the valves are closed at rest, the devices

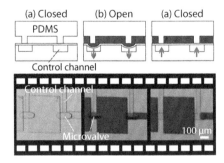

FIGURE 3.40 Metering of nanoliter volumes with PDMS "doormat" microvalves. (From Nianzhen Li, Chia-Hsien Hsu, and Albert Folch, "Parallel mixing of photolithographically defined nanoliter volumes using elastomeric microvalve arrays," *Electrophoresis* 26, 3758–3764, 2005. Figure con-tributed by Greg Boggy.)

can be unplugged and transported to a different location with all the fluids safely stored in microchambers.

Note that the doormat microvalve is limited to a monolithic PDMS architecture—unless the valves are implemented on a different level. This mandate for an "all-PDMS architecture" can be a hurdle for biological applications, especially cell culture studies, in which it is desirable to have a glass surface. Glass offers better control over the surface chemistry, has superior optical clarity, and is rigid even when it is very thin (an advantage for high-resolution optical microscopy). The Quake microvalve, on the other hand, can be used either as a stand-alone device, that is, it can be *applied to a surface*, or be fabricated entirely in PDMS. Also, devices that can be applied to a surface can, in principle, be used to deposit patterns on the surface. The Mathies laboratory has reported a doormat microvalve design made in PMMA (except the PDMS membrane); to bond the PDMS membrane to the PMMA layers, both the PMMA and the PDMS surfaces were exposed to a UV ozone treatment, which generated –OH groups on their surface.

3.8.1.6 PDMS Microvalve: The "Sidewall Design"

Andrew Berlin and colleagues at the University of Wisconsin (Madison) have presented a PDMS valve design that places the pneumatic actuators on the lateral walls of the microchannels (**Figure 3.41**). This design decision incurs a considerable cost in chip real estate, making high-density microvalve arrays very difficult, but enables the fabrication of valves with a single-mask process. Here, one or more of the sidewalls of the microchannel is fabricated thin enough that, when pressure is applied on the other side, the wall deflects into the volume of the microchannel; this scheme cannot produce strong seals but, nevertheless, alternatively actuated "leaky" valves were shown to produce pumping.

3.8.1.7 PDMS Microvalve: The "Curtain Design"

The "curtain design" by the Toner group from Harvard University in 2006 overcomes the main limitation of the doormat microvalve design: here, the membrane constitutes the microchannel's roof (as in the Quake design), which lifts a wall like a curtain to open a normally closed channel (unlike the Quake microvalves). The curtain design is fabricated entirely in SU-8 and can be assembled as a stand-alone device on a glass surface (**Figure 3.42**), making it ideal for cellular applications. It is to be expected that, as the PDMS membrane in the curtain design must displace a larger mass than in the doormat design, the response time of the doormat design should be faster.

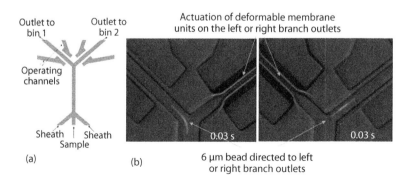

FIGURE 3.41 PDMS microvalves implemented in the sidewall. (From Narayan Sundararajan, Dongshin Kim, and Andrew A. Berlin, "Microfluidic operations using deformable polymer membranes fabricated by single layer soft lithography," *Lab Chip* 5, 350–354, 2005. Reproduced with permission from The Royal Society of Chemistry.)

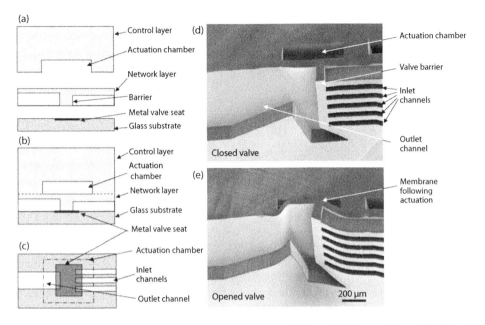

FIGURE 3.42 PDMS "curtain" microvalves. (From Daniel Irimia and Mehmet Toner, "Cell handling using microstructured membranes," *Lab Chip* 6, 345, 2006. Reproduced with permission from The Royal Society of Chemistry.)

3.8.1.8 PDMS Microvalve: The "Plunger Design"

If you have peeked in a toilet tank while you pull on the flush lever, you will have noticed how, in most designs, it pulls a valve which looks like a radially symmetric plunger that is sealed against a circular orifice. Sang Hoon Lee and coworkers of the University of Korea seem to have found inspiration in macro-world components of this type when they designed a microfluidic valve that works in a similar way (Figure 3.43)—except that the force that closes the valve is

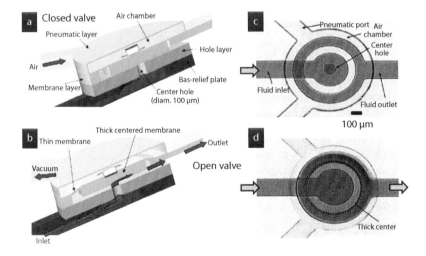

FIGURE 3.43 PDMS "plunger" microvalve. (From Ju Yeoul Baek, Ji Young Park, Jong Il Ju, Tae Soo Lee, and Sang Hoon Lee, "A pneumatically controllable flexible and polymeric microfluidic valve fabricated via in situ development," *J. Micromech. Microeng.* 15, 1015, 2005. Reproduced with permission from the Institute of Physics. Figure contributed by Sang Hoon Lee.)

not exerted by the weight of the water on top of the valve (as in the toilet tank) but by a PDMS membrane, which also acts as the seat of the valve.

TO SEAL OR NOT TO SEAL

THE ADVANCED STUDENT MIGHT HAVE NOTICED an exasperating fabrication challenge, common to all closed-at-rest PDMS valves, but particularly acute in the curtain design: how can one irreversibly seal the device (to avoid leaks) without sealing the valve? In the doormat design, the challenge is simply solved by applying vacuum to the valve during sealing (the surface oxidated state of PDMS is lost in about 30 minutes of keeping the valve open in air), but in the curtain design, this trick may not be possible if a portion of the sidewall is lifted when the end wall is lifted (as in **Figure 3.42**). One possibility around this problem is to limit the lifting to only a portion of the end wall, but this option will generally require larger actuation pressures and may not be possible (there is a limit, yet to be determined, as to how much suction can be applied from an air-filled PDMS microchannel, because PDMS itself is full of air and permeable to air). The other option is to limit the bonding by micropatterning an inert pad on the surface (such as a metal pad, as in **Figure 3.42**); this pad defines the area of PDMS that can be lifted (and not the lateral dimension of the control channel as **Figure 3.42** seems to imply). In the fabrication of the plunger design, to avoid bonding at the valve seat, salt (yes, NaCl) and poly(vinyl alcohol) was dried onto the valve—it was removed after the bonding step by applying water. In doormat designs, whether the membrane overlaps with both the end wall and the sidewall or only with the end wall (e.g., **Figure 3.40**) is irrelevant to the existential question of the seal. The Takayama group has recently reported a clever trick that consists of deactivating selected oxygen plasma–activated areas with (yes, you guessed it) a PDMS-wetted stamp! This way, the device bonds only on the areas not contacted by the "deactivator" stamp.

3.8.1.9 PDMS Valve: Latch-On Design

In computer memory chips, each memory unit in a given row is addressed by the *same physical wire* while the column wires read or write all the units on the same row. This information storage/retrieval scheme is possible because the information can be "latched" onto the physical medium (with electrical charges), saving an enormous amount of space in wiring real estate. It has been difficult to build latches in microvalves: a control channel cannot switch the state of a subset within a larger set of microvalves without switching the state of the whole set. (Note that the multiplexers that we will see in Section 3.8.3 solve a different problem—they minimize the number of *control channels* required to open/close all the fluidic channels, but at the expense of increasing the total number of *valves*.) In an effort in this direction, Mathies and colleagues have devised a scheme implemented on a PDMS membrane with a doormat design that adds two additional valves to the latching valve: one to hold the valve open (continues to supply vacuum) and one to help it close (to break the vacuum), as schematized in **Figure 3.44**. In essence, this scheme adds one or more nonlatching valves to hold the state of the valve. The duration that the valves stay latched are, unfortunately, on the order of minutes—with PDMS, this duration will never be infinite because PDMS is permeable to air so the walls leak: it is not possible to form a hermetic seal that holds a pressure differential.

A different approach based on the thermal expansion of a polymer introduced in the control channel cavity may be more promising (see Section 3.8.1.10) because the wires for heating occupy very little real estate on the chip and very little power (so both states can be considered "latched").

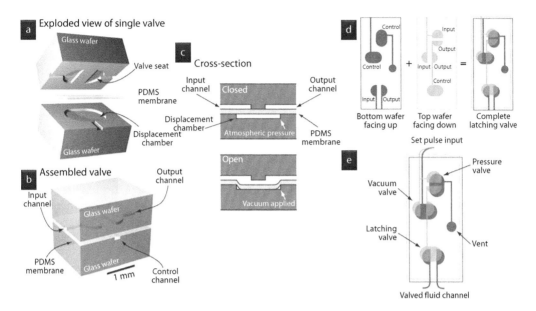

FIGURE 3.44 Latching microvalves. (From William H. Grover, Robin H. C. Ivester, Erik C. Jensen and Richard A. Mathies, "Development and multiplexed control of latching pneumatic valves using microfluidic logical structures," *Lab Chip* 6, 623, 2006. Reproduced with permission from The Royal Society of Chemistry.)

3.8.1.10 Electrically Actuated PDMS Microvalves

Note that all the previous designs require the connection of the device to a vacuum supply (and sometimes also a pressurized air source), so they are not portable while functioning. Ideally, one would like to generate the pressure with minimal power and using electrical circuits, which can be readily patterned and addressed with computers. With that in mind, Christopher Backhouse's group at the University of Alberta in Canada has introduced a temperature-sensitive polymer, which could be PEG or wax, into the control channel of a doormat design valve (**Figure 3.45**). To flow, the polymer has to be introduced in the molten phase (expands in volume), so when it cools down to room temperature it opens the valve. Upon local heating of the valve seat, the polymer in the control channel expands, causing closure of the valve. Note

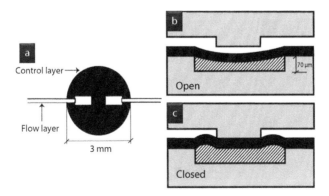

FIGURE 3.45 Microvalve based on thermal expansion of PEG. (From Govind V. Kaigala, Viet N. Hoang, and Christopher J. Backhouse, "Electrically controlled microvalves to integrate microchip polymerase chain reaction and capillary electrophoresis," *Lab Chip* 8, 1071, 2008. Reproduced with permission from The Royal Society of Chemistry.)

that here the doormat design is open at rest, which may be an interesting variation on its own for certain applications.

For biochemical reactors (e.g., PCR devices) or cell culture systems, which are operative only on a very limited range of temperature, heat-based actuation naturally raises serious concerns. These concerns may be addressable by placement of the valves away from critical sites and by designing heat sinks.

3.8.1.11 PDMS Microvalves Actuated by "Braille" Pins

These valves also belong to the pinch valve class, but with two key differences compared with the PDMS valves reviewed previously: (a) instead of being based on the deformation of a thin PDMS membrane next to the fluid-carrying microchannel, they are based on the deformation of the bulk of PDMS that forms the device; and (b) instead of being based on the remote generation of pressure—which keeps the Quake, doormat, curtain, and plunger designs tethered to a vacuum-supplying wall—they are based on the local generation of pressure via mechanical pins. An inexpensive, easily programmable source of pins is found in Braille displays universally used by the blind to communicate, hence, the name of the valves. One limitation is that the pinching points cannot be very close, with the smallest distance ultimately dictated by the thickness of the device (even if pin arrays smaller than Braille displays can be fabricated). However, the pinching points need not be the valving points. As shown in **Figure 3.46**, each pin presses onto a liquid-filled reservoir (~150 µm high, ~900 µm diameter), which acts as a "piston" that transmits the pressure to a membrane-based pinch valve (in this case, a Quake microvalve) at a remote location, even centimeters away. **Figure 3.46c** shows a top-down view of four intersections of pressurized control (red) and fluidic (green) microchannels (9 µm high and 100 µm wide except for the 40-µm-wide control channel on the bottom right, which is not pinched because it is narrower.

This "remote hydraulic piston" could, in principle, be implemented with actuation stimuli other than displacement, such as light or temperature, that can also cause expansion and retraction in certain materials but are often cumbersome to apply at the site of the valve.

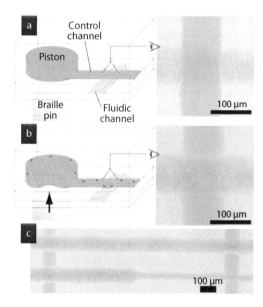

FIGURE 3.46 Braille-actuated microvalves. (From Wei Gu, Hao Chen, Yi-Chung Tung, Jens-Christian Meiners, and Shuichi Takayama, "Multiplexed hydraulic valve actuation using ionic liquid filled soft channels and Braille displays," *Appl. Phys. Lett.* 90, 033505, 2007. Reprinted with permission from the American Institute of Physics. Figure contributed by Shuichi Takayama.)

3.8.1.12 Smart Polymer Microvalves

There are polymers that undergo reversible phase transitions (such as a change in volume or solubility) on application of a stimulus (such as a change in pH, temperature or light); as the state is reversed on removal of the stimulus, these polymers can be used as a sensor for that particular stimulus, hence, they have been marketed as "**smart polymers**." (Water, with its transition to ice, is just as smart in that regard, but it is not a polymer.) In 2000, a team led by Jeffrey Moore and David Beebe, then at the University of Illinois at Urbana-Champaign, was able to photopolymerize smart-polymer features inside a microfluidic device (Figure 3.47), thus creating functional microarchitectures that change the flow patterns in response to the stimulus. Step responses of 8 seconds to pH changes were demonstrated with a variety of devices, including valves. Polymer posts that expand in diameter at low pH (~5, obstructing the channel) and contract at high pH (~8, allowing flow) could be fabricated in one channel, whereas posts made of a polymer of inverse behavior could be fabricated in a branching channel, thus acting as a sorter of flow depending on pH. As the photopolymerization process is generally applicable to many polymers, it is not difficult to envision other types of "smart valves."

To demonstrate light-induced switching (i.e., an optomechanical valve), the Beebe laboratory teamed up with Jennifer West's laboratory at Rice University. The West laboratory provided nanoparticles with distinct and strong optical absorption profiles (gold colloids and nanoshells), which, when mixed with a thermally responsive polymer (poly[N-isopropylacrylamide-co-acrylamide]), formed a hydrogel that collapsed/swelled in response to green light (the gold-colloid nanocomposite hydrogel) or near-infrared light (the gold-nanoshell nanocomposite hydrogel). As shown in Figure 3.48, when hydrogel microstructures were fabricated around cylindrical posts inside a microfluidic T-junction, switching of a fluid stream at the "T" was observed within approximately 5 seconds when the light source was switched from 532 nm (green) to 832 nm (near-infrared).

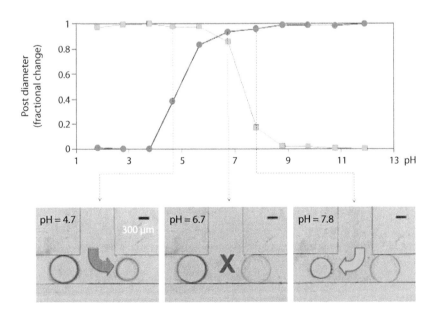

FIGURE 3.47 Smart-polymer microvalves. (From David J. Beebe, Jeffrey S. Moore, Joseph M. Bauer, Qing Yu, Robin H. Liu, Chelladurai Devadoss, and Byung-Ho Jo, "Functional hydrogel structures for autonomous flow control inside microfluidic channels," *Nature* 404, 588–590, 2000. Adapted with permission from the Nature Publishing Group. Figure contributed by David Beebe.)

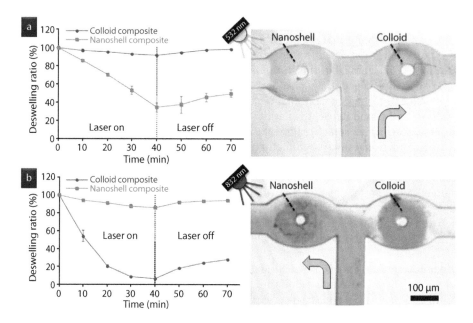

FIGURE 3.48 Optomechanical microvalves. (From Scott R. Sershen, Glennys A. Mensing, Marie Ng, Naomi J. Halas, David J. Beebe, and Jennifer L. West, "Independent optical control of microfluidic valves formed from optomechanically responsive nanocomposite hydrogels," *Adv. Mater.* 17, 1366–1368, 2005. Adapted with permission from John Wiley and Sons.)

3.8.1.13 Single-Use Microvalves

For some applications, to ensure that the valve never closes again, it may be beneficial to design single-use valves that are closed at first, open once, and stay open forever afterwards. In 1999, Robert Langer and colleagues at MIT designed a silicon-based "drug-release" microchip containing etched cavities that were capped with "sacrificial" gold membranes (**Figure 3.49**). The cavities could be filled with drug-containing solutions before being sealed with the membranes. The membranes could be electrochemically dissolved on-demand with the application of a small voltage (~1 V) pulse, so that each voltage pulse controlled the release of one dose (~25 nL) of "drug" for each cavity present in the microchip.

Sacrificial membranes composed of SU-8 and silicon nitride have also been reported. As with Langer's sacrificial gold membrane, these microvalves are opened when heat or electrical current is applied to the sacrificial membrane, resulting in a change in membrane integrity and rupture of the membrane. As the microvalve is opened, debris from the disintegration of the sacrificial membrane may be released into the fluid, potentially contaminating the system. In addition, actuation of the microvalve may induce localized heating near the electrodes if power and timing are not properly calibrated.

A sacrificial physical barrier may not be needed if all that is needed is a single-use microvalve—surface tension and wetting may do the job. "**Capillary burst microvalves**" passively control fluid flow by increasing capillary resistance inside the microchannel. This is typically accomplished by an abrupt change in the geometry or surface chemistry of the microchannel. The high surface energy as a result of the abrupt channel change traps the liquid meniscus at the interface of the valve. For example, as seen in "Lab-CD" platforms (see **Figure 3.36** in Section 3.8.1.2), a straight microfluidic channel may empty into a valve area with diverging sidewalls. The fluidic resistance then greatly increases at the interface between the straight channel and the angled valve sidewall, pinning the liquid meniscus at the straight capillary channel. The liquid meniscus forms a bulge into the valve area characterized by the equilibrium contact angle, θ_e,

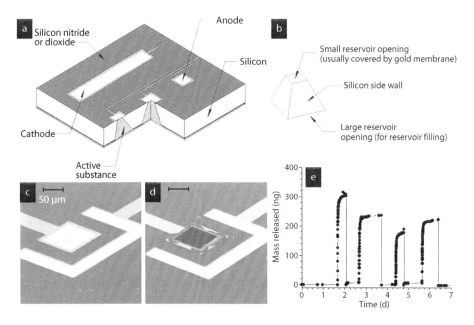

FIGURE 3.49 Sacrificial membranes for single-use microvalves. (From John T. Santini Jr., Michael J. Cima, and Robert Langer, "A controlled-release microchip," *Nature* 397, 335–338, 1999. Reprinted with permission from the Nature Publishing Group.)

with respect to the straight channel, which is dependent on the surface tension of the meniscus. Applying a driving force (such as pneumatic pressure or centripetal force) to the fluid channel will increase θ_e. When θ_e exceeds the advance critical contact angle, θ_A, the liquid meniscus bursts, thereby opening the valve (**Figure 3.50**). The driving pressure needed to open the valve can be derived from the Young–Laplace equation. Once opened, capillary microvalves cannot be recovered unless controlled slug volumes are used, such that the valve oscillates between filled and unfilled states.

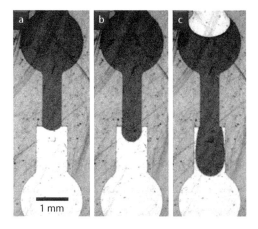

FIGURE 3.50 Capillary burst microvalve. (From Hansang Cho, Ho-Young Kim, Ji Yoon Kang, and Tae Song Kim, "How the capillary burst microvalve works," *J. Colloid Interface Sci.* 306, 379–385, 2007. Reprinted with permission of Elsevier.)

3.8.1.14 "Valve-Less" Approaches

It is possible to route the passage of fluids in many other ways, even *without* valves (at some cost). It is worth noting that the first demonstration of valving in a PDMS device was done in 1999 using…bubbles! Isao Endo's group from RIKEN in Saitama (Japan) used dedicated microchannels to inject air into the fluid-carrying channel so as to obstruct the passage of fluids on demand—an approach that is now considered impractical because the "bubble" is much more stable when it is covered by a PDMS membrane. Many other designs of valves have been invented and many more will continue to be invented, some with less applicability to biomedical applications than others. For example, bubbles can be thermally generated with microfabricated resistors to obstruct flow (albeit with low flow resistance) and reabsorbed to allow flow; however, changes in the concentrations of dissolved gases can affect cells and pH. There is a whole industry that produces micropumps based on piezoelectric elements for inkjet printers—they have been tried for printing protein solutions but their nozzles tend to clog (maybe people have not tried hard enough). Miniature screws have been used to pinch flow in predetermined locations, an approach that has high portability and low cost (two precious features for deploying diagnostic assays in resource-poor settings) but does not allow for automation. For applications that can afford large volumes of reagents, it may be more practical to switch flow…using flow! (a very old trick). Most often, these "flow valves" consist of using one stream to control the position of another stream on the same plane. Whitesides and colleagues have experimented with juxtaposing a stream *on top* of another one at a "crossroads" (**Figure 3.51**); these "tangential microchannels" behave in curious ways depending on their aspect ratio ("*A*" in **Figure 3.51**) and can be operated as fluidic switches.

Recently, Rustem Ismagilov's group at the University of Chicago demonstrated an ingenious scheme to move fluids from one microchamber to another within microfluidic devices without using pumps or valves. The scheme, called the "**SlipChip**," works because the fluids to be mixed are in different planes, and the device is formed of two silanized glass plates that can slide along the dividing plane (**Figure 3.52**). A lubricating layer of fluorocarbon facilitates the relative motion of the two plates. No cross-contamination between wells is observed. The

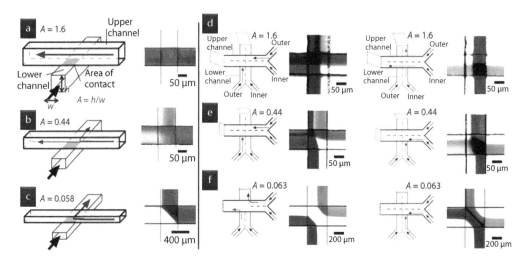

FIGURE 3.51 Tangential microchannels for switching flows. (From Rustem F. Ismagilov, David Rosmarin, Paul J. A. Kenis, Daniel T. Chiu, Wendy Zhang, Howard A. Stone, and George M. Whitesides, "Pressure-driven laminar flow in tangential microchannels: An elastomeric microfluidic switch," *Anal. Chem.* 73, 4682–4687, 2001. Reprinted with permission of the American Chemical Society. Figure contributed by George Whitesides.)

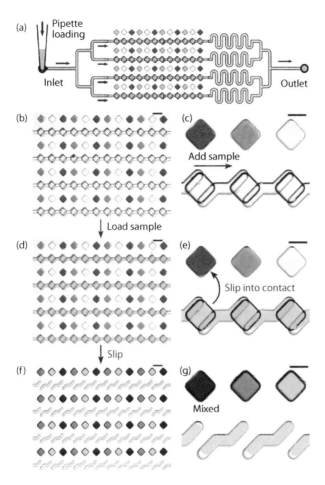

FIGURE 3.52 The SlipChip. Scale bars for (b), (d), and (f) are 500 μm and for (c), (e), and (g) are 250 μm. (From Wenbin Du, Liang Li, Kevin P. Nichols, and Rustem F. Ismagilov, "SlipChip," *Lab Chip* 9, 2286–2292, 2009. Reproduced with permission from The Royal Society of Chemistry.)

SlipChip has already been successfully used for protein crystallization screening and digital PCR.

3.8.2 Microfluidic Resistors

Fluid flow in microchannels, microvalves, and micropumps have many parallels with electrical current through wires, switches, and current sources of microelectronic circuits. It is surprising that the equivalent of a variable resistor, a very common electrical component, was not developed until fairly recently. In 2006, the author's laboratory presented "**microfluidic resistors**," microchannel segments whose height can be constricted to increase flow resistance (**Figure 3.53**). When the pneumatic lines are not pressurized (the microfluidic resistor is "inactive"), the top channels have a normal rectangular cross-section (**Figure 3.53a**). When the pneumatic lines are pressurized, the membrane deflects upward (the microfluidic resistor is "activated"), reducing the cross-sectional area of the microchannel to two narrow fluid paths at the top corners (**Figure 3.53b**). The outputs of a three-inlet mixer (**Figure 3.53c–e**) can be predicted using a purely resistive electrical circuit analogue.

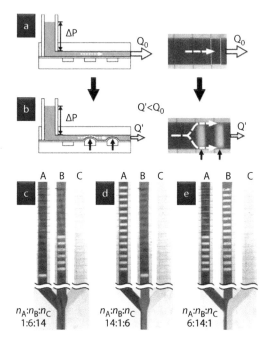

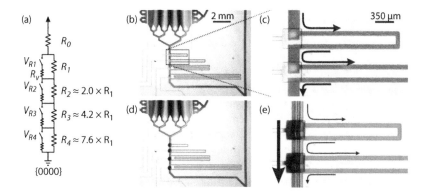

FIGURE 3.53 Microfluidic resistors using inflatable elements. (From Eric W. Lam, Gregory A. Cooksey, Bruce A. Finlayson, and Albert Folch, "Microfluidic circuits with tunable flow resistances," *Appl. Phys. Lett.* 89, 164105, 2006. Figure contributed by Eric Lam.)

A limitation of the design of the microfluidic resistors in **Figure 3.53** is that they need to be calibrated because it is not possible to accurately predict the flow resistance of the channel with deflected membranes (the uncertainties in both the deflected geometries and in the PDMS Young's modulus are too large). A more predictable design uses microvalves to divert flow (or not) through highly resistive channels (**Figure 3.54**). Depending on the number of side channels that are open, the circuit becomes more or less resistive; because all the channels are made with rectangular cross-sections, the flow resistances can be predicted with straightforward formulas (see **Equation 3.12**).

FIGURE 3.54 Microfluidic resistors using microvalves. (From Gregory A. Cooksey, Christopher G. Sip, and Albert Folch, "A multi-purpose microfluidic perfusion system with combinatorial choice of inputs, mixtures, gradient patterns, and flow rates," *Lab Chip* 9, 417–426, 2009. Reproduced with permission from The Royal Society of Chemistry.)

Controlling a large numbers of inlets with dedicated microvalves (i.e., one microvalve per inlet) requires, for each inlet, a separate line on the valve layer and an additional piece of tubing, thus taking up a large fraction of real estate in the device and adding additional tubes and switches connected to the device. As the number of valves grows, so does the number of channels required to control them. Microelectronic circuits face a similar problem with the number of transistors, each of which is addressed by three wires; typically, the state of the transistor is controlled by one of these wires (the gate). Electrical engineers, however, have invented wiring schemes termed *multiplexers* that allow for transistors to share wires, and the transistors only get activated if the *sum charge* of *two* wires gets activated—so they can be addressed in rows and columns, thus saving huge amounts of space in wiring. In 2002, Stephen Quake's group (then at Caltech) invented a similar concept that allowed for multiplexing microfluidic valves, although, unfortunately, nothing yet exists that is equivalent to the sum charge for microfluidics.

3.8.3.1 Multiplexer with Binary Valves

To reduce the number of valves necessary to control many inlets, Quake's group developed a clever binary multiplexer scheme, which can control N PDMS microchannels with only $X = \log_2 N$ pairs of pneumatic lines (or $2X$ total pneumatic lines). Using this scheme, more than 1000 flow channels could be controlled with only 20 pneumatic lines. In the Quake "binary" design, each pneumatic line is connected to a group of valves controlling half of the channels, and a "complementary" pneumatic line controls the other half of the valves/channels. (Although the words "valve" and "pneumatic line" are used interchangeably, strictly speaking, in a multiplexer, a pneumatic line opens a group of valves, each of which controls flow in a different channel.) The basic concept of Quake's multiplexer is schematized in Section 3.8.1.4 (Figure 3.38c). We recall that Quake's valves are essentially "binary" (on or off, using two pressure settings), and that the control channel pinches the fluid-carrying channel below only if it overlaps with it significantly. If the area of overlap is too small, the control channel cannot exert enough pressure. This feature is necessary for "wiring" a control channel over a fluid-carrying microchannel without actuating it. Note that the layout shown in Figure 3.38c is prone to cross-contamination, a problem for assays such as PCR which are sensitive to very small amounts of residue. A new design that eliminates cross-contamination has been proposed (Figure 3.55).

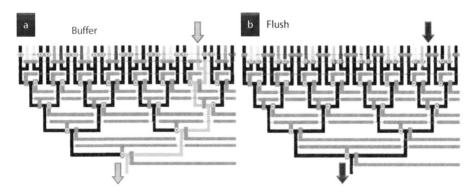

FIGURE 3.55 Binary multiplexer with "anticontamination" layout. (From Jessica Melin and Stephen R. Quake, "Microfluidic large-scale integration: The evolution of design rules for biological automation," *Annu. Rev. Biophys. Biomol. Struct.* 36, 213–231, 2007. Figure contributed by Stephen Quake.)

3.8.3.2 Combinatorial Operation of a Binary Multiplexer

The author's laboratory at the University of Washington in Seattle has extended the design of the Quake multiplexer to doormat microvalves and shown that its use can be broadened to create *combinations* of compounds. (In Quake's original implementation, only one input was selected.) This "combinatorial operation" is depicted in **Figure 3.56**. The device has 16 inlets (named A, B,...through P) controlled by eight multiplexing lines of valves (named V_1, V_2,...through V_8). Line V_1 must be opened to turn on any of the eight leftmost channels A through H and line V_2 must be opened to turn on any of the eight rightmost channels I through P; the V_1 and V_2 valves are said to be a "complementary pair" of valves; V_3 (which is necessary to open the first and last four channels) is complementary of V_4 (which is necessary to open the middle eight channels), and so on. The "state" of the multiplexer is denoted by the state of its valves, $V = \{V_1, V_2, V_3, V_4, V_5, V_6, V_7, V_8\}$, with V_i taking the values of 0 (closed) or 1 (open); note that, for clarity, a space is written between each pair of complimentary bits. To open a single inlet, one (and only one) valve from each complementary valve pair needs to be opened. For example, to open inlet *M*, valves V_2, V_3, V_6, and V_8 need to be opened, but V_1, V_4, V_5, and V_7 are "forbidden" and must be closed ($V = \{01\ 10\ 01\ 01\}$; **Figure 3.56b**).

It is crucial to note that the original implementation of Quake's multiplexer (i.e., selection of a single input) is based on respecting the "complementarity rule," in other words, none of the forbidden valves may be open. However, this operation is not a design constraint—it is only an operational constraint to ensure that the multiplexer is maintained in a "single-inlet state." If,

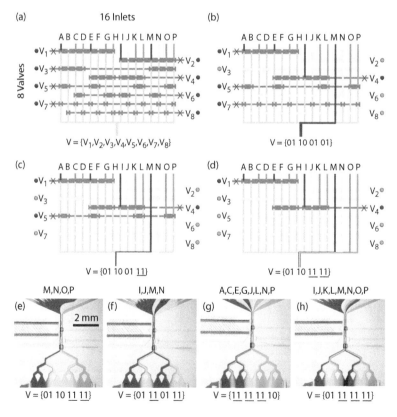

FIGURE 3.56 Combinatorial operation of a binary multiplexer. (From Gregory A. Cooksey, Christopher G. Sip, and Albert Folch, "A multi-purpose microfluidic perfusion system with combinatorial choice of inputs, mixtures, gradient patterns, and flow rates," *Lab Chip* 9, 417–426, 2009. Reproduced with permission from The Royal Society of Chemistry.)

in any single-inlet state, a previously closed forbidden valve is opened, then an additional inlet opens. For example, in the single-inlet state shown in Figure 3.56b in which inlet M is open ($V = \{01\ 10\ 01\ 01\}$), if forbidden valve V_7 is opened (i.e., the multiplexer state is changed to $V = \{01\ 10\ 01\ \underline{11}\}$) then inlet N opens, too (Figure 3.56c). As a rule, there will be two open inlets for any multiplexer state that has one complementarity violation. When there are two complementarity violations, then four inlets open; for example, in Figure 3.56d, $V = \{01\ 10\ \underline{11}\ \underline{11}\}$ opens inlets MNOP. With three complementarity violations, eight inlets open; and so on. Optical micrographs for several multiplexer states are shown in Figure 3.56e to h: $V = \{01\ 10\ \underline{11}\ \underline{11}\}$ (four inlets MNOP; Figure 3.56e), $V = \{01\ \underline{11}\ 01\ \underline{11}\}$ (four inlets IJMN; Figure 3.56f), $V = \{\underline{11}\ \underline{11}\ \underline{11}\ 10\}$ (eight inlets A,C,E,G,J,L,N,P; Figure 3.56g), and $V = \{01\ \underline{11}\ \underline{11}\ \underline{11}\}$ (eight inlets I through P; Figure 3.56h).

It is straightforward to see that there are (a) 32 possible two-inlet combinations; (b) 24 possible four-inlet combinations; (c) 8 possible eight-inlet combinations; (d) only one way of producing four complementarity violations, which is all eight valves (and consequently all 16 inlets) open; and (e) 16 possible ways of producing zero complementarity violations (single-inlet states). In sum, there are 81 possible combinations.

We emphasize that not all combinations of inlets can be selected and only combinations of $R = 2^n$ open inlets (with n = the number of complementarity violations) are possible with this multiplexer design (e.g., combinations of 3, 5, 6, 7, 9, 10, etc., inlets are not possible). But who needs more than 81 combinations of mixtures? For multiplexers having numbers of inlets other than 16 (and thus different requirements for the number of pneumatic lines), here, we provide a general expression for the number of possible combinations of R inlets (number of allowed "R-tuplets") that can be obtained with a multiplexer. Assume a multiplexer with $2X$ valves (i.e., X is the number of pairs of complementary valves), which can control up to $N = 2^X$ inlets. Thus, $R = 2^n$ with $n = 1 \ldots X$. Note that each pair of valves, for example, $\{V_1,V_2\}$, has two possible states that respect the complementarity rule, $\{V_1,V_2\} = \{1,0\}$ and $\{V_1,V_2\} = \{0,1\}$, and one state that violates it, $\{V_1,V_2\} = \{1,1\}$. From a set of X pairs of valves, there are obviously X possible ways of producing $n = 1$ complementarity violations ($R = 2$). In general, there are ^XC_n possible ways of producing n complementarity violations [i.e., of choosing n pairs in the (1,1) state], in which

$$^XC_n = \frac{X!}{(X-n)!n!}$$

For the $X - n$ pairs of valves that do not violate complementarity, there are 2^{X-n} possible states. Thus, the number of allowed R-tuplets is

$$^XC_n \times 2^{X-n}$$

where $R = 2^n$ and $n = 1 \ldots X$ is the number of complementarity violations.

A summary of the allowed combinations using this multiplexer design is shown in Table 3.4, along with the number of combinations that would be possible if each inlet were actuated by a unique valve (i.e., no multiplexing). Nevertheless, even with a moderate number of inlets, a large quantity of combinations becomes possible; for cell culture applications in which one wants to measure the response of cells to multiple factors (one in each inlet), addressing a large number but incomplete set of combinations is likely currently sufficient, given the complexity of analyzing the data from the responses of mixtures of just two factors.

3.8.3.3 Combinatorial Multiplexer

In 2006, an even more efficient multiplexing scheme was reported by Erdogan Gulari's group at the University of Michigan that uses only M pneumatic lines to address $M!/(M/2)!$ flow channels (Figure 3.57). The Gulari multiplexer is cunningly clever. To understand how it works, consider the schematic of Figure 3.57. Each fluidic channel (purple) is controlled by three valves, in ascending order: channel 1 by valves 1, 2, and 3, channel 2 by the next possible set up (valves 1, 2, and 4),

Table 3.4 Tabulation of the Number of Required Pneumatic Lines and Allowed Combinations for a Given Number of Inlets (C Denotes the Mathematical Symbol for Combinations: $^aC_b = a!/[(a-b)!b!]$)

No. of Inlets	No. of Required Pneumatic Lines (=2X)	No. of Allowed Pairs	Total Number of Possible Pairs	Number of Allowed Quadruplets	Total No. of Possible Quadruplets	Number of Allowed Octuplets	Total No. of Possible Octuplets
N	$2X = 2\log_2 N$	$X \times 2^{X-1}$	$^NC_2$	$^XC_{X-2} \times 2^{X-2}$	$^NC_4$	$^XC_{X-3} \times 2^{X-3}$	$^NC_8$
2	2	1	1	N/A	N/A	N/A	N/A
4	4	4	6	1	1	N/A	N/A
8	6	12	28	6	70	1	1
16 (Figure 3.56)	8	32	120	24	1820	8	12,780
32	10	80	496	80	35,960	40	1.05×10^7
64	12	192	2016	240	635,376	160	4.43×10^9
128	14	448	8128	672	1.07×10^7	560	1.43×10^{12}

channel 3 by the next one (valves 1, 2, and 5), channel 4 by valves 1, 3, and 4, and so on. One can see that applying pressure to any three of the six air (yellow) channels will block 15 of the 16 fluidic channels and leave only one channel open; conversely, if the six air channels are initially pressurized, removing pressure from any three of them will open a certain fluidic channel. Releasing pressure from additional air channels will open additional fluidic channels, which will give rise to mixtures. Using six pneumatic lines to select 16 inlets, it is possible to achieve 37 combinations (16 singles, 8 triplets, 6 quadruplets, 2 sextuplets, 4 nonuplets, and 1 with all 16 inlets on) compared with the 81 combinations of the multiplexer shown in the previous section (**Figure 3.56**). However, the Gulari design is very powerful in the number of single-inlet states that it can generate.

To visualize the number of combinations, it suffices to see the number of ways P in which M valves can be placed in N channels:

$$P = N!/[M!(N-M)!]$$

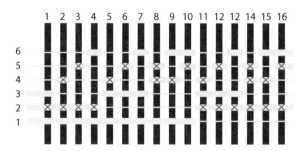

FIGURE 3.57 Combinatorial multiplexer. (From Zhishan Hua, Yongmei Xia, Onnop Srivannavit, Jean-Marie Rouillard, Xiaochuan Zhou, Xiaolian Gao, and Erdogan Gulari, "A versatile microreactor platform featuring a chemical-resistant microvalve array for addressable multiplex syntheses and assays," *J. Micromech. Microeng.* 16, 1433–1443, 2006. Figure contributed by Erdogan Gulari.)

Note that P reaches its maximum value when $M = [N/2]$, which is what the Gulari multiplexer design chooses by default. Thus, we have:

$$P = N!/[(N/2)!]2$$

The factorial scaling of this multiplexing algorithm is much more powerful than the logarithmic ($2 \log 2N$) scaling of the Quake multiplexing algorithm. For example, the Quake multiplexer requires 20 pneumatic channels to control 1024 fluidic channels, but the Gulari multiplexer can control a very similar number (924) of fluidic channels with only 12 pneumatic channels. As the number increases, the difference in performance is even more noticeable. With the Gulari multiplexer, 16 pneumatic channels can address up to $8!/(4!4!) = 12,870$ channels (!), whereas the Quake multiplexer can only address 256 fluidic channels with the same number of pneumatic channels.

3.8.3.4 Multiplexers with Ternary and Quaternary Valves

We have seen that the implementation of the Quake multiplexer itself is based on realizing that a very small area of overlap results in negligible valve deflections. In general, in the Quake microvalves, the area of overlap between the fluidic channel and the pneumatic channel is usually a constant for all the valves in the device—but it does not need to be so. Young-Ho Cho's group from the Korea Advanced Institute of Science and Technology recently demonstrated Quake microvalves with different depths/widths, each requiring different pressures to actuate (**Figure 3.58**). The scheme works well because the deflection of the PDMS membrane is rather nonlinear and each valve has a well-defined threshold. Valves opening at 50, 100, or 150 kPa were designed. At the highest pressure setting (150 kPa), a pneumatic line controlling three differently thresholded

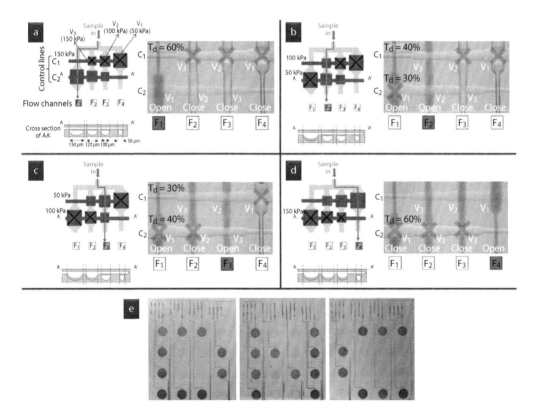

FIGURE 3.58 Multiplexer with quaternary valves. (From Dong Woo Lee and Young-Ho Cho, "High-radix microfluidic multiplexer with pressure valves of different thresholds," *Lab Chip* 9, 1681–1686, 2009. Reproduced with permission from The Royal Society of Chemistry.)

valves (each valving three different fluidic channels) closed all three channels, routing the fluid through a fourth channel (**Figure 3.58a**). Different pressure settings allow for selecting different valves (**Figure 3.58b–d**). The terms "ternary valve" and "quaternary valve" are somewhat misleading—each valve is inherently binary (their operation is thresholded), but they function as if they had three states (ternary) or four states (quaternary) when they are part of an ensemble of valves.

3.8.4 Micropumps

In Section 3.6.2, we have already introduced the basic practical concerns about how to drive flow into microchannels (using syringe pumps, gravitational flow, etc.), and we have covered ways to drive flow using electrical fields (i.e., electro-osmotic flow in Section 3.3.2 and electrowetting in Section 3.3.4). In this section, we will focus on nonelectrical strategies that have been designed specifically for pumping fluids on the microscale.

3.8.4.1 Micropumps Driven by Surface Tension

Small fluid volumes in contact with microstructured surfaces move spontaneously as a result of an interplay between the liquid's surface tension and both the surface's chemical composition and topography—always in the direction that minimizes the free energies between the vapor, fluid, and solid interfaces. This behavior is most strikingly revealed in the spontaneous wetting of small channels or capillaries and has also been exploited in valve and pump fluids. Early in 2002, Emmanuel Delamarche's group at IBM Zurich (Switzerland) reported a **capillarity pump** that moves fluid from one area of the device onto another (**Figure 3.59**). Eventually, the capillarity pump drains the liquid out of the service port until it is pinned at the capillary retention valve ("CRV" in **Figure 3.59c**); a second solution can then be dispensed into the drained service port, a process that can can be repeated as many times as desired (16 sequential steps were demonstrated), as long as the channels of the capillarity pump are not entirely filled. Flow rates of 220 nL/s and average flow speeds of 55 mm/s were observed (the device occupies $100 \times 100 \ \mu m^2$).

Note that Delamarche's pumping scheme is inherently limited to small footprints (which limits the designs to which it can be applied) and by the fact that the flow must be powered at all times by the surface tension of the meniscus at the end of the microchannel (so there is a point in which it runs "out of steam"). A variation on this concept, presented in 2002 by David Beebe's group from the University of Wisconsin at Madison, does not suffer from these limitations because it uses the surface tension of droplets placed at inlets/outlets of microchannels to drive the flow (**Figure 3.60**). The flow rates are dictated by the curvature of the droplets, which in turn are controlled by the amount of fluid dispensed. This droplet-based strategy allows for

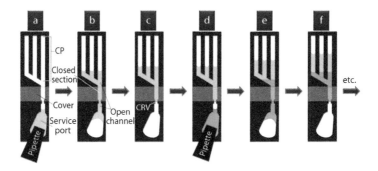

FIGURE 3.59 Capillarity pump. (From David Juncker, Heinz Schmid, Ute Drechsler, Heiko Wolf, Marc Wolf, Bruno Michel, Nico de Rooij, and Emmanuel Delamarche, "Autonomous microfluidic capillary system," *Anal. Chem.* 74, 6139, 2002. Reprinted with permission of the American Chemical Society. Figure contributed by Emmanuel Delamarche.)

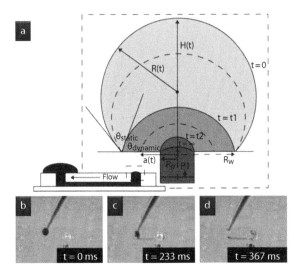

FIGURE 3.60 Surface tension–driven passive micropump. (a, from Erwin Berthier and David J. Beebe, "Flow rate analysis of a surface tension driven passive micropump," *Lab Chip* 7, 1475–1478, 2007; b–d, from Ivar Meyvantsson, Jay W. Warrick, Steven Hayes, Allyson Skoien, and David J. Beebe, "Automated cell culture in high density tubeless microfluidic device arrays," *Lab Chip* 8, 717–724, 2008. For the first paper on passive pumping see G. M. Walker and D. J. Beebe, "A passive pumping method for microfluidic devices," *Lab Chip* 2, 131–134, 2002. Reproduced with permission from The Royal Society of Chemistry.)

high-density, large-area addressing with robotic multipipetters and even multistream laminar flows for short periods of time, but (unlike Delamarche's scheme) it is sensitive to evaporation and it is not possible to load an arbitrarily large number of units at once.

The obvious advantages of the two surface tension–based pumping schemes shown in this section are their low cost, straightforward implementation, and compatibility with "tubeless microfluidics" (in which inlets are already integrated into the device in the form of reservoirs, see Figure 3.22), which makes them extremely valuable for resource-poor settings. However, the pumping rate decays with time (in microchannels; paper wicking is a form of capillarity-based pumping whose pumping rate stays constant) and it is not amenable to computerized control (i.e., the rate cannot be changed).

3.8.4.2 Gas-Permeation Micropumps

PDMS is extremely permeable to gases—one can inject air into a dead-ended microchannel with a syringe and the air goes into … the PDMS! This same concept, in reverse, has been applied to pump fluids: remove all the residual gas that is present in PDMS (by first placing the device in a vacuum jar for 15–20 minutes) and it will create a local vacuum in the microchannels. This simple idea was first applied by Mizuo Maeda's group from RIKEN in Japan to run biochemical assays (Figure 3.61) in 2004 but it is now used by many groups elsewhere. Unfortunately, the pump's pumping rate decays exponentially with time as the air reservoir gets gradually refilled (operation times of >15 minutes were reported).

In 2006, Bruce Gale's group at the University of Utah designed a different pump based on similar principles. Here, the idea was to inject air into microchannels at very small rates such that the air can in turn push plugs of fluids around (Figure 3.62). To add a high resistance to air flow, they simply placed a PDMS membrane of the right thickness (the study compared membranes of 100, 45, and 25 μm thicknesses). This pumping scheme is very robust but is incompatible with applications that are sensitive to bubbles (e.g., cells die if exposed to a bubble or even in the proximity of an air bubble in which the CO_2 concentration is not adequate).

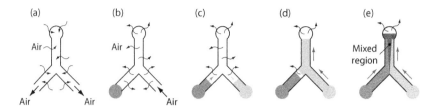

FIGURE 3.61 A PDMS vacuum pump. (From Kazuo Hosokawa, Kae Sato, Naoki Ichikawa, and Mizuo Maeda, "Power-free poly(dimethylsiloxane) microfluidic devices for gold nanoparticle-based DNA analysis," *Lab Chip* 4, 181–185, 2004. Reproduced with permission from The Royal Society of Chemistry.)

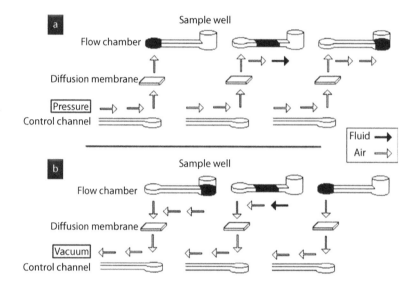

FIGURE 3.62 Gas-permeation micropump. (From Mark A. Eddings and Bruce K. Gale, "A PDMS-based gas permeation pump for on-chip fluid handling in microfluidic devices," *J. Micromech. Microeng.* 16, 2396–2402, 2006. Figure contributed by Bruce Gale. Reproduced with permission from the Institute of Physics.)

3.8.4.3 Three-Valve PDMS Peristaltic Micropumps

The Quake microvalves (**Figure 3.38a**) were first reported in 2000. The same *Science* article also reported the operation of three valves in series to obtain a peristaltic pump (see **Figure 3.38b** in Section 3.8.1.4) that achieved a pumping rate of 2.35 nL/s at zero backpressure; these pumps could be operated at 25 Hz (75 Hz each valve). In 2003, the Mathies group presented peristaltic pumps constructed with their "doormat" microvalves (**Figure 3.63**), featuring pumping rates up to 50 nL/s at zero backpressure. (The pumping rates do not necessarily reflect the performance of the pump, because it can be easily increased by increasing the area of the valve seal; the Mathies group investigated diaphragm pump dimensions between 1000 and 6000 μm, which is equivalent to displacement chamber volumes ranging between 67 and 2050 nL.)

An interesting variation of the "doormat" design (it can be constructed with an all–SU-8 mold) has been presented by Satoshi Konishi's group of Ritsumeikan University in Kyoto (Japan). Here the valve seat is circularly symmetric, which facilitates the ejection of the fluid out of the valve seat as the PDMS membrane attempts to seal with the valve seat in the valve-closing phase of the pumping cycle (**Figure 3.64**). The reported pumping rates were low (~1.2 nL/s at

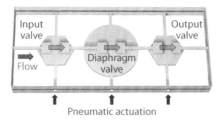

FIGURE 3.63 Peristaltic micropumps constructed with "doormat" PDMS microvalves. (From W. H. Grover, A. M. Skelley, C. N. Liu, E. T. Lagally, and R. A. Mathies, "Monolithic membrane valves and diaphragm pumps for practical large-scale integration into glass microfluidic devices," *Sens. Actuators B Chem.* 89, 315–323, 2003. Figure contributed by Rich Mathies.)

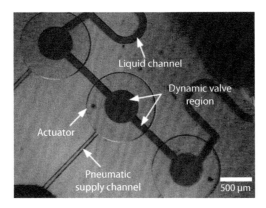

FIGURE 3.64 PDMS peristaltic micropump featuring circularly symmetric microvalves. (From O. C. Jeong and S. Konishi, "Fabrication of a peristaltic micro pump with novel cascaded actuators," *J. Micromech. Microeng.* 18, 025022, 2008. Reproduced with permission from the Institute of Physics.)

zero backpressure), likely because of the large size of the actuators (indeed, the maximum flow rate was observed at a very low frequency; ~2 Hz).

Regardless of the microvalve design of choice, peristalsis is typically actuated by the pattern 101, 100, 110, 010, 011, 001, in which 0 and 1 indicate open valves and closed valves, respectively. Note that this operation, although sufficient for many applications, requires three pneumatic control channels, which consumes a lot of real estate on the chip and puts heavy constraints on chip layout design.

3.8.4.4 "Single-Stroke" PDMS Peristaltic Micropumps

It was proposed early on by several groups that it should be possible to produce a peristaltic motion using a single pneumatic channel that crosses the fluidic channel several times in a serpentine path: the time-delay introduced by the pressure wave traveling down the serpentine channel is used as a substitute of the traditional time sequence used to produce peristalsis (Figure 3.65). We stress that, although this design uses the same number of valves as the Quake or the Mathies designs shown previously that featured one control line per valve, here, all three valves are activated (at different times) with a single stroke. The concept was first published in 2006 by Gwo-Bin Lee's group from National Cheng Kung University in Taiwan, who reported large pumping rates (124 nL/s using a seven-membrane device operated at 9 Hz). This group studied the delay lines and modeled them as electrical circuits containing resistors and capacitors. On first approximation, one could be tempted to model the transmission of the pressure wave inside the pneumatic channel as sound through a rigid pipe, traveling at 343 m/s for dry air

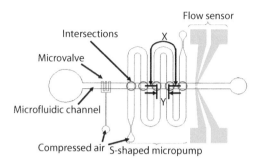

FIGURE 3.65 Serpentine channel micropumps. (From C.-H. Wang and G.-B. Lee, "Pneumatically driven peristaltic micropumps utilizing serpentine-shape channels," *J. Micromech. Microeng.* 16, 341–348, 2006. Reproduced with permission from the Institute of Physics.)

at 20°C, but the reality is that the pneumatic channel expands, so the walls behave like a capacitor (and so do the membranes), and the channel itself offers some aerodynamic resistance, R—all of which slows down the response of the line by more than two to three orders of magnitude. (The capacitance, C, is a function of volume, V, and pressure, P, expressed as $C = dV/dP$, and the resistance, R, is expressed by **Equation 3.12**.) In 2008, Gwo-Bin Lee studied how serpentine pumps (with 1, 5, and 10 mm pneumatic channel lengths) behave using high-speed cameras and determined that the total time needed to complete the activation and deactivation processes was 26, 38, and 106 ms (for the 1, 5, and 10 mm-length pumps, respectively); by comparison, sound in a rigid pipe would have taken a mere 2.9, 14.5, and 29 μs to travel the same lengths, respectively, and thus, clearly, the flexibility of PDMS plays a big (advantageous) role in slowing down the pressure wave. If PDMS were rigid, the serpentines would have to be three to four orders of magnitude longer. Still, serpentine design occupy a large amount of chip real estate.

The author's group at the University of Washington (Seattle) has investigated a design principle that is based on the universal mechanical principle that small membranes respond faster than larger membranes, so they can be connected to the same pneumatic line (**Figure 3.66**), eliminating delay lines. On arrival of the pressure pulse, the first valve to close is always the

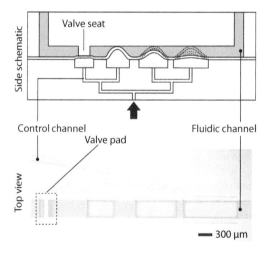

FIGURE 3.66 Compact PDMS peristaltic pump actuated by a single pneumatic channel. (From Hoyin Lai and Albert Folch, "Design and characterization of "single-stroke" peristaltic PDMS micropumps," *Lab Chip* 11, 336–342, 2011. Reproduced with permission from The Royal Society of Chemistry. Figure contributed by Hoyin Lai.)

smallest one and the last valve to close is always the largest one, thus producing a peristaltic wave. It matters very little whether the membranes are connected in series or in parallel: serially connected membranes that are connected backward (such that the pressure pulse arrives first at the largest membrane) still pump forward (i.e., the small membrane is activated first).

3.8.4.5 Diaphragm Micropumps

Miniature diaphragm pumps are commercially available and used routinely for pumping small fluid volumes, such as in toys, decorative displays, and so on. These pumps universally incorporate two check valves opening toward opposite sides of the diaphragm chamber, so that one of the valves opens when the chamber is at negative pressure (when the diaphragm is deflected away from the chamber, increasing its volume) and the other valve opens when the chamber is at positive pressure (when the diaphragm is deflected into the chamber, reducing its volume). By alternating between positive and negative pressure to the diaphragm, the check valves open in alternating sequence, making it possible to draw fluid from a given reservoir into an outlet. Interestingly, the microfabricated version of this principle was not fabricated in PDMS until 2006 (although the technology for building it existed for several years), probably because of the technical difficulty of fabricating check valves. A group from Korea University led by SangHoon Lee fabricated them with a UV-polymerized hydrogel (introduced into the device via dedicated inlets), and the diaphragm was straightforwardly fabricated as a PDMS membrane (Figure 3.67).

3.8.4.6 Piezoelectric Micropumps

Certain piezoelectric materials, such as lead zirconate titanate (PZT), undergo significant shape changes when they are loaded with an electrical charge. Stress exerted by the piezoelectric material, coupled to a thin diaphragm over a microchannel or a microchamber, can be used to valve and pump fluids. Some of the earliest designs of piezoelectric micropumps were micromachined from silicon wafers and were developed in the 1980s. In these micropumps, piezoelectric disks were glued on top of a diaphragm (usually glass) directly above a pumping chamber. Pumping strokes were generated by alternating deflection of the piezo-coupled diaphragm into and away from the pumping chamber. Additionally, check valves, diffusers, and nozzles were commonly incorporated into the flow channels to better rectify flow direction.

More recently, the development of piezoelectric micropumps has shifted toward using low-cost, optically transparent materials such as PDMS and PMMA, instead of micromachining

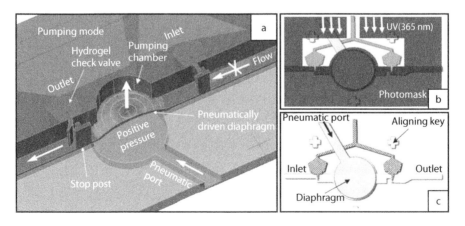

FIGURE 3.67 PDMS diaphragm pump with integrated hydrogel check valves. (From Jeong Yun Kim, Ju Yeoul Baek, Ki Hwa Lee, Yong Doo Park, Kyung Sun, Tae Soo Lee, and Sang Hoon Lee, "Photopolymerized check valve and its integration into a pneumatic pumping system for biocompatible sample delivery," *Lab Chip* 6, 1091–1094, 2006. Reproduced with permission from The Royal Society of Chemistry.)

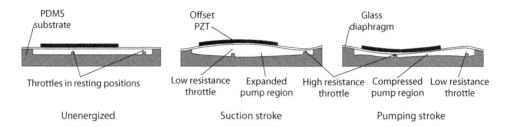

Unenergized   Suction stroke   Pumping stroke

FIGURE 3.68 Piezoelectric micropumps. (From M. C. Tracey, I. D. Johnston, J. B. Davies, and C. K. L. Tan, "Dual independent displacement-amplified micropumps with a single actuator," *J. Micromech. Microeng.* 16, 1444–1452, 2006. Reproduced with permission from the Institute of Physics.)

glass or silicon substrates. Dynamic modulation of flow resistance (termed "throttling") has been discussed as a potential method for flow rectification. Flow rate is inversely proportional to the fourth power of the hydraulic radius of a microchannel, so even a small reduction in a channel's cross-sectional area can produce a significant increase in flow resistance and consequently a drop in flow rate. Christabel Tan and coworkers from the University of Hertfordshire (England) have reported a micropump that manipulates the flow resistance using a pair of "microthrottles" molded into a PDMS substrate (**Figure 3.68**). An offset in PZT disk placement with respect to the throttles results in bimorphic flexion of the glass diaphragm such that the throttles operate antiphasically. When the PZT is deflected away from the channel, the inlet throttle has a lower resistance and thus fluid is drawn from the inlet. Conversely, when the PZT disk is compressed into the pump region, the outlet throttle has a lower resistance and thus fluid is driven toward the outlet. Pumping performance comparable to valved piezoelectric micropumps is observed at the throttling ratio (the high-to-low flow resistance ratio) of 8:1.

3.8.4.7 Ultrasound-Based Micropumps

We have seen that the air–liquid interface of a trapped air bubble can be vibrated at its resonance (ultrasound) frequency using an external piezoelectric transducer to produce acoustic microstreaming and rapid mixing around the bubble (see **Figure 3.10b** in Section 3.4.2). Abe Lee's group at the University of California (Irvine) has been able to put acoustic microstreaming to use for propelling fluids (**Figure 3.69**). The pump is formed by trapping a set of bubbles in

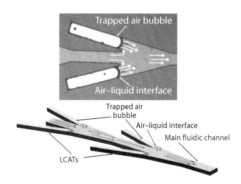

FIGURE 3.69 Micropump based on acoustic microstreaming. (From Armando R. Tovar and Abraham P. Lee, "Lateral cavity acoustic transducer," *Lab Chip* 9, 41–43, 2009. Reproduced with permission from The Royal Society of Chemistry.)

dead-end arms of the microchannel in which a solution is to be pumped; vibration of the bubbles at ultrasound frequencies generates acoustic microstreaming and forward motion. Acceptable flow rates of up to 250 nL/min were achieved when the PZT buzzer was powered at 40 V_{pp} (45 kHz), but only around 5 nL/min when it was powered at 5 V_{pp}. Pumps based on acoustic microstreaming have the enormous advantage that they can be powered with just electricity; however, they still require a wave function generator and an amplifier (they cannot be powered from a computer only). Most importantly, these pumps are fundamentally lossy, in that a lot of the energy generated by the PZT transducer is dissipated into heat and device vibrations (acoustic noise), and the bubbles can slowly dissolve into the PDMS or the fluid, so their performance may not be as scalable as other pumps.

3.8.5 Microfabricated Flow Gauges

Measuring instantaneous flow rates is notoriously difficult in a microchannel. Ideally, one would like to know the distribution of flow speeds in every point of the channel, but this goal is almost impossible to achieve and one must compromise for an average flow rate or a local measurement. But how can we locally measure flow rates, in such small spaces? One old trick is to use **particle imaging velocimetry** (PIV), whereby one tracks the position of beads in the stream and from that one infers the velocity and streamlines (see Section 3.6.3); however, this approach requires the contamination of the sample with beads and a careful treatment of focal depth (e.g., with a confocal microscope), especially in complex 3-D flows.

Stop-flow lithography (see Section 1.6.4) enables the in situ polymerization of mechanical elements into a microfluidic channel. PEG-diacrylate is photopolymerized using a transparency mask through the microscope objective to create a spring-like force sensor (**Figure 3.70**). At the PDMS surfaces (roof and floor of the microchannel), oxygen inhibition blocks the photopolymerization reaction so the microstructure is free to "slide" back and forth within the channel. The elastic modulus of the spring can be finely tuned by changing the monomer to cross-linker ratio or by lowering the photoinitiator concentration. The elongation of the spring could be used as a measure of the flow rate, Q.

However, because not everyone has access to stop-flow lithography, fluids may (once again) do the trick if one is willing to get by knowing the *pressure differential* between the unknown and a reference pressure. Frieder Mugele and coworkers at the University of Twente in the Netherlands have presented a clever microfluidic comparator that translates differences in pressure into displacements of an interface (which can be measured with a microscope). The comparator has been used to demonstrate the measurement of the hydrodynamic resistance of moving drops in real time (**Figure 3.71**).

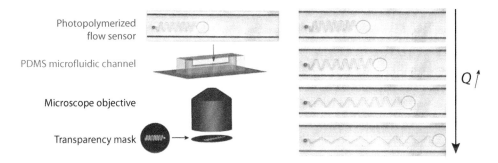

FIGURE 3.70 Microfluidic flow gauge fabricated by stop-flow lithography. (From Rafaele Attia, Daniel C. Pregibon, Patrick S. Doyle, Jean-Louis Viovy, and Denis Bartolo, "Soft microflow sensors," *Lab Chip* 9, 1213–1218, 2009. Reproduced with permission from The Royal Society of Chemistry.)

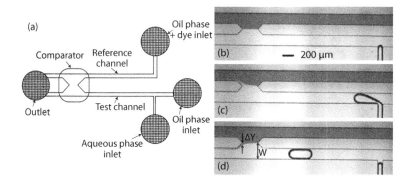

FIGURE 3.71 Microfluidic flow comparator. (From Siva A. Vanapalli, Arun G. Banpurkar, Dirk van den Ende, Michel H. G. Duits, and Frieder Mugele, "Hydrodynamic resistance of single confined moving drops in rectangular microchannels," *Lab Chip* 9, 982–990, 2009. Reproduced with permission from The Royal Society of Chemistry.)

3.9 Micromixers

In microfluidics, the word "micromixer" is loosely used to mean three distinct types of devices: devices in which the fluids are mixed to generate dilutions of one or more compounds ("**dilution generators**"), others in which the goal is to generate a gradient over a surface ("**gradient generators**"), and others in which the goal is to homogenize the mixture of a number of inlets as efficiently as possible ("**homogenizers**"). The terminology can be confusing because many gradient generators incorporate dilution generators (which are covered in Section 3.9.2), but not all. There are also a number of devices that make good use of microscale mixing phenomena for specific (e.g., cellular) applications, but their design is usually also application-specific, and thus they are covered elsewhere in this text. A newer generation of micromixers that use microvalves and micropumps to achieve mixing is described in Section 3.10.2.

The field of micromixers is perhaps where microfluidics has reached its tallest heights in inventiveness. There is a simple, very fundamental reason for that: fluids do not mix well on a microscale, so scientists have had to put all their creativity at work. In your bathtub or in your coffee mug, mixing of fluids occurs by diffusion, advection (i.e., by flow-assisted transport), or turbulence. In a microchannel, turbulence is ruled out because of the "imprisonment" of fluids within the channel walls that only allow for laminar flow motions. Hence, all the existing micromixer designs have been conceived to exploit either diffusion or advection, or both, to make up for the absence of turbulence in the microscale.

3.9.1 T- or Y-Mixer

Let us first understand in detail what happens when two streams (in the simplest case, two aqueous solutions) meet at an intersection between two channels in the shape of a "T" or a "Y"—the simplest possible mixer (also known as the "**T-mixer**" or "**Y-mixer**"). Both streams have a parabolic flow profile and homogeneous concentrations before merging, and some distance after merging the merged flow will also have a parabolic flow profile but the concentration will display an approximate step profile. How about further downstream? Understanding the downstream evolution of the step profile in the T-mixer is a great exercise for the microfluidics student, even if done qualitatively. It is actually not straightforward, indeed, it was not fully explained until the year 2000, when a team led by George Whitesides and Howard Stone at Harvard University went through the trouble of *imaging* what was going on. As shown in **Figure 3.72**, they used a fluorogenic reaction, the binding of the calcium indicator fluo-3 (introduced in one inlet) to $CaCl_2$ (introduced in the other inlet), to visualize

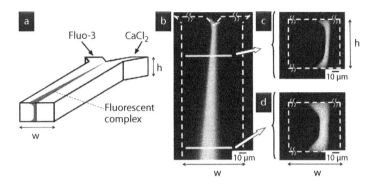

FIGURE 3.72 The "butterfly effect." (From Rustem F. Ismagilov, Abraham D. Stroock, Paul J. A. Kenis, George Whitesides, and Howard A. Stone, "Experimental and theoretical scaling laws for transverse diffusive broadening in two-phase laminar flows in microchannels," *Appl. Phys. Lett.* 76, 2376, 2000. Adapted with permission of the American Institute of Physics. Figure contributed by George Whitesides.)

the mixing in three dimensions with a confocal microscope. As expected, the profile diffuses downstream due to diffusion (Figure 3.72b). However, it broadens more closer to the walls and it broadens asymmetrically (as if diffusion happened faster toward the fluo-3 side). Why?

Let us first see why the broadening is asymmetrical. Ions do not have an intrinsic preference to diffuse in one direction or another, so what makes the profile become asymmetrical? As the calcium ions diffuse into the fluo-3 stream, they bind to fluo-3 and form a fluorescent $[Ca^{+2}]$-fluo-3 complex, which we see as a band. The complex also forms as the fluo-3 molecules diffuse into the $CaCl_2$ stream (the calcium ion stream), but at a much slower pace because the fluo-3 molecules are much larger and consequently diffuse much slower than the calcium ions.

Now let's see why the broadening is more pronounced at the roof and floor of the microchannel. Recall that, at the walls, the solution always flows slowest (see Section 3.2.4), so in any vertical cross-sectional image of the channel (such as Figure 3.72c), the $[Ca^{+2}]$-fluo-3 molecules at the top and bottom of the images have been in the channel for considerably longer than those traveling (toward you, the observer) along the center of the image. Therefore, the $[Ca^{+2}]$-fluo-3 molecules near the roof and floor of the microchannel have had much more time to diffuse than those near the center of the channel—hence, the diffusion profile appears wider near the walls in Figure 3.72c and d, giving the appearance of half of a butterfly wing. The microfluidics student will find it easy to remember that this effect, which is simply a particular manifestation of the Taylor dispersion, is popularly known as the "**butterfly effect**."

3.9.2 Dilution and Gradient Generators

In biotechnology, many assays require producing a set of dilutions of a given solution, and the dilutions must be painstakingly done by hand with a pipetman. To automate this procedure, and to enable the miniaturization of such assays, many groups have reported micromixer designs that generate microfluidic dilutions. These devices, importantly, can also be used to generate a gradient simply by merging all the outlets into one channel. (However, not all gradient generators work by merging the outputs of a dilution generator, and these will be covered in subsequent sections.) Conversely, the output of a gradient generator can often (but not always) be split into many outputs, yielding that many number of dilutions. Hence, we cover dilution and (some) gradient generators together in the same section. The difference between gradient generator and dilution generator is simply in their application: the gradient generator pursues the creation of a continuum of concentrations, typically in one channel or chamber, whereas the dilution generator pursues the creation of discrete concentrations in separate channels or chambers.

In 2001, Stephan Dertinger and George Whitesides at Harvard University demonstrated a symmetric 2-D microfluidic network that continuously generates a stepwise range of dilutions of two compounds. The concept is based on splitting and recombining the flows in a tree-like manner (Figure 3.73). Although the method is straightforward to implement and, not surprisingly, has become very popular, it has serious limitations: the gradient depends on the flow rate and it is not constant in space (it becomes smoother and flatter downstream), which limits its cellular applications to a narrow window.

In 2006, Mehmet Toner and colleagues at Massachusetts General Hospital implemented a generalization of the Dertinger design that allows for producing arbitrarily shaped gradients (Figure 3.74). In the Dertinger design, the vertical "squiggles" are designed to increase the fluid's residence time such that each pair of branches has time to recombine by diffusion; the squiggles, however, are always the same length and the amount of recombination is always the same. The Toner design *looks* very different but can be analyzed very similarly. In the "universal" gradient generator, the (topological equivalent of the) squiggles are channel sections of various widths and lengths, and depending on the sequence of lengths, gradients with various mathematical shapes are achieved. It can be proven mathematically that, for any desired output concentration profile, at least one configuration of channel sections exists that produces the output profile from only two inputs. Obviously, only monotonically varying gradients are possible, in other words, a sinusoid with two maxima is not possible and would require more inlets.

Toner's universal gradient generator assumes perfect mixing in every channel section and uses simple algebraic relationships to determine the configuration of the channel dividers that produce the desired output concentration curve. An alternative strategy is to regard the set of channel sections as electrical resistors and use electrical circuit theory (Kirchhoff's current and voltage laws) to solve the circuit (assuming complete mixing by diffusion at every segment). This approach, presented by Kwang W. Oh and colleagues from the State University of New York at Buffalo in 2009, allows for creating monotonic concentration profiles of any mathematical function (Figure 3.75).

The universal gradient generators shown previously are clever in that they present a set of alternatives to the fluids—and the fluids, based on the flow resistance of the network, distribute themselves spontaneously, which then automatically produces a downstream gradient by

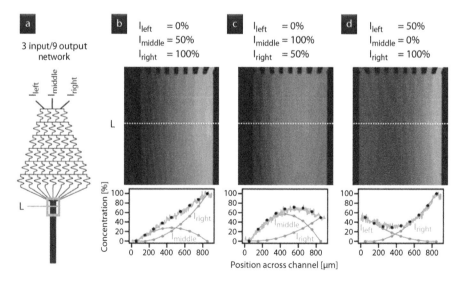

FIGURE 3.73 The Dertinger gradient generator. (From Stephan K. W. Dertinger, Daniel T. Chiu, Noo Li Jeon, and George M. Whitesides, "Generation of gradients having complex shapes using microfluidic networks," *Anal. Chem.* 73, 1240, 2001. Adapted with permission of the American Chemical Society. Figure contributed by George Whitesides.)

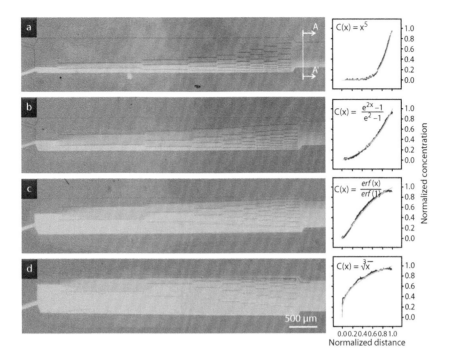

FIGURE 3.74 Universal gradient generator. (From Daniel Irimia, Dan A. Geba, and Mehmet Toner, "Universal microfluidic gradient generator," *Anal. Chem.* 78, 3472–3477, 2006. Reprinted with permission of the American Chemical Society. Figure contributed by Mehmet Toner.)

diffusion. This principle can be exploited in virtually unlimited combinations, of which we will highlight only a few here. Minoru Seki and coworkers at the Osaka Prefecture University in Japan, for example, have shown an elegant flow distributor that splits the flow from two distribution channels into many outlet channels (**Figure 3.76**). The amount of flow received by each outlet channel is a function of the ratio of distances between the outlet channel and the distribution channels, so the performance of the mixer is easy to predict. Note that the conceptual simplicity comes at a price: the outlets must cross over the distribution channels and, therefore, the device requires 3-D multilayer fabrication/assembly.

Similarly, Mengsu Yang's group at the City University of Hong Kong have presented a device in which a channel containing constant flow communicates through "microtunnels" with the area in which the gradient is desired (**Figure 3.77**). The shape of the gradient is determined simply by the length of the microtunnel.

Both the Dertinger and the universal gradient generators shown previously allow for exquisite spatial control of the gradients but are highly flow rate–dependent. Adrian O'Neill's group at North Carolina State University has presented a clever design that produces linear dilutions and preserves the linearity over a wide range of flow rates (0.5–16 µL/min). The ratio of the dilutions is controlled by the volumetric flow rate of the solution of interest and the diluent, which mix in a two-layer network (**Figure 3.78**). The flow rate is modulated by making use of the fact that the channel resistance varies linearly with length (see **Equation 3.12**).

Because of its simplicity, the Dertinger device has been used by several groups to generate gradients for cell culture applications. Indeed, it can generate quantifiable, steady-state gradients, but with several limitations: (1) it can only generate gradients under fluid flow, which induces shear and drag forces that can alter intracellular signaling or cause changes in cell shape and attachment that may lead to migrational bias; (2) the gradient evolves as the fluid flows downstream such that no two cells in the microchannel experience the same concentration

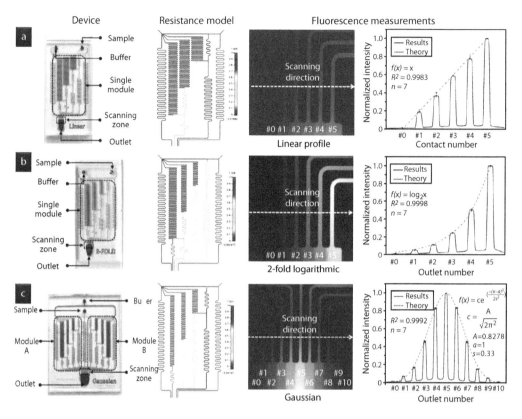

FIGURE 3.75 Microfluidic gradient generators for arbitrary gradients using electrical-circuit analogues. (From Kangsun Lee, Choong Kim, Byungwook Ahn, Rajagopal Panchapakesan, Anthony R. Full, Ledum Nordee, Ji Yoon Kang, and Kwang W. Oh, "Generalized serial dilution module for monotonic and arbitrary microfluidic gradient generators," *Lab Chip* 9, 709–717, 2009. Reproduced with permission from The Royal Society of Chemistry.)

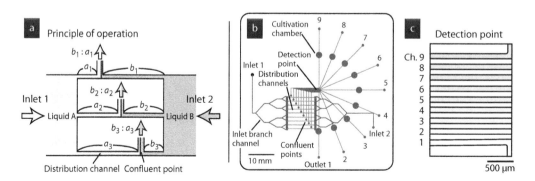

FIGURE 3.76 Dilution generator based on ratiometric distribution of flow resistance. (From Masumi Yamada, Takaya Hirano, Masahiro Yasuda, and Minoru Seki, "A microfluidic flow distributor generating stepwise concentrations for high-throughput biochemical processing," *Lab Chip* 6, 179, 2006. Reproduced with permission from The Royal Society of Chemistry.)

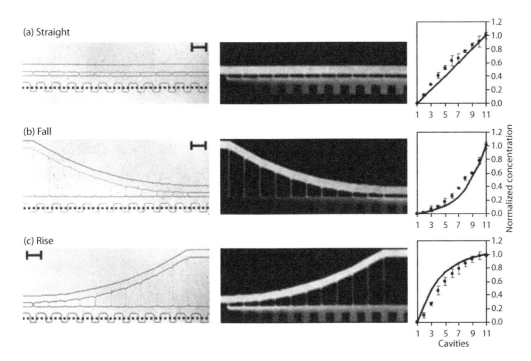

FIGURE 3.77 Gradient generator based on transport through microtunnels. Scale bar is 200 µm (From Cheuk-Wing Li, Rongsheng Chen, and Mengsu Yang, "Generation of linear and non-linear concentration gradients along microfluidic channel by microtunnel controlled stepwise addition of sample solution," *Lab Chip* 7, 1371–1373, 2007. Reproduced with permission from The Royal Society of Chemistry.)

gradient; (3) downstream cells are exposed to higher concentrations of cell-secreted molecules (e.g., metabolites and growth factors) than upstream cells, which precludes true redundancy in single-cell data; and (4) the observation area is an enclosed microfluidic channel, which limits gas and nutrient exchange for long-term cell viability and hinders physical access to single cells (e.g., patch-clamp recording, intracellular injection, atomic force microscopy, etc.) during the

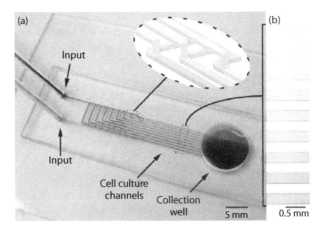

FIGURE 3.78 Linear dilution generator. (From Glenn M. Walker, Nancy Monteiro-Riviere, Jillian Rouse, and Adrian T. O'Neill, "A linear dilution microfluidic device for cytotoxicity assays," *Lab Chip* 7, 226–232, 2007. Reproduced with permission from The Royal Society of Chemistry.)

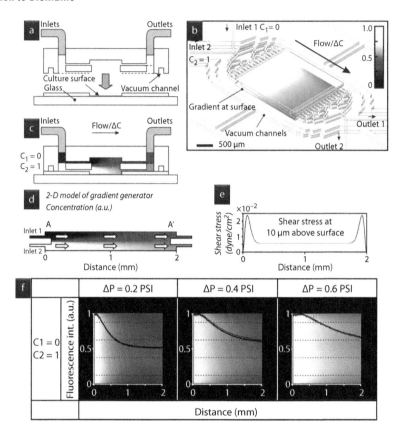

FIGURE 3.79 Stacked-flow gradient generator. (From Christopher G. Sip, Nirveek Bhattacharjee, and Albert Folch, "A modular cell culture device for generating arrays of gradients using stacked microfluidic flows," *Biomicrofluidics* 5, 022210, 2011. Figure contributed by Chris Sip.)

course of the experiment. Several groups have developed alternative designs that overcome these limitations.

One approach consists of using stacked flows (as opposed to parallel streams) to produce the gradient (**Figure 3.79**). The device can be sealed to a wet surface, washed, and reused (**Figure 3.79a–c**). The strategy is topologically equivalent to turning a T-mixer sideways (**Figure 3.79d**). Shear stresses exerted by the flow for a typical protein gradient are two orders of magnitude smaller than the smallest shear stresses reported to bias the migration of neutrophils toward the downstream direction in a Dertinger device (**Figure 3.79e**). The gradients can be shifted in position and tuned in shape (**Figure 3.79f**) simply by changing the pressure at the inlets (switching the inlets makes the gradient increase against the flow instead of decrease with the flow). Note that all the points at the same distance from the wall experience the same gradient (hence, it should be ideal to produce redundant cell data) and that there is no fundamental limit as to how much area can be perfused with this device.

3.9.3 Gradients Delivered through Microjets

In 2006, the author's laboratory at the University of Washington in Seattle presented a microfluidic device that generates gradients of diffusible molecules in an open reservoir with none of the aforementioned limitations for cell culture applications (**Figure 3.80**). PDMS is exclusion-molded from the tallest features of a three-level mold by compressing the mold against a polyester sheet (**Figure 3.80a**). The resulting device contains an open reservoir 66 μm deep and 200 μm

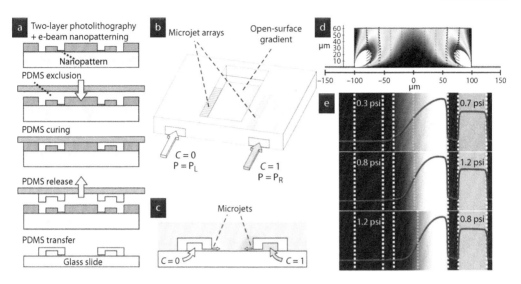

FIGURE 3.80 Gradient generator using "microjets." (From Thomas M. Keenan, Chia-Hsien Hsu, and Albert Folch, "Microfluidic "jets" for generating steady-state gradients of soluble molecules on open surfaces," *Appl. Phys. Lett.* 89, 114103 (2006). Figure contributed by Tom Keenan.)

wide (Figure 3.80b). Two 45-μm-tall and 100-μm-wide microchannels serve as manifolds to deliver fluids to each array of 25-μm-long channels termed "microfluidic jets" or "**microjets**" (1.5×1.5 μm cross-section). Pressurization of the manifolds causes fluid to be ejected (~40–150 pL/min) at the microjet locations, effectively "pinning" the microjet outlet concentration to that inside the manifold (Figure 3.80c). Because the air–fluid interface is open, the fluid lines are directed upwards and the shear stress (parallel to the surface) seen by the cells is negligible (Figure 3.80d). Because mass transport is established mostly by convection, not by diffusion, a stable gradient can be established within minutes and their position and slope can be modulated by the user simply by changing the manifold pressures (Figure 3.80e). Importantly, the established gradient is very insensitive to the diffusion coefficient, so the gradient (e.g., of a protein) can be visualized by adding a tracer of dissimilar diffusivity (e.g., fluorescein). The device has been applied to study neutrophil chemotaxis and axon guidance.

The microjets also work in combination with closed microchambers, which are easier to microfabricate. A team led by Sarah Heilshorn from Stanford University has presented a gradient generator based on microjets that incorporates a closed cell culture chamber, which communicates with a source channel and a sink channel through the microjets (Figure 3.81). Perfusion presents a problem because the flow needs to go somewhere after it enters the chamber, so the way the perfusion is set up influences the flow lines inside the cell culture chamber (Figure 3.81d). Nevertheless, because of the high flow resistance of the microjets (they are 4 μm high), the shear stress in the cell culture chamber during gradient generation is several orders of magnitude lower than the physiological levels encountered in blood capillaries.

Also using a closed microchamber, Laurie Locascio and colleagues have developed a symmetrical version of the microjets that reduces to practically zero the amount of outflow expelled by the microjets. In their design, each microjet is connected to the center of a loop, which is perfused with the solution of interest; because of design symmetry considerations, flow cannot go through the microjet into the chamber (it is closed), but solute can diffuse into it, producing a gradient that stabilizes in approximately 15 minutes (for a 1.5 mm diameter chamber) and that can be set to rotate (Figure 3.82). For convection to be truly cancelled, the pressures (both hydrostatic and pneumatic) at all the inlets and outlets need to be carefully balanced. The device has been applied to study bacterial chemotaxis.

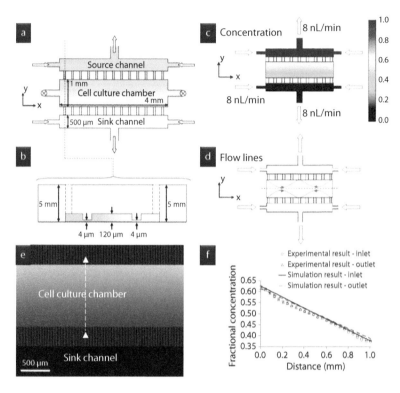

FIGURE 3.81 Microjets in a closed microchamber. See Amir Shamloo, Ning Ma, Mu-ming Poo, Lydia L. Sohn, and Sarah C. Heilshorn, "Endothelial cell polarization and chemotaxis in a microfluidic device," *Lab Chip* 8, 1292–1299 (2008). Reproduced with permission from The Royal Society of Chemistry.

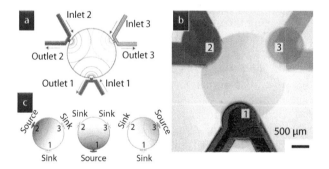

FIGURE 3.82 A diffusive gradient generator using microjets. (From Javier Atencia, Jayne Morrow, and Laurie E. Locascio, "The microfluidic palette: A diffusive gradient generator with spatio-temporal control," *Lab Chip* 9, 2707–2714, 2009. Reproduced with permission from The Royal Society of Chemistry.

3.9.4 Microfluidic Pens

Microjet devices and, in general, multistream laminar flows are limited in the types of patterns that can be created with a static layout of microchannels. Emmanuel Delamarche's group at IBM Zurich has developed a positionable "**microfluidic pen**" to deliver solutions (in fact, sharp gradients of them) at arbitrary positions of cell culture surfaces (**Figure 3.83**). The critical

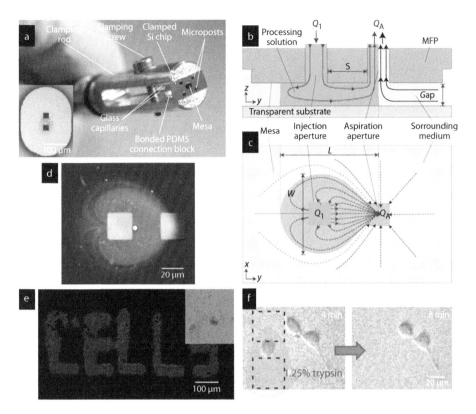

FIGURE 3.83 Microfluidic pen. (From Juncker, D., Schmid, H., and Delamarche, E., "Multipurpose microfluidic probe," *Nat. Mater.* 4, 622–628 (2005). Reprinted with permission of the Nature Publishing Group. Figure contributed by Emmanuel Delamarche.)

components of the device are its two apertures (20 × 20 μm), which are positioned close (~10 μm) to the surface to be patterned: flow is ejected out of the device through the "injection aperture" (at flow rate Q_I, typically Q_I = 0.44 nL/s) and is aspirated into the device through the "aspiration aperture" (at flow rate Q_A). To maintain a stable plume, it is crucial that $Q_A > Q_I$; typically, Q_A = 2.5 Q_I.

The five-letter word "CELLS" was written in DiI (a lipophilic membrane dye) on fixed fibroblast cells in just 3 minutes (**Figure 3.83e**); to produce the space between the letters, the probe was moved very quickly without interrupting the outflow of DiI. Selective treatment of live cells was demonstrated by trypsinizing target cells without affecting nearby ones (**Figure 3.83f**). More recently, David Juncker, from Delamarche's group and now directing his own group at McGill University in Canada, has produced a transparent, cheaper version of the fountain pen entirely in PDMS; the apertures on the flat surface are created simply by slicing a set of microchannels with a razor blade.

To simplify the fluidic controls—which are required to keep diffusion under control—Shuichi Takayama's group (University of Michigan at Ann Arbor) has produced a much simpler microfluidic probe (**Figure 3.84**) that keeps diffusion at bay simply by a clever choice of the solutions: the cells are bathed in an aqueous polymeric solution of PEG (molecular weight, 8000) and the solutions locally delivered to the cells are dissolved in an aqueous dextran solution (molecular weight, 500,000). Both polymers are highly immiscible and nontoxic, with dextran always at the bottom due to its higher density. (Cells were cultured in serum containing DMEM culture medium with 4% PEG or 5% dextran for 24 hours, resulting in viabilities >96%.)

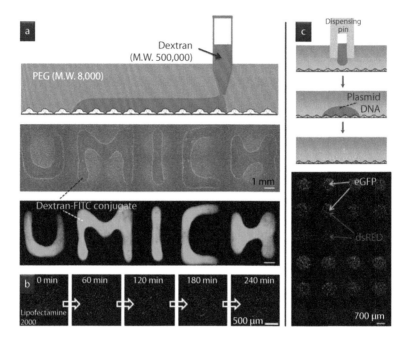

FIGURE 3.84 Local delivery of fluids onto cells using a two-phase system. (From H. Tavana, A. Jovic, B. Mosadegh, Q. Y. Lee, X. Liu, K. E. Luker, G. D. Luker, S. J. Weiss, and S. Takayama, "Nanolitre liquid patterning in aqueous environments for spatially defined reagent delivery to mammalian cells," *Nat. Mater.* 8, 736–741, 2009. Reprinted with permission from the Nature Publishing Group. Figure contributed by Shu Takayama.)

3.9.5 Gradients Delivered through a Semipermeable Barrier

We note that most of the previous devices rely on flow (and some on *fast* flow) to generate gradients or dilutions. For some applications, in particular, those that include cells in the microchannels, flow itself is a concern. Thus, there have been many efforts directed at developing gradient generators capable of producing gradients with negligible flow by interposing a semipermeable barrier between the area containing cells and the channel delivering the flow.

3.9.5.1 Gradient Generators that Incorporate Porous Membranes

One design option, first proposed by Matthew DeLisa and colleagues at Cornell University in 2006, is to build the walls of the device in … nitrocellulose paper! (Of note, this article can be considered the first contribution to the field of paper microfluidics.) This is a clever option because the paper has a defined thickness (140 μm) and pore size (0.45 μm) and can be laser-cut to define 400-μm-wide channels. The device consists of three channels, capped with glass. The gradient is created in the middle channel (where the cells reside) and the outer channels act as the (infinite) source and the (infinite) sink for the chemoattractant. Thus, the cells are never exposed to flow and are exposed to a linear gradient of chemoattractant (Figure 3.85).

A different configuration (in which diffusion occurs in the vertical direction) has been proposed by Michel Maharbiz and colleagues from the University of California at Berkeley. As semipermeable barriers, they used thin polyester membranes (containing 0.4 μm pores, commercially available as filters) that allow for diffusive but not convective transport. (This is technically incorrect: there is always some amount of convection as long as there is a pressure differential, an inlet, and a path to an outlet; in the Maharbiz design, convection is minimized by both the smallness of the pores, which maximizes resistance, and by the symmetry of the

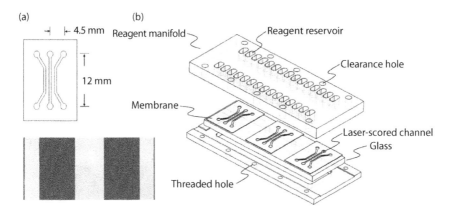

FIGURE 3.85 Gradient generator based on diffusion through nitrocellulose paper. (From Jinpian Diao, Lincoln Young, Sue Kim, Elizabeth A. Fogarty, Steven M. Heilman, Peng Zhou, Michael L. Shuler, Mingming Wu, and Matthew P. DeLisa, "A three-channel microfluidic device for generating static linear gradients and its application to the quantitative analysis of bacterial chemotaxis," *Lab Chip* 6, 381–388, 2006. Reproduced with permission from The Royal Society of Chemistry.)

design, which minimizes the pressure differential.) The device consists of two PDMS chambers, with the membrane acting as the roof of the bottom chamber and the floor of the top chamber (see **Figure 3.86**); the cells are cultured atop the membrane and any gradient introduced into the bottom chamber is "seen" by the cells by diffusion through the pores—even though flow rates in the bottom chamber can be quite fast, allowing for subsecond-scale switching of the gradient.

3.9.5.2 Gradient Generators that Incorporate Hydrogels
Hydrogels are the most readily available source of semipermeable barriers because they can be molded into the microchannels from the liquid state. Importantly, many hydrogels, such as collagen gel and Matrigel, are suitable for 3-D cell culture; hence, a number of devices exploit the use of biological hydrogels to both shield the cells from flow and to provide 3-D anchorage for the cells.

In 2007, a Cornell University team led by Michael Shuler and Mingming Wu used an idea introduced a few years earlier by Joe Tien's group (see Section 3.5.1.5): instead of engineering a device that contains a hydrogel, why not make the device in hydrogel *entirely*? They molded a simple three-channel device in agarose, such that the two peripheral channels acted as source and sink of the gradient, and the walls of the central channel—where the cells were cultured—acted as a flow barrier (**Figure 3.87**). Periodic recirculation of the source and the sink channels refreshed the solution at the walls and thus ensured that the gradient was time-invariant. (A note for historians: the footprint of this design was almost identical to the design by DeLisa made with laser-cut nitrocellulose shown in **Figure 3.85**; indeed, Shuler and Wu were coauthors of that DeLisa article, so they were reusing a successful idea!)

The Cornell agarose device has three important limitations. First, agarose does not have good optical clarity so it interferes with some microscopic measurements, such as phase-contrast microscopy, that require that light crosses the roof of the device. (That is not the case of fluorescence microscopy with an inverted microscope setup, which only collects light emitted from the sample through the floor of the device.) Second, although agarose is very cheap, the device is not reusable. Lastly, it cannot incorporate microvalves or micropumps for automation.

Richard Zare's group at Stanford University first demonstrated in 2006 an arbitrary gradient generator that does not require any flow, whereby the channel is sealed against the plane of a hydrogel slab preloaded with the gradient of interest (**Figure 3.88**). The idea is that the fluid in the (arbitrarily shaped) channel quickly equilibrates in contact with the hydrogel (which gets continuously loaded by two reservoirs); however, the hydrogel gradient is stationary only on first approximation because

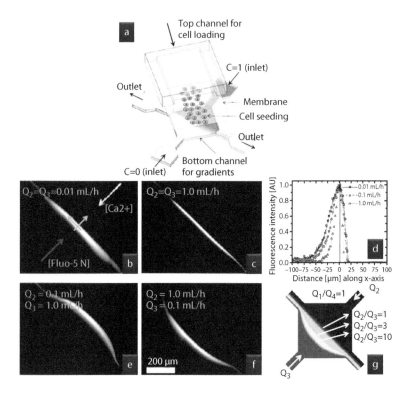

FIGURE 3.86 Gradient generator based on transport through a porous polyester membrane. (From Taesung Kim, Mikhail Pinelis, and Michel M. Maharbiz, "Generating steep, shear-free gradients of small molecules for cell culture," *Biomed. Microdevices* 11, 65–73, 2009. Adapted with permission from Springer. Figure contributed by Taesung Kim and Michel Maharbiz.)

the fluid in the channel slowly "smears" the features of the gradient by diffusion (it acts as a local short circuit) and, if the PDMS does not seal well against the slab, the hydrostatic pressure of the fluid column may wick in between the slab and the PDMS, confounding the values of the gradient.

The problem of spilling has an easy solution, as proposed by several groups: the hydrogel can be formed directly inside PDMS microchannels by injecting its precursor in liquid form (gelation is a process that typically takes a few minutes, with a small temperature change). In its simplest implementation, the sink and the source of the gradient are located at the inlets of the device, as demonstrated by David Beebe's group at the University of Wisconsin at Madison (**Figure 3.89**). This configuration, although simple, requires a careful gradient maintenance protocol because the source needs to be periodically replenished with a concentration *smaller* than the initial concentration (because the source is already equilibrated with the top part of the gel, which differs from the bottom). Thus, the height of the gel, the diffusivity of the factor in the gel, the volume of solution in the source, and the cross-sectional area at the channel entrance influence the equilibrium time—and the maintenance protocol.

Noo Li Jeon and coworkers (then at the University of California, Irvine) have proposed a hydrogel-in-channel design, termed a "ladder chamber," whereby the source and sink solutions can themselves be perfused through microchannels (**Figure 3.90**). The hydrogel precursor can be introduced through the same (source or sink) channels used to deliver the gradient (**Figure 3.90a–c**) or through a dedicated inlet (**Figure 3.90d**). A number of gel materials such as collagen type I, Matrigel, and fibrin were successfully polymerized in the device. Because of the high resistance of the fluid–hydrogel contact on either side of the hydrogel, exchanges of fluid on the sink or source channel do not perturb the hydrogel microenvironment. The shape of the gradient equilibrates

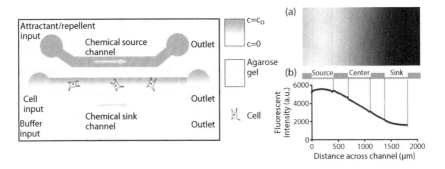

FIGURE 3.87 Gradient generator made in agarose. (From Shing-Yi Cheng, Steven Heilman, Max Wasserman, Shivaun Archer, Michael L. Shuler, and Mingming Wu, "A hydrogel-based microfluidic device for the studies of directed cell migration," *Lab Chip* 7, 763–769, 2007. Reproduced with permission from The Royal Society of Chemistry.)

in approximately 15 to 30 minutes as long as the sink/source stay constant. Importantly, varying the design of the gradient region (**Figure 3.90c** and **d**) can be used to change the gradient's shape (**Figure 3.90e** through **g**) and gradients through dissimilar gels are also possible (**Figure 3.90h**).

In most chemotaxis experiments, the cells are presented with a "control" condition at some point—a condition that does not contain the chemotactic factor, to verify that cells do not respond

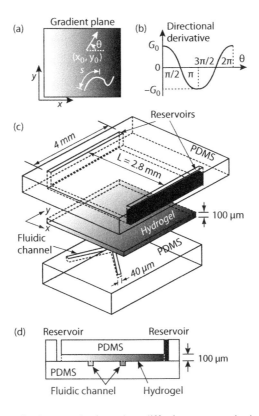

FIGURE 3.88 Arbitrary gradient generator based on diffusion across a hydrogel slab. (From Hongkai Wu, Bo Huang, and Richard N. Zare, "Generation of complex, static solution gradients in microfluidic channels," *J. Am. Chem. Soc.* 128, 4194–4195, 2006. Reprinted with permission of the American Chemical Society.)

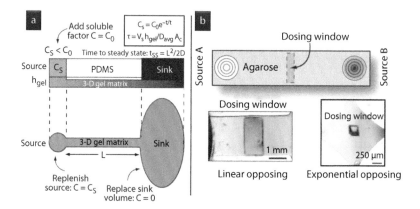

FIGURE 3.89 Agarose-filled microchannels as gradient generators. (From V. Abhyankar, Michael W. Toepke, Christa L. Cortesio, Mary A. Lokuta, Anna Huttenlocher, and David J. Beebe, "A platform for assessing chemotactic migration within a spatiotemporally defined 3D microenvironment," *Lab Chip* 8, 1507–1515, 2008. Reproduced with permission from The Royal Society of Chemistry.)

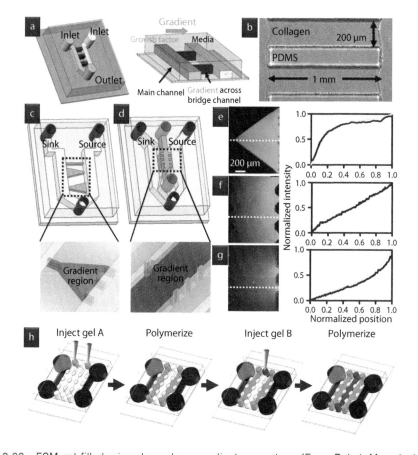

FIGURE 3.90 ECM gel-filled microchannels as gradient generators. (From Bobak Mosadegh, Carlos Huang, Jeong Won Park, Hwa Sung Shin, Bong Geun Chung, Sun-Kyu Hwang, Kun-Hong Lee, Hyung Joon Kim, James Brody, and Noo Li Jeon, "Generation of stable complex gradients across two-dimensional surfaces and three-dimensional gels," *Langmuir* 23, 10910–10912, 2007. Reproduced with permission from the American Chemical Society. Figure contributed by Noo Li Jeon.)

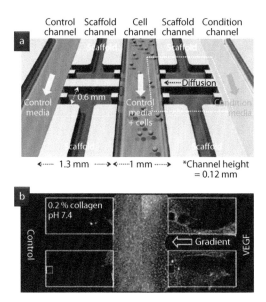

FIGURE 3.91 Gradient generator incorporating collagen gels as diffusional barriers. (From Seok Chung, Ryo Sudo, Peter J. Mack, Chen-Rei Wan, Vernella Vickerman, and Roger D. Kamm, "Cell migration into scaffolds under co-culture conditions in a microfluidic platform," *Lab Chip* 9, 269–275, 2009. Reproduced with permission from The Royal Society of Chemistry.)

to it. However, the control experiment is usually done with a different set of cells, not with the same set of cells because there is no way to expose the cells to two gradients simultaneously (to see which one the cells prefer). Roger Kamm's group at MIT has demonstrated a design that simultaneously presents the cells with both the control condition and the gradient condition—and the cells choose (Figure 3.91). The clever design also takes advantage of the fact that the hydrogel precursor is actually a viscous solution, so that when it is introduced in a channel that has small side-openings that connect to other channels, the hydrogel precursor will not spill into those other channels—this point is crucial because the small side-openings are the same openings through which the gradient is later established and cells are allowed to migrate. The channel that contains the cells is in the center and can be filled with cell culture medium or ECM hydrogel (potentially allowing for 3-D cell migration experiments). Ultimately, this device could benefit from all the automation technology that already exists for PDMS (microvalves, micropumps, etc.).

Hydrogels are inexpensive to produce in a research setting but are not adequate as commercial devices—they are not easily assembled nor manufactured in large quantities. In addition, they cannot be produced thin, so they are not suitable for fast-varying gradients or, equivalently, for steep gradients of small molecules—the hydrogel "blurs" them.

3.10 Combinatorial Mixers

In many applications, it is desirable to produce not only several titrations of a given compound but also *mixtures of two or more* compounds, including their titrations. The author's laboratory at the University of Washington in Seattle has devised a microfluidic mixer design that produces all the mixture combinations of four dilutions of (two) input compounds and delivers the sixteen mixture combinations in separate outlet microchannels (Figure 3.92). The device features four different flow levels made by stacking nine laser-cut Mylar laminates. The fluidic network has a symmetric design that guarantees that the flow rates are nearly identical at all the outlets. Such systems should find uses in cancer and toxicology screening.

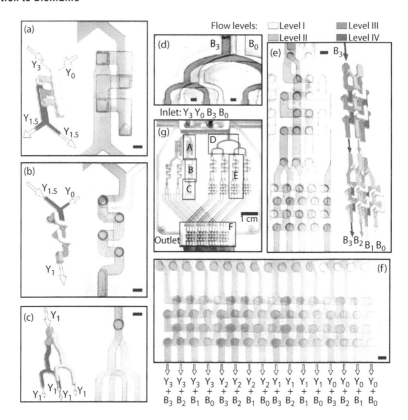

FIGURE 3.92 Combinatorial micromixer. Scale bars are 1 mm. (From Christopher Neils, Zachary Tyree, Bruce Finlayson, and Albert Folch, "Combinatorial mixing of microfluidic streams," *Lab Chip* 4, 342, 2004. Reproduced with permission from The Royal Society of Chemistry.)

3.10.1 Homogenizers

We have seen in Section 3.9.2 how laminar flow can be a powerful tool for producing gradients in microfluidic chambers. In many applications of biotechnology and cell biology, however, we are interested in mixing (i.e., homogenizing) the inlet solutions rather than establishing gradients with them. Obviously, this goal presents a challenge under laminar flow, because in laminar flow the only mechanism available for mixing is diffusion—unless one is really smart about it. The challenge, which essentially consists of homogenizing the solutions in as short a time and space as possible, has attracted very bright engineers to the field, so much so that probably the cleverest designs in microfluidics are seen in the area of homogenizers.

3.10.1.1 Homogenization Directed by Pulsatile Flow: The "Dahleh Micromixer"

In 1999, Mohammed Dahleh (1961–2000) and Igor Mezic at the University of California in Santa Barbara proposed in a theoretical article a novel scheme to homogenize solutions in a very short amount of space based on shearing flows orthogonally to each other—that is, quickly alternating the injection of two flows into the channel. The idea is brilliant in that (a) it potentially reduces the diffusive path (depending on the pulsatile frequency) and (b) by transposing the direction in which the fluids need to diffuse to homogenize the solution (the flow direction) by 90 degrees, Taylor dispersion (see **Figure 3.2b**) greatly contributes to enhancing homogenization. The scheme was put to practice a few months later, both by Dahleh's group and independently by another team led by Chih-Ming Ho at the University of California (Los Angeles). Since then, the idea has been perfected by several groups. For example, Nadine Aubry's group at the New Jersey Institute of Technology (Newark) has applied out-of-phase sinusoidal voltages to pump the fluids into the two inlets of a Dahleh micromixer, such that the total flow rate is kept constant (**Figure 3.93**).

Recently, Igor Mezic's group has developed a sophisticated version of the Dahleh micromixer whereby the main flow (containing the two solutions to be homogenized) is sheared sideways by several secondary channels, which are actuated with an oscillating pump at a frequency on the order of 2 Hz (**Figure 3.94**). The two fluids are mixed within 200 μm in less than 10 ms, which represents an improvement of five orders of magnitude over diffusion-limited mixing.

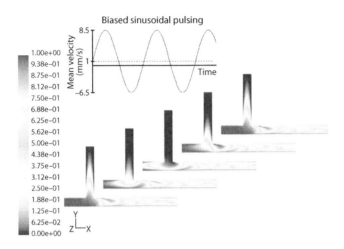

FIGURE 3.93 The Dahleh micromixer driven with sinusoidal pulsing. (From Ian Glasgow and Nadine Aubry, "Enhancement of microfluidic mixing using time pulsing," *Lab Chip* 3, 114–120, 2003. Reproduced with permission from The Royal Society of Chemistry.)

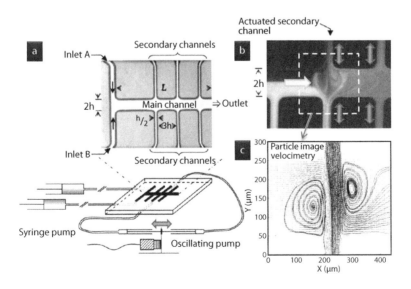

FIGURE 3.94 Shear superposition micromixer. (From Frédéric Bottausci, Caroline Cardonne, Carl Meinhart, and Igor Mezic, "An ultrashort mixing length micromixer: The shear superposition micromixer," *Lab Chip* 7, 396–398, 2007. Reproduced with permission from The Royal Society of Chemistry.)

3.10.1.2 Homogenization by 3-D Serpentines

Homogenization by pulsatile flow is very fast but produces a nonuniform flow (unless it is sinusoidally driven) and requires extra energy consumption and hardware. Several groups have researched designs that passively homogenize the solutions without any energy input. Because solutions only mix by diffusion when flowing in parallel flow lines, a successful strategy has been to force the flow lines to cross each other using "serpentine" designs with 3-D turns, a strategy inspired on macroscale "twisted-pipe" configurations. This strategy was first implemented on the microscale in 2000 by a collaborative effort between the groups of Hassan Aref (who had shown chaotic advection in twisted pipes) and David Beebe, then both at the University of Illinois at Urbana-Champaign (**Figure 3.95**). Characteristically of these serpentine mixers, the mixing effect is enhanced as the Re increases (**Figure 3.95c**, triangles). If the device is built with planar (2-D) turns, then mixing is not enhanced as much (**Figure 3.95c**, squares). It should be emphasized that, whereas the flow lines cross by an effect of the geometry of the channel, the regime is still laminar—there is no turbulence (**Figure 3.95d**).

There has been an intense debate about how to build the best possible serpentine—what should each unit (each turn) look like? The earliest implementation, by David Beebe's group, was limited by the use of silicon micromachining in the range of geometries and turns that could be implemented—only overlapping "U"-shaped turns were possible (see **Figure 3.95a** inset). With the advent of PDMS micromolding, it has been possible to investigate various other types of turns, including overlapping "C" and "L" turns (**Figure 3.96**). In **Figure 3.96d**, we see

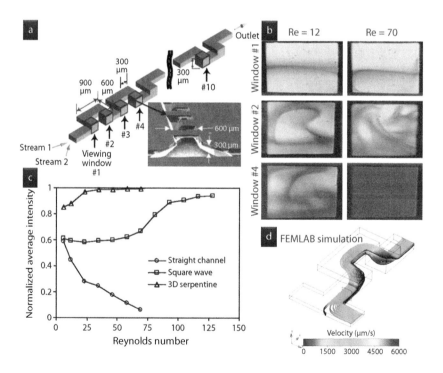

FIGURE 3.95 Passive micromixer based on chaotic advection. (a–c, from Robin H. Liu, Mark A. Stremler, Kendra V. Sharp, Michael G. Olsen, Juan G. Santiago, Ronald J. Adrian, Hassan Aref, and David J. Beebe, "Passive mixing in a three-dimensional serpentine microchannel," *J. Microelectromech. Syst.* 9, 190, 2000; d, FEMLAB simulation of flow in 3-D serpentine turns, courtesy of Paul Yager. Figure contributed by David Beebe.)

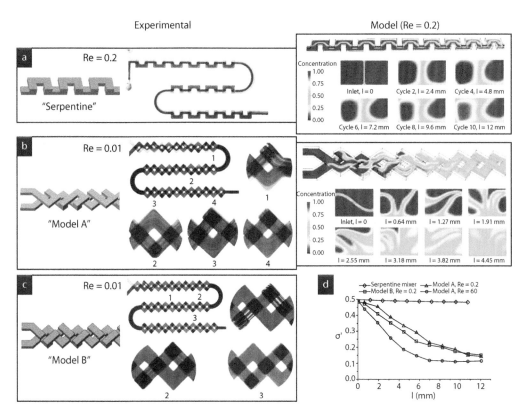

FIGURE 3.96 Microfluidic homogenizer with complex 3-D architecture. (From H. M. Xia, S. Y. M. Wan, C. Shu, and Y. T. Chew, "Chaotic micromixers using two-layer crossing channels to exhibit fast mixing at low Reynolds numbers." *Lab Chip* 5, 748–755, 2005. Reproduced with permission from The Royal Society of Chemistry.

again that a nonobvious and gratifying feature of these mixers is that they perform better as the Re…increases!

We note that one drawback of these 3-D serpentine mixer designs is that they consume large amounts of chip real estate. However, using lamination technologies, it is possible to "fold" back a serpentine channel over itself in a multilayer device, as shown in the example of **Figure 3.92a**.

An interesting feature about serpentine mixers is that a small change in their architecture can drastically change their performance. In 2004, a team led by Jun Keun Chang from Seoul National University in Korea presented a serpentine composed of overlapping "sigmoids"—a small difference, one might think, from overlapping square "C"'s—(see **Figure 3.97**). Yet the improvement with respect to C-segment-type serpentines was significant, especially at high Re. The mechanism of operation of this mixer is best understood by looking at where the fluid is forced to go through (see cross-sectional planes in **Figure 3.97d**), and realizing that there must occur a rotational motion (which enhances mixing).

In 2005, Chong Ahn's laboratory in Pohang University of Science and Technology in Pohang (South Korea) introduced a mixer with overlapping "F" turns (**Figure 3.98**). Although it looks very similar to a serpentine mixer, it is based on a rather different principle. The novelty of the design is that the "F" essentially behaves like a fork that splits the flow in the horizontal dimension and recombines (stacks) it in the vertical dimension, yielding a layered flow at the outlet that significantly shortens the diffusive path between the inlet solutions. The characteristic

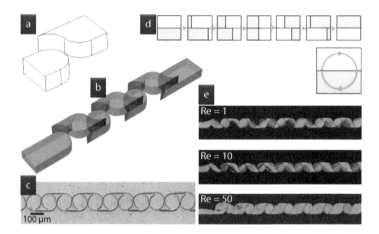

FIGURE 3.97 A passive micromixer that induces fluid rotation. (From Sung-Jin Park, Jung Kyung Kim, Junha Park, Seok Chung, Chani Chung, and Jun Keun Chang, "Rapid three-dimensional passive rotation micromixer using the breakup process," *J. Micromech. Microeng.* 14, 6–14, 2004. Reproduced with permission from the Institute of Physics. Figure contributed by Jun Keun Chang.)

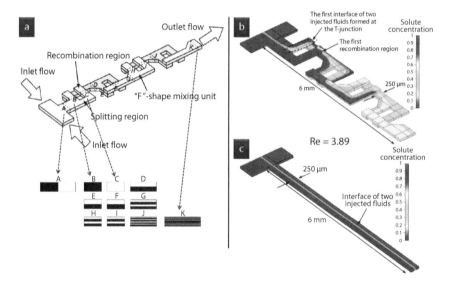

FIGURE 3.98 Microfluidic homogenizer with 3-D "F" splitter-recombiners. (From Dong Sung Kim, Se Hwan Lee, Tai Hun Kwon, and Chong H. Ahn, "A serpentine laminating micromixer combining splitting/recombination and advection," *Lab Chip* 5, 739–747, 2005. Reproduced with permission from The Royal Society of Chemistry.)

mixing length is approximately 5 to 7 mm within the range of Re = 0–12; therefore, mixing in this design is very insensitive to Re at low Re, although it works better for faster flow rates.

Of course, splitting and recombining can also be achieved with crisscrossing 3-D microchannel architectures, but the drawback is that these can only be manufactured with two-photon absorption stereolithography. Dong-Pyo Kim's group from KAIST the Korea Advanced Institute of Science and Technology (in Daejeon, South Korea) has used this "brute force" approach to manufacture highly efficient mixers in SU-8. At Re = 1 and Pe = 1000, complete mixing of water and ethanol is achieved in just 250 μm (**Figure 3.99**).

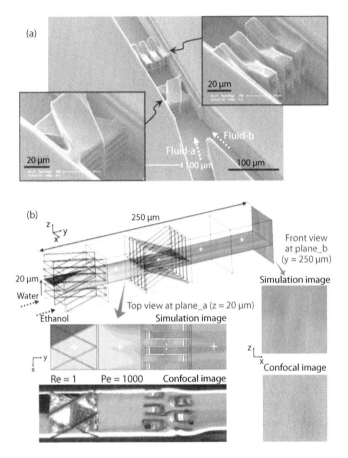

FIGURE 3.99 A crisscrossing 3-D micromixer. (From Tae Woo Lim, Yong Son, Yu Jin Jeong, Dong-Yol Yang, Hong-Jin Kong, Kwang-Sup Lee, and Dong-Pyo Kim, "Three-dimensionally crossing manifold micro-mixer for fast mixing in a short channel length," *Lab Chip* 11, 100–103, 2011. Reproduced with permission from The Royal Society of Chemistry.)

3.10.1.3 Homogenization by Tesla Mixer

In 2005, the Ahn laboratory reported the use of a well-known microfluidic component—the "**Tesla valve**," an asymmetric structure that offers more fluidic resistance in one direction than in the other—as a homogenizer (**Figure 3.100**); as it turns out, the valve causes transverse dispersion when a heterogeneous laminar flow crosses it. The researchers reported impressive insensitivity to flow rates from 0 to 100 μL/min, a range in which they obtained more than 95% homogenization; the device also worked better at faster flow rate. The device essentially exploits the **Coandă effect**, also known as "boundary layer attachment" (here, occurring at the walls of the end of the loops; Romanian aircraft inventor Henri Coandă discovered in 1910 that a stream of fluid flowing next to a convex surface has a tendency to stay attached to the surface, contrary to the expectation that, by inertia, the fluid would detach from the surface and follow the tangent).

3.10.1.4 Homogenization Directed by Surface Topology

What we learn from the serpentine designs is that it is beneficial to adopt strategies that cause the fluid lines to twist and intermingle, because then, the diffusive path, and the total mixing length, is shortened. But is it really necessary to create a 3-D channel to make the flow somersault, so to speak? A group led by Howard Stone and George at Harvard University realized in

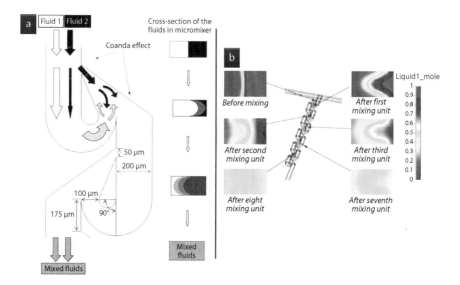

FIGURE 3.100 Microfluidic homogenizer based on Tesla mixer. (From Chien-Chong Hong, Jin-Woo Choi, and Chong H. Ahn, "A novel in-plane passive microfluidic mixer with modified Tesla structures," *Lab Chip* 4, 109, 2004. Reproduced with permission from The Royal Society of Chemistry.)

2002 that, in fact, you do not. Because fluids are incompressible, inasmuch the same way that a car at high speed can flip over if one of its front wheels hits a small post, a fluid will also start a rotating (helical) motion if it enters a channel that has slanted grooves on its floor (**Figure 3.101a** and **b**). More efficient swirling patterns can be achieved with an asymmetric "herringbone" or "Chevron" groove design whose center is shifted periodically (**Figure 3.101c** and **d**). In other words, the small amount of fluid that goes into the grooves transmits its momentum to the rest of the fluid in the channel, causing it to rotate; because the grooves are slanted, a recirculation occurs. This mixer (along with one of the early serpentine micromixers) was originally dubbed a "chaotic micromixer," which is a big word to express how a small deviation in the position of a particle at the inlet can result in a large deviation at the outlet.

Jing-Tang Yang and colleagues at National Tsing Hua University in Taiwan have done a thorough study of the herringbone mixer and several design variations and have found that by introducing a zigzag barrier on the roof of the channel(!), more recirculation (advection) is introduced, thus enhancing mixing (**Figure 3.102**). Not surprisingly, the performance of this design, which they call "circulation disturbance micromixer," is enhanced at higher Re. With specialized (tilted illumination) photolithography, it has been possible to fabricate mixers (in SU-8) that also have grooves…on the walls!

3.10.1.5 Homogenization Induced by Surface Charge Patterns

Patterns of surface electrical charges have been successfully exploited to create local recirculation of fluids. George Whitesides' group at Harvard University was able to deposit stripes of thin polymer layers with similar but reverse electrical charges, which locally modulates the sign of electro-osmotic flow in an orthogonally placed microchannel (**Figure 3.103**). Although clever, this design is sensitive to surface contamination and surface aging, which changes the charge density on the surface.

3.10.1.6 Homogenization Induced by Bubble-Based Acoustic Streaming

Ultrasonic waves have been investigated for a while as a means to improve mixing in microchannels by inducing pressure waves that disturb flow. Tony Jun Huang's group from

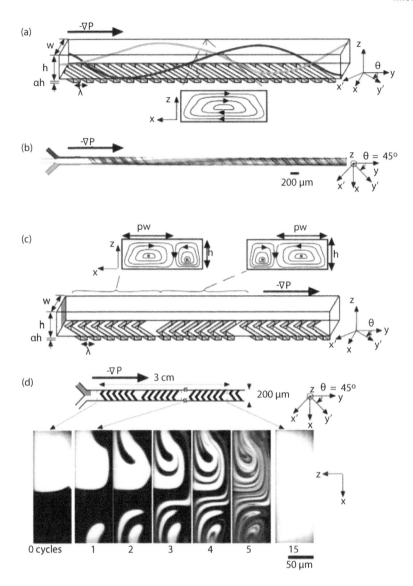

FIGURE 3.101 Homogenization directed by surface topology. (From Stroock, A. D., Dertinger, S. K. W., Ajdari, A., Mezic, I., Stone, H. A., and Whitesides, G. M., "Chaotic mixer for micro-channels," *Science* 295, 647–651, 2002; and Abraham D. Stroock and George M. Whitesides, "Controlling flows in microchannels with patterned surface charge and topography," *Acc. Chem. Res.* 36, 597–604, 2003. Reprinted with permission from the American Chemical Society.)

Pennsylvania State University, implementing a variation of an earlier design by Abraham Lee's group, trapped a bubble in a horseshoe structure in the middle of a channel and vibrated it at 70 kHz/8 V with a piezoelectric transducer glued to the side of the device (Figure 3.104). Because of the "acoustic streaming" phenomenon (see Section 3.4.2), which focuses acoustic energy on the oscillating air–fluid interface of the bubble, the surrounding flow is perturbed, resulting in mixing. It is not yet known whether the ultrasonic vibrations required for mixing are damaging to cells.

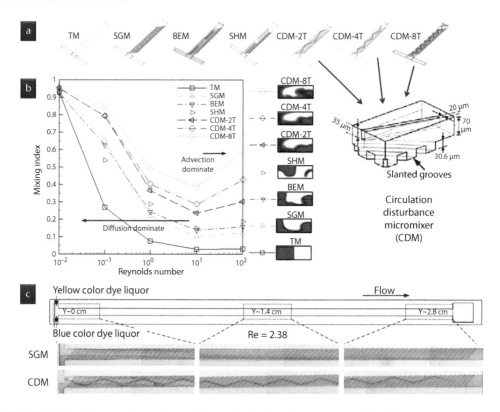

FIGURE 3.102 Homogenizer enhanced by a circulation-disturbance barrier. (From Jing-Tang Yang, Ker-Jer Huang, Kai-Yang Tung, I-Chen Hu, and Ping-Chiang Lyu, "A chaotic micromixer modulated by constructive vortex agitation," *J. Micromech. Microeng.* 17, 2084–2092, 2007. Reproduced with permission from the Institute of Physics. Figure contributed by Jing-Tang Yang.)

3.10.2 Micromixers Incorporating Dynamic Elastomeric Microelements

In Section 3.9, we have already seen micromixer designs that are "dynamic," in the sense that their performance could be changed with time—for example by changing the flow rate. However, that performance could not be changed unless the user changed the flow going into the mixer. In this section, we will see several designs in which the performance of the mixer can be tuned by controlling elastomeric microelements integrated into the device.

3.10.2.1 Microvalve-Based Mixers

A simple way to control mixing in a microdevice is to incorporate microvalves and micropumps that can be programmed to do the mixing on a desired schedule. Stephen Quake and coworkers at Caltech, then designed a microfluidic "formulator" that looks like a rotary mixer, which accepts plugs of several inputs through a multiplexer (**Figure 3.105**); note that if the plugs do not rotate, the diffusion path to achieve homogenization is the whole circle. However, as the plugs rotate, Taylor dispersion typical of the parabolic flow profile encountered in microchannels (see **Figure 3.2** in Section 3.2.4.2) greatly speeds up mixing, because the plugs get ensheathed within each other and the diffusion path is now limited to the width of the channel. Complete mixing takes an impressive 3 seconds.

A different concept, originally presented by Quake's group, also achieves mixing but uses only nanoliter-sized chambers that communicate via microvalves. The power of the approach is that groups of valves are operated simultaneously using the same pneumatic channel, so it is possible to create titrations in parallel on a very small scale. This system is now commercialized by

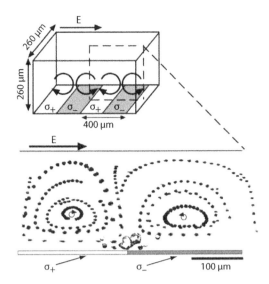

FIGURE 3.103 Homogenization induced by surface charge patterns. The patterned electro-osmotic flow is shown schematically with arrows on the top diagram. The bottom is a micrograph that shows the trajectories of fluorescent beads in a vertical plane through the channel. The micrograph was compiled from multiple series of images that were taken at intervals of 0.2 seconds. The direction of motion is indicated by the arrows. The applied electric field was $E = 95$ V/cm. The globular objects at the bottom of the image are beads that have stuck to the surface at the stagnation point in the flow at the transition between positively and negatively charged regions. (From A. D. Stroock, M. Weck, D. T. Chiu, W. T. S. Huck, P. J. A. Kenis, R. F. Ismagilov, and G. M. Whitesides, "Patterning electro-osmotic flow with patterned surface charge," *Phys. Rev. Lett.* 84, 3314–3317, 2000; and A. D. Stroock and G. M. Whitesides, "Controlling flows in microchannels with patterned surface charge and topography," *Acc. Chem. Res.* 36, 597–604, 2003. Reprinted with permission from the American Chemical Society.)

Fluidigm, who sells the chips for high-throughput screening of protein crystals. If implemented in the doormat microvalve design, as shown in **Figure 3.106**, the volumes of the chambers are fully predicted by photolithography and they do not distort transmitted-light microscopy modes (such as phase-contrast microscopy). The titrations are determined simply by the ratios of the volumes of the chambers.

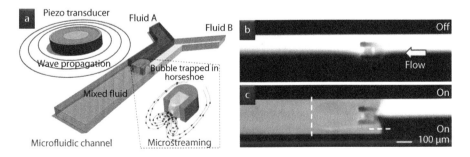

FIGURE 3.104 Bubble-based on–off millisecond homogenizer. (From Daniel Ahmed, Xiaole Mao, Jinjie Shi, Bala Krishna Juluri, and Tony Jun Huang, "A millisecond micromixer via single-bubble-based acoustic streaming," *Lab Chip* 9, 2738–2741, 2009. Reproduced with permission from The Royal Society of Chemistry.)

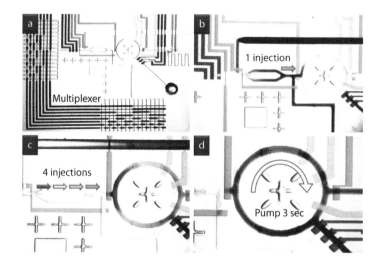

FIGURE 3.105 Automated combinatorial mixer based on microvalves. In all images, the diameter of the mixing ring is 1.5 mm. (a) The peristaltic pump pneumatic channels are filled with yellow dye and the multiplexer channels are filled with black dye. (b) Injection of ~250 pL of blue dye into the rotary mixer. (c) Color gradient formed by consecutive injections of blue, green, yellow, and red dyes into the rotary mixer. (d) Pumping around the ring for 3 seconds results in complete mixing. (From Carl L. Hansen, Morten O. A. Sommer, and Stephen R. Quake, "Systematic investigation of protein phase behavior with a microfluidic formulator," *Proc. Natl. Acad, Sci. U. S. A.* 101, 14431–14436, 2004. Copyright (2004) National Academy of Sciences, U.S.A. Figure contributed by Stephen Quake.)

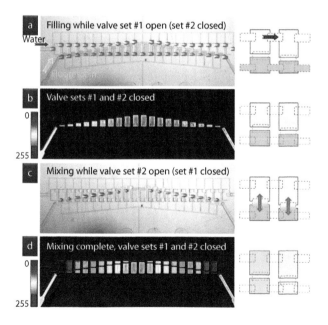

FIGURE 3.106 Metering of nanoliter-scale volumes in a micromixer using microvalves. (From Nianzhen Li, Chia-Hsien Hsu, and Albert Folch, "Parallel mixing of photolithographically defined nanoliter volumes using elastomeric microvalve arrays," *Electrophoresis* 26, 3758–3764, 2005. Figure contributed by Nianzhen Li.)

3.10.2.2 Tunable Microtopography

A PDMS microvalve is simply a rectangular pad of PDMS membrane that is deforming the wall of a microchannel or bridging the wall between two microchannels. What if the pad were smaller than the channel, becoming a variable topographical feature of the channel, such that it could be used to alter the flow? The author's laboratory at the University of Washington in Seattle has extended the concept of microvalves to such "tunable microtopographies" (Figure 3.107). We have seen in Section 3.10.1.4 how the surface topology of a microchannel, and in particular herringbone grooves, can be used to homogenize a mixture. Likewise, tunable microtopographies can be used to make grooves appear and disappear (Figure 3.107), so that the mixer can be turned on and off; this active micromixer can be used to create unique spatiotemporal concentration gradients downstream of the microchannel. In principle, it should be possible to implement similar systems with microtopographies that are made of smart polymers, so they could be responsive to stimuli (see Section 3.8.1.12).

The tunable microtopographies can also be used to form a microstructured membrane that traps picoliter-sized amounts of fluids inside a microchannel (Figure 3.108a) or that functionalizes the substrate within the microchannel in predetermined shapes (Figure 3.108b). We note that the concepts of microstructured membranes and microvalves can now be merged because there is a size for which it does not matter whether the features are defined on the PDMS bulk or on the membrane. For example, the microvalve-based nanoliter mixer shown previously (see Figure 3.106) can be further miniaturized down to the picoliter level (Figure 3.108c) simply by performing two-level photolithography (the photomask for the 3.5 pL chambers is a standard plastic mask—it is the height of the photoresist that determines the low volume of the chamber).

3.10.2.3 Vortex-Type Mixer

Gwo-Bin Lee and colleagues at National Cheng Kung University in Taiwan have presented a very creative device that imitates the influx and outflux that occurs in the wells of a 96-well plate when solutions are pipetted in and out for homogenizing the well's contents—except here, the mixing can be done in principle with hundreds of wells in parallel, in less than a second, in a totally automated way! As shown in Figure 3.109, the central unit consists of the well and, next to it, four flaps that can be actuated to cause the level of the fluid in the well to rise and lower very quickly in a vortex-type fashion. Homogenization is achieved in 0.7 seconds.

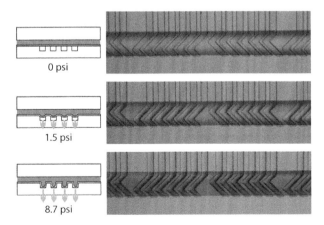

FIGURE 3.107 On/off chaotic micromixer. (From Hsu, C.-H. and Folch, A., "Spatiotemporally-complex concentration profiles using a tunable chaotic micromixer," *Appl. Phys. Lett.* 89, 144102, 2006. Figure contributed by Chia-Hsien Hsu.)

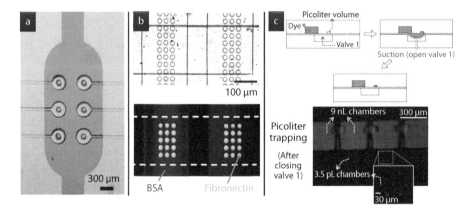

FIGURE 3.108 Microstructured membranes and microvalves for trapping fluids and for substrate patterning. (a; from Hsu, C.-H., Chen, C., and Folch, A. "Microfluidic devices with tunable micro-topographies," *Appl. Phys. Lett.* 86, 023508, 2005. Figure contributed by Chia-Hsien Hsu; b, from Siva A. Vanapalli, Daniel Wijnperle, Albert van den Berg, Frieder Mugeleac, and Michel H. G. Duits, "Microfluidic valves with integrated structured elastomeric membranes for reversible fluidic entrapment and in situ channel functionalization," *Lab Chip* 9, 1461–1467, 2009. Reproduced with permission from The Royal Society of Chemistry; c, from David Cate, Nianzhen Li, and Albert Folch, "Microvalve-addressable picoliter chambers for single-molecule enzymology," *Proceedings of MicroTAS 2008*, San Diego, 2008. Figure contributed by David Cate.)

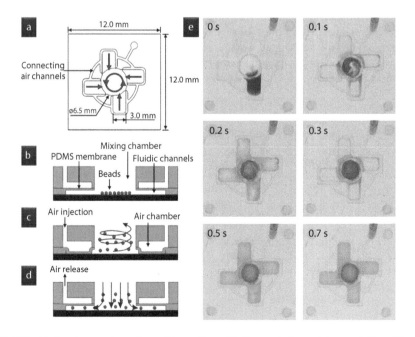

FIGURE 3.109 Vortex-type micromixer. (From Sung-Yi Yang, Jr-Lung Lin, and Gwo-Bin Lee, "A vortex-type micromixer utilizing pneumatically driven membranes," *J. Micromech. Microeng.* 19, 035020, 2009. Reproduced with permission from the Institute of Physics. Figure contributed by Gwo-Bin Lee.)

AND THE WINNER IS ...

IN ALL FAIRNESS, WHEN ASKING whether a given device gives "the best performance," we should first agree on a set of benchmark tests that allow us to compare it to its contemporary devices. The problem is that this set of benchmark tests, for homogenizers, keeps on changing in the BioMEMS field. It seems reasonable to assume that the best homogenizer should be the one that homogenizes solutions in the shortest amount of length; unfortunately, not all reports provide sufficient data to compare devices because the comparison is dependent on Reynolds number. Also, there are many applications (e.g., where cost is an issue) in which it is perfectly legitimate to sacrifice length for simplicity of operation or of fabrication. But for now, with scattered data, and with good sportsmanship, we can probably state that the brainchild of Tony Jun Huang, the "bubble micromixer" (Figure 3.104)—which achieves mixing practically instantly with minimal hardware, and can even be turned on and off—probably deserves the title of *The Micromixer Champion*.

3.11 Summary

Microfluidics is an enabling technology that allows for controlling fluids on the micrometer scale and in volumes down to the picoliter scale or lower. The laminar regime that governs the flow at these scales is a challenge for applications that require *homogenization* of an input mixture, but it is possible to accelerate mixing by breaking the pattern of parallel laminar flow—be it by inducing helical flow, by shearing the main flow transversally (using a secondary pump), or by using micropumps, among other strategies. On the other hand, laminar flow can be advantageous for mixing applications whose goal is to create a *gradient* and for keeping *droplets* (and cells) on a steady stream. An important feature of microfluidics is that for practically the same price of fabricating the microchannel, one can also integrate flow-control elements such as microvalves, micropumps, microfluidic resistors, and multiplexers, thus making its operation *amenable to automation*. This feature has enormous implications for the effect of microfluidics in health care and diagnostics because the end user (who pays for devices) demands user-friendly devices. Automation is also vital in cell biology and molecular biology research because it enables large-scale (hands-free, unassisted) experimentation.

Further Reading

Atencia, J., and Beebe, D.J. "Controlled microfluidic interfaces," *Nature* **437**, 648–655 (2005).

Brody, J., Yager, P., Goldstein, R., and Austin, R. "Biotechnology at low Reynolds numbers," *Biophysical Journal* **71**, 3430–3441 (1996).

Duffy, D.C., McDonald, J.C., Schueller, O.J.A., and Whitesides, G.M. "Rapid prototyping of microfluidic systems in poly(dimethylsiloxane)," *Analytical Chemistry* **70**, 4974–4984 (1998).

Nguyen, N.-T., and Wereley, S.T. *Fundamentals and Applications of Microfluidics*," Artech House (2002).

Purcell, E.M. "Life at low Reynolds number," *American Journal of Physics* **45**, 3–11 (1977).

Santiago, J. G., Wereley, S., Meinhart, C.D., Beebe, D.J., and Adrian, R.J. "A particle image velocimetry system for microfluidics," *Experiments in Fluids* **25**, 316–319 (1998).

Stewart, R.J., Ransom, T.C., and Hlady, V. "Natural underwater adhesives," *Polymer Physics* **49**, 757–771 (2011).

Stone, H.A., Stroock, A.D., and Ajdari, A. "Engineering flows in small devices: microfluidics toward a Lab-on-a-chip," *Annual Review of Fluid Mechanics* **36**, 381–411 (2004).

Teh, S.-Y., Lin, R., Hung, L.-H., and Lee, A.P. "Droplet microfluidics," *Lab on a Chip* **8**, 198–220 (2008).

White, F.M. *Fluid Mechanics*, McGraw-Hill (1979).

4

Molecular Biology on a Chip

THERE ARE MANY POSSIBLE MOTIVATIONS for developing microfluidic systems. Research can certainly take advantage of new tools for manipulation of small samples and single cells, as well as the potential for high-throughput processing of multiple samples in integrated micro-fluidic systems. By extension, the high (and still rising) costs of conventional molecular biology techniques has made microfluidics an almost-imperative tool. Not surprisingly, virtually every molecular biology assay has already been miniaturized. (If it has not been miniaturized as of this writing, it would be safe to bet that several groups are already working on it.)

LAB ON A CHIP VERSUS μTAS

ANDREAS MANZ AND COLLEAGUES at Ciba-Geigy in Switzerland first proposed in the early 1990s the integration into a single, miniature system of all the wet chemical and electrical detection technologies required to make laboratory measurements; hence, the term "microfabricated total analysis systems," or μTAS (read "micro-tas"), was coined. It made sense: microfabricated sensors (at the time, mostly silicon-based) were faster and cheaper (cost per device) than their macroscopic counterparts. So μTAS started as the natural, biomedical outgrowth of Si microfabrication technologies, and soon spread to many other countries (a little bit like a disease). The greeting "see you at μTAS" now refers to the international conference held every Autumn and that attracts more than 1000 participants from around the world. Throughout the 1990s, there was a search by those in the μTAS community for "the killer app"—the application that was going to establish microfluidics as a vital tool and practical product. One of those "killer apps" was point-of-care diagnostics, as shown later. It gradually became apparent that miniaturization had an enormous potential for tissue engineering and cell biology studies, so the more general term (not restricted to "analysis" only) "lab on a chip" was coined. It is a term that, probably because of its descriptive power, has stuck with the popular press—unlike our old, beloved μTAS: after all, we must admit that "total" and "system" mean different things to an engineer, a chemist, and a biologist. Now *Lab on a Chip* is also the name of a high-impact scientific journal.

4.1 The Importance of Miniaturizing Molecular Biology

We have already seen the conceptual benefits of miniaturization in general (Section 1.1) and microfluidics in particular (Section 3.1). To reiterate and focus our message on the benefits of miniaturization as applied to molecular biology applications, we may distinguish two classes of advantages:

A. Molecular biology performed on a microscale is *more efficient*; the devices allow for:

    a. Portability of the assay to the point-of-care

    b. Reduced reagent consumption

    c. Reduced animal use (a growing concern in our society)

    d. Expanded use of rare samples (stem cells, biopsies, circulating tumor cells, etc.)

B. Microscale molecular biology assays deliver *increased information content per unit cost*, thus they have *increased biological relevance*.

    a. Microscale assays provide new biological insights.

    b. Microscale assays can have more in vivo-like functionality (cocultures, gradient cultures, 3-D cultures), which increases the ability to predict in vivo efficacity when the assays are used as cell culture models for developing new drugs.

4.2 The Importance of Point-of-Care Diagnostics: Where Is Cost Really, Really, Really Important?

Our modern hospitals are the culmination of centuries of development of the best technology to help the most specialized physicians. Engineers, scientists, and doctors have labored to produce fabulously sensitive technologies like computerized tomography scanners, magnetic resonance imaging instruments, laparoscopic surgical tools, and the vast array of chemical testing equipment found in the centralized laboratories in such hospitals. With these tools, the physicians can diagnose almost anything, and do so rapidly and efficiently. Unfortunately, most of this comes at great cost, and is, therefore, only available to the most well-off societies in our world. For those not born into a family, society, or country that has great wealth, most of the remarkable diagnostic technologies developed in the last century have been unavailable. This is an inequity that may be balanced, at least in part, by microfluidics.

In the same sense that computation and the rapid transmission of information has been "democratized" by the microprocessor and the wireless data network, the information necessary for the best medical care could also be made more accessible. Medical emergencies do not begin at the emergency room. Voice and image transmission is now available even on inexpensive cellular telephones. What is missing is the ability to generate detailed chemical information at the point-of-care. With such information, the developed world could streamline medical care, reduce costs through less expensive medical testing, and allow diagnostics and theranostics (combining therapy with diagnostics) not possible today unless one had the luxury to stay in a hospital bed for one's whole life. For the developing world, effective rugged and inexpensive portable point-of-care diagnostics promise the delivery of an unprecedented level of diagnosis and therapy.

The difference between this new set of technologies and the great achievements in medical diagnostics of the last century is that one must *begin* by considering costs as one of the most important factors in instrument design. For this, microfluidics can be an excellent choice, but we must always be looking at the simplicity of the system, the robustness of the design, and the simplicity of operation for the end user, the health care worker or patients themselves.

NOT TOO SMALL!

AN IMPORTANT POINT WAS MADE to the µTAS community in 1990 by Andreas Manz. He pointed out that the concentrations of medically significant analytes in blood were not infinite. As a consequence, there was no point in using samples at lower than particular volumes because the number of pathogens or analyte molecules in very small volumes of sample could easily drop to less than statistically significant levels. For every particular range of analyte concentrations in blood, there was a minimum sample volume that must be obtained to get meaningful diagnostic results. Microfluidics, not nanofluidics, may well be the best format for many medical measurements.

4.3 Sample Preparation: A Bloody Example

A point-of-care diagnostic system ideally would take a readily accessible biological sample (saliva, urine, or blood, for example) and perform all necessary sample preconditioning, sample fractionation, signal amplification, analyte detection, data analysis, and possibly result display. Most of these steps have a microfluidic component, as we will describe in the next section.

There are many different samples one could imagine studying. For the purposes of this chapter, let us focus on a sample of human blood; it is both very rich in information and fluidically complex. Blood samples the entire body, reaching all corners every few minutes, so its contents can report on all aspects of human health. It is a non-Newtonian fluid (its viscosity depends on the shear rate), is heavy in particles (nearly 50% cells by volume), and is very high in protein content.

Cleaning up a "dirty" sample has always been a challenge. In a centralized laboratory, the first step for a sample of blood is usually a centrifuge, in which the plasma is separated for further study from the cellular components. Many different assays require that they be performed in a clear solution that is low in hemoglobin, so removal of all cells from whole blood is usually the first required step in processing whole blood in analytical systems. Centrifugation does not scale well to small sizes, although some approaches do rely on inertial forces and the difference in momentum of particles and the surrounding medium for sample preconditioning. In general, however, other methods have prevailed.

4.3.1 Fluid Conditioning for Cell-Free Analysis

Several different approaches have been made in the problem of removing whole cells from the plasma of blood. The goal is generally the complete removal of any cells of any size from the sample, so the cells range in size from the 10-µm-diameter white blood cells to 1-µm platelets and bacteria. The approaches have fallen into a few simple classes: sieves, weirs, inertial confinement, and flow diversion devices.

Sieves have been manufactured by making high-aspect-ratio columns or posts in Si or moldable materials using deep Reactive Ion Etch (RIE) or SU-8 molding. The spacing between columns must be set small enough so that no cells can penetrate at a given flow rate. This two-dimensional filter approach can work very well, but has the disadvantage of clogging as cells are captured. Filters of this sort have a maximum capacity, so it only works well when the loading is well defined and low, or it is possible to build in a back-flush cycle that resets the filter by pushing the cells off the columns of the sieve.

A macro-solution to the clogging problem of such filters has been to flow the sample across the filter at a relatively high velocity, so that thick layers of cells that would pile up on the filter would be washed off. This requires recycling of the retentate, which adds considerable complexity to the system. Even recycling has its limits.

A slightly different type of filter uses a **weir**, or shallow mesa-like section, to restrict flow to a narrow slot between layers, rather than using a set of posts. Such a system has many of the advantages and disadvantages of a conventional sieve with posts, but the slot between the mesa and the top of the channel can be made arbitrarily small with conventional microfabrication methods, so very good selectivity is possible, and the absence of discrete posts makes the recycling of the retentate more effective to prevent clogging. Weir-shaped cell traps are considered in Section 5.3.6.

A particularly successful device for separating plasma components for cells (that is based on microfluidic-specific physics) is known as the **H-filter** (see **Figure 4.1**). This device relies on the fact that at low Reynolds numbers, the flow lines are stable, and particles moving in those flow lines can only move from one line to another by diffusion (or some other force). In the absence of gravity, which can cause settling of blood cells in plasma even at 1 g, diffusion is the only other force that causes particles to cross those lines. As a consequence, in an H-shaped device such as that shown, blood cells, bacteria, and even large molecules can be made to stay on the initial side of the device, whereas smaller particles (from small molecules on up) can be partially equilibrated across the channel. The degree to which the molecules equilibrate depends on the flow rate through the device, the distance they have to diffuse (the width of the channel), and their time of residence in the channel (controlled by both the channel geometry and the flow rate). Under the right circumstances, it is possible to completely exclude cellular components (which diffuse much slower than molecules, for example) from the filtrate, whereas allowing most of the molecules to reach nearly 50% of their concentrations in the sample. Best of all, this device does not clog at all, and can run indefinitely. It has some advantages over operating with a membrane between the two solutions, in that there is no selectivity for the transport process except the diffusion coefficient—differently charged species can transport at equal rates. The throughput of the device depends only on the depth of the device—only the channel width need be "microfluidic enough" to maintain laminar flow.

No device is perfect; the H-filter requires careful control of three flow rates to maintain the expected rates of transfer from sample to filtrate. Particles need to be small enough not to clog the device, of course, and if one does have particles like blood cells in the device that will sediment, one must be sure that the sample solution is either on the downsides of the device, or the device is oriented so that sedimentation is along the flow direction in the common channel. Analytes are inevitably diluted by a factor of 2 in the H-filter, an important consideration if the ratio of input flows is adjusted to something other than 1:1.

A recently developed technique could also be used to precondition particle-laden samples, even though it was designed for a different purpose. This method, dubbed **pinched-flow fractionation** (see Section 5.2.4), relies on the restriction of a particle-laden fluid to one side of the channel by pumping in another particle-free solution at a higher rate into a common channel. If the sample flow is so low that its flow lines are restricted to a width smaller than the diameter of particles in that stream, the particles will collide with the wall, and so must be displaced from their original flow lines during passage through the pinched segment. When the flow expands

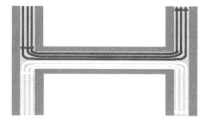

FIGURE 4.1 The H-filter. Microfluidic device with two inlets and two outlets that exploits laminar flow and diffusion to separate components that diffuse across the interface between the two inlet streams. (Figure contributed by Paul Yager.)

after the pinched segment, these particles will move along flow lines that are shifted toward the particle-free fluid in proportion to their size. Subsequent splitting of the exiting flow allows the recovery of essentially particle-free solutions. Some dilution of the sample is also inevitable, as in the H-filter, and there is a risk of mechanical damage to the particles in the pinched region, but this remains a very promising new approach to sample preconditioning.

4.3.2 Fluid Conditioning for Cell Analysis

Often, the aim of the microfluidic system is the analysis of cell populations. There are generally two approaches to such analysis—one in which the cell can be studied intact, and one in which the cell is lysed. A further advantage of single-cell analysis is that the heterogeneity of signals contributed by cells of different ages and phenotypic states can be directly assessed.

In point-of-care assays of blood, the most common form of blood cell analysis is counting one or more of the several types of cells present. This differential count can be used for diagnosis and staging of a wide range of congenital and acquired infectious diseases, such as AIDS and cancer, as well as monitoring chemotherapy. The primary centralized laboratory method for such analysis is flow cytometry (having displaced the Coulter counter some decades ago because of higher throughput and more general sensitivity). In flow cytometry, a suspension of cells or similar small particles is surrounded by a sheath fluid and forced through a nozzle at high velocity, precisely aligning the cells along the center of a free-standing jet. In the jet, the cells are interrogated optically, generally by a focused laser to stimulate light scattering or fluorescence (or both). Because the fluid stream that originally held the cells is focused down to a stream narrower than the cells themselves, the cells can be very well aligned into a single-sell stream.

Flow cytometry and similar techniques have been adapted repeatedly to microfluidic formats (see Section 5.1). The drag on the constrained stream can create high pressures, so microfluidic implementations of flow cytometry have generally used lower flow velocities than their macroscopic counterparts, and the number of cells that can be analyzed in a given period has typically been smaller. However, the advantages of being able to analyze the cells with a smaller and simpler instrument are very attractive for point-of-care purposes. Considering the utter impracticality of taking a conventional flow cytometer into the field in developing nations to stage AIDS, the need is particularly acute.

Today there is a growing interest in lysing cells before analysis, and a variety of methods for chemical lysis have been developed that rely on bringing a stream of buffer containing a lytic agent adjacent to a stream containing cells, followed by rapid lysis of the cells as the lytic agent diffuses into the cell-carrying stream. By controlling the flows, the timing of the cell lysis before an analytical process can be quite precise.

4.4 The Problem with Microfluidic Sample Separation

The biochemical analysis of complex samples often involves the separation of the sample into its components. A typical batch separation can be done by means of filtration, centrifugation, chromatography, or electrophoresis—all of which can be implemented in a microfluidic format. The problem with batch procedures is that they require the precise injection of minute amounts of samples into the separation channel (Figure 4.2a). An example of a technique that has followed this strategy is **capillary electrophoresis** (**CE**), as explained in Section 4.4.1. An entirely different strategy, dubbed **continuous-flow separation** (Figure 4.2b), involves the application of a force field at an angle to the direction of flow so as to deflect the path of the sample with respect to the flow of the buffer. Then, the sample is collected in a different channel and the output is visualized with a microscope, which can be done at very high resolution with standard optical microscopes (i.e., no temporal resolution is required). This approach is clever in that it better exploits differences in the microscale diffusive behavior between the sample and the buffer in

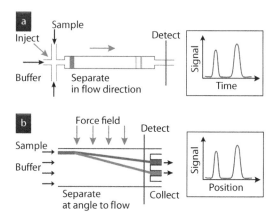

FIGURE 4.2 Microfluidic separations in time and in space. (a) Batch separation procedure, in which separation occurs in the flow direction and is detected as a function of time; (b) continuous flow separation procedure, in which a force field acting at an angle to the flow spatially separates the sample from the buffer. (From Nicole Pamme, "Continuous flow separations in microfluidic devices," *Lab Chip* 7, 1644–1659, 2007. Reproduced with permission from The Royal Society of Chemistry. Figure contributed by Nicole Pamme.)

the microchannel. The many modalities of continuous-flow separations, depending on the force field applied, will be covered in Section 4.4.2.

4.4.1 Capillary Electrophoresis on a Chip

Chromatography has long been the mainstay of analytical chemistry, and was one of the first methods to be implemented in microfluidics. CE, developed in the 1960s, is based on the use of a long glass capillary (thin tube) for separation of chemical species by voltage; because of the length of the capillary, a high-voltage power supply is typically needed. The analytes separate as they migrate by electro-osmotic flow because of differences in their electrophoretic mobility. Generally, the method involves creating a small bolus of sample, and running it through a long column in which some species travel slower than others as they are retarded by some interaction with a stationary species in the column. The presence of different compounds is detected by their arrival at an expected time after the beginning of the run at the outlet of the column. Exquisite differentiation between very closely related molecules is possible using a host of different methods.

In the 1990s, it became apparent that CE seemed excellently suited to microfabrication because the only negative consequence of making the column thinner is a reduction in the size of the bolus of sample that can be loaded onto the column to produce a separation of the same resolution. In theory, making the column thinner can actually enhance operation because of allowing faster removal of unwanted byproducts of the separation process (such as heat in the case of electrophoretic separations). As long as small samples contain adequate numbers of analyte molecules for detection, going small should improve matters.

MICROTAS: FROM BIRTH TO ADOLESCENCE—JUMPING PLANETS IN ONE HUMAN GENERATION

TWO MAJOR ENTICEMENTS DROVE the pace of research in the 1990s. One was the fact that drug discovery had been limited in recent decades by the time and cost of running many chromatographic analyses in parallel on macroscopic equipment. The potential for manufacturing high-throughput chromatographic chips for drug discovery (particularly using capillary electrophoresis) powered the creation of most of the first microfluidics-based companies in the mid-1990s. In 1992, a group from Ciba-Geigy in Switzerland, led by Michael Widmer, wrote an influential article in the *Journal of Chromatography*, bearing a long, forgettable title and a much more memorable subtitle: "Capillary electrophoresis on a chip." The three junior authors of this article, Andreas Manz, Jed Harrison, and Elisabeth Verpoorte, are now widely considered microfluidics pioneers. In fact, Andreas Manz and Michael Widmer had already coauthored an article in 1990, in which they had outlined the theoretical advantages of miniaturizing "total chemical analysis systems" (without demonstrating a prototype)—this 1990 article, published in *Sensors and Actuators B*, is universally regarded as the foundational article of microTAS as a field.

As it happens, they were not the first to have conceived the idea that biochemical separations would be improved by miniaturizing fluid manipulation—most of the electrophoresis field had noticed by then: Philpot had invented free-flow electrophoresis using parallel-plate devices in 1940 and Hannig had invented a continuous-flow paper electrophoresis device in 1950 (see Section 4.4.2). Seeking no publicity, a team of engineers at McDonnell Douglas worked with Hannig through the mid-1970s to design a continuous-flow device for performing electrophoretic separations in space. The device was even successfully tested in various shuttle flights (the first of them in June 1982). One of the first microTAS experiments was performed…away from our planet! (The subtle difference is that, in these pioneering experiments, the solutions are confined in fluid sheets whereas in microTAS systems, as sketched by Manz and Widmer, the fluids would be confined in microchannels—which would require a new set of fabrication technologies and would impose a new set of engineering and scientific constraints on the investigator.)

The other great driver was the Human Genome Project and its seemingly insatiable demand for enhanced DNA sequencing capacity. Fortunately, DNA can be very efficiently separated by CE on a chip, and relatively intense labeling allows easy detection using laser-induced fluorescence. Great effort was put into making fused silica, glass, and plastic chips that could carry out sensitive and highly parallel laser-induced fluorescence detection–based CE chips.

In 2005, Richard Mathies' group (University of California, Berkeley) reported in *PNAS* the development of the Mars Organic Analyzer, a CE chip for amino acid analysis to be deployed on the rover in the ExoMars Mission in 2016 to 2018. Consider the pace of progress, starting from Manz and Widmer's 1990 vision: it will have taken the field slightly more than 25 years (the measure of a human generation) to take a microfluidic chip to another planet!

Hundreds of complete CE channels can now routinely be fabricated on a single glass wafer. Richard Mathies' group at the University of California (Berkeley) has been the main powerhorse in this field, continuously pushing the technology. In 1999, they had already developed a 96-channel system (110 μm wide, 50 μm deep channels) on a 10-cm-diameter glass wafer (Figure 4.3a) that could be loaded via 96 capillaries (Figure 4.3b) and scanned with a radial laser scanner (Figure 4.3c), achieving a separation resolution of one base pair in just 10 minutes (Figure 4.3d). The improved version of this device, which will go into the Mars rover (see previous text box), now contains on-chip microvalves and micropumps activated by a pneumatically deflectable PDMS membrane. Another group actually manufactured a glass chip nearly a meter long to maximize the number of channels that could be run in parallel.

Two technical challenges cropped up in the development of highly parallel CE-on-a-chip systems. To load a sample bolus small enough to take advantage of the small column volumes,

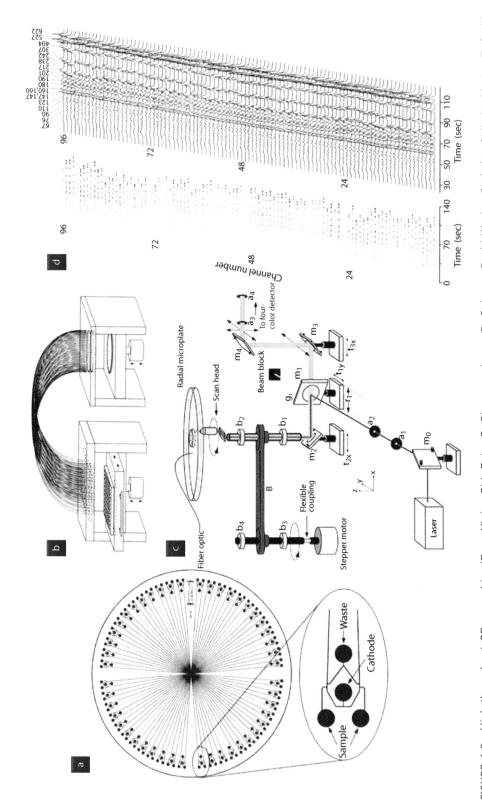

FIGURE 4.3 High-throughput CE on a chip. (From Yining Shi, Peter C. Simpson, James R. Scherer, David Wexler, Christine Skibola, Martyn T. Smith, and Richard A. Mathies, "Radial capillary array electrophoresis microplate and scanner for high-performance nucleic acid analysis," *Anal. Chem.* 71, 5354–5361, 1999. Figure contributed by Rich Mathies.)

conventional loading loops were not practical. Instead, the crossed-T injector and appropriate high-voltage electrophoretic control systems evolved. This allowed many separate picoliter samples to be accurately positioned and sent down the CE channels under computer control. The other problem was one of geometry. To put one or more long chromatographic separation columns onto a finite-sized (perhaps 5-in.-diameter) glass wafer, the channels cannot be straight. To pack detection zones into a small region, which is also critical, it proved necessary to use a radial packing of the chromatographic columns (e.g., see **Figure 4.3a**). However, the radius of the wafers was not adequate for sufficient separation, so the channels had to typically have a few hairpin (180 degree) turns. These turns badly smeared out the bands, costing as much in resolution as had been gained by allowing the channels to be more than one wafer radius in length. There were two reasons, one of which was a concentration of current on the inner edge of the turn, and the other was the longer path taken by analyte molecules on the outer edge of the channel. It was ultimately shown by the Mathies and Santiago groups that (some rather unintuitive) modifications of the geometry of the turn itself allowed restoration of almost all the resolution.

4.4.2 Continuous-Flow and "Free-Flow" Electrophoresis

This electrophoresis modality has found less effective uses than CE but was historically developed much earlier than CE. In fact, this idea can be traced back to the dawn of microfluidics, well before the first microtechnologies had been invented. (Note that it is possible to construct rudimentary microfluidic devices with anything that confines fluid flow to a small space—a thin glass tube, a sheet of paper, or two glass plates separated by a spacer will do.) Electrophoretic separation does not need to be performed along the length of the capillary. For certain applications (in particular, where the sample can be guaranteed to flow continuously for as long as data needs to be gathered), it may be advantageous to apply the field perpendicular to the channel (**Figure 4.4a**). In July 1939, J. St. L. Philpot, a biochemist from Oxford University, submitted a report entitled "The use of thin layers in electrophoretic separation" to the *Transactions of the Faraday Society* in which he used laminar flow principles to layer five input flows (the central one containing protein solution); the voltage deflected the protein solution and separated the "wanted" fractions to one output and the "unwanted" fractions to other outputs. To describe laminar flow, he reported that "the five solutions flow quite smoothly on top of each other." The idea of **continuous-flow electrophoresis** was picked up by a few researchers after Philpot, most notably by Kurt Hannig from the Max Planck Institute (Munich) who in 1950 designed an electrophoresis device in which gravity-fed flow was sustained by filter paper while subjected to

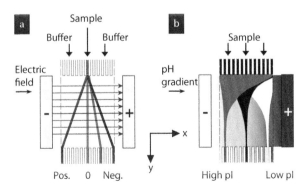

FIGURE 4.4 Continuous-flow electrophoresis and continuous-flow isoelectric focusing. (From Nicole Pamme, "Continuous flow separations in microfluidic devices," *Lab Chip* 7, 1644–1659, 2007. Reproduced with permission from The Royal Society of Chemistry. Figure contributed by Nicole Pamme.)

an orthogonal electrical field—Hannig here not only invented the first **paper electrophoresis** device but also the first paper microfluidic device! Not quite satisfied with the uneven transport properties of filter paper, in 1961 he reported a continuous-flow electrophoresis device in which flow was injected between two parallel plates (essentially, redesigning the Philpot device) and dubbed the technique "**free-flow electrophoresis**" (**Figure 4.4a**) because the fluid was flowing "freely" in contrast with inside the paper. (I find the original term "continuous" more intuitive and inclusive of both modalities, and should have the benefit of historical precedence.) In the 1970s, when NASA became interested in techniques for automating sample analysis in space, they looked into who had developed electrophoresis devices that could fit their automation and small-sample requirements—and they worked with Kurt Hannig and a team of McDonnell Douglas engineers to develop the first outer space microfluidic device, which was successfully tested on the shuttle in June 1982 to separate a mixture of proteins at varying concentrations.

Because of the large surface to volume ratio and the continuous perfusion, Joule heating can dissipate quickly, so application of higher fields enables very fast separations compared with traditional CE. Also, collection of the separations and integration of this system with other microfluidic devices is straightforward, which simplifies automation substantially.

4.4.3 Isoelectric Focusing

Isoelectric focusing is a technique that allows for separating proteins with different isoelectric points (pIs, or the pH at which a particular molecule or surface carries no net electric charge). The setup for continuous-flow isoelectric focusing is similar as that for continuous-flow electrophoresis, as shown in (**Figure 4.4b**). A pH gradient is first created perpendicular to the flow direction, usually by means of voltage application (e.g., through local electrolysis at the electrodes). A sample protein entering at a location in which the pH is different from its pI exhibits a net charge, and thus experiences an electrophoretic force. The protein migrates perpendicular to the flow until it reaches a point in the pH gradient in which the pH equals its pI, where it no longer has a net charge and it ceases to migrate perpendicular to flow. The protein now only follows the direction of flow because it is focused at its isoelectric point. Reported separation times vary widely, ranging from subseconds to minutes, depending on design and the sample.

4.4.4 Continuous-Flow Magnetic Separations

Devices that require electric fields consume power and tend to be bulky, with complex controls and lots of wires. However, electric fields are not the only means to exert forces

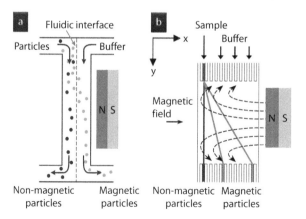

FIGURE 4.5 Free-flow magnetic separations. (From Nicole Pamme, "Continuous flow separations in microfluidic devices," *Lab Chip* 7, 1644–1659, 2007. Reproduced with permission from The Royal Society of Chemistry. Figure contributed by Nicole Pamme.)

in nature; how about magnetic fields? Many researchers have remembered the days when they played with little magnets as kids and dreamed of devices that might put that convenient, portable force into play. Unfortunately, proteins, DNA, and cells are not magnetic—however, they can be attached to magnetic particles by fairly standard biochemistry. Then, under flow, particles of different sizes or different magnetization can be separated if an inhomogeneous magnetic field is applied perpendicular to flow. Two different setups are shown in Figure 4.5. Magnetic particles or magnetically-labeled cells are attracted into the field and thus deflected from the direction of laminar flow. The magnetic force depends on the gradient of the applied magnetic field as well as on the volume and magnetization of the particle.

4.4.5 Molecular Sieving

The vast majority of microfluidic devices for molecular separation aim at counteracting diffusion (which acts isotropically) by imposing an external force (in the desired direction of separation). In 2002, a Princeton University team led by Edward Cox, Robert Austin, and James Sturm introduced a new key concept: if the molecules were forced to pass through an array of posts, and the spacing between posts were small enough, then at some point the larger molecules would have a harder time flowing through the post array (which acted, in essence, as a microfabricated sieving matrix). When DNA fragments (61–209 kb) were flowed through the array and an electric field was applied (Figure 4.6), separation at a resolution of approximately 13% was possible in just 15 seconds (compared with 10 to 200 hours for traditional pulsed-field gel electrophoresis).

The concept of DNA sieving can be extended to large particles, such as beads (which behave in highly predictable ways), without the requirement of the electric field for separation applications. The same Princeton team elegantly demonstrated that beads that are small compared with their lanes tend to stay in their lanes, whereas beads that are large effectively invade another lane and get displaced with respect to the average flow velocity vector (Figure 4.7). Note that, because the array is at an angle with the flow, to advance straight the small beads have to zigzag through the posts (Figure 4.7c).

In 2007, Jongyoon Han's group at MIT designed a more sophisticated version of the DNA-sieving posts discussed previously. Here, the posts were substituted by pillars, such that the only space for the molecules to diffuse through was the spaces between the top of the pillars and the ceiling of the chamber (Figure 4.8). Because the critical dimension was dictated by the pillar-to-ceiling spacing, which could be fabricated to submicrometer specifications, the device allowed for the study of small proteins (which had remained a challenge for previous post-based devices) as well as for separation of DNA (down to 2000-bp fragments).

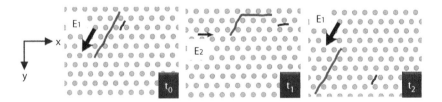

FIGURE 4.6 DNA prism. (From Nicole Pamme, "Continuous flow separations in microfluidic devices," *Lab Chip* 7, 1644–1659, 2007; and L. R. Huang, J. O. Tegenfeldt, J. J. Kraeft, J. C. Sturm, R. H. Austin, and E. C. Cox, "A DNA prism for high-speed continuous fractionation of large DNA molecules," *Nat. Biotechnol.* 20, 1048, 2002. Reproduced with permission from The Royal Society of Chemistry. Figure contributed by Nicole Pamme.)

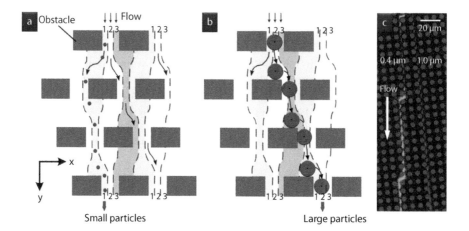

FIGURE 4.7 Deterministic lateral displacement. (From Nicole Pamme, "Continuous flow separations in microfluidic devices," *Lab Chip* 7, 1644–1659, 2007; and Lotien Richard Huang, Edward C. Cox, Robert H. Austin, and James C. Sturm, "Continuous particle separation through deterministic lateral displacement," *Science* 304, 987, 2004. Reproduced with permission from The Royal Society of Chemistry. Figure contributed by Nicole Pamme.)

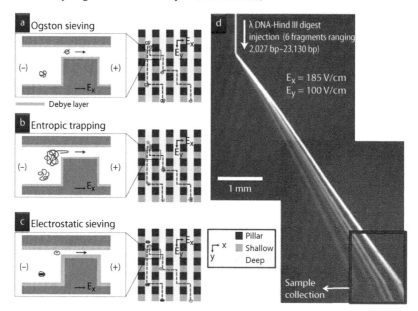

FIGURE 4.8 Anisotropic sieving. (From Nicole Pamme, "Continuous flow separations in microfluidic devices," *Lab Chip* 7, 1644–1659, 2007; and Jianping Fu, Reto B. Schoch, Anna L. Stevens, Steven R. Tannenbaum, and Jongyoon Han, "A patterned anisotropic nanofluidic sieving structure for continuous-flow separation of DNA and proteins," *Nat. Nanotech.* 2, 121, 2007. Reproduced with permission from The Royal Society of Chemistry. Figure contributed by Nicole Pamme.)

4.5 Microfluidic Immunoassays

Molecular biology has given us a new set of nucleic acid–based assays that are unsurpassed for measuring small quantities of nucleic acids themselves. However, they are no use for measuring most other types of chemicals. One of the most powerful and versatile biomedical diagnostic tools, the immunoassay, is used today to monitor almost everything else: the levels of drugs and hormones in body fluids, to diagnose infectious and autoimmune diseases, and to both diagnose and monitor treatment of cancer. All immunoassays rely on the ability of properly designed or selected antibody molecules to strongly and selectively bind small and large molecules.

THE WORLD'S MOST FAMOUS MICROFLUIDIC IMMUNOASSAY: THE PREGNANCY TEST

MODERN PREGNANCY TESTS BASED ON THE STRIP ASSAY that can be dipped in urine detect the presence of human chorionic gonadotropin (hCG). This pregnancy marker was discovered in 1930 and is produced by the trophoblast cells of the fertilized ovum (blastocyst). Home pregnancy testing hit the American market by the end of 1977, but then it consisted of mixing the woman's urine with solutions in a rather complex procedure that took hours to produce a result. The first lateral-flow test (based on a strip of nitrocellulose paper, with flow driven by wicking, which reads as "two strips, you are pregnant," see Figure 4.9) was the ClearBlue commercialized by UniPath (now SPD) in 1985. By the same token, it can also be considered the first microfluidic product to have ever reached the market. Many of us have made important planning decisions after seeing the results displayed by those two bars. It is a clever piece of paper microfluidics, invented before the first article on paper microfluidics!

Although some *qualitative* immunoassays have been reduced to simple test strip assays (like pregnancy tests), the performance of *quantitative* immunoassays is a relatively complex sequence of processes today largely restricted to centralized laboratories because of the need for long assay times, complex and expensive equipment, and highly trained technicians. If a wider range of the 700 million immunoassays performed annually in the United States alone could be run more inexpensively, more frequently, and at the point-of-care, the health of millions of patients could be improved. The inexpensive availability of such assays for diagnosis in the developing world would have an enormous effect on global health.

There are many different detection modes for immunoassays, some of which rely on the binding of the analyte to change a property of the surface, whereas others rely on some form of amplification process that may or may not rely on a surface. We will break our discussion of immunoassays into two sections, but first, we will explain in detail the most ubiquitous immunoassay ever deployed.

4.5.1 The Pregnancy Test

Human chorionic gonadotropin (hCG) is a hormone secreted by the developing placenta after fertilization and it can be found in blood and urine. Modern pregnancy tests use a "sandwich" ELISA (short for "enzyme-linked immuno-sorbent assay") method involving three different antibody preparations (located in three different strips) on a porous paper sheet (Figure 4.9). The porosity of the paper serves to pump the urine by capillarity from one end of the strip to the other. The reaction strip "R" is the first to be exposed to urine flow and contains unbound mouse monoclonal anti-hCG antibody–enzyme conjugates. Note that the urine carries these conjugates upstream whether or not there is hCG in the urine, but if there is hCG in the urine, a fraction of the conjugates will, in addition, bind to hCG (there is great excess of conjugates with respect to the normal concentrations of hCG). The test strip ("T") is the next to be exposed to the urine flow and contains immobilized polyclonal anti-hCG antibodies (designed to bind to a different epitope on the hCG molecule than the one that is already bound to the monoclonal antibody) and a dye substrate (shown in red). The control strip ("C") contains immobilized goat

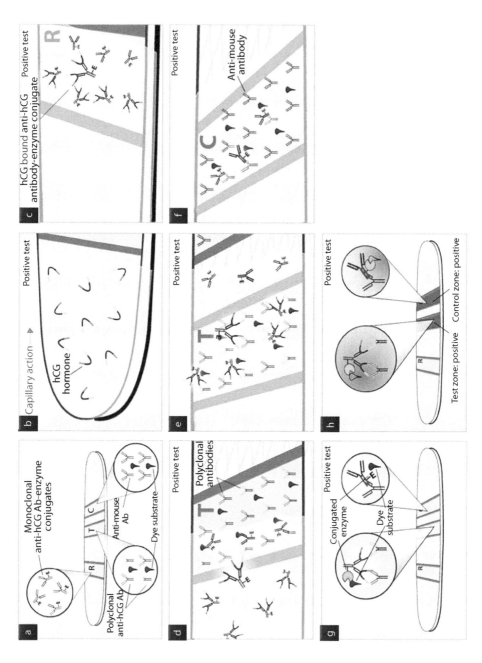

FIGURE 4.9 Home pregnancy test (Adapted from "*The HCG Pregnancy Test: How It Works*," a video developed by Susana Maria Halpine for the *Kuby Immunology* Web site. Copyright (2000) W.H. Freeman and Sumanas, Inc.)

anti-mouse antibodies and the same dye substrate as the T strip. The immobilized enzymes are now able to activate the dye substrate in both the test and control strips, confirming the pregnancy.

The pregnancy strip assay has now been generalized to a variety of immunochromatographic strip (ICS) tests. The nonprofit organization Program for Appropriate Technology in Health (PATH; Seattle, WA), which focuses on global health applications, has developed ICS tests for diphtheria toxin and a number of sexually transmitted diseases (gonorrhea, syphilis, chancroid, and chlamydia), among others. These devices require minimal training by health workers but often only provide a yes/no readout answer (rather than an analyte level) and sometimes still need sample preconditioning.

4.5.2 Homogeneous Phase Immunoassays

Because proteins come with a wide range of surface charges and surface chemistries, they tend to adhere more to surfaces than the relatively homogeneous nucleic acids. Therefore, implementing protein-based assays is generally more problematic, requiring efforts at nonfouling as discussed previously. However, microfluidics has been applied to implementing immunoassays in several formats. Successful CE on a chip approaches involve the electrophoretic separation of bound and unbound labeled analytes, or monitoring the change in electrophoretic migration speed of labeled antibodies bound or unbound to their target ligands. These methods are sensitive and highly reproducible.

A completely different approach involves the use of the T-sensor to perform a "diffusion immunoassay" (DIA; Figure 4.10). In the original embodiment of the T-sensor, two fluids interact during parallel flow until they exit the microchannel. Large particles, such as blood cells, do not diffuse significantly within the time the flow streams are in contact. Small particles diffuse rapidly between streams, whereas larger polymers diffuse more slowly and equilibrate between streams further from the point of entry to the device. As interdiffusion proceeds (Figure 4.10a–c), interaction zones are formed in which sample and reagents may bind and react (Figure 4.10d). If an indicator solution is used in the detection solution, the diffusion interaction zones will be optically detectable (Figure 4.10f). The positional variation in intensity of that signal is a complex function of the concentration of the indicator and analyte. However, it is straightforward to calibrate the optical response to analyte concentration.

In the DIA, the transport of molecules perpendicular to flow in a microchannel is affected by binding between antigens and antibodies. The DIA is based on the difference in diffusion coefficients of antigens and antigen–antibody complexes. Like other T-sensor assays, it relies entirely on the fluid dynamics of solutions and chemical interactions of components in solution—no interactions with immobile phases such as the channel walls are required or desired. By imaging

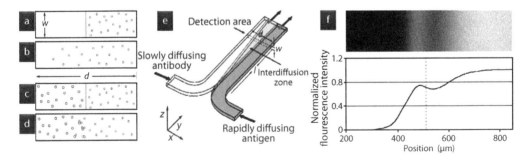

FIGURE 4.10 The DIA in a T-sensor. (From A. Hatch, A. E. Kamholz, K. R. Hawkins, M. S. Munson, E. A., Schilling, B. H. Weigl and P. Yager, "A rapid diffusion immunoassay in a T-sensor," *Nat. Biotechnol.* 19, 461, 2001. Reproduced with permission from the Nature Publishing Group. Figure contributed by Paul Yager.)

the steady-state position of labeled components in a flowing stream, the concentration of very dilute (<1 nM) analytes can be measured in a few microliters of sample in seconds. This assay has been demonstrated in the format of a small molecule analyte competition immunoassay using fluorescence imaging detection, and for larger analytes as well. The DIA can also monitor concentrations of analytes as large as proteins, at the cost of increased assay times. Most exciting is that this assay can operate almost completely without sample preconditioning, greatly strengthening its applicability in point-of-care diagnostics.

4.5.3 Heterogeneous Phase (Surface-Bound) Immunoassays

The utilization of the walls of a microchannel allows access to the greater complexity and sensitivity of assays like ELISA. This strategy requires very good control of surface binding (k_{on} and k_{off}), as illustrated in **Figure 4.11**, because mass transport to and from the surface is affected by diffusion (a "force" that works isotropically), convection (which works in the direction of flow), and the presence of electrical fields (if any).

The first microfluidic immunoassay, consisting of a set of PDMS channels that delivered antibodies to micron-sized lines of a glass substrate for fluorescence detection, dates from 1997 and is credited to Hans Biebuyck's group (then at IBM Zurich), as explained in Section 2.4.3 (see **Figure 2.17**). Recent implementations of immunoassays using surface plasmon resonance detection have also been made in microfluidic implementations.

Microfluidics is generally appealing because it provides low-cost automation solutions, so it is slowly becoming the technological substitute of fluidic robotics. For a much lower price, it also consumes fewer reagents. (The main drawbacks are that it tends to operate on nonstandard platforms and requires some uncommon expertise, but those barriers are being lowered every day.) A very common problem in immunoassays is to establish a serial dilution curve to test antibody binding response. These dilutions are cumbersome to produce, requiring several pipetting steps with traditional multiwell pipettors. Several microfluidic titrators have been developed (see Section 3.9.2). As an example, **Figure 4.12** shows a serial dilution fluorescent immunoassay device reported in 2003 by George Whitesides' group from Harvard University. The assay can measure the concentrations of multiple antibodies in parallel in one experiment using a network of microchannels to achieve serial dilution. The branching structures of the microfluidic network serially dilute one stream by half by connecting it to the adjacent stream (so long as proper mixing occurs at each stage, which is ensured with the incorporation of "chevron" grooved mixers on the surface of the microchannels). To illustrate the assay, the concentration of antibodies in HIV + human serum (anti-gp41 and anti-gp120) was determined by secondary

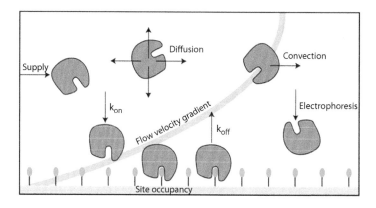

FIGURE 4.11 Mass transport in microfluidic immunoassays. (Figure contributed by Paul Yager.)

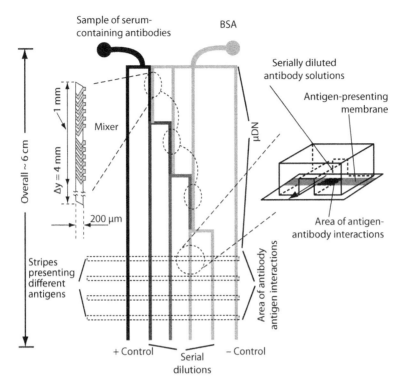

FIGURE 4.12 Serial dilution immunoassay. (From Xingyu Jiang, Jessamine M. K. Ng, Abraham D. Stroock, Stephan K. W. Dertinger, and George M. Whitesides, "A miniaturized, parallel, serially diluted immunoassay for analyzing multiple antigens," *J. Am. Chem. Soc.* 125, 5294–5295, 2003. Reprinted with permission of the American Chemical Society.)

immunofluorescence on a polycarbonate membrane presenting strips of the antigens gp41 and gp120 (HIV ENV proteins).

4.5.4 Capture and Enrichment of Biomolecules

Biofluids often contain precious quantities of rare proteins (or other biomolecules), so biochemists have devised a variety of (macrofluidic) purification methods to enrich biological samples. However, these methods are not always easy to automate in microfluidic format, as they entail several steps and reagents. Recently, a team led by Patrick Stayton from the University of Washington (Seattle) has reported a microfluidic device that can reversibly capture and enrich biomolecules from solution. They utilized a phase-transition polymer, poly(N-isopropylacrylamide) (PNIPAAm) to facilitate capture and release of protein–polymer conjugates. PNIPAAm undergoes a reversible hydrophilic-to-hydrophobic phase transition when heated above a lower critical solution temperature (LCST). PNIPAAm can be conjugated to a protein (to enable reversible conjugate aggregation) and be grafted to PDMS channel walls (to enable conjugate immobilization through hydrophobic interactions when heated, and to hasten the release of biomolecules from the surface with cooling), as shown in **Figure 4.13a**. Streptavidin was conjugated to PNIPAAm through a biotin linkage and introduced into a PDMS microfluidic device containing a PNIPAAm-grafted recirculator (consisting of a herringbone mixer and a three-valve peristaltic pump, see **Figure 4.13c** and **e**, respectively). After washing the recirculator with warm buffer, and subsequently releasing conjugates by cooling to a temperature below the

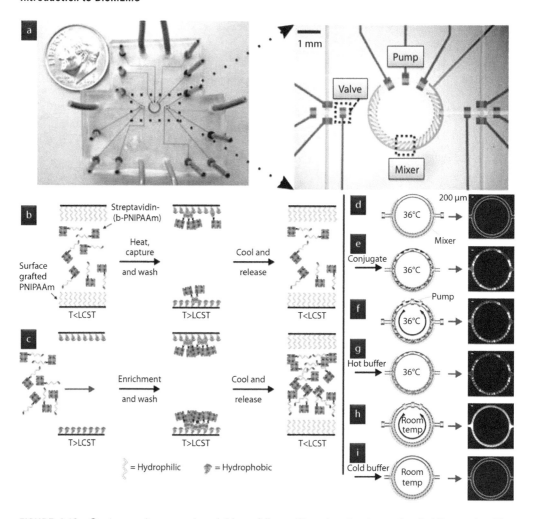

FIGURE 4.13 Capture, release, and enriching of "smart" conjugates in a microfluidic reactor. The mixer and the pump were operated at 5 Hz. The LCST for PNIPAAm is 36°C. (From John M. Hoffman, Mitsuhiro Ebara, James J. Lai, Allan S. Hoffman, Albert Folch, and Patrick S. Stayton, "A helical flow, circular microreactor for separating and enriching "smart' 'polymer–antibody capture reagents," *Lab Chip* 10, 3130, 2010. Reproduced with permission from The Royal Society of Chemistry. Figure contributed by John Hoffman.)

LCST, the fluorescence intensity of labeled streptavidin–polymer conjugates was measured and compared with the original bolus intensity. Under the correct conditions, 60% of a conjugate bolus was captured and subsequently released. Enrichment of protein–polymer conjugates was achieved by continuous conjugate flow into the recirculator heated above the LCST, followed by mixing, cooling and release of conjugates. Conjugate concentration was increased nine times within 30 seconds of continuous flow.

4.6 Chips for Genomics and Proteomics

Animals have thousands or even hundreds of thousands of genes. In 1990, the U.S. Department of Energy and the U.S. National Institutes of Health formally founded a $3 billion project to sequence the Human Genome, and was expected to take 15 years, but was declared "complete" in 2000 (the central parts of the chromosomes, the centromeres, and the endings of the chromosomes, the telomeres, are highly repetitive and remain for the most part unsequenced, as well as occasional gaps, amounting to a total of ~7%–8% of the genome). Here, we review how microfabrication technology has contributed to making genomics (and its cousin proteomics) technology and research both cheaper and faster.

4.6.1 Microarrays of DNA-Based Molecules

In these last couple of decades, microfabrication technology has been paramount to the overall vision of making genome analysis faster and cheaper. (Nanotechnology approaches, such as sequencing by passing DNA through nanopores, are beyond the scope of this textbook.) The goal is to know which genes are being activated and, for any given nucleic acid sample, what its sequence is—either by sequencing it from scratch or by comparing it to a known sequence. Most of the challenges associated with depositing DNA in microarrays are directly extrapolatable to proteins; obviously, the synthesis of proteins and the dynamics of adsorption of proteins to surfaces differ radically from those of DNA, but that usually does not change the equipment used to perform the deposition.

4.6.1.1 Oligonucleotide Chips

In 1991, a team from Affymax Research Institute (a company based in Palo Alto, CA) led by Stephen Fodor presented a novel approach to find out a gene sequence. (This technology became the basis of a new company, Affymetrix, which was formed as a division of Affymax and began operating independently in 1992.) The novelty of the technique resided in that it used light to micropattern the chemical synthesis of oligonucleotides, nucleotide by nucleotide, on a surface (**Figure 4.14**). To add one nucleotide length to all nucleotides on every spot, it required four exposures (through four different photomasks) in the right nucleotide solution followed by four washes. Because approximately 25-mer oligonucleotide fragments are necessary to provide unambiguous sequencing, this means that the chips would require...100 photomasks! At the beginning, the chips were prohibitively expensive but, fortunately, Affymetrix soon adopted Texas Instruments' Digital Micromirror Device, which projects virtual photomasks (see Section 1.3.6.1 and **Figure 1.8**). This approach is obviously not suited for creating protein microarrays.

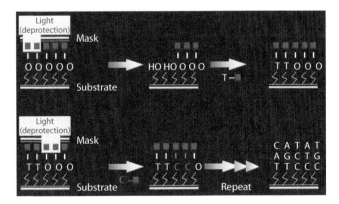

FIGURE 4.14 Photochemical synthesis of oligonucleotide chips. (From Stephen P. A. Fodor, J. Leighton Read, Michael C. Pirrung, Lubert Stryer, AmyTsai Lu, and Dennis Solas, "Light-directed, spatially addressable parallel chemical synthesis," *Science* 251, 767–773, 1991. Figure contributed by Steve Fodor.)

4.6.1.2 DNA Microarrays

In 1995, Patrick Brown's group from Stanford University presented a new approach for detecting which genes get expressed at any given time in a cell or group of cells. The first step is to collect mRNA from the cells, convert it into complementary DNA (cDNA) by reverse transcription, and separate it into different vials so that the cDNA can be printed. The technique requires depositing approximately 100-µm-diameter spots (each corresponding to one gene) of cDNA fragments on glass by means of slotted metallic pins (the slot serves to wick in fluid, see **Figure 4.15a**), which Pat Brown manufactured in a machine shop. This "sense cDNA," which is in a known location of the array, is simply used as a sensor to hybridize to the fluorescently labeled "antisense" single-strand cDNA from the sample (**Figure 4.15b**). A differential approach is used: an untreated sample (labeled in a different color, say green) as a control is analyzed as well as a treated sample (labeled red), so that the differences in activated spots tells the investigator which genes have been activated by the treatment (**Figure 4.15c**). A red dot means that the treatment caused the gene to be expressed at higher levels than in the control. A green dot means that the corresponding gene is expressed at higher levels in the control compared with the treatment (i.e., it is repressed in normal conditions), and a yellow dot reveals genes that are expressed

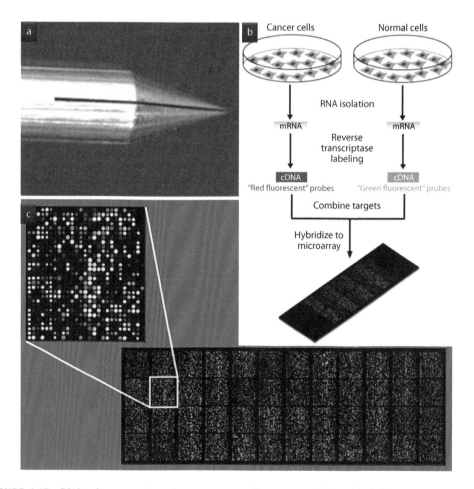

FIGURE 4.15 DNA microarrays to probe gene expression patterns. (From Mark Schena, Dari Shalon, Ronald W. Davis, and Patrick O. Brown, "Quantitative monitoring of gene expression patterns with a complementary DNA microarray," *Science* 270, 467, 1995. Images courtesy of the Wikimedia Project.)

at equal levels in both samples (dark spots indicate negligible expression in both the treatment and the control). The spotting technique is suitable for proteomics because the metallic pins can deposit any biomolecule from solution.

"DNA" CHIPS VERSUS DNA "MICRO"-ARRAYS

IT IS UNFORTUNATE THAT THE TWO MAIN CHIP-BASED PLATFORMS for genomic analysis have misleading names. The so-called "DNA" chips are actually microarrays of ~25-mer oligonucleotides synthesized and photochemically immobilized on the chip, not microarrays of full-length DNA. And the so-called DNA "micro"-arrays are not really microfabricated because the arrays are made by spotting (yes, full-length) DNA molecules with a robotic pin arrayer. The terminology is confused so often that even *Wikipedia* had it wrong until very recently: the entry for Pat Brown, the Stanford professor and inventor of the DNA microarrays, reads that "his research uses **DNA microarrays** to study the gene expression patterns", but the "**DNA microarrays**" link would take you to…the oligonucleotide microarrays invented by his competitors, Affymetrix! (sigh). The error has now been (partially) corrected. Clearly, Pat Brown does not buy Affymetrix chips for his research.

4.6.1.3 DNA Chips versus DNA Microarrays

In this section, we will point to the differences between the two main chip-based platforms for DNA analysis. The most salient features of the Affymetrix "DNA" chips, which are really microarrays of oligonucleotides photochemically immobilized on glass, are:

- Fabricated by photolithography.
- Resolution of approximately 1 μm.
- Very expensive: $4N$ masks for N-mers! + photochemistry reagents.
- Synthesis has a poor (95%) yield, resulting in aberrant sequences (only 28% of any given 25-mer oligomer will have the targeted sequence).
- Sequencing requires approximately 40 different unique oligonucleotides per gene (to compensate for overlapping motifs).
- Qualitative (not useful for detecting levels of gene expression).
- Allows for detection of single-base mutations.
- Limited to patterning of oligonucleotides.

The DNA microarrays invented by Pat Brown, on the other hand, are characterized by:

- cDNA solutions printed on a glass slide with an array of pins.
- Resolution: approximately 100 μm (limited by "spotting/drying" process).
- Relatively inexpensive.
- Highly specific (full gene).
- Quantitative (levels of actively expressed genes compared with those of control).
- Cannot be used for sequencing.
- Can be extended effortlessly to proteins and other biomolecules.

Because DNA microarrays are not really microfabricated, they will not be discussed further here. However, there have been a few notable attempts to miniaturize this technology. The challenge is essentially to deposit small volumes of large numbers of DNA solutions (each originally separated in a macroscopic vial) in a microarray format. Several solutions have been proposed in addition to those by Stephen Fodor and Pat Brown, but among those, two deserve a special creativity mention: the ingenious bead arrays invented by David Walt (now marketed by Illumina) and the quasimagical electroaddressable DNA deposition conceived by Susan Brozik.

4.6.1.4 Self-Assembled Microarrays of Beads

In 1998, David Walt, a chemistry professor at Tufts University (Cambridge, Massachusetts), pondering about the high cost of Affymetrix's chips, had a revolutionary idea: Why did the DNA-sensing spots need to be printed? It occurred to him that they could be prefabricated in solution as *beads*(!) and then deposited into approximately 3-μm-deep, 3.6-μm-diameter microwells that fitted just one approximately 3.1-μm-diameter bead each (**Figure 4.16**). At this scale, the viscosity of water makes it virtually impossible for a bead to be dislodged from a microwell once it is inside. Beads can be easily derivatized with DNA (or any other biomolecule) as well as with dyes (to distinguish them from one another)—by the millions. Of course, every time that a mixture of beads is deposited on an array, the arrangement of beads changes (unlike with Affymetrix' arrays), but as long as the beads are encoded with a dye, their identity can be decoded. The microwell array can potentially be built in many ways, but Walt's group (in the same article!) came up with a cunningly clever design: take a bundle of glass fiber optics and dip the end of it in hydrofluoric acid (HF) for a few minutes (the core of the fibers is pure quartz, and thus etches faster in HF than the cladding, which is doped glass)—this results in a set of glass wells, as shown in the top image of **Figure 4.16**; the reason this design is so clever is that it is cheap to fabricate (does not require access to a clean room for photolithography) and, most importantly, it is already *optically wired*: each fiber optic can be used to project as well as collect light onto the beads, and it is straightforward to interface the other end of the bundle with a digital camera for data acquisition.

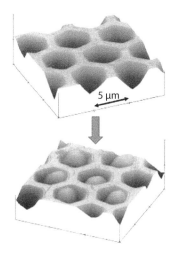

FIGURE 4.16 Optically addressable, self-assembled microarrays of beads. (From Karri L. Michael, Laura C. Taylor, Sandra L. Schultz, and David R. Walt, "Randomly ordered addressable high-density optical sensor arrays," *Anal. Chem.* 70, 1242–1248, 1998. Figure contributed by David Walt.)

BEADS, THE NEW PDMS?

IN FACT, DAVID WALT'S BEAD-BASED APPROACH can hardly be categorized as a "MEMS-based" approach (none of its components require miniaturization), so why is it featured in a BioMEMS book? Precisely because this approach manipulates miniature components for biological applications (the stated subject of the book), and it was revolutionary at that.

Historically speaking, David Walt's story has an interesting parallel in that, once more in the field of BioMEMS, another "outsider" (also a chemist working in Cambridge, Massachusetts) brought a radical new solution that lowered costs and enabled new frontiers, as George Whitesides had done with the introduction of PDMS in a silicon-based arena. And now, every engineer must at least consider this sensing design for high-throughput applications, just like every engineer will consider PDMS as a material for building a microdevice.

This sends a clear message to our young researchers (as creativity declines with age, there is little point in reminding the older generation, myself included): please think outside of the box!—whether the box is inside or outside Cambridge.

4.6.1.5 Electroaddressable Deposition of DNA and Protein

In 2007, Susan Brozik's group at Sandia National Laboratories developed what seems to be the ultimate convenient scheme for depositing DNA and detecting hybridization (Figure 4.17): the deposition is electroaddressable and hybridization is electrochemically detectable (so it requires neither optical equipment nor biochemical reagents). For the same price, in addition, it also allows for the deposition of proteins. Simultaneous electrochemical detection of a DNA sequence related to the breast cancer BRCA1 gene and the human cytokine protein interleukin-12 was demonstrated.

In certain settings (such as a biological laboratory), however, electrical addressability may not be as important an advantage as optical readability (which can be performed with microscopes).

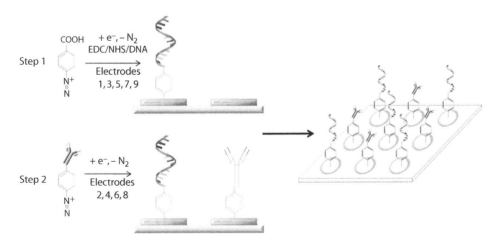

FIGURE 4.17 Electroaddressable deposition of DNA and protein. (From Jason C. Harper, Ronen Polsky, David R. Wheeler, Shawn M. Dirk, and Susan M. Brozik, "Selective immobilization of dna and antibody probes on electrode arrays: Simultaneous electrochemical detection of DNA and protein on a single platform," *Langmuir* 23, 8285–8287, 2007. Figure contributed by Susan Brozik.)

Richard Crooks' group from the University of Texas at Austin has developed arrays of "bipolar electrodes," which detect DNA hybridization by an electrochemical reaction that produces light. The electrodes are rectangular and aligned parallel to the channel; because of the voltage drop along the solution in the channel, for any given electrode, one end of the electrode is at a different voltage than the other, making it possible to bias the electrode without wires (the "wire" is the solution). DNA is detected by exposing the array to a complementary target bearing a 4-nm platinum nanoparticle, which acts as a label. Upon hybridization, the Pt label catalyzes O_2 reduction; the electrons required for this reaction originate from oxidation of *tris*(2,2′-bipyridyl) ruthenium(II) $\left[Ru(bpy)_3^{2+} \right]$ and tri-*n*-propylamine (TPrA). $Ru(bpy)_3^{2+}$ is the light-emitting species that needs an amine (here, TPrA) as a co-reactant. Although electrochemiluminescent bipolar electrodes have been used for more than a decade (Andreas Manz' group already used them in CE devices in 2001), Crooks' work made it possible to extend it to large molecules such as DNA. The simultaneous imaging of 1000 bipolar electrodes (each 1 mm long, 200 μm wide)— reporting 1000 DNA hybridization events—has been demonstrated.

4.6.2 Automated DNA Purification

Genomics and proteomics applications demand the development of devices that automate the most tedious and repetitive sample preparation procedures. Automation enables the

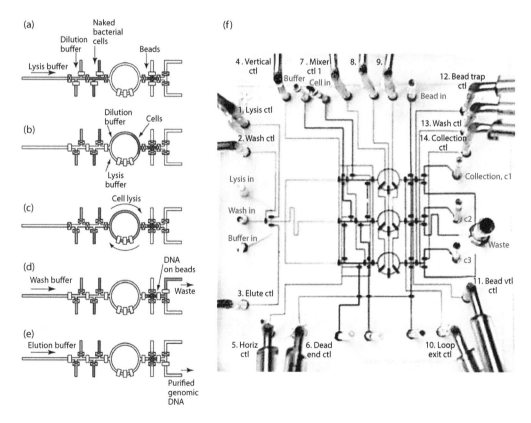

FIGURE 4.18 A microfluidic device to purify genomic DNA. (From Jong Wook Hong, Vincent Studer, Giao Hang, W. French Anderson, and Stephen R. Quake, "A nanoliter-scale nucleic acid processor with parallel architecture," *Nat. Biotechnol.* 22, 435–439, 2004. Copyright (2004) Nature Publishing Group.)

preparation of large batches of samples in parallel as well as the manipulation of samples as small as single cells without error. Microfluidics is optimally suited for the task, in terms of the cost associated to develop a device and the ability to manipulate small samples. In 2004, Quake's group presented a "nucleic acid processor" that could lyse cells (bacterial or mammalian) and extract their DNA or mRNA (Figure 4.18). The mixing and metering of the cell suspension with the lysis and dilution buffers was performed by means of microvalves and a "ring micromixer." The beads trapped downstream of the ring micromixer are derivatized with DNA-binding molecules and thus act as a DNA affinity column: as the lysed-cell solution is passed through the beads, DNA retention occurs, and the rest of the lysate gets eluted to the waste port. Purified DNA is recovered from the chip by introducing an elution buffer (which releases the DNA from the beads).

4.6.3 A Microfluidic cDNA Synthesizer

As detection methods become so sensitive that the contents of even a single cell can be meaningfully analyzed, interest in lysing cells one-at-a-time has grown. Thus far, this approach has been taken primarily in research studies, that is, in basic research and in drug discovery, but it is possible that it will soon be of clinical utility as well. The primary methods for single-cell lysis involve the use of a powerful voltage pulse (electroporation) or a rapid pulse by a high-powered laser, which causes rapid heating and cavitation to explode the cells. If these methods are designed to occur at the mouth of a chromatographic separation column, the contents of a single cell can be rapidly labeled and then separated for analysis at the other end of the separation column. Recently, a team from University of Southern California and Stanford University presented a complex microfluidic device (Figure 4.19) that extracts total mRNA and synthesizes the cDNA on the same device with high mRNA-to-cDNA efficiency. Their results indicate notable differences between population-averaged gene expression data and the expression levels in individual cells, which is only revealed (for a reasonable cost) when using such high-throughput microfluidic approaches.

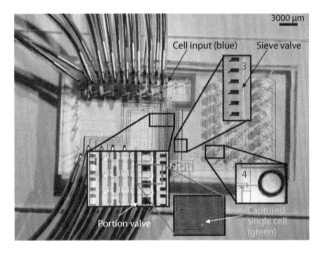

FIGURE 4.19 Microfluidic device for extracting mRNA and synthesizing cDNA on the same chip from single cells. (From Jiang F. Zhong, Yan Chen, Joshua S. Marcus, Axel Scherer, Stephen R. Quake, Clive R. Taylor, and Leslie P. Weiner, "A microfluidic processor for gene expression profiling of single human embryonic stem cells," *Lab Chip* 8, 68–74, 2008. Reproduced with permission from The Royal Society of Chemistry.)

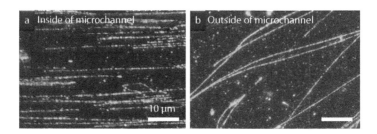

FIGURE 4.20 Flow-induced DNA elongation inside microchannels. (From J. M. Sidorova, N. Li, D. C. Schwartz, A. Folch, and R. J. Monnat, Jr., "Microfluidic-assisted analysis of replicating DNA molecules," *Nat. Protoc.* 4, 849, 2009. Figure contributed by the author.)

4.6.4 Microfluidic Elongation of DNA to Produce "Optical Maps"

In 1999, David Schwartz and colleagues (then at New York University) presented the famous "optical maps" of single DNA molecules, essentially whole-genome restriction maps constructed by single-molecule microscopy on charged glass surfaces without using DNA libraries, polymerase chain reaction (PCR), or electrophoresis. The immobilized DNA molecules are typically digested in situ with methylation-insensitive restriction endonuclease *Swa*I (its moderate average restriction fragment size balances good restriction map resolution with accurate fragment sizing), and then stained with the fluorescent dye YOYO-1, which can be visualized with standard fluorescence microscopy. In 2004, Schwartz' team (at the University of Wisconsin) implemented the technique in PDMS microchannels, which makes the deposition more reproducible (the force gradient caused by flow, or shear, that uncoils the DNA can be modeled and changed to the experimenter's specifications) and allows for high-throughput experiments, as hundreds of flows can be fed by capillarity or with a pump simultaneously (**Figure 4.20**). As bioinformatics techniques have become more sophisticated, the Schwartz group has recently been able to generate high-resolution single-molecule restriction maps of the human genome.

4.6.5 PCR Chips

One of the central molecular biology techniques of the last few decades has been the selective amplification of sequences of DNA using PCR. In 1993, a group from Lawrence Livermore Laboratories led by R. T. Watson announced in a conference on *Sensors and Actuators* that they had achieved DNA amplification with a microfabricated silicon-based reaction chamber. The first peer-reviewed report was published in 1994 by Larry Kricka's team, from the University of Pennsylvania. Shortly thereafter, many groups (several in industry) published similar results and started incorporating various improvements such as on-chip heaters, temperature probes and, later, microvalves.

In PCR, the temperature of a DNA sample is cycled between three temperatures in the presence of a thermally stable DNA polymerase, doubling the number of copies of the target sequence of DNA with each cycle. Various methods can be used to detect the amplified DNA sequence after perhaps 35 cycle steps. Some of these methods allow real-time detection by fluorescence within the thermal cycler itself. Several schemes exist for amplifying DNA without thermal cycling, for example using DNA helicase, and enzyme that unwinds DNA.

4.6.5.1 Chip Substrates and Surface Treatments

The choice of surface is critically important because of both the high surface-to-volume ratio (dictating adsorption) and the sensitivity of PCR to nucleic acid contamination. Because PCR yields enormous amplification factors (10 orders of magnitude or more), any trace of DNA contaminants or amplicons (reaction products) from preceding uses of the device can easily

Table 4.1   Materials Used as Substrates for PCR Chips

Material	Advantages	Disadvantages
Silicon	Thermal conductivity (rapid heating and cooling for rapid cycling) Microfab processes available	Bare Si inhibits PCR (adsorption of DNA?) Thermal conductivity (thermal insulation) Opaque (limits fluorescence readouts) Expensive processing Micropumps difficult
Glass	Transparent (optical readout) Allows for EOF for integration with CE (DNA separation)	Etching difficult, expensive processing No micropumps Adsorbs PCR sample
PDMS	Inexpensive Transparent Molded ⇒ fab inexpensive Flexible ⇒ micropumps and valves Less adsorption than glass	Permeability to water (concentration changes → can be avoided by presoaking or "priming" PDMS for 24 hours) Permeability to air (air bubbles when cycling → can be avoided by pressurization, by degassing the PCR sample, and by adding a solvent such as glycerol that increases the boiling point)
PC	Inexpensive fab (hot embossing) Inexpensive material High glass transition temperature (150°C)	Some autofluorescence
PMMA	Inexpensive fab (CO_2 laser) Inexpensive material Little adsorption of DNA or protein Little autofluorescence	Glass transition temperature 105°C (close to 95°C for strand separation) limits heat treatments

dominate the input copy number. The trends have followed the general trends in microfluidics: the first devices were fabricated in silicon and glass (mostly due to convenience, as fabrication was performed in MEMS facilities) and slowly, as the advantages of new polymeric materials such as PDMS, polycarbonate (PC), and poly(methyl methacrylate) (PMMA) have been unveiled and characterized, designs in these materials have become available. Fluidigm sells a reusable PDMS PCR chip (for digital PCR) that can process 36,960 reactions simultaneously from up to 48 samples. Table 4.1 summarizes the advantages and disadvantages of the materials commonly used in the fabrication of PCR chips.

4.6.5.2 PCR Chip Architectures

PCR chip architecture can be classified as stationary (chamber-based) PCR chip or dynamic (continuous flow-based) PCR chip. Chamber-based architectures, which were the first to be developed, are the result of shrinking down the traditional macroscopic PCR reactors—a strategy that does not scale well because of the importance of the surface-to-volume ratio in such critical biochemical reactions as PCR (or the devices and the solutions must be kept proportionately cleaner of nucleic acid contaminants, which increases processing costs). One of the leading laboratories in this area is Richard Mathies' laboratory at the University of California

(Berkeley), which has been able to integrate glass PCR chips with PDMS microvalves and other processing functions such as CE (**Figure 4.21**). A four-layer glass-PDMS-glass-glass hybrid device integrated microvalves, microheaters, temperature sensors, 380-nL reaction chambers, and the channels for CE where the amplicons (reaction products) are separated and detected by fluorescence after 30 cycles of RT-PCR (in 45 minutes, allowing for attomolar detection sensitivity, equivalent to ~11 template RNA molecules).

In the continuous-flow architecture, introduced in 1998 by Andreas Manz's group, then at London's Imperial College of Science, the nucleic acid solution flows through a microfluidic channel with alternating hot and cold regions, thereby producing amplification (**Figure 4.22**). A rich variety of designs (ranging from serpentines to spirals to fit as many meandering channels

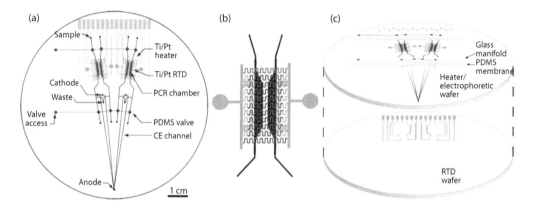

FIGURE 4.21 Integrated RT-PCR and CE on a chip. (From Nicholas M. Toriello, Chung N. Liu, and Richard A. Mathies, "Multichannel reverse transcription-polymerase chain reaction microdevice for rapid gene expression and biomarker analysis," *Anal. Chem.* 78, 7997-8003, 2006. Reprinted with permission of the American Chemical Society.)

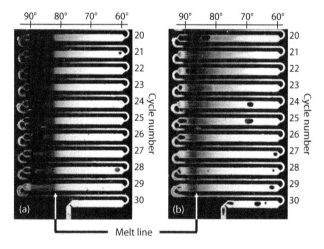

FIGURE 4.22 Continuous-flow PCR. The "melt line" shows the point where the sample has a relatively high amplicon concentration (a, DYZ1; and b, CYP2C9*3 targets). The random dark ellipsoids within the channels are slow-moving bubbles. (From Niel Crews, Carl Wittwer, Robert Palais, and Bruce Gale, "Product differentiation during continuous-flow thermal gradient PCR," *Lab Chip* 8, 919, 2008. Reproduced with permission from The Royal Society of Chemistry.)

as possible through the heating and cooling elements) have been presented, but no "optimal design" has emerged. This strategy features high throughputs but suffers from large reagent consumption, gas bubbles (during thermal cycling, gas solubility changes), and Taylor dispersion.

Reactions can also be confined in microdroplets or microchambers ("**digital PCR**") whereby the sample is subdivided into many small volumes (so quantification is done by counting the volumes that become fluorescent, rather than the total fluorescence levels). The term *digital PCR* was coined in 1999 by Bert Vogelstein and Kenneth Kinzler from the Johns Hopkins Oncology Center, who demonstrated the concept in 7-μL volumes using 96-well polypropylene PCR plates. The first implementation of microfluidic digital PCR was in 2006 by Stephen Quake's group, then at Caltech, and consisted of an array of 1176 PDMS chambers (6.25 nL each) actuated by microvalves (Figure 4.23).

HARD TO BEAD

IN TERMS OF REDUCING COST, as it turns out, microfluidics has probably been already superseded as a technology in the PCR chip race (although microfabrication will always be superior in speed). The PDMS microchamber array concept can be extended to an emulsion of microdroplets, each containing a bead bound with primers. In this method, called BEAMing (for beads, emulsion, amplification and magnetics) and developed by Bert Vogelstein of Johns Hopkins University, the beads are confined with preamplified templates into individual compartments (the microdroplets) of a water-in-oil emulsion in a test tube. The single molecules are amplified on the bead by PCR. Next, the emulsion is broken magnetically, such that a population of molecules has been transformed into a population of beads by a ~1:1 ratio. It will be hard to beat millions of beads using microchannels and microvalves.

The latest implementation of microfluidic digital PCR, by Rustem Ismagilov's group at the University of Chicago, features 1280 chambers (of ~2.6 nL each) and...zero valves(!) using the SlipChip (see Section 3.8.1.14 and Figure 3.53), as shown in Figure 4.24. DNA concentrations as low as 1 fg/μL were detectable, and no evidence of contamination was observed because the control (loaded with no DNA) did not show any activated chambers (Figure 4.24l).

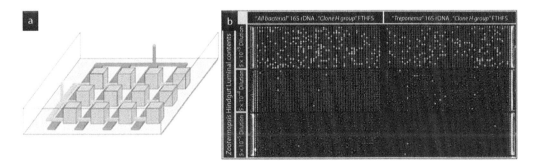

FIGURE 4.23 Digital PCR. (From Elizabeth A. Ottesen, Jong Wook Hong, Stephen R. Quake, and Jared R. Leadbetter, "Microfluidic digital PCR enables multigene analysis of individual environmental bacteria," *Science* 314, 1464–1467, 2006. Figure contributed by Stephen Quake.)

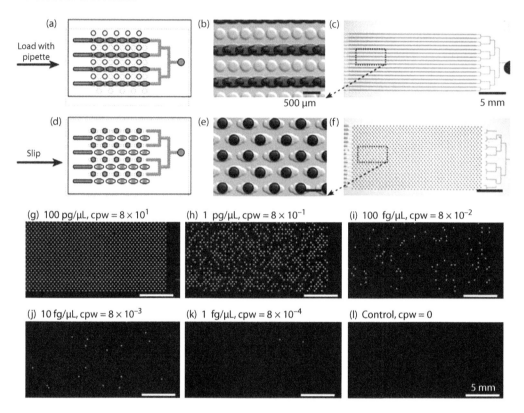

FIGURE 4.24 Digital PCR using the SlipChip. (From Feng Shen, Wenbin Du, Jason E. Kreutz, Alice Fok, and Rustem F. Ismagilov, "Digital PCR on a SlipChip," *Lab Chip* 10, 2666–2672, 2010. Reproduced with permission from The Royal Society of Chemistry.)

4.6.5.3 PCR Reaction Volume, Temperature Control, and Speed

Miniaturization of the reaction volume is not always desirable, as effects such as loss of sample/solution on or into the channel walls become relatively more severe. The smaller the thermal mass (including the reaction volume), the faster the cycling. (World record of stationary PCR, as of 2006, is 40 cycles in less than 6 minutes using a silicon chip with integrated heater.) Fabricating the heater as a separate reusable unit makes it cheaper to assemble, optimize, and reuse but increases cycling time. Continuous-flow systems are faster because the heat inertia is limited mostly to the sample thermal mass; a 500 base pair DNA fragment was amplified in 1.7 minutes in 2005. Also, note that the time dependence of fluorescence acquisition typical of real-time PCR is eliminated because the cycles and temperatures are spread over space instead of over time. Incorporating microfabricated temperature sensors is feasible but increases device thermal mass and, anyways, reads only temperature at a few points; a better option is to add a thermochromic liquid crystal or fluorescence dye indicator or monitor thermal radiation in the infrared (IR) spectrum ("IR Thermography") or by laser-based Raman spectroscopy so temperature can be read optically. Integration of several platforms into one presents intellectual property challenges that hinder commercialization.

Although there are other amplification schemes possible, PCR remains the standby. One of the earliest biochemical procedures to be embodied in silicon was PCR amplification. Clearly, one would like to speed up the amplification process as much as possible. This can be accomplished by increasing the speed of thermal cycling. However, there is a minimum time required for the creation of the new copies of DNA that depends on the rate at which the enzyme turns over. Thermal cycling faster than that cannot improve the process, and can even make it worse. It was thought that the excellent thermal conductivity of Si would allow rapid temperatures changes.

Very uniform temperature within the sample are also required to make sure that amplification process proceeds with predictable fidelity. As it turned out, Si itself poisoned the polymerase reaction, so some form of chemical insulator had to be interposed between the reaction solution medium and the device itself. Thermal cycling consisting of a static cell apposed to a single Peltier heater and cooler has been used, as well as a serpentine channel that runs a sample sequentially through zones on a chip held at three different temperatures, and even a system that consisted of a single loop that caused cycling of the contents by buoyancy. Ultimately, the apposition of a polymeric chamber to a high-powered Peltier element has proven to be one of the most efficient strategies, allowing 35 full extension cycles for short sequences in less than 8 minutes.

In sum, there are still several practical challenges in the field: many designs are expensive (they are not disposable, so cross-contamination is difficult to avoid), most can only perform DNA/RNA amplification, and detection methods such as gel electrophoresis are not as amenable to miniaturization as the device volume is shrunk further.

4.6.6 High-Throughput Protein Immunoblotting on a Chip

Given a sample of tissue extract or homogenate, specific proteins can be detected by a technique known as **protein immunoblot** or **Western blot**. The first step in the technique uses electrophoresis to separate the proteins in a gel matrix, usually a polyacrylamide gel; often the proteins are denatured with the addition of a detergent (such as sodium dodecyl sulfate) so that protein migration inside the gel is only a function of its size and not its charge or shape. The second step in the technique consists of transferring (or "blotting") the proteins to an antibody-coated membrane (typically made of nitrocellulose or polyvinylidene fluoride (PVDF)) that captures the protein(s) of interest.

Until recently, Western blots have essentially been a manual, extremely slow technique that lasts between 5 and 12 hours because of the long incubation steps. Obviously, this low throughput precludes Western blots from being used in proteomics analysis. There are large cancer tissue and sample repositories that have grown over several decades (containing tens of thousands of frozen samples from patients whose disease outcomes are now known) which are waiting to be analyzed with a high-throughput immunoblotting approach.

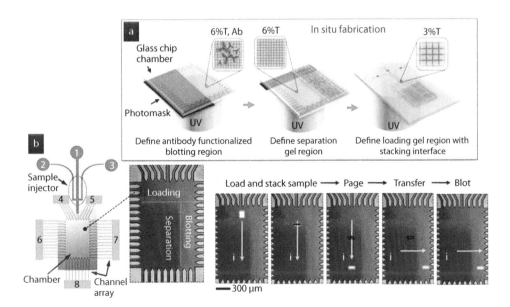

FIGURE 4.25 Microfluidic platform that integrates electrophoretic separation and immunoblotting of proteins. (From Mei He and Amy E. Herr, "Automated microfluidic protein immunoblotting," *Nat. Protoc.* 5, 1844 (2010). Reprinted with permission from Nature Publishing Group.)

Amy Herr's laboratory at the University of California (Berkeley) has developed a microfluidic glass chip platform that integrates polyacrylamide gel electrophoresis with immunoblotting (**Figure 4.25**). The main chamber is filled with gels that are photochemically patterned to provide either separation or immunoblotting functionality (**Figure 4.25a**). Transport of fluids within the chip is achieved by electrophoresis; to achieve homogeneous current injection and parallel electric field lines, the main chamber is designed with many electrical contacts (connected to ports numbered "4" through "8" in **Figure 4.25b**). The sample (5 µL) is loaded onto reservoir 2 and moved sequentially down and right by applying voltage (50–500V) sequentially to the various ports (**Figure 4.25c**). The yellow band stops at (i.e., is captured by) the antibody-functionalized blotting region despite the continued application of field: note that the blue band (not captured by the antibodies) keeps going out of the field of vision. An assay takes typically 1 minute and the chips can be reused after cleaning with piranha solution.

4.6.7 Protein Crystallization Chips

Elucidating the relationships between protein structure and protein function has been one of the most intense areas of study in the field of molecular biology for many decades, and it remains so in this decade of high-throughput experimentation where it has become one of the goals of proteomics. The determination of a protein's structure is presently done by X-ray crystallography and requires the formation of a crystal of the protein so that an X-ray spectrum (the set of X-ray dots scattered by the crystal onto a screen) can be recorded and analyzed. Protein crystal formation depends on a number of biophysical parameters that are typically optimized empirically, for example, temperature, salt and protein concentration, and others. Not all conditions produce good crystals. Screening this combinatorial parameter space may involve running on the order of hundreds of experiments, which has slowed the progress of several fields that rely heavily on protein structure determination, such as structural biology, molecular biology, and pharmacology. In 2002, Stephen Quake's group (then at Caltech) produced a chip with 480 microvalves capable of simultaneously performing 144 mixing reactions (each using only 10 nL of protein sample). After incubation at 25°C (for up to 3 weeks), the resulting crystals (**Figure 4.26**) could be removed from the chip and analyzed. The microfluidic chip is now commercially available.

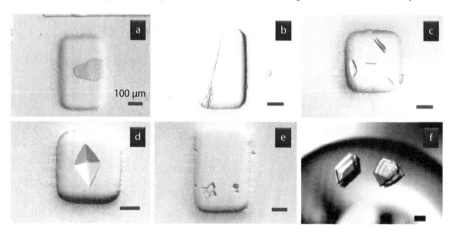

FIGURE 4.26 Protein crystallization on a chip. (a-e; From Carl L. Hansen, Emmanuel Skordalakes, James M. Berger, and Stephen R. Quake, "A robust and scalable microfluidic metering method that allows protein crystal growth by free interface diffusion," *Proc. Natl. Acad. Sci. U. S. A.* 99, 16531–16536, 2002; f, from Carl L. Hansen, Morten O. A. Sommer, and Stephen R. Quake, "Systematic investigation of protein phase behavior with a microfluidic formulator," *Proc. Natl. Acad. Sci. U. S. A.* 101, 14431–14436, 2004. Copyright (2002 and 2004) National Academy of Sciences, U. S. A. Figure contributed by Stephen Quake.)

It only took Rustem Ismagilov's group from the University of Chicago a couple of years to develop the droplet version of Quake's device. (And an even simpler one in the SlipChip format.) In the droplet version, the combinations are achieved by a mixer upstream that mixes three inlets in various proportions to obtain all the desired combinations (**Figure 4.27**). To form droplets containing different concentrations of reagents, the relative flow rates of the three streams are constantly changed to combine streams in several ratios (black arrows in **Figure 4.27c**). At the same time, to index these concentrations, the sizes of droplets are changed by changing the flow rate of the carrier fluid. By using just 10 µL of protein solution, a researcher could set up approximately 1300 crystallization reactions in 20 minutes. After crystallization in various conditions at 23°C for 9 days, the plugs (carrying crystals) were slowly transferred into paraffin oil for freezing. The advantages, with respect to Quake's device, are that there is no disposable device to deal with and that the throughput is only limited by the length of the Teflon tube to which the microfluidic device is connected to (rather than by the area of the device).

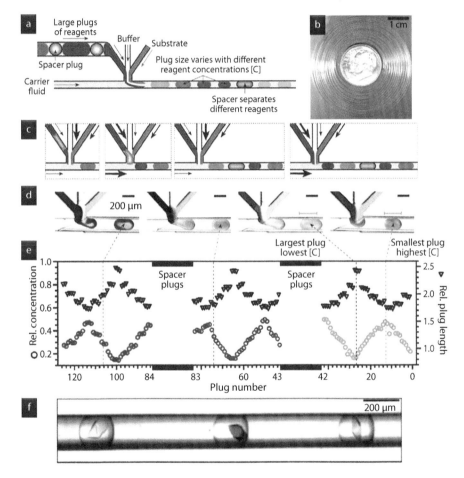

FIGURE 4.27 Protein crystallization in a droplet (a–e). Thaumatin crystals in microdroplets visualized under polarized-light (f). (From See Liang Li, Debarshi Mustafi, Qiang Fu, Valentina Tereshko, Delai L. Chen, Joshua D. Tice, and Rustem F. Ismagilov, "A nanoliter microfluidic hybrid method for simultaneous screening and optimization validated with crystallization of membrane proteins," *Proc. Natl. Acad. Sci. U. S. A.* 103, 19243–19248, 2006. Copyright (2006) National Academy of Sciences, U. S. A.; Bo Zheng, Joshua D. Tice, and Rustem Ismagilov, "Formation of arrayed droplets by soft lithography and two-phase fluid flow, and application in protein crystallization," *Adv. Mater.* 16, 1365–1368, 2004. Reprinted with permission from John Wiley and Sons.)

4.6.8 Measuring DNA–Protein Interactions Using PDMS Mechanical Traps

In eukaryotes (unlike bacteria), RNA polymerase cannot initiate transcription on DNA without the help of a set of proteins called *transcription factors*. Hence, studying the binding characteristics (on and off rates) of transcription factors has been an important thrust to understand and model one of the most fundamental biochemical reactions of life. An important technological challenge has been that these molecular interactions are transient and exhibit nanomolar to micromolar affinities, so by the time the experimenter is ready to measure the bound molecules, there are no molecules bound. Recently, Stephen Quake's group at Stanford University has developed a new high-throughput microfluidic device (**Figure 4.28**) that enables the detection of low-affinity transient binding events on the basis of mechanically induced trapping of molecular interactions. The device uses pneumatically actuated PDMS membranes (i.e., Quake

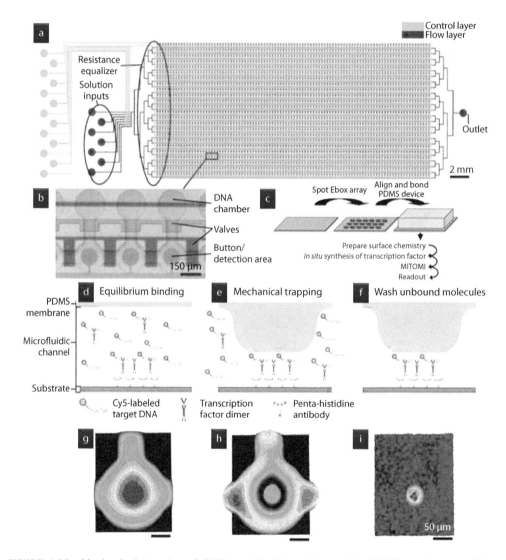

FIGURE 4.28 Mechanical trapping of DNA–protein interactions using PDMS membranes. (From Sebastian J. Maerkl and Stephen R. Quake, "A systems approach to measuring the binding energy landscapes of transcription factors," *Science* 315, 233–237, 2007. Figure contributed by Stephen Quake.)

microvalves) to apply pressure against the DNA–transcription factor complexes as soon as they are bound, thus preventing their dissociation into solution. The approach can, in principle, be extended to just about any two low-affinity biomolecules.

4.7 Electrospray Mass Spectrometry

Mass spectrometry is a technique used to determine the composition and chemical structure of a sample or a chemical compound. After vaporizing and ionizing the sample (e.g., with electron beam irradiation), the charged molecule fragments are accelerated by an electric field and sorted spatially according to their mass-to-charge ratio (m/z). The m/z of the particles is computed

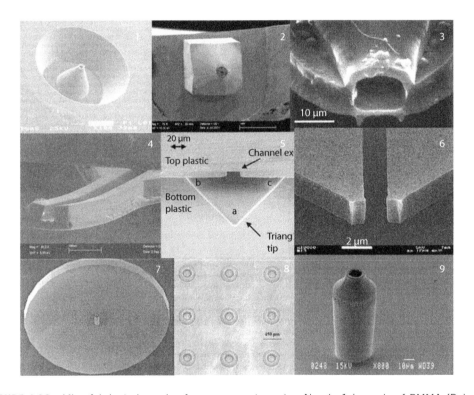

FIGURE 4.29 Microfabricated nozzles for mass spectrometry. Nozzle 1 is made of PMMA (Roland Hergenröder laboratory, ISAS, Dortmund, Germany, 2004), nozzle 2 is made of polycarbonate/PMMA by injection molding (Karin Markides laboratory, Uppsala University, Sweden, 2003), nozzle 3 is a parylene emitter (Terry Lee's laboratory, Beckman Research Institute of the City of Hope, California, 2000), nozzle 4 is a PDMS emitter made by soft lithography (Fredrik Nikolajeff's laboratory, Uppsala University, 2004), nozzle 5 is an emitter formed from a triangular piece of parylene sandwiched between two Zeonor substrates structured by hot embossing (Harold Craighead laboratory, Cornell University, New York, 2002), nozzle 6 is made of SU-8 using standard clean room processes (Christian Rolando laboratory, Univ. des Sciences et Technologies de Lille, France, 2005), nozzle 7 is a commercial emitter made of silicon using deep reactive-ion etching (Advanced Bioanalytical Services, Inc., 2000), (the array of) nozzle 8 is made of polycarbonate by laser ablation (Richard Smith laboratory, Pacific Northwest National Lab, US, 2001), and nozzle 9 is made of silicon dioxide using deep reactive-ion etching (Johan Roeraade laboratory, Royal Institute of Technology, Sweden, 2003). (From Sander Koster and Elisabeth Verpoorte, "A decade of microfluidic analysis coupled with electrospray mass spectrometry: An overview," *Lab Chip*, 7, 1394–1412, 2007. Reproduced with permission from The Royal Society of Chemistry.)

based on the details of the motion of the ions as they transit through the fields. In most mass spectrometers, the sample molecules are converted into ions from gas phase by means of an ion source. The use of electrospray ionization allows for using liquid samples.

The first reports of microfluidic devices that implemented electrospray ionization-mass spectrometry used electrospray from channels terminating at the edge of the device, such as the glass device reported in 1997 by Barry Karger's group at Northeastern University. Other groups have used conventional electrospray emitters (e.g., fused-silica capillaries) attached to the device. However, these approaches are nonoptimal because the spray at the edge of the device is troubled by droplet formation. Importantly, electrospray nozzles can be miniaturized using micromachining techniques so that the nozzles can be integrated with the device (**Figure 4.29**), producing reliable sprays. The first integrated micromachined emitter, made of silicon nitride, was reported by Terry Lee's group from the Beckman Research Institute of the City of Hope (Duarte, CA) at the *Transducers '97* conference but the design was quickly abandoned because of unreliable performance. The first micromolded (PDMS) emitter tip integrated as part of a microfluidic device was demonstrated by Daniel Knapp's team from the Medical University of South Carolina in 2001. The performance of the nozzle is logically a function of its shape and materials, and thus, many groups have devoted great efforts not only to develop better electrospray emitters but also to understand the fundamentals of electrospray emission.

The diameter of the nozzle and the flow rate at which fluid is ejected are the critical parameters for measuring the performance of a microfabricated electrospray mass spectrometer. Present-day nozzles with inner dimensions on the range of a few microns are capable of generating a "nanospray," resulting in higher ionization efficiency (because the droplets are smaller, the number of charges available per analyte molecule is much higher in a nanospray, which enhances the probability of ionization), lower electrospray voltages (which allows the nozzle to be positioned closer to the mass spectrometer), and less sample consumption compared with traditional electrosprays.

In 2007, a team led by Jonathan Sweedler and Ralph Nuzzo from the University of Illinois at Urbana-Champaign was able to capture the release of neuropeptide by single *Aplysia californica* bag cell neurons using mass spectrometry imaging. Peptide release was not analyzed online—the peptides were chemically captured by the surface (a self-assembled monolayer of octadecyl alkyl chains, which happen to adsorb the neuropeptides) and the surface was interrogated by scanning it, performing a mass spectrometric analysis at each "pixel" of the sample. Imaging resolution ranged from 100 to 200 μm.

4.8 Biochemical Analysis Using Force Sensors

Another very successful strategy for detecting binding or affinities between biomolecules is based on measuring the force of adhesion between them. There are many devices that can be used to measure forces, and many do not use microfabricated components (e.g., laser tweezers), but the simplest is probably the spring, or its cousin, the spring board (also known as cantilever) because it translates force to displacement.

Cantilevers have been used to detect forces for centuries: the force applied at the tip produces a displacement, which can be measured; the relationship between the force, F, and the tip displacement, d (usually linear, $F = kd$) can be calibrated with known applied forces or weights, so that when an unknown force is applied, the value can be known by interpolation. Without going into the details of how the formula is derived, for small deflections, the relationship between d and F for a cantilever of width, w, thickness, t, and length, L, can be expressed as:

$$F = \frac{1}{4} E \frac{wt^3}{L^3} d \qquad (4.1)$$

where:

E, the magnitude that needs to be calibrated for each material is called **Young's modulus** and the rest, w, t, and L, are constants *for each cantilever.*

Thus, for small deflections, the cantilever follows essentially Hooke's law ($F = kd$), like a spring. It is straightforward to see that the best strategy to improve sensitivity, that is, to maximize deflection for any given force, is to use very thin or very long cantilevers: a cantilever of half the thickness (or double the length) of another one will deflect eight times more than the latter. Why, then, microfabricate the cantilever, if doubling the length improves the force sensitivity by the same factor as halving the thickness?

The answer lies in the **resonance frequency**, which is the frequency at which the cantilever oscillates with a maximum amplitude (for a given driving amplitude). We worry about the resonance frequency because the world surrounding the instruments introduces vibrations, and if these vibrations are on the order of the resonance frequency, then the cantilever will "resonate"—as if "in tune," much like a guitar string would. Usual disturbances come from people walking, acoustic noise, traffic, and so on—all in the range lower than 100 kHz. If one could build a cantilever that resonates at 1 MHz or higher, it would be pretty much insensitive to a child screaming next to it.

Recall that for a spring, the resonance frequency, ω_0, can be expressed as

$$\omega_0 = \sqrt{\frac{k}{m}} \tag{4.2}$$

where:

m would be the mass that hangs from the spring.

In a cantilever, however, the mass is distributed all over the spring, so the effective mass is approximately one-fourth of the mass of the cantilever, m_c:

$$\omega_0 \approx \sqrt{\frac{k}{0.24m_c}} \tag{4.3}$$

where:

m_c = (density of cantilever material) × (cantilever volume) = $\rho \times w \times t \times L$.

(Note that the units of ω_0 are radians, but most engineers prefer to use $f_0 = \omega_0/2\pi$, measured in Hz or cycles because 2π rad = 1 cycle.) Because $F = kd$, from **Equation 4.1**, we have

$$k = \frac{1}{4}E\frac{wt^3}{L^3} \tag{4.4}$$

Substituting **Equation 4.4** into **Equation 4.3**, we obtain:

$$\omega_0 \approx \sqrt{\frac{\frac{1}{4}E\frac{wt^3}{L^3}}{0.24\rho wt L}} \approx \sqrt{\frac{E}{\rho}\frac{t^2}{L^4}} = \sqrt{\frac{E}{\rho}}\frac{t}{L^2} \tag{4.5}$$

Note that k is proportional to t^3/L^3, whereas ω_0 is proportional to t/L^2, so they scale differently when the size and proportions change. Thus, if we halve the thickness of a cantilever, k is divided by 8 and ω_0 is halved, but if we double its length, k is also divided by 8 whereas ω_0 is divided by four. If we reduce all the dimensions w, t, and L by half, k decreases by half (i.e., the same applied force produces twice as much deflection) and ω_0…increases by 2!

From **Equations 4.4** and **4.5**, we see that if we want to increase the sensitivity of a cantilever (i.e., increase its sensitivity to forces but reduce its sensitivity to vibrations, so that small displacements can be resolved) what we want to do is scale down *all* its dimensions—which calls for microfabrication. That is why the most sensitive cantilevers are microfabricated.

Researchers have used microcantilevers as substrates to measure whether biological material has adhered to the surface. The most sensitive measurement uses a cantilever vibrated close to its resonance frequency: a small amount of material added to the cantilever will shift the resonance frequency and significantly change the vibration amplitude. Piezoelectric cantilevers are particularly powerful; these are made of a piezoelectric material that produces a voltage when the cantilever is deflected and, conversely, can be driven to vibrate by application of an AC voltage. First, the cantilever is coated with an antibody to the antigen of interest. Then, the cantilever is vibrated at its resonance frequency while it is exposed to the sample. When the antibody recognizes the antigen, the antigen presence can be readily detected as a change in vibration amplitude.

Using optical interferometry detection, Harold Craighead and colleagues from Cornell University have developed nanocantilevers with gold dots on the tip (Figure 4.30). The dots can be used for binding thiolated SAMs or antibodies, allowing for a minimum resolvable mass of 0.37 ag. A thiol SAM monolayer covering a typical 50 nm gold dot, for example, weighs just 6.3 ag, and a (dried) 1.43-µm-long *E. coli* bacterium was found to weigh 665 fg.

What about the detection of molecules in solution? The major problem is that the high viscosity of water dampens the vibration of the cantilever, impairing the detection sensitivity. Therefore, it is clear, from fundamental principles, that one cannot dip a cantilever in an aqueous solution

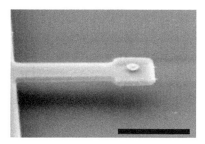

FIGURE 4.30 Attogram detection using nanocantilevers. Scale bar is 2 µm. Deflection of the cantilever was detected by optical interferometry. (From B. Ilic, H. G. Craighead, S. Krylov, W. Senaratne, C. Ober, and P. Neuzil, "Attogram detection using nanoelectromechanical oscillators," *J. Appl. Phys.* 95, 3694–3703, 2004. Reprinted with permission of the American Institute of Physics.)

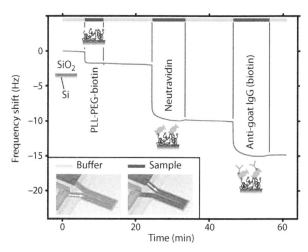

FIGURE 4.31 Liquid-filled cantilevers for biochemical detection. (From Thomas P. Burg, Michel Godin, Scott M. Knudsen, Wenjiang Shen, Greg Carlson, John S. Foster, Ken Babcock, and Scott R. Manalis, "Weighing of biomolecules, single cells and single nanoparticles in fluid," *Nature* 446, 1066–1069, 2007. Copyright (2007) Nature Publishing Group.)

and vibrate it and obtain the same mass balance sensitivities that can be obtained in vacuum. However, human inventiveness seems endless. In 2003, Scott Manalis' group at MIT managed to microfabricate a hollow cantilever (3-μm-thick fluid layer) that can be filled with fluids and vibrated in vacuum (without spilling the fluids into the vacuum!). Adsorption of the solutes onto the inner walls of the cantilever alters the mass of the cantilever and changes its resonance frequency (by a fraction of a percent, at ~200 kHz), which can be readily detected (Figure 4.31). This "suspended microchannel resonator" has also been used to measure the mass, density, and size of particles, including cells in their normal life cycle. Sensitivity is highest near the cantilever tip, enabling the measurement of particle masses with a precision of 300 ag (1 ag = 10^{-18} g).

4.9 Summary

Microfluidics is revolutionizing virtually every molecular biology technique. This revolution is inevitable for two overlapping reasons: (1) microfluidics allows for handling minute quantities of precious fluids (which could be done previously with robotic pipetters), and (2) provides dirt-cheap automation. Automated fluid handling is a requisite to reduce human errors and to perform high-throughput experimentation. Robotic pipetters are too costly for shrinking university and research budgets and too bulky to be deployed in home care and field settings. Clearly, microfluidics should become a winning technology.

FUTURE MICROFLUIDIC ASSAYS: THE "CAR TREND" VERSUS THE "TOY TREND"

WE SEE TWO DIVERGING TRENDS in the miniaturization of diagnostic devices (typically, microdevices that perform a few biochemical reactions), depending mostly on *where* they will be used (and less by *who* will use them). For those systems for which there is no incentive to be untethered from a laboratory wall, the trend is an increase in integration, similar to what we have seen in the car industry in the last 25 years with the incorporation of all the technologies from Formula 1: electronic chips now control almost every function of the car, from steering to braking, fuel injection, shifting, etc. Importantly, all these technologies are hidden from the driver, providing a more pleasant experience. Similarly, the end users of a microfluidic system do not want to be bothered with the difficulties of microvalves and micropumps, they want to be *helped* by them. When the device can be used in a laboratory setting, in which it can be powered up with electricity, fluids, and pressurized gases, and can be placed on a computerized microscope stage for inspection, the design can easily incorporate a range of power-hungry sensors and actuators. Ideally, these systems should work by pushing buttons only and loading of fluids into them should be through open ports ("top-loadable")—as an analogy, we would like their connections to be as simple as a USB connector.

However, there is also a large class of diagnostics devices that cannot be tethered to a wall—those that need to be portable so they can deployed to the field. These systems need to be made the size of a disposable plastic credit card or, even better, a paper stamp. They cannot consume energy, or at best, they must run on batteries or solar cells. The typical end user will be a patient who does not have a technical education and might even have a physical disability, so the system should be *really* easy to use, like a toy. Note that there is nothing wrong with the term "toy"—toys are safe, simple devices that help us learn.

Further Reading

Bashir, R., and Wereley, S. (editors). "Biomolecular sensing, processing and analysis," *Vol. IV of BioMEMS and Biomedical Nanotechnology Series*. Springer (2006).

Berthier, J., and Silberzan, P. *Microfluidics for Biotechnology*, 2nd ed. Artech House (2010).

Hardt, S., and Schönfeld, F. (editors). *Microfluidic Technologies for Miniaturized Analysis Systems*. Springer (2007).

Ozkan, M., and Heller, M. J. (editors). "Micro/nano technology for genomics and proteomics" *Vol. II of BioMEMS and Biomedical Nanotechnology Series*. Springer (2006).

Pamme, N. "Continuous flow separations in microfluidic devices," *Lab on a Chip* **7**, 1644–1659 (2007).

Saliterman, S.S. *Fundamentals of BioMEMS and Medical Microdevices*. SPIE Press (2006).

5

Cell-Based Chips for Biotechnology

THE PREVIOUS CHAPTER DEALT EXCLUSIVELY with microdevices that process cell-free samples or cell extracts—in other words, the biological material was *not alive*, which simplifies both its maintenance and its readout. A whole new set of challenges is presented to the experimenter when the measurement or question requires that the interrogated sample be kept alive, as we will see in the next three chapters. In this chapter, we will look at applications that have traditionally concerned the biotechnology industry, that is, primarily with a health care device focus.

5.1 Microfluidic Flow Cytometers

Possibly the most basic operation one can do with cells is to count them. **Flow cytometers** are devices that allow for counting particles (such as cells or beads) suspended in a stream of fluid. The device produces a stream (typically a single cell wide) carrying a cell suspension ensheathed in a larger stream, which acts as a carrier that places the cells in line with the detector. Traditional (nonmicrofabricated) flow cytometers are expensive, bulky, and tedious to operate, hence great research efforts have been deployed to miniaturize and automate flow cytometry since the early days of microfluidics.

Michael Ramsey's group, from Oak Ridge National Laboratory in Tennessee, was the first to demonstrate in 1997 the two-dimensional confinement or "focusing" of flow (of rhodamine 6G in Figure 5.1). The focusing is achieved by "pinching" the sample flow between two focusing flows; the width and position of the output focused flow is a function of the focusing flow rates relative to that of the sample. The device was built in glass and the flow was controlled electrokinetically; however, the same functionality can be achieved in a much simpler setup with a PDMS device and gravity-driven flows or syringe pumps.

However, achieving focus in two dimensions is generally not enough because of the "butterfly effect" (see Section 3.9.1 and Figure 3.73)—the particles that are traveling close to the roof or the floor of the channel will be flowing much slower than the particles traveling close to the midline of the channel, which will confound the detector. Ideally, a scheme that allows for focusing in three dimensions (3-D), creating a single-cell-wide, constant-velocity stream in the center of the channel, is desirable. Wanjun Wang's group at Louisiana Tech University used sophisticated tilted photolithography to create a real microfabricated nozzle that focuses flow in all three dimensions (Figure 5.2).

The SU-8 nozzle of Figure 5.2 is an example of miniaturization of a macroscale component. However, can a focused flow be achieved in a simpler way, one perhaps that does not require exotic photolithography equipment, that is compatible with PDMS molding, or that exploits

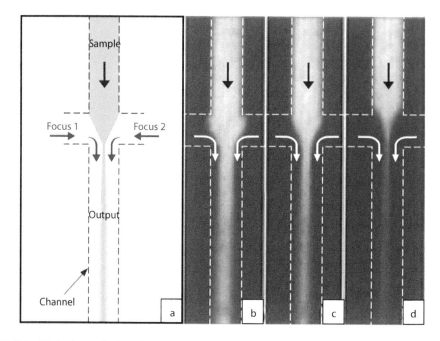

FIGURE 5.1 Hydrodynamic focusing using electrokinetic flows. The voltages applied to the inlets are 2.4 kV (focus 1), 2.6 kV (focus 2), 0 kV (output), 3.9 kV (sample in B), 2.9 kV (sample in C), and 1.9 kV (sample in D). (From S. C. Jacobson and J. M. Ramsey, "Electrokinetic focusing in microfabricated channel structures," *Anal. Chem.* 69, 3212, 1997. Adapted with permission of the American Chemical Society.)

some clever hydrodynamic phenomena on the microscale? As we see below, all of the above have already been invented—there are plenty of smart people!

In 2007, Ruey-Jen Yang and colleagues at National Cheng Kung University in Tainan, Taiwan, came up with a brilliantly simple way to sandwich the sample flow using two-layer PDMS microfluidics (Figure 5.3)—making the device entails only two masks, one in the shape of a cross ("ADE" in Figure 5.3) and the other in the shape of two lines ("BC" in Figure 5.3)!

Alex Groisman's group at University of California in San Diego has further shown that the cross-device can be "planarized" and fabricated in a single mold (i.e., "B" and "C" in Figure 5.3 would be joining channel A at the same plane as the floor of the channel, not under it); the mold requires triple-layer photolithography, but after that, fabrication of each device does not require alignment (Figure 5.4). The sample stream was focused to approximately 10 μm and an impressive 17,000 counts/s were achieved with this device, rivaling commercial nonmicrofabricated flow cytometers.

Note that the above cross-devices require a delicate balance of four steering streams to keep the sample stream stationary. Shouldn't it be possible to use microfluidic effects that automatically ensheath the sample, using only one inlet for the sample and two for the sheath (or only one for the sheath, if steering is not required)? The first thought along these lines came from Frances Ligler's group at the Naval Research Laboratory. She noticed that the presence of chevron grooves on the floor and roof of a microchannel causes 3-D hydrodynamic focusing, and implemented the principle into a flow cytometer (Figure 5.5).

Tony Jun Huang and colleagues at Pennsylvania State University have presented a clever single-layer microfluidic design that induces Dean flow in a curved microfluidic channel (Figure 5.6). The curvature, when particles are flown at high flow velocities (on the order of

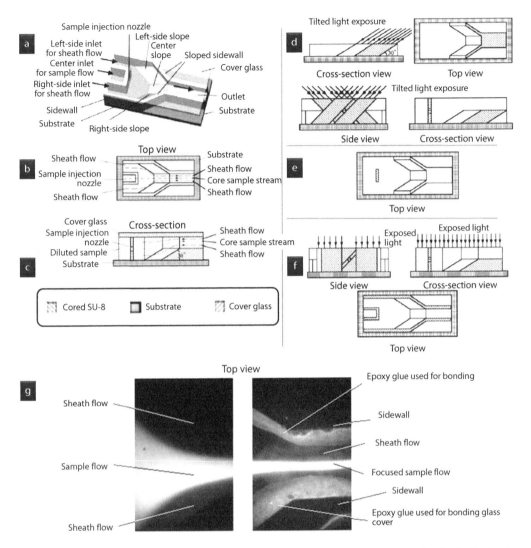

FIGURE 5.2 Microfluidic flow cytometer integrated in SU-8. (From Ren Yang, Daniel L. Feeback, and Wanjun Wang, "Microfabrication and test of a three-dimensional polymer hydro-focusing unit for flow cytometry applications," *Sens. Actuators A* 118, 259–267, 2005. Adapted with permission of Elsevier.)

meters per second), generates "drifting" that can be used to focus cells in 3-D at rates greater than 1700 cells/s.

To create Dean flow, this device requires flow velocities that are extremely high by microfluidic scales, which raises concerns of shear stress acting on the cells. Is there another way to create Dean flow, perhaps one that would be gentler to cells? Je-Kyun Park's group at Korea's Advanced Institute of Science and Technology (KAIST) has elegantly demonstrated that Dean flow can be induced by a series of contraction–expansion structures in the microchannel (Figure 5.7). Every time the fluid enters an expansion of the microchannel, it is forced to follow rotational lines (to conserve the momentum) that end up concentrating the sample flow in the 3-D center of the channel.

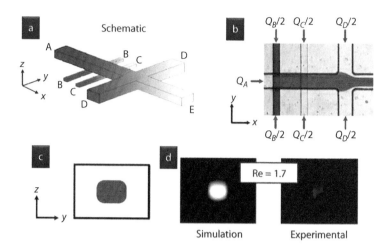

FIGURE 5.3 3-D hydrodynamic focusing using two-layer PDMS microfluidics. (From Chih-Chang Chang, Zhi-Xiong Huang, and Ruey-Jen Yang, "Three-dimensional hydrodynamic focusing in two-layer polydimethylsiloxane (PDMS) microchannels," *J. Micromech. Microeng.* 17, 1479–1486, 2007. Adapted with permission of the Institute of Physics.)

Note that in all these devices, one of the inlets is the sample carrying the cells and other inlet(s) carry solution simply to focus it in 3-D. However, because the cell solution also carries plain solution, shouldn't it be possible to design a device that focuses the cells with a single inlet, making use of inertial effects? Dino Di Carlo's group at the University of California, Los Angeles, has demonstrated a device of ultimate simplicity (see **Figure 5.8a**): the cells, introduced at the right flow speeds (~10 cm/s) into channels *of the right size*, experience inertial forces that bring them to equilibrium positions. There are two inertial lift forces acting on the cells: a "wall effect lift" and a "shear gradient lift" (**Figure 5.8b**). A design with 256 channels (**Figure 5.8c** and **d**) can easily count 1 million cells per second(!). The equilibrium positions for different channel aspect ratios are shown in **Figure 5.8e**.

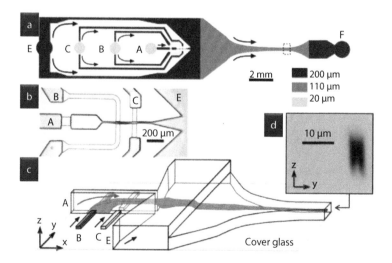

FIGURE 5.4 Single-layer 3-D flow focusing. (From Claire Simonnet and Alex Groisman, "High-throughput and high-resolution flow cytometry in molded microfluidic devices," *Anal. Chem.* 78, 5653–5663, 2006. Reprinted with permission of the American Chemical Society.)

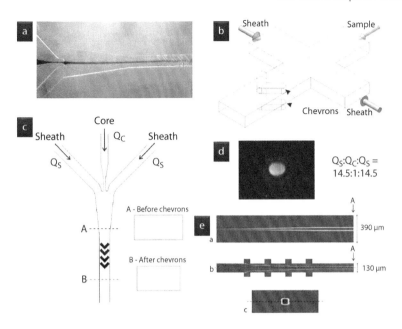

FIGURE 5.5 3-D hydrodynamic focusing using chevron grooves. (From Peter B. Howell Jr., Joel P. Golden, Lisa R. Hilliard, Jeffrey S. Erickson, David R. Mott, and Frances S. Ligler, "Two simple and rugged designs for creating microfluidic sheath flow," *Lab Chip* 8, 1097–1103, 2008; and Joel P. Golden, Jason S. Kim, Jeffrey S. Erickson, Lisa R. Hilliard, Peter B. Howell, George P. Anderson, Mansoor Nasir, and Frances S. Ligler, "Multi-wavelength microflow cytometer using groove-generated sheath flow," *Lab Chip* 9, 1942–1950, 2009. Reproduced with permission from The Royal Society of Chemistry.)

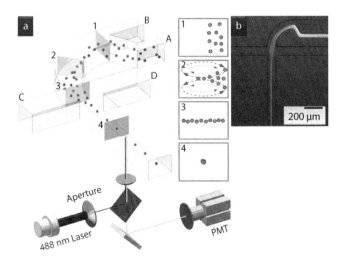

FIGURE 5.6 Single-layer microfluidic device producing 3-D focused flow. (From Xiaole Mao, Sz-Chin Steven Lin, Cheng Dong, and Tony Jun Huang, "Single-layer planar on-chip flow cytometer using microfluidic drifting based three-dimensional (3D) hydrodynamic focusing," *Lab Chip* 9, 1583–1589, 2009. Reproduced with permission from The Royal Society of Chemistry.)

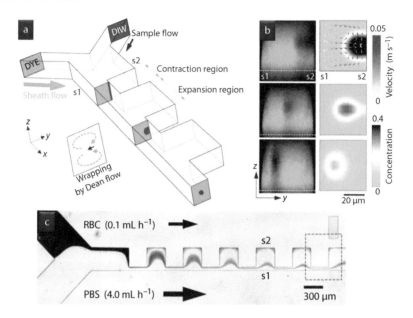

FIGURE 5.7 Contraction–expansion array resulting in 3-D focused flow. (From Myung Gwon Lee, Sungyoung Choi, and Je-Kyun Park, "Three-dimensional hydrodynamic focusing with a single sheath flow in a single-layer microfluidic device," *Lab Chip* 9, 3155–3160, 2009. Reproduced with permission from The Royal Society of Chemistry.)

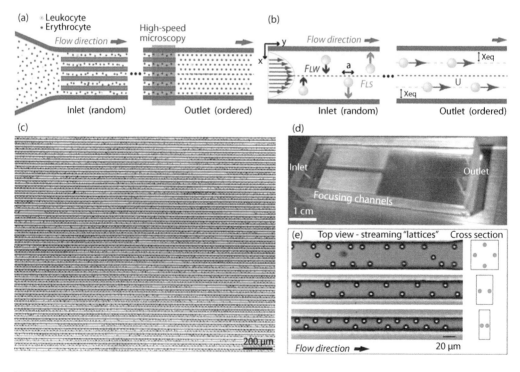

FIGURE 5.8 Extreme-throughput, sheathless flow cytometry: the "Di Carlo cell counter." (From Soojung Claire Hur, Henry Tat Kwong Tse, and Dino Di Carlo, "Sheathless inertial cell ordering for extreme throughput flow cytometry," *Lab Chip* 10, 274–280, 2010. Reproduced with permission from The Royal Society of Chemistry.)

THE SIMPLEST OUTSMARTS THEM ALL

THE DI CARLO CELL COUNTER sets a very high bar for improvement, and it sends a humbling note as well. How can it be that the brightest minds in the field spent so many years investigating complicated designs for flow cytometry—didn't it seem like a complicated problem?—and then a first-year professor, in one of his first articles, presented the *optimal* solution to the problem, which simply consists of using *straight* channels? Indeed, the device contains straight single-layer channels, uses fairly simple camera equipment to capture the data, and yet it is capable of the highest throughputs ever achieved in the field.

Moving forward, it is rather unlikely that anyone will want to perform flow cytometry at rates higher than a million cells per second (or using less than one inlet per device), so the challenge now will be the integration of inexpensive detection modules, such as optofluidics, lensless microscopy, and electrical detectors to acquire the data without the need of microscopes such that the samples can be analyzed in home or field settings.

5.2 Cell Sorting

If flow cytometry (counting cells) is the simplest operation one can do with cells, the next-simplest is to try to sort them (i.e., route them into separate paths using flow) for analysis. Next, we will consider the challenge of trapping cells, a particular case of sorting that involves immobilizing the cells, usually so that a static assay such as microscopy can be performed.

5.2.1 Red Blood Cell Assays

Blood is our most abundant fluid; hence, the diagnostics of many diseases take advantage of blood analysis. Automating and miniaturizing blood analysis into a microfluidic format is very attractive because it reduces required sample size, operator expertise, patient discomfort, assay time, reagents, and cost. As early as 1989, a team of researchers led by Kazuo Sato at Hitachi in Tokyo (Japan) presented in the journal *Biorheology* a microfluidic device for measuring mechanical properties of blood cells. Albeit simple, this often-forgotten article deserves the honorable mention of having pioneered the introduction of *any type of* live cells into microchannels. The device consisted of a large chamber with a central inlet hole through which blood was injected into the chamber; as blood entered the chamber floor, it was radially forced to exit through one of the 2600 silicon microgrooves (each 6 µm deep, 9 µm wide at the top, and 10 µm long) etched in one of the four walls of the chamber. Both preparations of white blood cells (before and after activation with **fMLP**) and of erythrocytes were pushed through the device to study differences in the suspension passage time; however, the results are difficult to interpret because of the triangular cross-section of the channels. In 1995, Adrian Barnes and colleagues from Hertfordshire University in England (**Figure 5.9**) designed a similar device consisting of eight square, cross-section microchannels connecting two glass-capped silicon-pit reservoirs. The microchannels were smaller than the red blood cells (mimicking the size of microcapillaries), which forced the cells to deform in the direction of the flow and allowed for discerning anemic cells from healthy cells by the speed at which they traveled through the microchannel (anemic cells are more deformable, so they travel faster).

Dan Chiu's group at the University of Washington in Seattle elaborated on this idea several years later to model malaria infection using PDMS microchannels that contained constrictions (see **Figure 5.10**). Indeed, in people afflicted by malaria, organ failure is observed due to an increase in the rigidity of red blood cells and the blockage of capillaries. Uninfected red blood cells (6 µm wide and highly elastic) were able to traverse channels of widths 2, 4, 6, and 8 µm. In their tropozoite stage, the cells could traverse only channels 6 µm wide or larger. Also, uninfected cells were able to squeeze through the blockages formed by immobile schizonts in a 6-µm-wide microchannel.

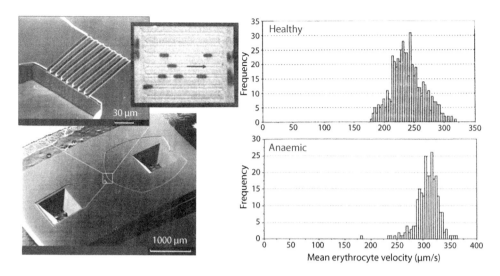

FIGURE 5.9 A microfluidic device to study erythrocyte deformability. (From Mark C. Tracey, Richard S. Greenaway, Arindam Das, Paul H. Kaye, and Adrian J. Barnes, "A silicon micromachined device for use in blood cell deformability studies," *IEEE Trans. Biomed. Eng.* 42, 751, 1995. Figure contributed by Mark Tracey and Paul Kaye.)

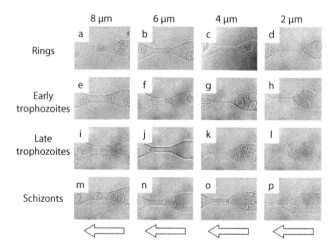

FIGURE 5.10 Modeling malaria infection in microchannels. (From J. Patrick Shelby, John White, Karthikeyan Ganesan, Pradipsinh K. Rathod, and Daniel T. Chiu, "A microfluidic model for single-cell capillary obstruction by Plasmodium falciparum-infected erythrocytes," *Proc. Natl. Acad. Sci. U. S. A.* 100, 14618–14622, 2003. Copyright (2003) National Academy of Sciences, U. S. A.)

5.2.2 Electrokinetic Routing of Cells

Sorting cells necessarily requires the ability to switch the fluids in real time, a nontrivial task. One of the pioneering articles in microfluidics, presented in 1997 by Jed Harrison's group at the University of Alberta, Canada, demonstrated the simple task of transporting cells (yeast, bacteria, and erythrocytes) from one location of the chip to a choice of two other locations; that chip, predating soft lithography, was made of glass and, to move the fluids and the cells, electrokinetic

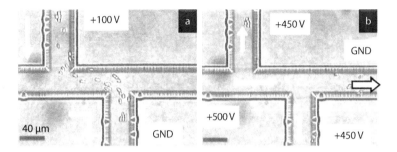

FIGURE 5.11 Switching cells and biochemical reagents using electrokinetic flow. (From Paul C. H. Li and D. Jed Harrison, "Transport, manipulation, and reaction of biological cells on-chip using electrokinetic effects," *Anal. Chem.* 69, 1564, 1997. Reprinted with permission of the American Chemical Society.)

flow was used—requiring voltages on the order of 400 to 500 V (**Figure 5.11**). (Note: the term "**electrokinetic**" refers to all varieties of electrohydrodynamic mechanisms for moving fluids and particles; in this work, the fluids were likely moving by electro-osmosis and, in addition, the cells might also have moved by dielectrophoresis.) On-chip chemical treatment (cell lysis) was also demonstrated.

Because of the required large voltages, electrokinetic flow turns out to be impractical, raising concerns about cell viability (the electrodes can generate pH gradients and electrolysis), researcher safety, and equipment size and cost (high-voltage amplifiers/switches are bulky and expensive). Therefore, a variety of alternatives has been proposed.

5.2.3 Dean Flow in Spiral Microchannels

In 1997, Ashutosh Kole's team at the Palo Alto Research Center introduced a novel concept for particle separation: subject the particles to a curvilinear laminar flow by forcing them through a spiral microchannel at high speeds—similar to passing them through a centrifugation machine. Since then, the designs have been improved by many groups, using spiral as well as serpentine designs. Interestingly, the separation effect is amplified not only by the centrifugal force but also by the so-called Dean's vortex. The vortex generates the recirculation of fluid close to the walls to satisfy fluidic continuum when the fast-moving core moves outward at the center of the channel due to centrifugal force. Using a five-loop spiral microchannel 100 μm wide and 50 μm high, Ian Papautsky's group succeeded in separating 7.3 μm beads from 1.9 μm beads at Dean number $De = (\rho V D_h/\mu)(D_h/2R)^{1/2} = Re\,(D_h/2R)^{1/2} = 0.47$, where ρ is the density of the fluid medium, V is the average fluid velocity, D_h is the channel's **hydraulic diameter** $[D_h \equiv 2HW/(H + W)]$, μ is the fluid viscosity, and R is the radius of curvature of the path of the channel (**Figure 5.12**).

5.2.4 Pinched-Flow Fractionation

In 2004, Minoru Seki and colleagues from Osaka Prefecture University realized that, as a liquid containing particles is "pinched" against a wall of a microchannel by another particle-free liquid, the particles exit the microchannel sorted by sizes (**Figure 5.13a**). This effect, dubbed "**pinched-flow fractionation**," is generated because the particle-free liquid forces all the particles against the wall, thus homogenizing their speed (in the edge of the parabolic flow profile) as a function of their size; as they exit the microchannel into the broadened segment, the particles follow lines of equal flow velocity, so they are sorted by size with submicron resolution (**Figure 5.13b**).

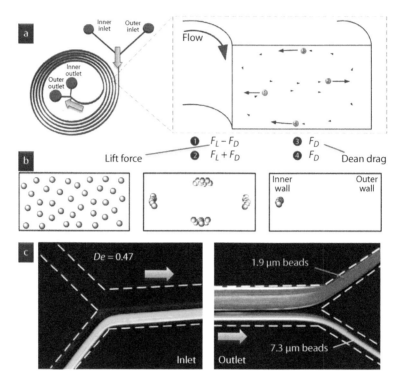

FIGURE 5.12 Particle separation using spiral microchannels. (From Ali Asgar S. Bhagat, Sathyakumar S. Kuntaegowdanahalli, and Ian Papautsky, "Continuous particle separation in spiral microchannels using dean flows and differential migration," *Lab Chip* 8, 1906–1914, 2008. Adapted with permission from The Royal Society of Chemistry.)

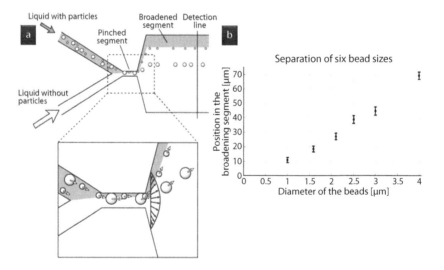

FIGURE 5.13 Pinched-flow fractionation of beads. (a, from Masumi Yamada, Megumi Nakashima, and Minoru Seki, "Pinched flow fractionation: Continuous size separation of particles utilizing a laminar flow profile in a pinched microchannel," *Anal. Chem.* 76, 5465–5471, 2004. Reprinted with permission from the American Chemical Society; b, from Asger Vig Larsen, Lena Poulsen, Henrik Birgens, Martin Dufva, and Anders Kristensen, "Pinched flow fractionation devices for detection of single nucleotide polymorphisms," *Lab Chip* 8, 818–821, 2008. Reprinted with permission from The Royal Society of Chemistry.)

5.2.5 Tunable Hydrophoretic Focusing

None of these approaches can be turned off or tuned in real time (except by changing the flow rate). Recently, Je-Kyun Park's group from KAIST (Korea) has reported a device that is reminiscent of Ligler's flow cytometer (see **Figure 5.5**), but implemented in PDMS so that the roof can be pushed up and down. Slanted obstacles in the roof of the microchannel induce rotational flow and (if the roof is at the right height) produce size-dependent ordering of the microparticles (**Figure 5.14**). The effect has been put to use to demonstrate the separation of blood cells from plasma (**Figure 5.14c**).

5.2.6 Cell Sorting Using Surface Acoustic Waves

Surface acoustic waves (SAWs) are devices that contain interdigitated transducers that convert electrical signals into acoustic waves using piezoelectric materials. Recently, David Weitz's group at Harvard used a SAW device to produce acoustic streaming (see Section 3.4.2) orthogonally to the cell-containing stream, which quickly displaced the stream within the channel; switching speeds on the order of 1 kHz were achieved (**Figure 5.15**).

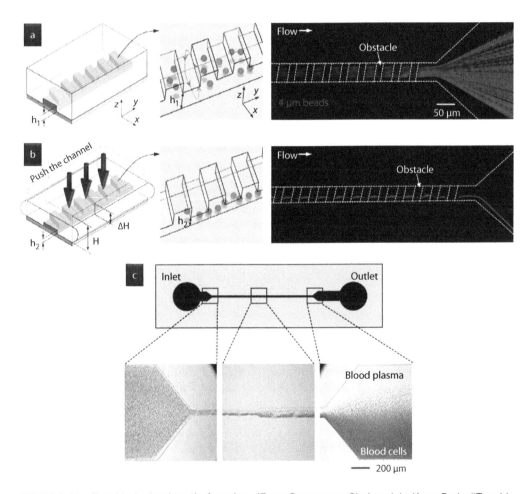

FIGURE 5.14 Tunable hydrophoretic focusing. (From Sungyoung Choi and Je-Kyun Park, "Tunable hydrophoretic separation using elastic deformation of poly(dimethylsiloxane)," *Lab Chip* 9, 1962–1965, 2009. Reproduced with permission from The Royal Society of Chemistry. Figure contributed by Je-Kyun Park.)

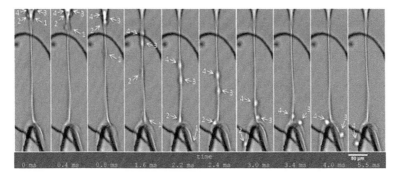

FIGURE 5.15 Cell sorting using acoustic streaming. (From T. Franke, S. Braunmüller, L. Schmid, A. Wixforth, and D. A. Weitz, "Surface acoustic wave actuated cell sorting (SAWACS)," *Lab Chip* 10, 789–794, 2010. Reproduced with permission from The Royal Society of Chemistry.)

5.3 Cell Trapping

Here we consider technologies for trapping cells, which can be considered a special case of sorting whereby the cells end up being immobilized such that they can be observed or assayed for a long time.

5.3.1 Neuro-Cages

Since the late 1990s, Jerome Pine's group at Caltech has used micromachined "neuro-cages," initially made in silicon, to confine neurons. The idea behind confining the neurons in cages is that neurons tend to migrate onto chemically untreated silicon, so it becomes necessary to physically constrain their soma in a "cage" with "tunnels" that allow the cells to grow processes out of the cage to synapse with other (also confined) neurons. In the initial silicon cages (produced by KOH etching), the cages were actually wells and the cells were supported by a metalized 20-μm-thick membrane at the bottom of the well, which allowed for extracellular recordings. In a more recent design, the cages have been constructed on an electrode-patterned wafer by two-layer photolithography in parylene, a biocompatible photosensitive polymer (**Figure 5.16**). The device, however, is nontransparent (although, in principle, it could be built on quartz), so it is incompatible with many microscopy modalities based on light transmission such as phase-contrast microscopy.

5.3.2 PDMS Microwells

The author's group at the University of Washington in Seattle has demonstrated an inexpensive alternative to the problem of trapping cells based on PDMS molding. A master made by photolithography contains an array of cylindrical posts, each slightly wider than the size of a single cell. The PDMS replica thus contains an array of circular wells. The cell-trapping procedure exploits the high apparent viscosity of water on the microscale: once a cell falls into a well, it is very hard for fluid to dislodge the cell out of the well. With repeated seedings, it is possible to obtain more than 90% microwell occupancy rates even with low seeding densities (**Figure 5.17**). In principle, the microwells can be designed to accommodate more than one cell (of more than one cell type) for studies of cell–cell communication. Adherent cells conform to the curved walls or edges of the bottom of the microwells, adopting shapes that may produce confounding results. With a large-chip cooled-CCD camera and an optimized array, it is possible to image more than 100,000 cells (in particular, calcium transients in olfactory sensory neurons) simultaneously at 4× magnification.

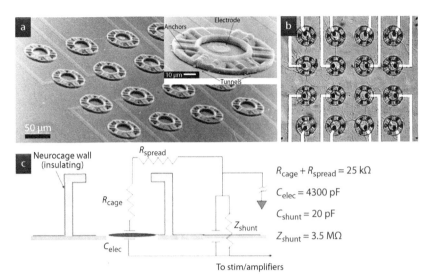

FIGURE 5.16 "Neuro-cages" for extracellular recording from in vitro neural networks. (From Jonathan Erickson, Angela Tooker, Y.-C. Tai, and Jerome Pine, "Caged neuron MEA: A system for long-term investigation of cultured neural network connectivity," *J. Neurosci. Meth.* 175, 1–16, 2008. Reprinted with permission from Elsevier.)

Microfluidics can be used in combination with microwells to selectively "dock" cells to designated microwells or groups of microwells. In the simplest configuration possible, Robert Langer's group at MIT devised simple linear microchannels to address rows of microwells (a different cell type on each row), then turning the microchannels orthogonally addressed the array by columns (Figure 5.18). However, this approach does not allow for individually addressing each microwell. Recently, Joel Voldman and colleagues at MIT have devised a laser-levitation method to extract selected cells from microwells: the laser is focused onto the cells through the microscope objective from underneath so as to counteract gravity, and the photons' momentum is sufficient to lift the cell out of the well, at which point it is taken away by the flow.

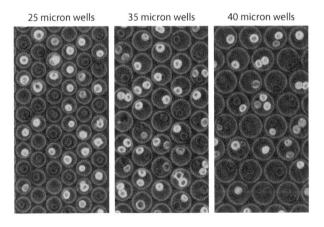

FIGURE 5.17 Cell trapping in PDMS microwells. (From Rettig, J. R. and A. Folch, "Large-scale single-cell trapping and imaging using microwell arrays," *Anal. Chem.* 77, 5628, 2005. Figure contributed by Jackie Rettig.)

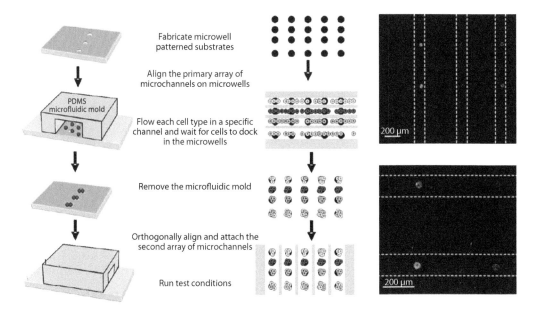

FIGURE 5.18 Microwells addressed with microfluidics. (From Ali Khademhosseini, Judy Yeh, George Eng, Jeffrey Karp, Hirokazu Kaji, Jeffrey Borenstein, Omid C. Farokhzad, and Robert Langer, "Cell docking inside microwells within reversibly sealed microfluidic channels for fabricating multiphenotype cell arrays," *Lab Chip* 5, 1380–1386, 2005. Reproduced with permission from The Royal Society of Chemistry.)

5.3.3 PEG Microwells

The microwells need not be made of PDMS. As a matter of fact, for applications in which the cells are adherent and grow over time, it is best if they are made of a cell-repellent material. This is exactly the reasoning behind making the microwells out of PEG (**Figure 5.19**). The images in **Figure 5.19b** and **c** show a culture of embryonic stem (ES) cells inside 100-μm-diameter PEG wells that were seeded with ES cells and washed.

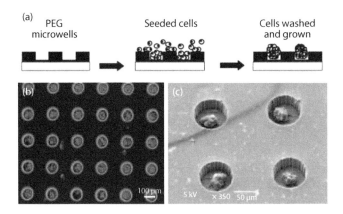

FIGURE 5.19 Cell trapping with PEG microwells. (From Ali Khademhosseini, Robert Langer, Jeffrey Borenstein, and Joseph P. Vacanti, "Microscale technologies for tissue engineering and biology," *Proc. Natl. Acad. Sci. U. S. A.* 103, 2480–2487, 2006. Copyright (2006) National Academy of Sciences, U. S. A.)

5.3.4 Dielectrophoretic Traps

In 1995, Peter Gascoyne's group from the University of Texas Anderson Cancer Center (Houston, Texas) built the first device capable of trapping *and* sorting suspended cells, and to achieve the feat, they used dielectrophoresis (DEP, see Section 3.3.3). They constructed a 60-μm-deep glass plate chamber with interleaved gold electrodes at its bottom surface and introduced a diluted blood suspension containing tumor cells (normal/cancerous cell ratio = 3:1) into the chamber in the direction orthogonal to the electrodes. Upon applying a 200 kHz 5V peak-to-peak sinusoidal signal to the electrode array for 30 seconds, the tumor cells were observed to concentrate by positive DEP on the electrode tips (Figure 5.20a), whereas all the blood cells were removed with the eluate (Figure 5.20b).

In 1997, Tomokazu Matsue and coworkers first applied DEP to micropattern cells. An array of microelectrodes was placed approximately 50 μm away from the cell culture surface, with cellophane tape acting as the spacer. When a dense cell suspension (~10^6 cells/mL) was allowed into the gap, voltages (~7 V_{rms}) applied to the microelectrodes repelled or attracted the cells depending on the applied frequency. Water electrolysis was avoided by applying high frequency (>3 kHz) fields. Micropatterning of mouse myeloma (nonadhesive) cells was demonstrated. The frequency dependence of the sign and magnitude of the dielectrophoretic force was also used by van den Berg and Lammerink to selectively position cells on substrates. The field, applied perpendicularly to the cell surface, repels cells at frequencies on the order of 10^6 Hz and higher and attracts them at lower frequencies. At high electric field magnitudes of approximately 50 kV/m, local heating of a few degrees centigrade was observed. Because fibroblasts only grow and attach between 30°C and 40°C, selective attachment in cold (30°C) medium and selective detachment in hot (40°C) medium occurred on the locally heated areas.

A collaborative team from the Center of Engineering in Medicine (Massachusetts General Hospital) and MIT led by Martin Schmidt has presented arrays of DEP traps that allow for selectively trapping and sorting single cells in suspension (Figure 5.21). The cylindrical shape of the electrodes increases the trapping efficiency of the electrodes compared with a normal planar tetrode. In this design, the DEP force is used to counteract the drag forces from the fluid, which is continuously flowing; as soon as a given trap is turned off, the cell is taken away by the flow (see Figure 5.21d and e) and can be collected in a separate channel.

DEP forces have been used not only to trap cells in aqueous media but also cells encapsulated in gel microspheres, within a gelling medium, to create "3-D tissue constructs" (see Figure 7.4 in Section 7.1.1).

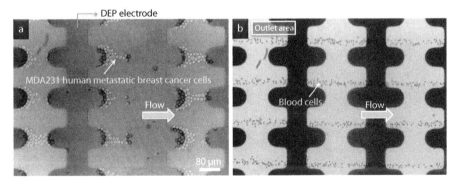

FIGURE 5.20 Dielectrophoretic trapping of cells. (From Frederick F. Becker, Xiao-Bo Wang, Ying Huang, Ronald Pethig, Jody Vykoukal, and Peter R. C. Gascoyne, "Separation of human breast cancer cells from blood by differential dielectric affinity," *Proc. Natl. Acad. Sci. U. S. A.* 92, 860–864, 1995. Copyright (1995) National Academy of Sciences, U. S. A.)

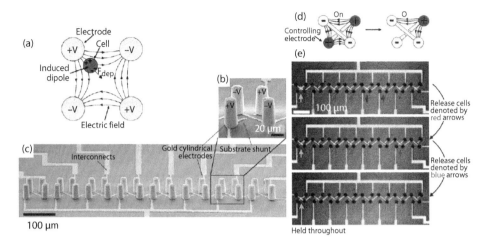

FIGURE 5.21 Addressable dielectrophoretic traps for single cell sorting. (From Joel Voldman, Martha L. Gray, Mehmet Toner, and Martin A. Schmidt, "A microfabrication-based dynamic array cytometer," *Anal. Chem.* 74, 3984–3990, 2002. Adapted with permission of the American Chemical Society. Contributed by Joel Voldman.)

5.3.5 Micromagnetic Traps

As a general rule of thumb (so we learn in textbooks), magnetism does not miniaturize well, because most forms of magnetism arise from the cooperation between groups of atoms—so the smaller you try to make a device, the less magnetic poles per unit volume one has, and the magnetization or force one will be able to generate with the device will necessarily be smaller. Despite this seemingly losing proposition, Robert Westervelt from Harvard University fabricated complementary metal-oxide semiconductor (**CMOS**) arrays that can power small coils, each of which generates a magnetic field powerful enough to trap a cell decorated with magnetic beads (**Figure 5.22**). The advantage of these traps is that the CMOS array of microelectromagnets

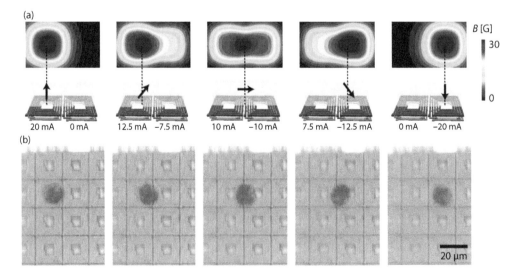

FIGURE 5.22 CMOS-based micromagnetic traps cell manipulation. (From Hakho Lee, Yong Liu, Donhee Ham, and Robert M. Westervelt, "Integrated cell manipulation system—CMOS/microfluidic hybrid," *Lab Chip* 7, 331–337, 2007. Reproduced with permission from The Royal Society of Chemistry.)

is easily programmed to change the magnetic field patterns, so the cells can be pushed from one trap to the next.

5.3.6 Hydrodynamic Traps

In 2003, a collaborative team led by Richard Zare at Stanford and Antoine Daridon at Fluidigm Corporation introduced a new powerful approach to cell trapping: instead of devising complicated devices that use external force fields such as electrical or magnetic force fields to trap the cells, why not make use of the forces generated by the flow itself? The high apparent viscosity of water at these scales plays in our favor: as with the microwells, once the cell is inside the trap, the flow lines change and they contribute to keeping the cell confined to the trap because they cannot penetrate the trap. Their device contained a single trap (Figure 5.23), and fast assays (such as methanol-induced cell death in Figure 5.23b and staining within 5 seconds in Figure 5.23c) on single Jurkat T-cells were demonstrated; complete solution changes were achieved in approximately 100 ms (Figure 5.23d).

Several groups have scaled up this concept into an array format. Luke Lee's group from the University of California at Berkeley, for example, has devised arrays of traps that are cleverly shaped as inverted "weirs" (a structure commonly used in dams and rivers to provide even flow and facilitate navigation) that hang from the ceiling of the microchannel (Figure 5.24). As with the device by Zare and Daridon shown previously, the presence of a cell trapped in any given weir diverts the streamlines, and subsequent incoming cells progress around loaded traps (Figure 5.24b). Longer/wider weirs that trap several cells at a time were also demonstrated. This array was applied to measure single-cell enzyme kinetics for three different cell types (HeLa, 293T, and Jurkat) using fluorogenic substrates.

Similar weirs have been used by Joel Voldman and colleagues at MIT to trap pairs of cells. In the traps, they were able to demonstrate cell fusion between dissimilar cell types (a procedure of frustratingly low throughput and of great biotechnology interest) at extremely high throughputs (Figure 5.25).

A different, also promising cell manipulation device to control cell–cell contacts has been proposed by Luke Lee's group at UC Berkeley. The device consists of a microchannel (approximately two cells wide) with multiple lateral channels (much smaller than a cell), as shown in Figure 5.26. When the large microchannel is filled with cells, suction applied to the small channels causes the cells to plug them. Each opening thus acts as a "suction cup" for a cell. The researchers were

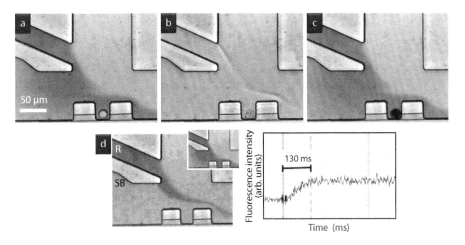

FIGURE 5.23 Hydrodynamic traps for high-speed single-cell analysis. (From Aaron R. Wheeler, William R. Throndset, Rebecca J. Whelan, Andrew M. Leach, Richard N. Zare, Yish Hann Liao, Kevin Farrell, Ian D. Manger, and Antoine Daridon, "Microfluidic device for single-cell analysis," *Anal. Chem.* 75, 3581–3586, 2003. Adapted with permission of the American Chemical Society.)

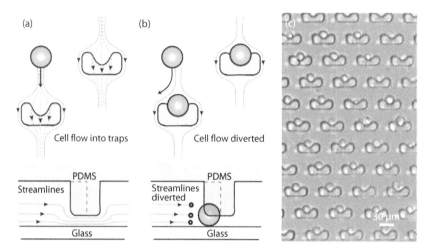

FIGURE 5.24 Arrays of hydrodynamic cell traps. (From Dino Di Carlo, Nima Aghdam, and Luke P. Lee, "Single-cell enzyme concentrations, kinetics, and inhibition analysis using high-density hydrodynamic cell isolation arrays," *Anal. Chem.* 78, 4925–4930, 2006. Reprinted with permission of the American Chemical Society.)

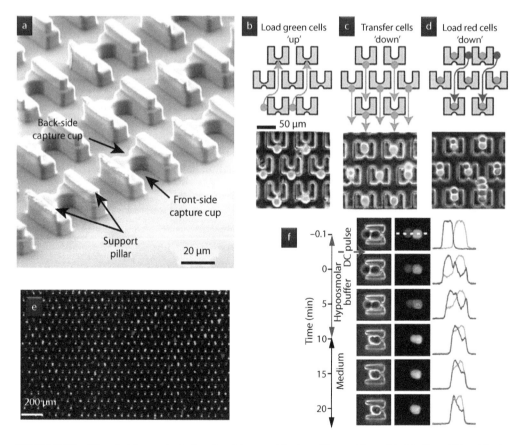

FIGURE 5.25 Microfluidic control of cell pairing and fusion. (From Alison M. Skelley, Oktay Kirak, Heikyung Suh, Rudolf Jaenisch, and Joel Voldman, "Microfluidic control of cell pairing and fusion," *Nat. Methods* 6, 147–152, 2009. Adapted with permission of the Nature Publishing Group.)

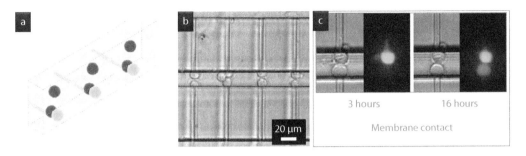

FIGURE 5.26 Cell traps for studying cell–cell contacts. (From Philip J. Lee, Paul J. Hung, Robin Shaw, Lily Jan, and Luke P. Lee, "Microfluidic application-specific integrated device for monitoring direct cell-cell communication via gap junctions between individual cell pairs," *Appl. Phys. Lett.* 86, 223902, 2005. Figure contributed by Luke Lee.)

able to demonstrate gap-junctional communication (transfer of fluorescent dye) between pairs of cells. This approach should find uses in a variety of cell–cell communication studies; for example, antigen-presenting cells and T cells could be brought together to study—or engineer—the immunological synapse.

These approaches all suffer from the drawback that most cells (80%–90%) in the cell suspension do not get trapped in the device, so an excess of cells is required. What if one required the trapping of all the cells (or the large majority of the cells), or only very few cells were available? We would like to have a design that takes care of trapping every cell that goes through it. That is precisely the problem that Shoji Takeuchi's group, from the University of Tokyo, solved in 2007

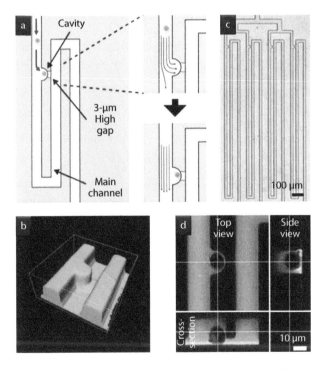

FIGURE 5.27 A high-efficiency hydrodynamic trapping device. (From Stefan Kobel, Ana Valero, Jonas Latt, Philippe Renaud, and Matthias Lutolf, "Optimization of microfluidic single cell trapping for long-term on-chip culture," *Lab Chip* 10, 857–863, 2010. Reproduced with permission from The Royal Society of Chemistry.)

(but with beads). Takeuchi's "self-regulating" clever trapping design was recently optimized for cell capture by Matthias Lutolf and colleagues at the École Polytechnique Fédérale de Lausanne (**Figure 5.27**), resulting in cell trapping efficiencies near 100%, 100% single-cell trap occupancy, and long-term cell survival of 95% (for a nonadherent cell type). The principle behind the high-efficiency capture is based on a system of straight, small channels (with a 3-μm gap too small for the cells to go through) and loops; the small channels have less fluidic resistance than the long loops, so most of the flow goes through them—until they get plugged by a cell: when this happens, the rest of the cells get rerouted through the loops and can get trapped by plugging other small channels.

5.3.7 Trapping Cells Using Antibodies

Biologists have used antibodies on surfaces to purify cell populations for a long time, a technique called "immunopanning": as the antibodies recognize and bind molecules on the surface of the cells, only the targeted cells will be immobilized on the surface and the other ones will be washed away. Mehmet Toner's laboratory at Harvard Medical School has implemented an immunopanning procedure in a microfluidic format for the immunocapture of circulating tumor cells (CTCs). CTCs are very rare, around one cell in a billion normal blood cells, and are thought to be the seed of incurable metastatic cancers, so their early detection is crucial to start a life-saving treatment. Toner's CTC chip improves the efficiency of capture by including 78,000 microposts within a surface area of 970 mm^2 (**Figure 5.28**). The microposts are coated with antibodies against cancer cell markers for various cancers (lung, colorectal, breast, head, and neck cancers). The chip, which is now in clinical trials, correctly detected CTCs in more than 99% of cancer patients with metastases.

In its present design, Toner's chip cannot retrieve the cells so it does not allow for gene sequencing or proteomics analysis of the cells. But why is it necessary to capture the cells at all? Shouldn't it be possible to detect the binding between the cell's surface marker protein and the immobilized antibodies as the cells "fly by?" Lydia Sohn and colleagues at the University of California (Berkeley) built an ingenious Coulter counter with a narrow (15 μm × 15 μm) fluidic pore and coated its surface with antibodies against CD34, a cancer cell surface marker (**Figure 5.29**). As each cell passes through the pore, a current pulse is detected (independently

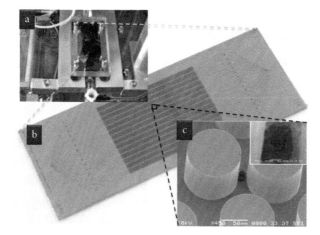

FIGURE 5.28 Microfluidic immunocapture of rare cancer cells. (From Sunitha Nagrath, Lecia V. Sequist, Shyamala Maheswaran, Daphne W. Bell, Daniel Irimia, Lindsey Ulkus, Matthew R. Smith, Eunice L. Kwak, Subba Digumarthy, Alona Muzikansky, Paula Ryan, Ulysses J. Balis, Ronald G. Tompkins, Daniel A. Haber, and Mehmet Toner, "Isolation of rare circulating tumour cells in cancer patients by microchip technology," *Nature* 450, 1235–1239, 2007. Reprinted with permission of the Nature Publishing Group. Figure contributed by Mehmet Toner.)

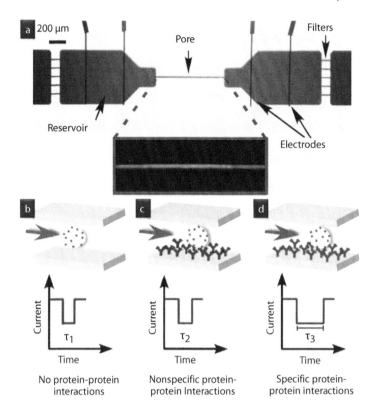

FIGURE 5.29 A Coulter counter for screening cancer cells. (From Andrea Carbonaro, Swomitra K. Mohanty, Haiyan Huang, Lucy A. Godley, and Lydia L. Sohn, "Cell characterization using a protein-functionalized pore," *Lab Chip* 8, 1478–1485, 2008. Reproduced with permission from The Royal Society of Chemistry.)

of whether it is cancerous or not). When cancer cells bearing the CD34 protein on their surface were passed through the antibody-functionalized pore, they spent on average 4 ms more transit time than normal cells or than cancer cells through an uncoated pore. In principle, this tool could be interfaced downstream with a sorter to select the CD34+ cells for analysis.

5.3.8 Trapping and Culturing Microfabricated Cell Assemblies

In many tissue engineering, biotechnology, and cell biology applications, the ability to produce, culture, and manipulate *assemblies* of cells (rather than single cells) is desirable because many cell types do not grow well unless they "feel the company" of their kin. This "companionship" is signaled to them through secretion of growth factors that have an effect only when they reach a certain concentration threshold, so cell density (and time) plays an important role in the local accumulation of growth factors. Nancy Allbritton's group from the University of North Carolina has devised a method to produce small, transferrable units of solid cell culture substrates that support cell attachment and culture (**Figure 5.30**). These (~100 μm-side square) "rafts," which can be micromolded in a number of biocompatible polymers (such as polystyrene or epoxy) from a PDMS template, can be mechanically released from their template by a pin and transferred onto a different substrate to amplify the cell culture.

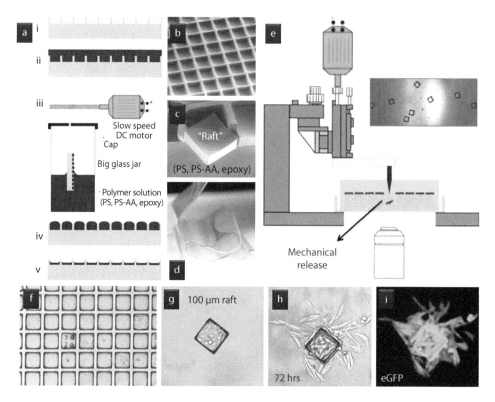

FIGURE 5.30 Micromolded rafts for cell culture. (From Yuli Wang, Colleen Phillips, Wei Xu, Jeng-Hao Pai, Rahul Dhopeshwarkar, Christopher E. Sims, and Nancy Allbritton, "Micromolded arrays for separation of adherent cells," *Lab Chip* 10, 2917–2924, 2010. Reproduced with permission from The Royal Society of Chemistry.)

What if the cells were not adherent? David Gracias' laboratory at Johns Hopkins University has developed microactuators that change shape in response to thermal or chemical changes. The microactuators can take the shape of microcages (**Figure 5.31**) or microgrippers when they fold up. The actuators contain trilayer joints or hinges composed of photoresist and a stressed bimetallic thin film. The assembly of the microcages is thermally triggered, without the need for any external connections, so that the highly parallel self-loading of glass beads, fibroblasts, and tadpole shrimp embryos could be demonstrated.

5.3.9 Microdroplet Cultures and Assays

And how about using *dissimilar* fluids to confine the cells—that is, confining the cells into oil-surrounded aqueous droplets, which can be produced at very fast rates in microchannels? Rustem Ismagilov's group at the University of Chicago was the first to pioneer what are now intensive investigations in this direction. Oil-suspended droplets filled with single bacterial cells could be grown and split for parallel tests (cellulase assay, cultivation, cryopreservation, and Gram staining); the plugs could be deposited on a plate for fluorescent in situ hybridization (**FISH**) or on an agar plate for culture (**Figure 5.32**).

To generalize this concept to mammalian cells, the cells will need to be pre-encapsulated in beads made of a matrix that supports cell attachment. Still, a team from RainDance Technologies (Lexington, MA) and Harvard Medical School led by Michael Samuels has

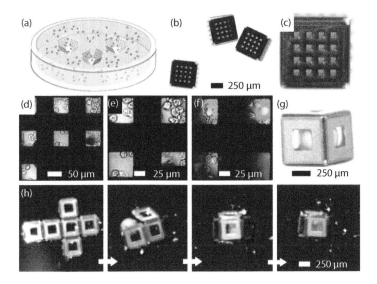

FIGURE 5.31 Microcontainers loaded with biological contents. (From Timothy G. Leong, Christina L. Randall, Bryan R. Benson, Aasiyeh M. Zarafshara and David H. Gracias, "Self-loading lithographically structured microcontainers: 3D patterned, mobile microwells," *Lab Chip* 8, 1621–1624, 2008. Reproduced with permission from The Royal Society of Chemistry.)

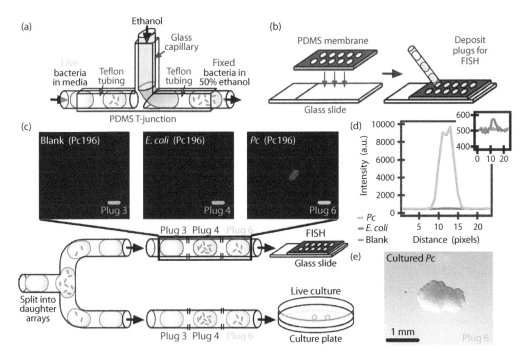

FIGURE 5.32 Bacterial cell cultures and assays in microdroplets. (From Weishan Liu, Hyun Jung Kim, Elena M. Lucchetta, Wenbin Du, and Rustem F. Ismagilov, "Isolation, incubation, and parallel functional testing and identification by FISH of rare microbial single-copy cells from multi-species mixtures using the combination of chemistrode and stochastic confinement," *Lab Chip* 9, 2153–2162, 2009. Reproduced with permission from The Royal Society of Chemistry.)

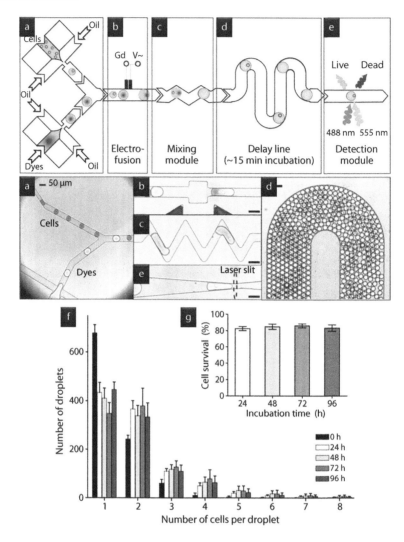

FIGURE 5.33 High cellular viability in microdroplets. (From Eric Brouzes, Martina Medkova, Neal Savenelli, Dave Marran, Mariusz Twardowski, J. Brian Hutchison, Jonathan M. Rothberg, Darren R. Link, Norbert Perrimon, and Michael L. Samuels, "Droplet microfluidic technology for single-cell high-throughput screening," *Proc. Natl. Acad. Sci. U. S. A.* 106, 14195, 2009. Figure contributed by Eric Brouzes.)

reported a high-throughput cytotoxicity screening system capable of encapsulating human monocytic U937 cells in oil droplets over a period of approximately 4 days—a surprisingly high viability (Figure 5.33). David Weitz laboratory at Harvard University has also demonstrated microfluidic droplet systems capable of encapsulating single cells in droplets and sorting the droplets at high throughput based on fluorescent indicators.

5.4 Microfluidic Cell Culture Laboratories

This section reviews the technologies that have been developed for culturing cells in microfluidic environments, mostly geared toward cell biology and tissue engineering research, but also toward stem cell biotechnology and high-throughput pharmaceutical testing.

THE FIRST MICROFLUIDIC CELL CULTURE

In 1948, K. K. Sanford, G. D. Likely, and W. R. Earl were interested in the growth of tissues from single cells, but single cells did not seem to want to grow in large baths. (The importance of dilution of growth factors had not been yet identified.) In an influential article in the *Journal of the National Cancer Institute* that year, they successfully argued that, if they reduced the bath to the level of single cells, the cells would be able to "adjust" and would reproduce—hence, they cultured single cells within 6- to 8-mm-long glass capillaries. The cell culture medium was renewed by diffusion through an open end connected to a larger reservoir. The first microfluidic cell culture, as it turns out, predates microfluidic technology!

5.4.1 Limitations of Traditional Cell Culture Technology

The ability to culture cells outside of their natural organism—pioneered by Harrison and Carrel—has, despite constituting a simplistic simulation of the organism's inner workings, revolutionized hypothesis testing in basic cell and molecular biology research and become a standard methodology in drug testing and toxicology assays. Indeed, cell culture ("in vitro") systems inherently lack the 3-D, multicellular architecture found in an organism's tissue but offer precious advantages over whole-animal ("in vivo") experimentation: (a) the parameters necessary for cell function can be isolated without interference from more complex, whole-organism or whole-organ responses; (b) because many experimental conditions can be tested with the cells from only one sacrificed animal—or a small portion of it—it reduces animal care expenses, human labor costs, and animal suffering; (c) because the cells are distributed in a thin layer, optical observation under a microscope is unobstructed by other cell layers; and (d) with cell lines, the researcher effectively circumvents the time necessary to raise the animal and its very sacrifice. A wide range of sophisticated medium formulations and cell lines from almost any type of tissue are now commercially available; cell culture equipment is becoming increasingly ubiquitous: dissociating cells from their organ using enzymes, culturing them in humidified CO_2 incubators, and time-lapse imaging them with fluorescence microscopes, to name a few techniques, are now common procedures even in some undergraduate laboratories. This accumulated wealth of knowledge has brought in vitro experimentation closer to real animal research and has led to many important discoveries in fields ranging from basic biology to pharmaceutical screening.

However, the technology of cell culture is falling behind in the pace of progress. Genes can now be probed simultaneously by the thousands on a **DNA chip** as animal genomes are being fully sequenced. Biochemists can synthesize a combinatorial variety of drug candidates as well as a myriad of reporters that tag specific biomolecules and organic compounds that mimic the function of other biomolecules. Our molecular understanding of cell behavior is materializing as a picture of pathways of biochemical reactions intricately entangled with each other. Clearly, the biomedical field is entering an era in which data retrieval and analysis has to deal with complex systems. Not surprisingly, social and political concerns are being raised which reflect the proportions that genetic engineering and animal experimentation are taking in modern science. In cell culture, because of recent advances in drug discovery and molecular cell biology, there is an increasing pressure for testing even more complex medium formulations that include putative drug candidates, growth factors, neuropeptides, genes, and retroviruses, to name a few. Because these various factors have nonlinear effects that may change when combined with other factors, the complexity of testing increases exponentially as new components are added to the cell culture medium.

Yet cell culture methodology has remained basically unchanged for almost a century: it consists essentially of the immersion of a large population of cells in a homogeneous fluid medium. This requires at least one cell culture surface (such as a petri dish, a slide, or a well) for each cell culture condition to be investigated; in general, a few surfaces are typically used for each condition to account for sample variability and measurement errors. Hence, as cells need to be fed periodically with fresh medium, testing of a few medium conditions already involves many cell culture surfaces, bulky incubators, large fluid volumes (~0.1–2 mL per sample), and expensive human labor or equipment (**Figure 5.34**).

5.4.2 The Cell-on-a-Chip Revolution

Microfluidic devices—and, more generally speaking, microsystems—promise to play a key role in circumventing the above limitations for several reasons:

- Devices consume small quantities of precious/hazardous reagents (thus reducing cost of operation/disposal).

- Devices can be straightforwardly integrated with other microfluidic devices.

- Devices can be mass-produced in low-cost, portable units.

- The dimensions of their microchannels can be comparable to or smaller than a single cell.

- Because of the large surface-area-to-volume ratio of these microchannels, liquids flow in sheets or "laminar flow" (see Section 3.2.3), that is, without turbulence. Thus, flow and diffusion around the cells can be accurately predicted, if not analytically, using

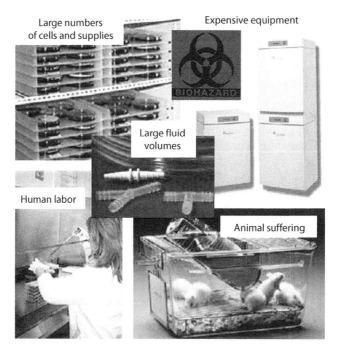

FIGURE 5.34 Limitations of traditional cell culture technology.

finite element modeling—hence, mass transport to cells and shear forces can be systematically incorporated into the experimental design of the device.

As we saw in Chapter 4, there has been, in the last 5 years, an eruption of microfluidic implementations of almost every known traditional bioanalysis technique. Considerable efforts are underway to integrate two or more of the above functions in "micro-total analysis systems" (MicroTAS). By comparison, the integration of microfluidic systems with live cells, despite its great potential for cell culture and biosensor technologies, is still in its infancy.

A big leap ahead in the progress of cell-based microfluidic devices has been the advent of soft lithography, based on the mold-replication of a master in PDMS (see Chapter 1). Soft lithography has become an enabling technique for cell-based microfluidics because of the optimal properties of PDMS. PDMS is inexpensive (both the material and its replica-molding process, which requires no expertise except for the one-time fabrication of the master mold), transparent (i.e., compatible with optical microscopes), and biocompatible (cells may be cultured on top of a PDMS surface). Importantly, PDMS microstructures can be sealed against almost any smooth dry surface (and also against wet surfaces, with some precautions), which allows for fabricating microfluidic devices for a small fraction of the cost of traditional methods.

Here we will show how microfluidic cell cultures promise to yield *higher throughputs* than their macrofluidic counterparts because the latter are often limited by fluid-handling issues. We will see in the next chapter how microtechnology is also enabling a new type of cell culture studies based on the study of cells in *small areas and volumes*, with increased throughput as a secondary advantage.

Many scientists regard the issues related to cost and throughput as merely practical because they can, in principle, be overcome with increased resources or funding. In the interest of time and cost, however, most will make a compromise on the number of conditions (different factors, substrates, or cell types) that are tested before reporting a finding, resulting in poor statistics (particularly when obtaining single-cell data) and semiquantitative conclusions. Indeed, the microfluidic implementation of a cell culture experiment often represents simply a *practical improvement*, in the sense that it can be done by other means, but it is an improvement that can yield not only lower overall cost but also higher throughput and more quantitative descriptions of single-cell behavior variability. When the conclusion of an experiment is not a new model of cell behavior but a measurement, the advantage of large numbers is that averaging can be key for detecting small signals just above the noise. In that regard, very large increases in throughput may *enable* certain classes of experiments—such as statistically demanding ones—that would otherwise be unworkable with traditional cell culture methods and realistic budgets.

5.4.3 Seeding Cells in Microchannels

As depicted in **Figure 5.35**, seeding cells inside a microfluidic channel involves three steps that deserve a few comments on practical (but important) considerations: (1) a cell suspension is injected into the channel (i.e., there is flow); (2) cells are deposited while flow is stopped; and (3) cells must be retained in the channel when flow is resumed. During the elapsed time, the cells must attach or be physically captured in traps (for example, in wells or in sieves; cell traps are covered extensively in Section 5.2), a procedure that in either case is cell type-dependent. For very small channels, stopping the flow might hinder the cellular microenvironment (e.g., nutrient depletion and metabolite accumulation). For loosely attached or delicate cells, an uncontrolled pulse (i.e., acceleration) in restarting the flow can produce large forces that could shear the cells off the substrate. Thus, automated cell culture systems must incorporate a certain level of flow control to ensure proper seeding.

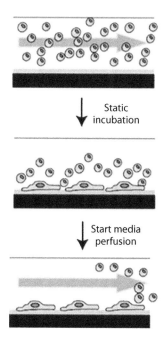

FIGURE 5.35 Seeding cells in microchannels. (From L. Kim, Y.-C. Toh, J. Voldman, and H. Yu, "A practical guide to microfluidic perfusion culture of adherent mammalian cells," *Lab Chip* 7, 681, 2007. Reproduced by permission of The Royal Society of Chemistry.)

5.4.4 From Serial Pipetting to Highly Parallel Micromixers, Pumps, and Valves

Fluid handling—ranging from reagent mixing to medium changes and supernatant sampling—is a major throughput-limiting step in cell culture technology. It currently involves interrupting incubation and subsequently pipetting the fluid in or out of the cell culture medium, which requires time-consuming manual labor or very costly automated multipipetters equipped with complex tubing and valve systems—both processes being prone to contamination and spills. The advent of multiwell plates represents an attempt to mitigate these limitations, but it is expensive to scale up because the required number of plates and fluid delivery steps grows geometrically as the number of cell culture parameters increases.

Let us consider two examples (note to young investigators: either of these would make for great microfluidics projects):

🌸 The optimization of a defined cell culture medium (for a particular cell type, a particular application, and even a particular phenotype), which may involve studying the nonlinear, nonadditive effects of various hormones, growth factors, amino acids, glucose, salts, and so on at different concentrations, as critical as it can be for the success of an experiment, can become a monumental task in its own that only a few cell culture laboratories can afford to undertake. As cell culture studies become increasingly sophisticated, the development of fast, inexpensive mixers that generate a combinatorial range of fluid mixtures becomes imperative.

🌸 Most cell culture protocols are presently based on changing the cell culture medium every 24 hours (or 48 hours). Biologists have not adopted this schedule based on first principles (i.e., this is what the cells need); they have adopted it because it is convenient for humans, who need to rest between experiments. If the schedule were being carried by a microfluidic automaton (i.e., by microvalves and micropumps), one might find that cells need to be fed more often, or continuously as it happens in vivo. As a matter of fact, it is *extremely unlikely* that the optimal feeding schedule for *all* cell types is 24 hours!

We are proposing a transition from dispensing fluids by pipetting (whether manual or robotic) to dispensing fluids through micropumps and microvalves. However, this transition raises the issue of fluid metering. In a microfluidic device, although the volumes of microchambers can be known with high precision, the measurement of flow rates is challenging because it requires specialized flow visualization techniques; as a result, continuous flow systems such as micromixers are not easily metered. However, stationary microscale volumes (e.g., microchambers closed by microvalves) are straightforwardly known independently (roughly) of their size and number. The volume of the chamber is easier to predict with "doormat" microvalves (see Figure 3.41) than with Quake's pinch microvalves (see discussion in Section 3.8.1.5). For example, the smallest chamber in Figure 3.107 is $500 \times 220 \times 55$ μm ≈ 6 nL; because it is possible to know the dimensions with an error of approximately 1% in each dimension, the volume can be known within approximately ±3% = ±20 pL. An example of a fully automated microfluidic cell culture system has been demonstrated by a collaborative team led by Christopher Chen at the University of Pennsylvania and Stephen Quake at Stanford University. The device creates arbitrary culture media formulations in 96 independent culture chambers and maintains cell viability for weeks. The team demonstrated osteogenic differentiation of human embryonic stem cells as visualized by on-chip alkaline phosphatase assays (Figure 5.36).

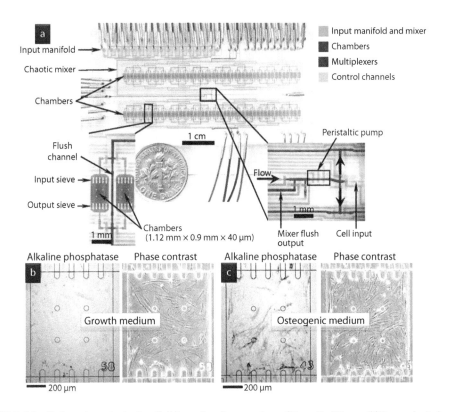

FIGURE 5.36 Fully automated microfluidic cell culture system. (From R. Gómez-Sjöberg, A. A. Leyrat, D. M. Pirone, C. S. Chen, and S. R. Quake, "Versatile, fully automated, microfluidic cell culture system," *Anal. Chem.* 79, 8557–8563, 2007. With permission. Contributed by Stephen Quake.)

5.4.5 From Incubators to "Chip-Cubators"

Cell culture equipment—comprising at least an incubator (required to preserve temperature, humidity, and gas concentration of the cell culture medium) and a tissue culture hood (required to preserve sterile conditions)—is expensive and bulky; usually, a dedicated room is recommended. Culturing cells in closed, microfluidic chambers circumvents the need for constant-humidity systems; however, it also calls for the development of gas exchange and temperature control systems on chip if the miniaturization of the cell maintenance equipment is to be fully realized. Although seemingly trivial from an engineering point of view, the full miniaturization of a cell maintenance system (without modifying the cell culture medium) is yet to be achieved; this may reflect not so much its technical difficulty but the fact that such a system has a small payoff for the average researcher, who still needs the traditional cell culture room for intrinsically "bulky activities" such as primary cell isolation and expansion of cell lines. Relatively small (yet macrofluidic) perfusion chambers featuring incubation capabilities for long-term live cell (e.g., time lapse) microscopy are commercially available. The laboratory of Shuichi Takayama at the University of Michigan (Ann Arbor) has developed a portable microfluidic cell culture incubator (i.e., capable of heating and pumping the cell culture solutions); by using a modified cell culture medium, osteoblast and myoblast differentiation was demonstrated outside a CO_2 incubator (**Figure 5.37**).

5.4.6 From High Cell Numbers in Large Volumes (and Large Areas) to Low Cell Numbers in Small Volumes (and Small Areas)

A typical cell culture experiment uses large numbers of cells, which causes substantial animal suffering—an increasing concern in our society—and large quantities of supplies (the culture surfaces as well as the fluids); this results in expensive experiments that take a lot of bench or incubator space. When the cultures are used for harvesting certain biochemicals, the need for large numbers of cells is most often not intrinsic to the experiment but mostly because of the need for collecting high concentrations of the biomolecule of interest (e.g., a cell-secreted product, DNA or mRNA content) or the need for collecting high volumes if the assay so requires (e.g., centrifugation, filling a well of a 96-well plate). In a microfluidic culture, compared with a traditional open-dish culture, both the area occupied by the cells and the volume bathing the cells are scaled down, but the volume can be (and typically is) scaled down by a larger factor than the area because of the reduced height of the fluid. Aside from considerations on the constancy of the cell culture environment, this scaling consideration means that, in microfluidic cultures compared with their nonmicrofluidic

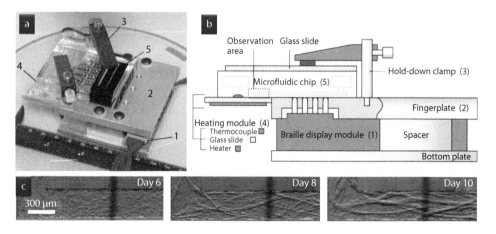

FIGURE 5.37 Handheld microfluidic incubator. (N. Futai, W. Gu, J. W. Song, and S. Takayama, "Handheld recirculation system and customized media for microfluidic cell culture," *Lab Chip* 6, 149–154, 2006. Reproduced by permission of The Royal Society of Chemistry.)

counterparts, the collected concentrations are higher (which contributes to increase the signal-to-noise ratio of the measurement), but the collected volumes are smaller (which often makes the fluid collection and the detection more challenging). Thus, depending on the sensitivity of the assay, the microfluidic collection of fluids for use in a macrofluidic assay may not be beneficial nor practical, and a microengineered version of the assay may have to be developed. Whenever possible, assays based on microscopic observation (e.g., calcium imaging using fluorescent markers, time-lapse imaging of changes in cell morphology), which can be done in situ, should prove more practical. However, microscopy assays have inherently low throughput (the microscope probes only a small field of view), so adopting a microscopy assay in a microfluidic cell culture hinders the high-throughput benefits of microfluidic perfusion; to achieve high throughputs, a programmable motorized stage can be used to automate image acquisition from multiple fields of view.

5.5 Gene Expression Cellular Microarrays ("Cellomics")

For many applications in biotechnology and cell biology, it is desirable to create microarrays of cells in which each array unit (either a single cell or a group of cells) is made genetically unique, either by the addition of a plasmid containing a bioluminescent protein, a **GFP** reporter system, or a plasma-permeable dye. In all three cases, the response of the cell(s) to the addition of an analyte can be detected optically and depends on their (artificially–added) genetic identity. David Sabatini's group from the Whitehead Institute (MIT) in Cambridge (Massachusetts) printed cDNA on a glass slide, coated it with the standard transfecting reagents, and observed that (surprisingly) cells cultured on top of the cDNA get spontaneously transfected by it (it is surprising that this "surface-bound transfection" works at all, because the reagents that package the cDNA were designed to work in solution, not on a surface). Because it is straightforward to print cDNA in microarrays using robotic pin arrayers, the technique effectively allows for producing "gene expression arrays" (Figure 5.38).

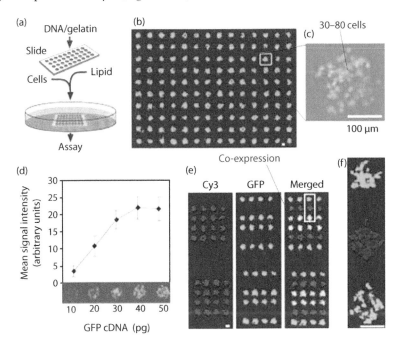

FIGURE 5.38 Gene expression cell arrays. (From Junaid Ziauddin and David M. Sabatini, "Microarrays of cells expressing defined cDNAs," *Nature* 411, 107–110, 2001. Adapted with permission from the Nature Publishing Group.)

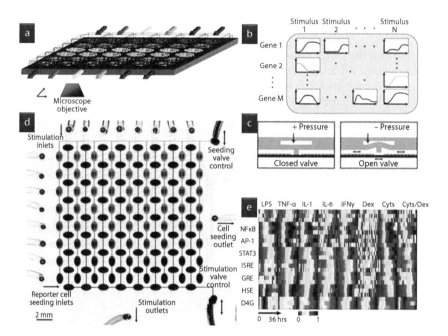

FIGURE 5.39 Microfluidic gene-expression cell array. (From Kevin R. King, Sihong Wang, Daniel Irimia, Arul Jayaraman, Mehmet Toner, and Martin L. Yarmush, "A high-throughput microfluidic real-time gene expression living cell array," *Lab Chip* 7, 77–85, 2007. Reproduced with permission from The Royal Society of Chemistry.)

These gene expression microarrays are limited in that all the cells, which form a monolayer culture, share the same cell culture medium. A Massachusetts General Hospital team led by Maish Yarmush has accomplished a gene expression platform in which cells are cultured in a square 16 × 16 array of circular microchambers (50 μm height and 420 μm diameter), connected to each other by microvalves; the cells are transfected with different genes by rows and exposed to different stimuli by columns, thus allowing for a rich combinatorial matrix of conditions of genes and stimuli that can be monitored noninvasively by time-lapse fluorescence microscopy (**Figure 5.39**). As a demonstration, hepatocyte inflammatory gene expression dynamics were profiled [see heat map in **Figure 5.39e**; each reporter was stimulated with bacterial endotoxin (LPS), inflammatory cytokines (TNF-α, IL-1, IL-6, and IFNγ), a synthetic glucocorticoid hormone (dexamethasone), and combinations thereof (Cyts ≡ TNF-α/IL-1/IL-6 or Cyts+Dex ≡ TNF-α/IL-1/IL-6/Dex), cellular fluorescence was measured from three cell chambers for each of the 64 stimulus–response pairs every 90 minutes for 36 hours to create the 192 time series composed of 4608 single time point measurements].

5.6 Micro-Bioreactors

Living systems are far from homogeneous. In fact, they are full of microscale (substrate-bound and soluble) gradients of small ions and a rich variety of growth factors, which require energy expenditure to create and maintain. It is not an exaggeration to say that, deep inside, our physiological systems are all microfluidic (**Figure 5.40**). If these microfluidic systems stop working, we develop a disease and eventually die.

Biologists and doctors have attempted to recreate "organ-like" conditions in petri dishes. In these conditions, cells dissociated from the organ are seeded on a homogeneous plastic surface (usually coated with protein) and homogeneously bathed in cell culture medium. These conditions do not reproduce the microscale gradients present in vivo, which can affect critical

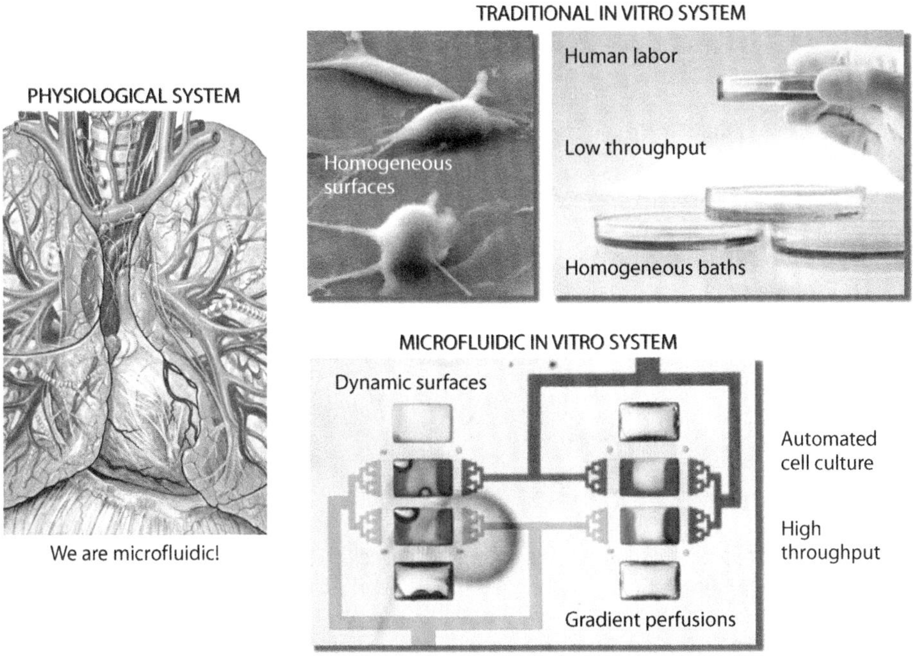

FIGURE 5.40 Increased throughput and physiological relevance of microfluidic systems.

cellular functions negatively and irreversibly. They also require considerable human labor, so they produce results at very low throughput. Hence, a number of groups have proposed the use of microfluidic technology for mimicking the physiological conditions on a cellular scale and for automating high-throughput cell culture experiments that can provide low-cost alternatives to animal and clinical studies (Figure 5.40).

In 2005, Andre Levchenko's group at Johns Hopkins University reported the design and operation of a microfluidic "chemostat," a bioreactor to and from which fresh medium is continuously added and removed while keeping the culture volume constant. Chemostatic conditions were demonstrated for bacteria and yeast colonies growing in an array of shallow microchambers (Figure 5.41). The walls of the chambers featured tunnels, so they were impenetrable to the cells but allowed diffusion of essential chemicals. The tunnels served to equilibrate the concentrations with an adjacent microchannel, which was continuously perfused. The device allows to monitor, control, and induce growth starting from a single cell.

As far as bioreactors are concerned, bacteria and yeast do not have very stringent needs. Mammalian cells represent the next challenge because they are much more difficult to maintain. Michael Shuler's group at Cornell University has presented a bioreactor capable of maintaining not just one but *three* mammalian cell types alive using the recirculation of the same cell culture medium. Cells embedded in 3-D hydrogels, such as Matrigel or alginate gel, are cultured in separate microchambers representing the liver (hepatoma cells), tumor (colon cancer cells), and marrow (myeloblasts), which are connected by channels mimicking blood flow (Figure 5.42). The gel protects the cells from flow. The device is able to reproduce the metabolism of Tegafur to 5-fluorouracil (5-FU, an anti–colon cancer drug) in the liver by p450 enzymes and consequent death of cancer cells by 5-FU, whereas the cultures in a 96-well microtiter plate were unable to do so.

Several groups have started questioning the need for dissociating the cells: Why not use pieces of intact tissue directly, which better preserve tissue architecture and cell–cell interactions? (See box below.) Recently, a group from the University of Groningen (Netherlands) led by Geny Groothuis and Elisabeth Verpoorte assessed interorgan interactions in a microfluidic device using

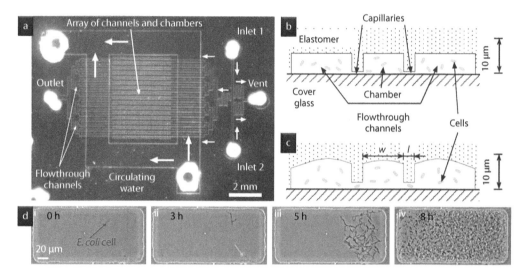

FIGURE 5.41 A microfluidic chemostat to maintain constant cell culture conditions. (From Alex Groisman, Caroline Lobo, HoJung Cho, J. Kyle Campbell, Yann S. Dufour, Ann M. Stevens, and Andre Levchenko, "A microfluidic chemostat for experiments with bacterial and yeast cells," *Nat. Methods* 9, 685, 2005. Adapted with permission from the Nature Publishing Group. Figure contributed by Andre Levchenko.)

precision-cut liver slices and intestinal slices (**Figure 5.43**). The slices were placed in adjacent microchambers and perfused sequentially, so that the metabolites excreted by the intestinal slice were directed to the microchamber containing the liver slice. Interplay between the two organs was demonstrated by exposure of the slices to chenodeoxycholic acid (i.e., bile), which induced the expression of fibroblast growth factor 15 (FGF-15) in the intestinal slice. FGF-15, in turn, caused down-regulation of the detoxification enzyme cytochrome p450 activity in the liver slice.

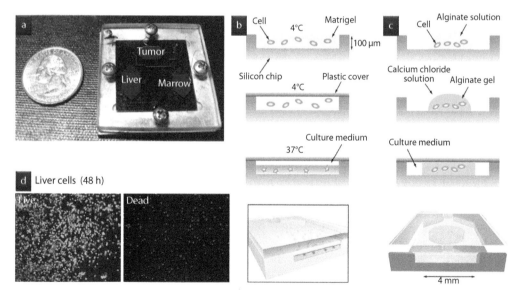

FIGURE 5.42 Probing interorgan interactions with a microfluidic cell culture analogue. (From Jong Hwan Sung and Michael L. Shuler, "A micro cell culture analog (μCCA) with 3-D hydrogel culture of multiple cell lines to assess metabolism-dependent cytotoxicity of anti-cancer drugs," *Lab Chip* 9, 1385–1394, 2009. Reproduced with permission from The Royal Society of Chemistry.)

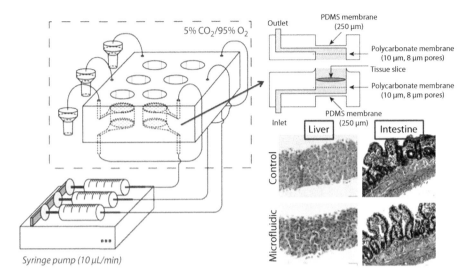

FIGURE 5.43 Sliced-organ-on-a-chip. (From Paul M. van Midwoud, Marjolijn T. Merema, Elisabeth Verpoorte, and Geny M. M. Groothuis, "A microfluidic approach for in vitro assessment of inter-organ interactions in drug metabolism using intestinal and liver slices," *Lab Chip* 10, 2778, 2010. Reproduced with permission from The Royal Society of Chemistry.)

TO DISSOCIATE OR NOT TO DISSOCIATE, THAT IS THE QUESTION

THE HISTORY OF DISSOCIATING CELLS has had its ups and downs. Biologists have always recognized that, in dissociating cells, essential components of the tissue architecture and that play key roles in cell physiology, such as the vasculature, cell–cell neighbor interactions, and signaling gradients cannot be preserved. The German biologist Otto Warburg, working at the Kaiser Wilhelm Institute for Biology in Berlin, first introduced the tissue-slicing technique in 1923 to study tissue metabolism; in 1931 he was awarded the Nobel Prize in Physiology or Medicine for his "discovery of the nature and mode of action of the respiratory enzyme." (Three of Warburg's disciples, among them Hans Adolf Krebs, went on to win Nobel prizes.) In 1957, Henry McIlwain at the Birmingham University (UK) was able to prepare viable brain slices, and in 1966, he published a seminal electrophysiology article that demonstrated synaptic transmission in slices for the first time. Toxicologists have also used tissue slices for decades since Klaus Brendel (from the University of Arizona) and Carlos Krumdieck (the developer of the Krumdieck slicer, from the University of Alabama) perfused thin tissue slices of the liver (and of many other organs) to show improved functionality with respect to dissociated cultures. By comparison, surprisingly, the cancer research community has barely made use of "tumor slices," perhaps because oncologists have long had the option of extracting needle biopsies—a less invasive option than slicing an organ. A biopsy is a approximately 0.4- to 1-mm-wide cylindrical section of tissue (usually less than 1 cm long) that is typically fixed and stained after extraction (and often embedded and sliced for imaging purposes); the biopsy is not usually cultured, but live needle biopsy screening should prove to be a highly viable format for applications as diverse as toxicology, pharmacology, and cancer.

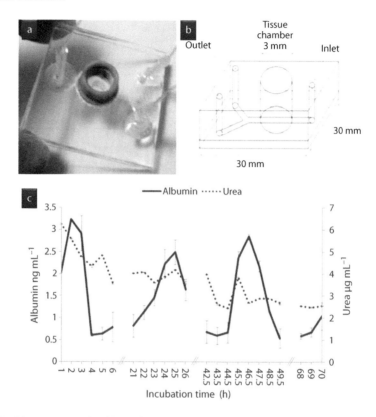

FIGURE 5.44 Biopsy-on-a-chip. (From Samantha M. Hattersley, Charlotte E. Dyer, John Greenman, and Stephen J. Haswell, "Development of a microfluidic device for the maintenance and interrogation of viable tissue biopsies," *Lab Chip* 8, 1842, 2008. Reproduced with permission from The Royal Society of Chemistry.)

However, tissue extraction in the form of a slice (requiring surgery) is too painful for many clinical applications. On the other hand, in clinical settings, the extraction of a needle biopsy is a well established, routine procedure (e.g., for cancer diagnostics). Stephen Haswell's group at the University of Hull (UK) has devised a simple microfluidic device that can accept a small needle biopsy and perfuse it. With the device, the group has demonstrated that liver biopsies can be kept alive for up to 70 hours with intact liver-specific functions such as albumin secretion and urea synthesis (**Figure 5.44**).

5.7 Cells on Microelectrodes

Clearly, the neuronal circuits of the vertebrate central nervous system are so complex that electrical recordings in vivo could provide only fragmentary information about the network's activity. Since the 1970s, in an effort to investigate basic network parameters, many groups have merged modern neuron culture techniques with microelectronics. In 1972, a team from Harvard Medical School and the University of Utah led by L. M. Okun produced microarrays of 30 platinum black-coated gold electrodes on glass. The platinized electrodes enabled good contact with chick heart cell sheets (although recordings from single beating cells were not possible). Many groups, to this day, still use metal electrodes (platinized or not) to record from electrically active cells such as neurons and muscle cells. Several designs of **microelectrode arrays** (**MEAs**) are commercially available.

However, metal electrodes are opaque, a feature that interferes with microscopic observation (when using transmitted light, which is necessary to obtain phase-contrast of subcellular organelles of live cells). For this reason, in 1985, a collaboration between Guenter Gross from North Texas State University and Jacob Lin from Polytronix, Inc., pioneered the use of **microelectrodes** made of indium-tin oxide (ITO), a transparent semiconductor, to record from ensembles of neurons (**Figure 5.45a** through **c**). A team, led by Markus Meister at Harvard University and Denis Baylor at Stanford University, first used this technique with great success in the early 1990s to record from excised retinas placed on top of 61 ITO microelectrodes (to reduce the impedance of the electrodes, they were coated with a thin layer of platinum black, resulting in an impedance of 100 kΩ at 1 kHz). The electrodes were 10 μm in diameter and approximately 70 μm apart, arranged in a 500-μm-wide hexagonal array whose spacing was designed as two to three times the distance between neighboring retinal ganglion cells in the amphibian or (peripheral) mammalian retina (**Figure 5.45d** and **e**). Since then, the combination of ITO microelectrodes and the retina has become a very productive system and has allowed for testing of critical hypotheses in vision research. For example, Meister's group was able to show that the retina (of salamanders

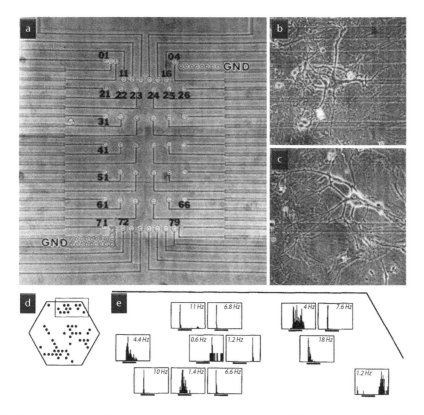

FIGURE 5.45 Transparent MEAs fabricated in indium tin oxide (ITO). (a–c; from G. W. Gross, W. Y. Wen and J. W. Lin, "Transparent indium-tin oxide electrode patterns for extracellular multisite recordings in neuronal cultures," *J. Neurosci. Methods* 15, 243–252, 1985. The electrodes are 10 μm in diameter. d and e; from M. Meister, J. Pine, and D. A. Baylor, "Multi-neuronal signals from the retina: Acquisition and analysis," *J. Neurosci. Methods* 51, 95–106, 1994. Reprinted with permission from Elsevier.)

and rabbits) has a remarkable capacity for processing motion information, a function previously thought to be restricted to the neurons of the visual cortex.

With hindsight, the success of the retina-on-MEA system is not surprising because it is a combination of good choices: (1) As neural circuits go, the retina is extremely well character-ized and its general function is also very well known; (2) the circuit can be removed from its organ without damaging its internal connections; (3) as the supporting cells are removed with the circuit and the layer of tissue is so thin that nutrients can be delivered to the cells by diffu-sion through the tissue, the circuit can be kept alive in simple saline solution for many hours; (4) there is no need to engineer complicated interfaces to provide an input to the circuit because the natural input is light (which can be projected in patterns through the objective of a micro-scope); and (5) the retina's output (the action potentials in ganglion cell axons) is clearly defined and organized in a single flat layer, which can be recorded with the planar MEA.

Note that the impedance of the cell–electrode contact is expected to vary from cell to cell, depending on how closely the cell is attached to the electrode, and to vary with time because focal attachments to the substrate are inherently dynamic and the electrode's adhesiveness (e.g., given by its protein coverage) may also be changing.

Recently, Bruce Wheeler's laboratory, formerly at the University of Illinois at Urbana-Champaign, devised a clever variation of MEAs designed for detecting action potential propa-gation from isolated axons in culture (**Figure 5.46**). The neurons are seeded in an open well that has microfabricated tunnels (10 µm wide, 3 µm high, and 750 µm long) through which the axons

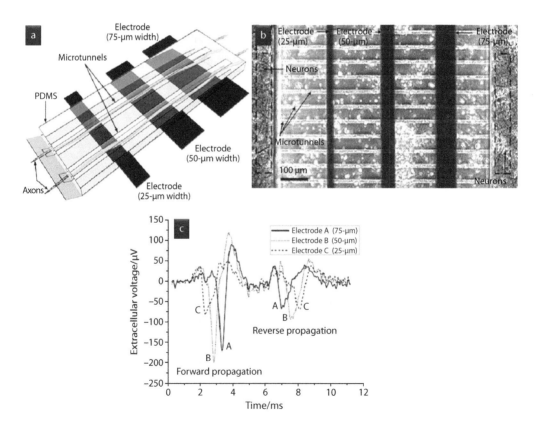

FIGURE 5.46 MEAs combined with microtunnels to measure axonal signal conduction velocity. (From Bradley J. Dworak and Bruce C. Wheeler, "Novel MEA platform with PDMS microtunnels enables the detection of action potential propagation from isolated axons in culture," *Lab Chip* 9, 404–410, 2009. Reproduced with permission from The Royal Society of Chemistry.)

grow over several days. At several points along the microtunnels, the floor of the tunnels contain metal electrodes that allow for recording large (up to 200 µV) electrical signals, including their direction and propagation speed.

MEAs, however, are passive signal detectors and require an extra amplifying step, which adds noise because of the distance between the MEA and the amplifier. How about using the exquisite voltage sensitivity of microelectronics components such as transistors (right below the cells) for detecting small bioelectric signals? This was precisely the thinking behind the work by Jürgen Weis' laboratory, then at the University of Ulm (Germany), who in 1991 pioneered the use of **field-effect transistors** (FETs) for bioelectric recordings (Figure 5.47). In microelectronics, FETs are used as switches; the voltage applied to the "gate" electrode is used to modulate the passage of current between the "source" and the "drain," two adjacent regions of differentially doped silicon. Not surprisingly, a custom-fabricated FET (without metal on the gate, to allow for contact between the neuron and the gate oxide) was able to detect changes in membrane potential of a Retzius neuron. The work raised many hopes, as the sensors are readily fabricated and integrated with other sensors in large microarrays. The fact that they are opaque did not seem to deter a wealth of research and development on cellular applications of FETs in the following two decades. The biggest obstacle seems to be the temporal stability of the recordings, which are impaired by the leaky seal between neuron and gate and (more fundamentally) by transistor breakdown (electrolyte ions diffuse into the silicon bulk and irreversibly alter the conductance of the device).

Recent developments on FET technology may enable new modalities of neuronal recordings. Charles Lieber's group at Harvard has shown that silicon nanowire FETs (Si-NWFETs) can be built on transparent substrates and interfaced with brain slices (Figure 5.48). Unlike traditional FETs or MEAs, Si-NWFETs allow for imaging cells in various optical microscopy modalities and provides highly localized measurements (their active surface is ~0.06 µm²), at submillisecond temporal resolution and 30 µm spatial resolution.

The biggest drawback of MEAs and FETs is, of course, that they are only capable of recording extracellular signals, that is, membrane voltages. However, electrophysiologists are even more interested in learning about the intracellular signals that give rise to those membrane voltages. Traditionally, the intracellular signals can only be recorded with patch clamp probes or patch clamp chips (see Section 5.8), but a new revolution in electrode technology has come to the rescue. Micha Spira's group from the Hebrew University of Jerusalem, in Israel, has developed a new type of MEA consisting of gold, mushroom-shaped microelectrodes (gMµEs) that allow for recording and stimulating intracellularly from arrays of neurons, while *the electrodes maintain*

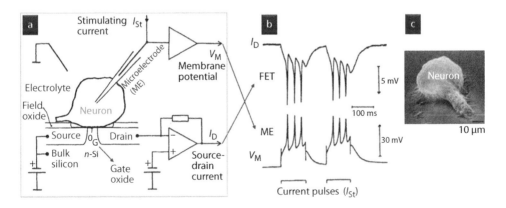

FIGURE 5.47 FETs to record the electrical activity of cells. (From Peter Fromherz, Andreas Offenhäusser, Thomas Vetter, and Jürgen Weis, "A neuron-silicon junction: A Retzius Cell of the leech on an insulated-gate field-effect transistor," *Science* 252, 1290–1293, 1991. Figure contributed by Peter Fromherz.)

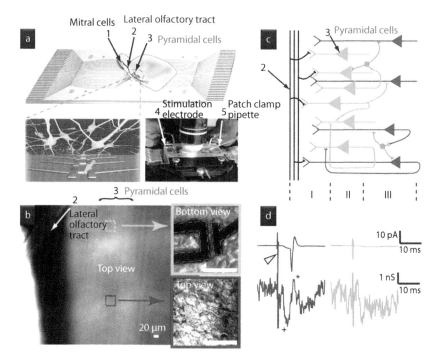

FIGURE 5.48 Nanowire FET arrays for recording from brain slices. (From Quan Qing, Sumon K. Pal, Bozhi Tian, Xiaojie Duan, Brian P. Timko, Tzahi Cohen-Karni, Venkatesh N. Murthy, and Charles M. Lieber, "Nanowire transistor arrays for mapping neural circuits in acute brain slices," *Proc. Natl. Acad. Sci. U. S. A.* 107, 1882–1887, 2010. Figure contributed by Venkatesh Murthy and Charles M. Lieber.)

an extracellular position. The "magic" is that the gMμEs are derivatized with a peptide that generates phagocytosis of the gMμE, so that the neuron engulfs it and forms a low-resistance actin ring around the stalk of the gMμE. Action potentials and subthreshold potentials were recorded with quality and signal-to-noise ratio matching that of conventional intracellular patch clamp pipettes for up to 2 days. It is not clear that the "mushroom" geometry is essential, or whether the engulfing peptides are the only strategy available to access the intracellular milieu: the group of Nicholas Melosh at Stanford University has recently presented a variation of this design whereby the post is cylindrical and the cell forms a seals with a hydrophobic 5 to 10 nm narrow band on the post that is functionalized with butanethiol, which causes it to fuse with the cell membrane (seal resistances of 3.8 ± 1.9 GΩ were recorded). In any case, these "in-cell" extracellular electrodes, as Spira's group calls them, are expected to revolutionize the technology of neuronal recordings both in vitro and in vivo.

5.8 Patch Clamp Chips

The patch clamp chip field has brought many interesting lessons for both biologists and engineers alike. Many biologists had predicted that, unless the surface of the chips were built with glass to imitate that of patch clamp pipettes, the chips would never provide the same quality of recordings—yet patch clamp recordings obtained with all-PDMS devices are now routine. Most engineers attempted designs that incorporated glass or silicon, but aperture fabrication (or fabrication throughput) was an immense challenge. The solutions were very creative and worth studying; however, the field has been somewhat outsmarted by Spira's (and others) molecular trickery that results in much more benign "in-cell" probes (see **Figure 5.49**).

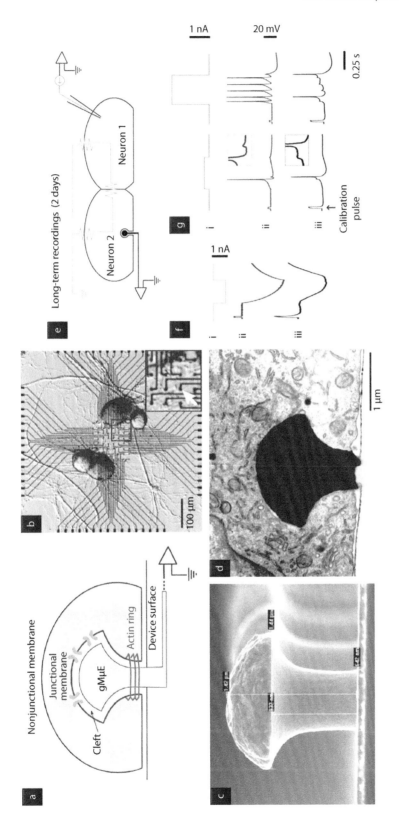

FIGURE 5.49 "In-cell" recording and stimulation by extracellular microelectrodes. (From Aviad Hai, Joseph Shappir, and Micha E. Spira, "Long-term, multisite, parallel, in-cell recording and stimulation by an array of extracellular microelectrodes," *J. Neurophysiol.* 104, 559–568, 2010. Copyright (2010) by The American Physiology Society. Figure contributed by Micha Spira.)

THE PATCH CLAMP TECHNIQUE

THE **PATCH CLAMP TECHNIQUE**, developed in the late 1970s and early 1980s by Erwin Neher and Bert Sakmann (who obtained the 1991 Nobel Prize for its development) is a method that allows for the direct measurement of currents through ion channels with submillisecond resolution and single-channel resolution using glass micropipettes sealed against the cell membrane (**Figure 5.50**). Furthermore, the technique allows to control the electrical and chemical environment of a membrane and the application of signaling molecules, drugs and others to both sides of the membrane. The electrical conductivity of a biological membrane comes not from the lipid, but from ion channels embedded in the lipid. The technique is the gold standard for screening for adverse effects in drug discovery research (the Food and Drug Administration, in the United States, for example, mandates that every new drug be screened for its effect on the hERG K^+ ion channel present in the heart muscle). However, the present patch clamp recording and stimulation technology is based on a delicate procedure that requires an experienced operator to position the glass pipette (with bulky micromanipulators under a microscope) onto the cell to obtain a "gigaohm (GΩ) seal" (or "**giga-seal**," for short), which results in very low throughput and inability to probe many cells simultaneously. Thus, the traditional pipette-based patch clamp technique represents a huge drug-testing bottleneck: drug companies can easily synthesize hundreds of thousands of compounds in a week by combinatorial chemistry, but they can only test 10 to 20 of these on cell cultures every day per operator (on a good day). Essentially, the difficulty relies in automating the giga-seal process. As we can see in **Figure 5.50**, a good giga-seal is created by minimizing the leakage current through the seal resistance R_s (path "B" in **Figure 5.50**), which is largely a function of the contact between the cell membrane and the insulator surface (the pipette or the recording device), and can be improved by applying suction. If the patch is not perforated, then most of resistance "seen" by the device (which collects current through path "A" in **Figure 5.50**) corresponds to the resistance of the area covered by the aperture ($R_{m\text{-in}}$), but if enough suction is applied, then the patch is perforated and "whole-cell access" is gained (i.e., $R_{m\text{-in}}$ is essentially zero and the total cell resistance is approximated by $R_{m\text{-out}}$). In whole-cell mode, the device records from all the ion channels in the cell rather than those limited to the patch, so the currents are much larger.

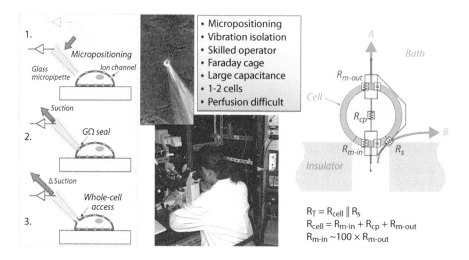

FIGURE 5.50 Patch clamp technique. (Figure contributed by Chihchen Chen and Albert Folch.)

Spurred by the biopharmaceutical interest in overcoming the drug-screening bottleneck, several laboratories have developed various procedures to build planar patch clamp–on-a-chip devices, whereby cells are deposited on top of a micron-sized hole on a thin insulating substrate over a cavity or channel filled with electrolyte solution (i.e., the hole and the cavity act as a "microfabricated, prepositioned pipette," see conceptual rendering in Figure 5.51).

The evolution of the field provides a very interesting lesson for materials scientists and biologists, as the device material (glass) was known by biologists to be critical for device function yet engineers often compromised on the material of choice due to fabrication challenges (building the structure of Figure 5.51b exactly as shown in glass is, to this day, impossible!)—until a new material (PDMS) was found to be suitable. These are the main pioneering contributions:

✳ In 2000, Horst Vogel's group from the Swiss Federal Institute of Technology made apertures (0.6–7 µm diameter) in silicon nitride membranes (0.1–1 µm thick) suspended over a silicon pit; they used a combination of anisotropic KOH silicon etching and reactive-ion etching (RIE), followed by building a SiO_2 layer with deposition and thermal oxidation. Lipid vesicles were electrophoretically positioned onto the aperture and seals with the membrane patches could be obtained with resistances up to 200 GΩ if the chip was coated with polylysine or chemically modified with 4-aminobutyl-dimethyl-methoxysilane.

✳ In 2002, James Heath and colleagues at UCLA reconstituted K^+ ion channels in lipid bilayers suspended in the 100- and 200-µm-diameter pores of silanized (hydrophobic) SiO_2 membranes created by RIE. Two years later, this group was able to demonstrate good seals with RAW 264.7, CHO-K1, HIT-T15, and RIN m5F cells, although only approximately 5% of the attempts yielded a giga-seal (with CHO-K1 and RIN m5F cells; with HIT-T15 and RAW 264.7 the resistances were ~100 MΩ).

✳ In 2002, a collaboration between James Klemic and Fred Sigworth's groups at Yale University produced micromolded arrays of planar PDMS patch electrodes with apertures of 2 to 20 µm. The apertures were formed by point-contact replication of epoxy pyramids (in contact with a hard surface), but the pyramids were fabricated from crystalline-silicon molds, so the points of the pyramids were necessarily formed by the coincidence of four crystalline planes (which never forms a perfect point—a point is defined by the intersection of three planes) and small irregularities in the etching conditions produced differences in the height of the pyramids (so during replication, some point-contacts received more pressure than others, resulting in irregular holes, as can be seen in Figure 5.52a). Nevertheless, by plasma oxidation of PDMS, they could temporarily improve the hydrophilicity of PDMS and obtain seals on oocytes expressing potassium channels. Two years later, the same group proposed a clever molding strategy based on a focused stream of pressurized air as the PDMS

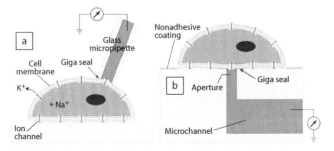

FIGURE 5.51   Conceptual comparison between (a) the traditional approach and (b) the planar chip approach to patch clamp recording. The nonadhesive coating in (b) is optional. (Figure contributed by the author.)

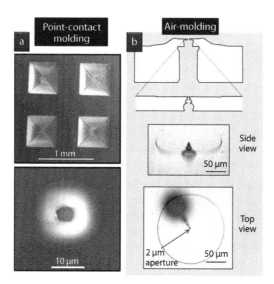

FIGURE 5.52 **Patch clamp chips** micromolded in PDMS. (a; from Kathryn G. Klemic, James F. Klemic, Mark A. Reed, and Fred J. Sigworth, "Micromolded PDMS planar electrode allows patch clamp electrical recordings from cells," *Biosens. Bioelectron.* 17, 597–604, 2002. Reprinted with permission from Elsevier; and b, from Kathryn G. Klemic, James F. Klemic, and Fred J. Sigworth, "An air-molding technique for fabricating PDMS planar patch-clamp electrodes," *Pflugers Arch. – Eur. J. Physiol.* 449, 564–572, 2005. Reprinted with permission from Springer.)

prepolymer was curing that produced very fine, smooth apertures (**Figure 5.52b**) and demonstrated giga-seals on oocytes, Chinese hamster ovary (CHO) cells, and rat basophilic leukemia (RBL) cells (which are three major cell types used as expression systems for ion channel research), although at relatively low yields (25%, 7%, and 10%, respectively; with glass pipettes and a skilled operator, these yields approach 90% for these cell types). This was the first time that a new material other than glass was seriously brought to the center stage of patch clamp electrophysiology.

In 2002, a collaborative team from the Swiss Federal Institute of Technology (Lausanne) and from Genion (Hamburg, Germany) led by Martin Gijs and Ulrike Bischoff, respectively, mimicked the 3-D shape of a glass micropipette by creating micron-sized (down to 2.5 μm), impressive-looking hollow SiO_2 nozzles on Si/SiO_2 wafer using photolithography, RIE, and thermal oxidation processes. Cells were reliably positioned on the nozzles by suction through the aperture and the underlying fluidic channels. Seal resistances of 100 to 200 MΩ were obtained on CHO cells. Unfortunately, as impressive as the microstructures were, silicon substrates are opaque, precluding microscopy or spectroscopy measurements. More importantly, because silicon is a semiconducting material, Si chips introduce capacitance because of the free charge carrier density in the substrate (which produces transient currents when voltage steps are applied).

The first report of giga-seal recordings on mammalian cells is credited to Jan Behrends' group from the Center for NanoScience in Munich (Germany), who in 2002 invented a process for sculpting submicron apertures in glass substrates (called "ion-track etching," which consists of shooting gold ions into glass with a particle accelerator, thus leaving a trace of the gold atom in the glass lattice that etches faster in certain etchants). They were able to record from **CHO cells** and N1E-115 neuroblastoma cells with typical seal resistances of 1 to 10 GΩ, allowing for recordings of single ion

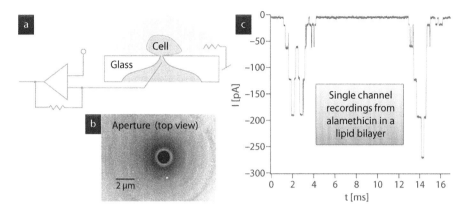

FIGURE 5.53 A glass patch clamp chip capable of detecting single-ion channel activity. (From Niels Fertig, Robert H. Blick, and Jan C. Behrends, "Whole cell patch clamp recording performed on a planar glass chip," *Biophys. J.* 82, 3056–3062, 2002. Adapted with permission of Elsevier.)

channels (in CHO cells as well as in lipid bilayers, see Figure 5.53). Unfortunately, although glass patch-on-a-chip devices are most promising because of their chemical surface similarity with micropipettes, the particle accelerator required for ion track etching is not available in most research institutions. However, this technology is now commercially available through Nanion Technology, which has reported a high-throughput system capable of obtaining whole-cell I-V curves (voltage-gated as well as ligand-gated channels in HEK-293 cells) from 96 cells simultaneously (the "SynchroPatch 96").

In 2005, bioengineer Luke Lee from the University of California at Berkeley and his colleagues proposed a radical new approach to patch-clamping: instead of letting the cells settle onto a planar aperture (which is the most obvious mimic of a pipette aperture), they reasoned that the cells should also be able to produce good seals against a cornered or "lateral" aperture (i.e., the end of a small microchannel, as depicted in Figure 5.54). For the patch array in cell-attached mode, the seal resistance was 150

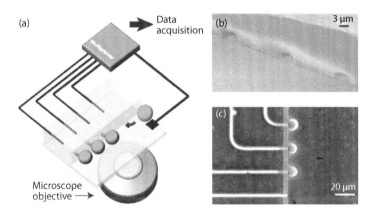

FIGURE 5.54 PDMS patch clamp chip with a lateral aperture. (From Cristian Ionescu-Zanetti, Robin M. Shaw, Jeonggi Seo, Yuh-Nung Jan, Lily Y. Jan, and Luke P. Lee, "Mammalian electrophysiology on a microfluidic platform," *Proc. Natl. Acad. Sci. U. S. A.* 102, 9112–9117, 2005. Copyright (2005) National Academy of Sciences, U. S. A.)

MΩ to 1.2 GΩ, with 27% of the seals at more than 250 MΩ and 5% at higher than 1 GΩ. In whole-cell mode, the cells had a seal resistance of 100 to 250 MΩ at −80 mV. These giga-seal yields and seal resistance qualities were not competitive with the best yields achievable by a human operator and a pipette, but the device offered straightforward automation and multiplexing for pennies, because it was the first device that could integrate multiple units with microfluidics using inexpensive polymer micromolded materials (PDMS). The lateral aperture design has several advantages over the planar aperture configuration: (a) it does not interfere with the microscopic observation of the cells; (b) aperture fabrication is simpler because it does not require thinning down of a thick substrate and it can be molded; (c) fluidic access to the reservoirs is already designed in the photomask, does not require additional packaging, and allows for integrated fluidic automation such as microvalves and micropumps, unlike in the planar patch clamp configuration; and (d) last but not least, in the planar geometry, the conducting buffers are separated by a relatively thin substrate, which produces a nonnegligible capacitance, whereas in the lateral patch clamp chip design, the two fluids are separated by an insulator wall.

The author's group at the University of Washington in Seattle conducted an investigation of the fundamental reasons why Luke Lee's PDMS design produced such low yields and concluded that a key factor was its inability to keep the aperture clean from cell debris and proteins. (In pipette-based recordings, to mitigate this problem, the pipette is positively pressurized so as to continuously perfuse the aperture rim with clean, nonproteic buffer until the very last moment, when the tip is very close to the cell; only then, is the pressure inverted.) A lateral patch clamp PDMS design that incorporated a dedicated "rinse line" to keep the aperture clean was demonstrated ("D" for "drain" in **Figure 5.55**). Using RBL cells, the device delivered high-stability giga-seals with success rates comparable with those of pipettes. The high stability enabled exchanges of both the extracellular solution (delivered through nanochannel nCh$_{EC}$ in **Figure 5.55**) and intracellular solution (delivered through nCh$_{IC}$) during whole-cell recordings. In a test of 103 different devices, 66 cells (64%) were successfully immobilized at the patch aperture; 38 cells (58% of immobilized cells, 37% of all cells) were successfully giga-sealed; and 25 cells (65% of giga-sealed cells, 34% of immobilized cells, 24% of all cells) were successfully perforated for whole-cell access. In the last group of 27 experiments, 79% of the cells could be immobilized, of which 68% could be giga-sealed and 46% perforated for whole-cell access, indicating that dexterity was still important.

How about if it were possible to create lateral apertures *in glass*? After all, glass seems to be unmatched in terms of giga-seal quality as a material (cell membranes just stick to clean glass more than any other material!). This is exactly what Levent Yobas' group from the Institute of Microelectronics in Singapore did in 2007. In fact, Yobas' "lateral" apertures are truly lateral, not cornered, as they are elevated from the bottom of the channel and the cell is patched against a flat surface. The process relies on the same physical principles as the preparation of conventional micropipette electrodes (heat pulling and fire polishing) but using phosphosilicate glass (PSG, 8 wt.% phosphorus content) deposited through plasma-enhanced chemical vapor deposition in microfabricated silicon trenches. A 2-μm-wide, 3.5-μm-deep Si trench was coated with a 4-μm layer of PSG (which "closed" the trench, leaving a keyhole void inside) and was heated to 1150°C for 30 minutes, resulting in reflow (melting) of the keyhole void into a cylindrical cavity. Apertures with a diameter of approximately 1.5 μm (variation of <10%) were demonstrated. One hundred apertures were tested on RBL-1 cells: 61% formed giga-seals (>1 GΩ) and of those, approximately 48% (29% of all) achieved whole-cell recordings (**Figure 5.56**).

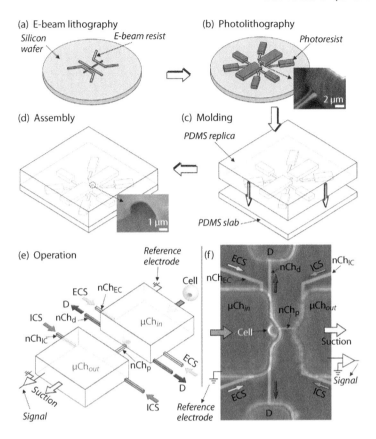

FIGURE 5.55 A PDMS lateral patch clamp design that delivers high-yield giga-seals. (From C. Chen and A. Folch, "A high-performance elastomeric patch clamp chip," *Lab Chip* 6, 1338, 2006. Reproduced with permission from The Royal Society of Chemistry. Figure contributed by Chihchen Chen.)

"GLASS OR POLYMER?": THAT'S THE PATCH CLAMP QUESTION

POLYMERS (MOST NOTABLY PDMS) provide a route for cheap microfabrication through replica-molding. However, if the price one has to pay is a suboptimal surface chemistry that compromises the throughput of the assay (e.g., the percentage of giga-seals), then it may no longer be the best option. The Yobas design may have found the optimal strategy because it uses glass (thus far unmatched as a patch clamp material, if we ask *any* cell type), a microscopy-friendly lateral design, and although the processing is expensive, the cost per device can be brought down using batch fabrication if the device is commercialized.

In 2010, two groups (Micha Spira's from the Hebrew University of Jerusalem and Nicholas Melosh's from Stanford University, see Section 5.7) report "in-cell" patch clamp recordings achieved with "self-impaling" nanofabricated electrodes. These electrodes can be fabricated by more traditional techniques (e.g., they do not require microfluidic protocols) and their operation does not dyalize the cells, so they are more benign than traditional perforating electrodes.

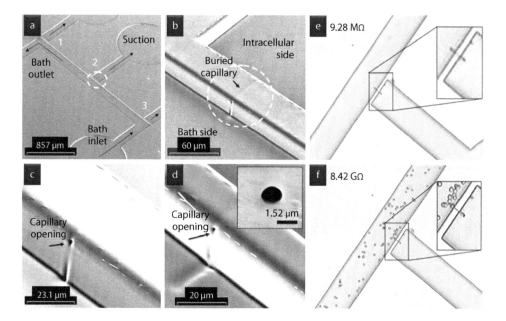

FIGURE 5.56 Microfabricated glass capillaries for patch clamp on a chip. (From Wee-Liat Ong, Kum-Cheong Tang, Ajay Agarwal, Ranganathan Nagarajan, Lian-Wee Luo and Levent Yobas, "Microfluidic integration of substantially round glass capillaries for lateral patch clamping on chip," *Lab Chip* 7, 1357–1366, 2007. Reproduced with permission from The Royal Society of Chemistry.)

As a variation on the patch clamp theme, several other groups have seized the opportunity to develop devices that are able to introduce compounds (i.e., DNA, proteins) into the cells, either using the patch clamp aperture or using very sharp needles that perforate the cell membrane. The needles, which can be deposited serially, surface-micromachined, or grown at random by a chemical process, are typically coated with the delivery agent or can be made of a biodegradable polymer mixed with it. As the cells are much bigger than the needles, the cell membrane self-heals rapidly around the perforation made by the needle, which releases the agent once inside the cell.

5.9 Cryopreservation

Cryopreservation is the process used to stop all biological activity by cooling the cells to low subzero temperatures. A recurrent problem in cryopreservation has been to find the right conditions to freeze biological tissue because ice crystals can form inside a cell and burst the cell membrane if cryoprotectants are not introduced into the cell. To find the optimal cryopreservation conditions of cells, it is necessary to better understand the properties of membrane transport and cell osmotic behavior. Dayong Gao and colleagues at the University of Washington have developed a microfluidic system for trapping isolated RBL cells (which, conveniently, live in suspension) in weir microstructures that let the fluid through but not the cells (**Figure 5.57**). This simple system allows for monitoring kinetic changes of cell volume under various extracellular conditions, either hypertonic solutions of nonpermeating solutes (e.g., NaCl) or solutions containing permeating cryoprotective agents (e.g., dimethylsulfoxide), from which it is possible to obtain the cell's osmotically inactive volume and the permeability coefficient of water and of dimethylsulfoxide for RBL cells.

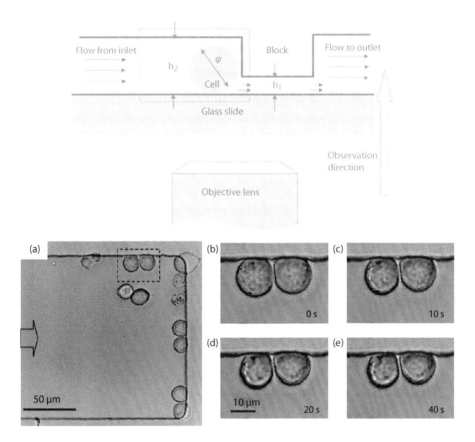

FIGURE 5.57 A microfluidic device for measuring cell osmotic properties. (From Hsiu-hung Chen, Jester J. P. Purtteman, Shelly Heimfeld, Albert Folch, and Dayong Gao, "Development of a microfluidic device for determination of cell osmotic behavior and membrane transport properties," *Cryobiology* 55, 200–209, 2007. Reprinted with permission from Elsevier. Figure contributed by Hsiu-hung Chen.)

5.10 Assisted Reproductive Technologies

Assisted reproductive technologies, including cryopreservation, cloning, nonsurgical embryo transfer, and the in vitro production of mammalian embryos, have become increasingly important in the last 15 years both to treat human infertility and to genetically improve livestock. The efficiency of these technologies, however, is still very low. This low efficiency is partially attributed to in vivo mammalian embryos residing in submicroliter amounts of fluid (within the crypts in the lumen of the female reproductive tract) whereas in vitro the embryos are bathed in large volumes of cell culture medium. Microfluidics offer opportunities for (a) mimicking in vivo microenvironments (i.e., topographical/biochemical mimics to which cells adhere, as well as minimization of dead volumes that mimic physiological cavities and mass transport), (b) straightforward integration of fluid-dispensing automation through microactuators (e.g., microvalves and micropumps), (c) inexpensive integration of sensors (e.g., pH, temperature) for real-time quality control (which reduces environmental stress and, in turn, increases efficiency), and (d) high-throughput testing (fabricating/operating 100 channels in parallel costs almost the same as fabricating/operating only one).

Sperm motility is an important measurement in the evaluation of male infertility. However, when the sperm is unconstrained on the surface of a slide, the motility is hard to quantify. In 1993, Peter Wilding and colleagues from the University of Pennsylvania were the first to apply microfluidic technology to evaluate sperm motility and perform sperm selection using silicon channels (**Figure 5.58**), which restrict sperm motion in the direction along the channel. The chemoattraction of sperm toward cervical mucus and toward hyaluronic acid was probed by filling a microchannel with semen and an adjacent microchamber with either cervical mucus or hyaluronic acid. Various concentrations of different spermicides (nonoxynol-9 and Cl 3G) were delivered through different microchannels to simultaneously assess in one assay the relative potency of each spermicide (while consuming minimal sample).

More recently, Shuichi Takayama's group from the University of Michigan at Ann Arbor has presented a cleverly simple device that allows for separating motile from nonmotile sperm based on microfluidic principles. As shown in **Figure 5.59**, the device is operated with two inlets at constant flow, one with sperm sample and one with saline or plain cell culture medium. The device exploits the simple principle that only motile sperm are able to escape from their laminar

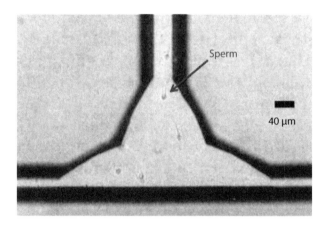

FIGURE 5.58 Evaluation of sperm motility using microfluidic channels. (From Larry J. Kricka, Osamu Nozaki, Susan Heyner, William T. Garside, and Peter Wilding, "Applications of a microfabricated device for evaluating sperm function," *Clin. Chem.* 39, 1944–1947, 1993. Figure contributed by Larry Kricka.)

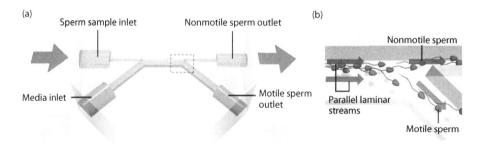

FIGURE 5.59 Sperm sorting based on sperm motility in a laminar flow. Schematic drawing of sperm sorting device. (a) Top view of device with two inlets and two outlet reservoirs. (b) Magnified view of outlet channels showing motile sperm separating from nonmotile sperm. (From Brenda S. Cho, Timothy G. Schuster, Xiaoyue Zhu, David Chang, Gary D. Smith, and Shuichi Takayama, "Passively driven integrated microfluidic system for separation of motile sperm," *Anal. Chem.* 75, 1671–1675, 2003. Figure contributed by Shuichi Takayama.)

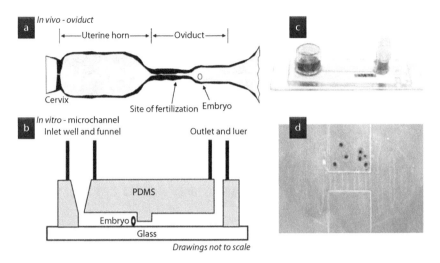

FIGURE 5.60 A biomimetic device for in vitro fertilization. (From Sherrie G. Clark, Kathyrn Haubert, David J. Beebe, C. Edward Ferguson, and Matthew B. Wheeler, "Reduction of polyspermic penetration using biomimetic microfluidic technology during in vitro fertilization," *Lab Chip* 5, 1229–1232, 2005. Reproduced with permission from The Royal Society of Chemistry.)

stream and go into the laminar stream that leads to the bottom outlet, whereas nonmotile sperm necessarily continue straight to the top outlet.

The penetration of in vitro–fertilized oocytes by more than one sperm ("polyspermy") results in unviable embryos and has haunted animal in vitro fertilization efforts for many years. In vivo, the oocyte is "parked" at the ampullary–isthmic junction (the site of fertilization) and the oviduct acts as a reservoir with a narrowing or "gate," which guides the sperm to flow to the junction (**Figure 5.60a**). A team led by Matthew Wheeler from the University of Illinois at Urbana-Champaign and David Beebe from the University of Wisconsin at Madison designed in 2005 a biomimetic microfluidic device featuring a microchannel constriction (**Figure 5.60b**) that allows for immobilizing pig embryos (**Figure 5.60c** and **d**) and demonstrated an increase in monospermy from approximately 22% (the controls, fertilized with the traditional "microdrop" technique) to approximately 55% or 64%, depending on whether 10 or 15 oocytes were present in the microchannel, respectively. There is additional evidence that this "biomimetic" effect is real, because separate results with mouse embryos also demonstrated that embryonic development of mouse embryos to the blastocyst stage was enhanced (exhibited a faster rate of cleavage, producing ~28%–24% more blastocysts) and showed a 2- to 3-fold reduction in degenerated embryos by 96 hours of culture compared with control microdrop cultures (the numbers varied depending on whether the microchannels were constructed in silicon or PDMS).

5.11 Whole Animal Testing

The roundworm *Caenorhabditis elegans* is an approximately 1-mm-long transparent nematode that has been widely used as a model organism for several decades now. The lineage of every cell is known, as well as the connectivity and position of every neuron in its 302-neuron nervous system. Seeing that it conveniently fits in microchannels, in 2004, Cornelia Bargmann's laboratory at Rockefeller University started using microfluidics to study *C. elegans* behavior (see Section 6.5.5). Recently, the laboratories of Nikos Chronis (who trained in the Bargmann laboratory) at the University of Michigan and of Fatih Yanik at MIT have incorporated PDMS microactuator technology to immobilize the worms at high throughput (a PDMS membrane pushes the worm down),

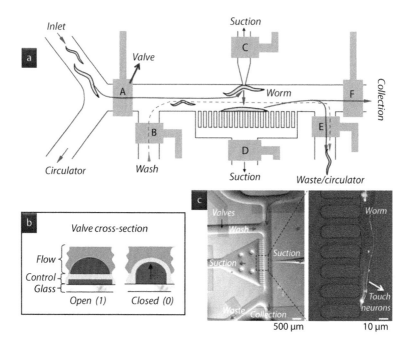

FIGURE 5.61 High-throughput screening of *C. elegans* with a microfluidic chip. (From Christopher B. Rohde, Fei Zeng, Ricardo Gonzalez-Rubio, Matthew Angel, and Mehmet Fatih Yanik, "Microfluidic system for on-chip high-throughput whole-animal sorting and screening at subcellular resolution," *Proc. Natl. Acad. Sci. U. S. A.* 104, 13891–13895, 2007. Copyright (2007) National Academy of Sciences, U. S. A. Figure contributed by Fatih Yanik.)

to exert forces on the worms, as well as to switch fluids. Yanik's laboratory has demonstrated an impressive worm-screening device (**Figure 5.61**) that is capable of isolating, immobilizing, imaging, performing femtosecond laser microsurgery, and sorting worms into multiwell plates—all within a fraction of a second per worm, an improvement of several orders of magnitude over previous manual procedures. A newer version of the device can manipulate zebrafish embryos. These systems have a great potential and are already being explored for drug screening applications.

5.12 Summary

The ability to handle and analyze single live entities (ranging from cells, embryos, to worms) at high throughput in small fluid volumes is revolutionizing the fields of molecular biology, biochemistry, and cell biology because it is providing a more quantitative description of cellular heterogeneity—which is crucial in understanding both physiological and pathophysiological phenomena such as differentiation, migration, reproduction, and cancer, among many others. In the same token, this ability is greatly benefitting many related efforts in biotechnology, such as in the development of cell analysis chips for PCR, patch clamp electrophysiology, and others.

Further Reading

Chen P., X. Feng, W. Du, and B. Liu. "Microfluidic chips for cell sorting," *Frontiers in Bioscience* **13**, 2464–2483 (2008).

Chung, T. D. and H. C. Kim. "Recent advances in miniaturized microfluidic flow cytometry for clinical use," *Electrophoresis* **28**, 4511–4520 (2007).

Dittrich, P. S. and A. Manz. "Lab-on-a-chip: microfluidics in drug discovery," *Nature Reviews Drug Discovery* **5**, 210–218 (2006).

Godin, J., C. H. Chen, S. H. Cho, W. Qiao, F. Tsai, and Y. H. Lo. "Microfluidics and photonics for Bio-System-on-a-Chip: a review of advancements in technology towards a microfluidic flow cytometry chip," *Journal of Biophotonics* **1**, 355–376 (2008).

Mairhofer, J., K. Roppert, and P. Ertl. "Microfluidic systems for pathogen sensing: a review," *Sensors* **9**, 4804–4823 (2009).

Meyvantsson, I., and D. J. Beebe. "Cell culture models in microfluidic systems," *Annual Review of Analytical Chemistry* **1**, 423–449 (2008).

Nilsson, J., M. Evander, B. Hammarstrom, and T. Laurell. "Review of cell and particle trapping in microfluidic systems," *Analytica Chimica Acta* **649**, 141–157 (2009).

Verpoorte, E. "Microfluidic chips for clinical and forensic analysis," *Electrophoresis* **23**, 677–712 (2002).

Walker, G. M., H. Zeringue, and D. J. Beebe. "Microenvironment design considerations for cellular scale studies," *Lab on a Chip* **4**, 91–97 (2004).

Wang, W. and S. A. Soper (editors). *Bio-MEMS: Technologies and Applications*. CRC Press (2007).

Wu, M.-H., S.-B. Huang, and G.-B. Lee. "Microfluidic cell culture systems for drug research," *Lab on a Chip* **10**, 939–956 (2010).

Xu, Y., X. Yang, and E. Wang. "Review: aptamers in microfluidic chips," *Analytica Chimica Acta* **1**, 12–20 (2010).

Yan H., B. Zhang, and H. Wu. "Chemical cytometry on microfluidic chips," *Electrophoresis* **29**, 1775–1786 (2008).

Zhang, C., D. Xing, and Y. Li. "Micropumps, microvalves, and micromixers within PCR microfluidic chips: advances and trends," *Biotechnology Advances* **25**, 483–514 (2007).

6

BioMEMS for Cell Biology

CELLS ARE PROGRAMMED TO BE EXQUISITELY SENSITIVE and responsive to their changing microenvironment. In a healthy organism as well as in a diseased one, cells respond to *local* variations (sometimes in space, sometimes in time, often in both) of dozens of biochemical and biophysical signals that intervene in multiple signaling pathways. As schematically depicted in Figure 6.1, biochemical signals can be soluble molecules—nutrients, enzymes, inorganic ions, and growth factors that are secreted by cells that can be adjacent or, for a large animal, as far as several meters—or immobilized molecules that are anchored to the membrane of adjacent cells (e.g., membrane receptors and cadherins) or bound to extracellular scaffolds, such as the extracellular matrix (ECM) or bone. Examples of biophysical signals can be membrane voltage, molecular conformation, incident light, temperature, and the rigidity or roughness of the substrate to which the cells are anchored. All these factors may vary locally and temporally in smooth gradients or in sharp steps.

This microenvironment is far from static. In fact, living systems are nonequilibrium systems with large fluxes of energy and mass flowing through them. In essence, dynamic microenvironments are what life is about: at the heart of any of life's many defining "macroscale" processes—reproduction, development, motility, immunological response, sensorial perception, wound healing, and so on—is a vibrant microenvironment orchestrated by cells. Tragically, this temporal component is necessarily lost in static cell culture systems, in which the cells are bathed in a homogeneous bath for prolonged periods of time. Hence, microfluidics (which allow for spatiotemporal modulation of the soluble environment) and other techniques that allow for micron-scale modulation of the substrate characteristics, offer a precious opportunity to recreate physiological conditions that cannot be attained with traditional cell culture substrates.

In this chapter, we will see how BioMEMS technology can be used to increase the biochemical and biophysical complexity of cell culture microenvironments—so as to trigger, control, or influence cellular processes such as adhesion, migration, growth, secretion, and overall gene expression in a physiologically relevant, quantitative way that *cannot* be achieved with traditional cell culture tools. For the same price, BioMEMS devices can also be designed to improve fluid handling (automate routing, reduce consumption and disposal of fluids) and increase experimentation throughput, which may simply facilitate research by lowering costs or enable new single-cell studies of rare-cell behavior.

6.1 An Enabling Technology: The Hurdles

In Chapter 5, we saw how microfluidic devices can be used to overcome some of the fluid handling limitations—low throughput, high cost—of traditional cell culture methodology. However, conventional cell culture techniques are limiting for two additional, even more fundamental,

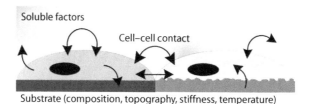

Substrate (composition, topography, stiffness, temperature)

FIGURE 6.1   The cellular microenvironment.

reasons: they are universally based on seeding the cells *on a static homogeneous substrate* and bathing them *in a static homogeneous bath*. This static homogeneity of the soluble and insoluble factors surrounding the cells is clearly a poor analogue of the dynamic signaling heterogeneity observed in vivo. Biologists have always been keenly aware of it, of course, and have indeed jumped at the opportunity to use probes such as micropipettes (e.g., iontophoresis), focused lasers (e.g., uncaging), and coated/soaked beads to locally stimulate cells at desired time points.

6.1.1 From Random Cultures to Microengineered Substrates

In traditional cell culture, cells are randomly seeded over the whole substrate; thus, cellular interactions that are strongly dependent on the proximity, distribution, and relative position of cells or biomolecules can become confounded by the presence of other, similar interactions. The advent of cellular micropatterning methods has offered the possibility of tailoring the cellular and biochemical neighborhood of cells with resolution down to single cells, in some cases, even with electrical addressability (see Sections 2.5 and 2.6).

Unfortunately, the implementation of cellular micropatterning approaches by cell biologists has faced, until recently, an important hurdle: most cell biologists do not have the required microfabrication expertise; such a technological gap has become narrower in recent years with the advent of straightforward, inexpensive micropatterning techniques (especially soft lithography) and with the increasing population of scientists with cross-training in both biology and engineering.

6.1.2 From "Classical" to "Novel" Substrates: The Cell Biologist's Dilemma

Micropatterning cells (and often also combining cells with microfluidic devices) naturally requires that the cell culture areas be contacted by chemical or physical obstacles that block the adhesion of cells or proteins on selected areas, so in the contact process undesired deposition of other materials may occur. For example, deposition of PDMS monomers may occur during microstamping of proteins or microfluidic patterning of cells, and methods based on photolithography processes may result in photoresist remains on the cell culture areas; these residues may be difficult to detect. This is a concern from a cell biologist's perspective because cells are exquisitely sensitive to submonolayer coverages of adsorbates (either directly or by affecting the adsorption of key proteins) and, as a result, contaminants may produce confounding results.

Importantly, this cell biologist also faces a second, often-overlooked subtle dilemma: typically, the cell biologist considering the implementation of a cellular micropatterning approach chooses to do so based on a previous line of research that, until then, used a randomly organized cell culture on homogeneously adhesive substrate. The micropatterned cell cultures thus represent the "next experiment" after the random cultures, and the random cultures serve as the "control experiments." Ideally, the control (random) cultures should use the same exact type of substrate and seeding/culture protocols as used previously for random cultures in the past—otherwise, only the cellular functions measured in that particular experiment are "controlled," and the available body of knowledge obtained on other cellular functions may not apply on the "novel" substrates. We stress that this preference is not merely for convenience; rather, given the high sensitivity of cells to minute changes in surface composition, changing the protocols and substrate may yield artifactual results in micropatterning experiments or affect future experiments probing different cellular functions.

This need for the random-culture experiment to be an appropriate control for the micropatterned-culture experiment places stringent constraints on the choices of micropatterning techniques that are appealing to the cell biologist. The substrates universally used for cell culture in molecular and cell biology research are either glass or polystyrene—either coated with proteins or bare. On the other hand, cell micropatterning approaches largely use one of two strategies to deposit cells on designated areas of the cell culture substrates: (1) selective cell attachment is guided by differential adhesiveness of the substrate (a very simple and widely used method to deter cell attachment consists of adsorbing albumin, a protein that lacks cell adhesion motifs, on the cell culture substrate); or (2) cell attachment to a homogenously adhesive substrate is blocked in selected areas with a removable physical barrier. For an extensive coverage of these methods, see Sections 2.5 and 2.6.

6.1.3 From Cells in Large Static Volumes to Cells in Small Flowing Volumes

As pointed out previously, in high-density microfluidic cell cultures, the microenvironment's nutrients, pH, and gas concentrations cannot be kept constant for long periods of time, and replenishing the volume of the microfluidic chamber becomes necessary. An extreme deviation of a "proper" cell culture environment results in obvious cell death. However, the effects can be much subtler because (a) cells are exposed to shear stress from the flow, and (b) at the cell membrane surface, the actual concentration of growth factors (that either bind to cells or are secreted by them) "seen" by the cell depends on the speed of the flow, which effectively "washes away" the growth factor molecules that happen to diffuse into the stream. Some researchers have argued that cells that normally do not grow well in culture, such as neurons and stem cells, should grow better in small microfluidic chambers where endogenous growth factors rapidly accumulate.

Given that flow/shear stress may produce unanticipated deleterious effects on the cells, several groups have been implementing "**open microfluidics**" systems whereby the cells are seeded on open pools or reservoirs, and the fluids are supplied to the cells via side channels (e.g., "microjets," see Section 3.9.3, or diffusion ports). These systems are inherently more benign to the cells because at least in their no-operation mode they are guaranteed to produce the same results as the control surface. (On the other hand, a closed-microchannel system in its no-operation mode is guaranteed … to kill the cells!).

6.1.4 From a Homogeneous Bath to Microfluidic Delivery

In traditional cell culture, cells are bathed in a homogeneous medium; thus, any cellular response that relies on graded or focal exposure of the cell to a given factor will not be observable, including many cell growth and motility phenomena. Traditionally, focal stimulation of cells with fluids has been possible only using micropipettes—which eject fluid on application of a pressure pulse ("puffing") or voltage pulse ("iontophoresis")—or using chemically-caged compounds that are "uncaged" (i.e., released) by a laser pulse. Clearly, these techniques result in very low throughputs and poorly characterized volumes or gradients, they require bulky equipment and substantial manual skill, and are unscalable (stimulation at more than three or four sites is not practical, at least with pipettes). Microfluidic systems, on the other hand, constitute a technology for directing many different fluids to a large number of (small) cell populations that is scalable, amenable to fluid dynamics modeling, and where the delivery system is prealigned with the cells. In 1999, a collaborative team led by Don Ingber at Children's Hospital (Boston) and George Whitesides at Harvard University first demonstrated the use of multiple laminar

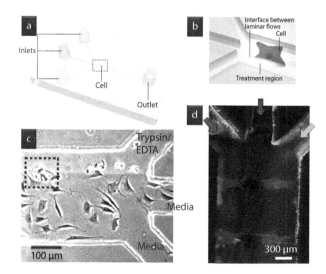

FIGURE 6.2 Partial cell stimulation with laminar flow streams. (a) Schematic drawing of setup for partial cell stimulation, with three inlets combining to form parallel streams in the main channel. (b) Close-up view of cell placed at the interface between laminar streams. (c) Cells partially stimulated with trypsin detach only partially. (From Shuichi Takayama, J. Cooper McDonald, Emanuele Ostuni, Michael N. Liang, Paul J. A. Kenis, Rustem F. Ismagilov, and George M. Whitesides, "Patterning cells and their environments using multiple laminar fluid flows in capillary networks," *Proc. Natl. Acad. Sci. U. S. A.* 96, 5545–5548, 1999. Figure contributed by Shuichi Takayama.) (d) Myotubes (muscle cells fused from one microchannel wall to another in the course of ~1 week) in laminar streams of Orange, Blue and Green CellTracker are differentially stained for several hours despite their cytoplasmic continuity. (Anna Tourovskaia, Xavier Figueroa-Masot, and Albert Folch, "Long-term microfluidic cultures of myotube microarrays for high-throughput focal stimulation," *Nat. Protoc.* 1, 1092, 2006. Figure contributed by Anna Tourovskaia.)

flows to stimulate and address various cell populations at once and also to partially treat a cell (**Figure 6.2a** through **c**). Many groups are now routinely able to culture cells, and even cellular micropatterns, for long periods of time (~2–3 weeks) in microfluidic systems (**Figure 6.2d**).

6.2 Cell-Substrate Signaling

Cell biologists have been aware for a long time that cell anchorage to its substrate plays a key role in its behavior. The first cell culture studies (the growth of frog nerve cells by Ross G. Harrison in 1907 at Johns Hopkins University) already addressed this issue by seeding cells on clotted lymph. Later, others experimented with a variety of materials, ranging from glass, ceramics, and plastics. The picture that emerged—even before it was concluded that (most) cells require ECM for anchorage—was that cells are critically sensitive to both the composition of the substrate and its microscale topography.

6.2.1 Cell Behavior Controlled by Cell Shape

The idea that cell shape influences cell behavior is as old as the field of BioMEMS: it is the reason why Carter started using microfabricated nickel stencils for his investigations of cell proliferation back in 1967 (see Section 2.5.1 and **Figure 2.24**). Unfortunately, Carter's reports are not very quantitative, but we do know that they inspired two excellent articles by a team led by Grenham Ireland at the Imperial Cancer Research Fund in London (first, in 1986) and at the University of Manchester (later, in 1990) 20 years after Carter, when immunofluorescence staining techniques for molecular imaging of cells were already in full swing. This group used microfabricated 5-μm-thick copper stencil mask containing feature linewidths of approximately 20 μm to deposit cell-adhesive Pd islands onto substrates previously coated with cell-repellent poly-2-hydroxyethyl methacrylate (poly-HEMA). The shape of the cells was confined to the shape of the islands (Figure 6.3a and b). In their first article, they demonstrated that cell proliferation was a strong function of island size (Figure 6.3c). For Nil-8 fibroblasts, the DNA synthesis (a more accurate measurement of cell proliferation) reached a cell proliferation minimum below the "zero contact" line (reference point obtained from suspended cells) for small islands and increased for larger islands (Figure 6.3d), a finding that was very similar on mouse embryo cells. However, 3T3 fibroblasts did not display a DNA synthesis minimum (Figure 6.3e), which suggested for the first time that this cell type had a major requirement for contact.

In a follow-up article, in 1990, Ireland and colleagues produced micropatterns of various shapes (circles, triangles, or lines). Using actin and vinculin immunofluorescence, they showed that cytoskeletal organization in 3T3 fibroblasts depends on island shape. In particular, the focal points tended to accumulate to the peripheries of circles, the apices of triangles, and the margins of lines (most obvious when the island reached ~1000 μm² in size), whereas the total focal adhesion area was independent of the shape of the cell. Cells tended to favor the formation of small focal points, independently of island size—with the caveat that, given enough space, on large islands, they would also form larger focal points; a good analogy, if we consider that the focal points are the "feet of cells," is that cells constrained to small islands are forced to tiptoe whereas cells free to spread on large islands have space to both tiptoe and stomp on their feet. Cell motility speed increased linearly with island size on linear islands. DNA synthesis (measured

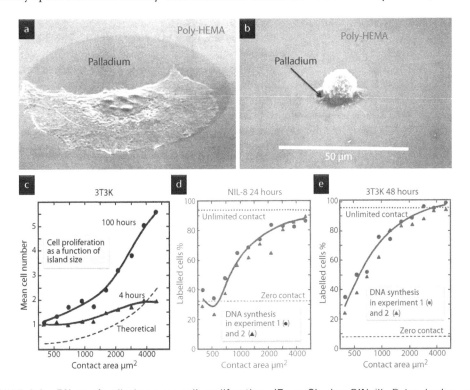

FIGURE 6.3 Effect of cell shape on cell proliferation. (From Charles O'Neill, Peter Jordan, and Grenham Ireland, "Evidence for two distinct mechanisms of anchorage stimulation in freshly explanted and 3T3 Swiss mouse fibroblasts," *Cell* 44, 489–496, 1986. Reprinted with permission from Elsevier.)

by thymidine incorporation for 2 days) was independent of island shape but increased with island area. This remarkable discovery (DNA synthesis occurs in the nucleus and the connection between cell shape and its effect on the nucleus could not be obvious at the time) was not emphasized particularly—how was the cell shape being "explained" to the nucleus? Proliferation rates (as measured by the number of cells in each island) were correlated with the number of focal adhesions that the cells made with the underlying ECM (as quantified with vinculin staining). Taken together, these experiments showed that islands of widely different shapes can produce similar anchorage stimulation and that the resulting proliferative stimulus is a function of the total focal adhesion area.

Several years later, in 1997, a team led by George Whitesides at Harvard University and Donald Ingber at Children's Hospital (Boston) confirmed the results of Ireland's group in two entirely different cell systems (hepatocytes and human and bovine endothelial cells) and a more sophisticated substrate chemistry (microstamped self-assembled monolayer (SAM) patterns, see **Figure 2.27**), which speaks for the general validity of the observed principle that gene expression is modulated by cell shape. This group also observed that, in general, albumin secretion rates decreased as the size of the island increased and DNA synthesis increased (in agreement with Ireland's results).

The elegance of these experiments resides in the fact that, unlike previous work that modulated cell shape and function by varying ECM density on the substrate or by limiting the area available for attachment on microspheres, cell spreading could be varied independently of ECM density and of total integrin–ECM contact area. Previously, it had been shown that endothelial cells attached to small spheres underwent apoptosis (owing to their small size, the small spheres did not induce spreading), whereas cells attached to larger spheres (>100 μm diameter) survived. However, microsphere cultures do not allow for discerning whether the critical parameter that switches cells between apoptosis and survival is the projected spread cell area or the total area of focal adhesions. To address this question, the Ingber-Whitesides team cultured human and bovine endothelial cells on micrometric fibronectin-coated islands surrounded by HEG-terminated alkanethiol areas (**Figure 6.4**). When cells were cultured on arrays of closely spaced 3- or 5-μm-diameter islands, they spread over several islands across the nonadhesive areas. Stress fibers were seen to anchor at the periphery of islands where focal adhesions formed and stretched from island to island across PEG regions as if they mapped out "tension field lines" within the cytoskeleton. Vinculin staining showed that cells formed focal adhesions only on the fibronectin-coated islands. DNA synthesis increased with increasing total projected area at

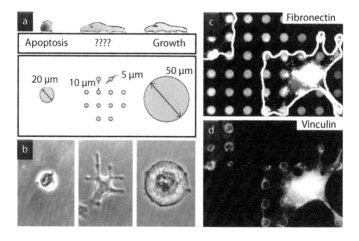

FIGURE 6.4 Subcellular control of integrin–ECM interactions. (From Christopher S. Chen, Milan Mrksich, Sui Huang, George M. Whitesides, and Donald E. Ingber, "Geometric control of cell life and death," *Science* 276, 1425, 1997. Figure contributed by George Whitesides.)

nearly constant integrin–ECM focal adhesion area. This finding indicates that cell shape, and not cell–ECM contact area, determines cell fate.

Using immunofluorescence microscopy, a team led by Martin Bastmeyer, then at the University of Konstanz (Germany), performed in 2004 a detailed molecular study of the focal adhesions that cells establish with the ECM during cell adhesion, spreading, and migration. Following the Whitesides–Ingber protocol (see Figure 2.27), this group used microstamping of alkanethiols on gold followed by PEG-thiol SAM formation on the background to create ECM-adhesive regions on the stamped areas; ECM only adsorbed on the alkanethiol-stamped areas, so cells attached on these ECM islands and they were unable to form contacts with the PEG-thiol background (Figure 6.5). Cells cultured on these substrata adhere to and spread on ECM regions as small as 0.1 μm², when spacing between dots was less than 5 μm. Spacing of 5 to 25 μm induces a cell to adapt its shape to the ECM pattern. The ability to spread and migrate on 1 μm² dots ceases when the dot separation is 30 μm. The extent of cell spreading is directly correlated to the total substratum coverage with ECM proteins, but irrespective of the geometrical pattern. An optimal spreading extent is reached at a surface coating of more than 15%. On homogeneous substrata, staining for molecules such as paxillin or vinculin are found in dot-like adhesions at the periphery of the cell, with most vinculin-positive foci connected to actin bundles (Figure 6.5a). When cells were forced to make contacts on 0.6 μm² fibronectin dots, they expressed vinculin normally (only on the fibronectin dots) and actin fibers terminated in the peripheral adhesion sites (Figure 6.5b). On a patterned substrate of 1 μm² fibronectin dots, cells stained focally for focal adhesion kinase (Figure 6.5c) and for phosphotyrosine (Figure 6.5d). On patterned substrata of 0.6 μm² vitronectin dots, cells expressing β3-integrin-GFP stained for paxillin (Figure 6.5e) and for actin (Figure 6.5f). The cell in Figure 6.5f is growing at the border between a uniform and a patterned substratum: note the redistribution of integrin receptors on the patterned substratum.

Because cell shape influences gene expression (protein secretion and cell proliferation) through the cell's cytoskeleton, it is not surprising that the positioning of the cell division axis itself (the spindle orientation) can be experimentally manipulated by micropatterning of the

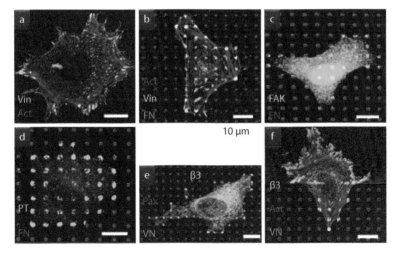

FIGURE 6.5 Molecular composition of focal adhesions on micropatterned substrates. Abbreviations: Vin, vinculin; Act, actin; FN, fibronectin; FAK, focal adhesion kinase; PT, phosphotyrosine; Pax, paxillin; β3, β3-integrin-GFP. (From Dirk Lehnert, Bernhard Wehrle-Haller, Christian David, Ulrich Weiland, Christoph Ballestrem, Beat A. Imhof, and Martin Bastmeyer, "Cell behaviour on micropatterned substrata: Limits of extracellular matrix geometry for spreading and adhesion," *J. Cell Sci.* 117, 41–52, 2004. Figure contributed by Martin Bastmeyer.)

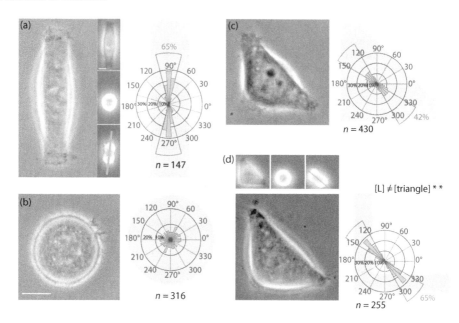

FIGURE 6.6 Micropatterned cell division. Scale bars represent 10 μm. The distributions of spindle orientation were significantly different on triangles and on "L." (From Manuel Théry, Victor Racine, Anne Pépin, Matthieu Piel, Yong Chen, Jean-Baptiste Sibarita, and Michel Bornens, "The extracellular matrix guides the orientation of the cell division axis," *Nat. Cell Biol.* 7, 947–953, 2005. Copyright (2005) Nature Publishing Group.)

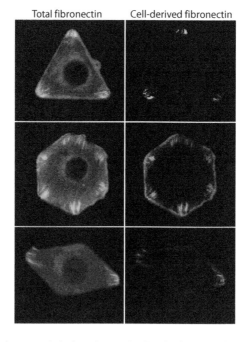

FIGURE 6.7 Deposition of new protein in polygonal adhesive islands. (From Amy Brock, Eric Chang, Chia-Chi Ho, Philip LeDuc, Xingyu Jiang, George M. Whitesides, and Donald E. Ingber, "Geometric determinants of directional cell motility revealed using microcontact printing," *Langmuir* 19, 1611–1617, 2003. Figure contributed by Don Ingber.)

ECM. Michel Bronens' group from the Institut Curie in Paris, France, microstamped islands of fluorescently tagged fibronectin onto a PEGylated-glass background and observed how HeLa and L929 cell lines divided when constrained to the various island shapes (rectangle, circle, triangle, or "L," as shown in Figure 6.6). Based on the analysis of the average distributions of actin-binding proteins during interphase and mitosis, it seems that the ECM controls the location of actin dynamics at the membrane, so it segregates the cortical components during interphase. This segregation is also maintained on the cortex of mitotic cells and used for the orientation of the spindle during cell division.

It is important to note that the cells are continuously remodeling (and interacting with) their underlying ECM substrate, disassembling the old contacts and forming new ones—especially during cell migration. In a cell culture experiment, the ECM coverage before the cells are seeded can be found by immunofluorescence and digitally subtracted from the total ECM coverage after the cells have remodeled the substrate to reveal in what parts of the cell most of the remodeling occurs. This experiment was done by a team led by Don Ingber (Harvard Medical School), who found that remodeling occurs mostly at the cell periphery, concentrating at the vertices of cytoskeletal high-stress points (Figure 6.7).

George Whitesides' group was able to "bias" the direction in which cells migrate on a cell culture substrate by micropatterning their shape (Figure 6.8). The cells' shapes were confined to

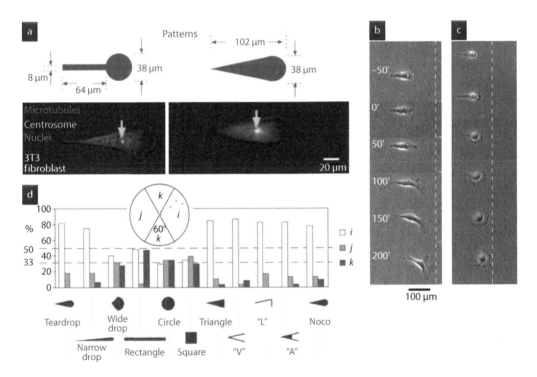

FIGURE 6.8 Cell migration induced by substrate asymmetry. In panel (a), note how the centrosomes (green arrows) are located closer to the blunt end's half of the 3T3 cell. Voltage pulses were applied at $t = 0$. In panel (d), the sample labeled as "noco" indicates that the cells were initially confined to a teardrop shape, then treated with nocodazole to disrupt their microtubules, and finally electrochemically released and the medium immediately replaced to allow for microtubule reassembly. (From Xingyu Jiang, Derek A. Bruzewicz, Amy P. Wong, Matthieu Piel, and George M. Whitesides, "Directing cell migration with asymmetric micropatterns," *Proc. Natl. Acad. Sci. U. S. A.* 102, 975–978, 2005. Copyright (2005) National Academy of Sciences, U. S. A. Figure contributed by George Whitesides.)

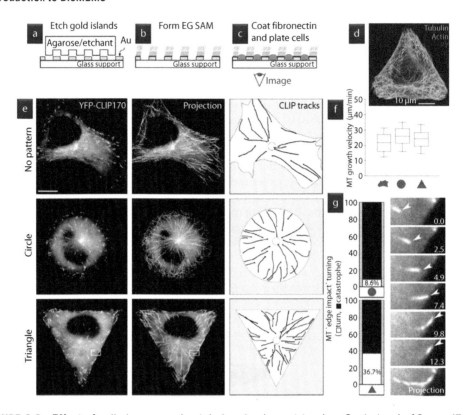

FIGURE 6.9 Effect of cell shape on microtubule edge impact turning. Scale bar is 10 µm. (From Kristiana Kandere-Grzybowska, Christopher Campbell, Yulia Komarova, Bartosz A. Grzybowski, and Gary G. Borisy, "Molecular dynamics imaging in micropatterned living cells," *Nat. Methods* 2, 739–741, 2005. Adapted with permission from the Nature Publishing Group.)

asymmetric "teardrop" or similar shapes using PEG-thiol surface chemistry (see Section 2.6.1.1 and **Figure 2.27**). The shape constraint was then released by electrochemical desorption of the thiol SAMs. The cells (whether 3T3 fibroblasts, endothelial cells, or COS-7 cells) always moved toward their blunt ends. This experiment demonstrates that morphological polarity itself can determine the direction of motility in the absence of chemoattractant gradients.

A team led by Gary Borisy at Northwestern University in Chicago has used molecular dynamics imaging and total internal reflection to visualize the motion of microtubules within live micropatterned cells in real time. This team used a patterning technique developed by Bartosz Grzybowski's group based on selectively etching gold with an etchant-soaked agarose stamp (**Figure 6.9a** through **d**; see also **Figure 1.35** in Section 1.7). Microtubule dynamics, in particular the "edge impact turning" (a measure of the change in microtubules growth direction when they hit the edge of the cell), was observed to depend on cell shape, because microtubules that reach the straight edges of triangular cells turn more frequently than those encountering convex edges of circular cells (**Figure 6.9e** and **g**). Restriction of cells to the islands did not inhibit microtubule growth velocity (**Figure 6.9f**).

6.2.2 Microtopographical Signaling

In the late 1980s, Donald Brunette at the University of British Columbia in Vancouver (Canada) pioneered the study of cell behavior on substrates containing micromachined grooves. He knew of previous studies that had observed that cells oriented along scratches and wanted to produce finer, more controlled "scratches." He coated micromachined silicon substrates with titanium and seeded them with fibroblasts as a model for wound healing over dental implants (he was in the Department of Oral Biology). He also used titanium-coated epoxy replicas of silicon templates as cell culture substrates. He showed that the grooves enhanced the elongation and the migration of the cells in the direction of the grooves in a manner that depended on the groove pitch and depth—sometimes, as seen in Figure 6.10a (depicting Brunette's more recent work), extending lamellipodia over several ridges ("r") and grooves ("g"). The cytoskeleton plays an important role in the overall mechanism of how cells align to the microtopographies because the addition of colcemid, a microtubule-depolymerizing drug, made the cells unable to obey the microtopography (Figure 6.10b and c). Brunette's laboratory has followed up this work producing dental implants that feature microfabricated topographies that outperform the traditional dental implants in terms of stimulating connective tissue and bone attachment and preventing epithelial migration (see Section 8.1).

The way substrate microtopography produces contact guidance deserves further attention. The elongation of the cell itself changes patterns of gene expression (probably by exerting direct forces on the nucleus, as its shape is elongated with the shape of the cell). Hence, it follows logically that the microtopography of the substrate should affect other aspects of cell behavior, not just contact guidance. In 1994, Andreas von Recum's group, then at Clemson University, observed that fibroblasts on 2-μm- and 5-μm-wide PDMS grooves proliferated faster than on smooth PDMS. In the late 1990s, Christopher Wilkinson, Adam Curtis and coworkers at Glasgow University (Scotland) showed that microfabricated grooves direct the migration of macrophages, an observation that was later reproduced on neutrophils by Mark Saltzman's group at Cornell University. More recently, Mogens Duch and collaborators from Aarhus University in Denmark have discovered that substrate microtopography influences the differentiation of stem cells in the absence of a layer of feeder cells (which are usually required to keep the stem cells alive). Although the mechanism is presently not clear, the implication is: it is possible to empirically screen for a variety of microtopographies (Figure 6.11a) and check which one produces the largest differentiation signal; in this case, mineralization (Figure 6.11b and c). The vision for this "**TopoChip**" is that soon petri dishes for stem cell culture will be sold with an engraved microtopography that will induce a certain differentiation depending on what pattern is engraved.

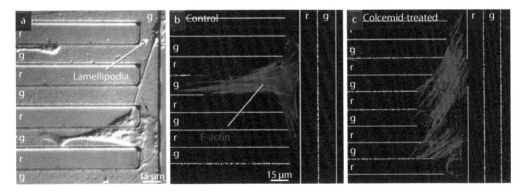

FIGURE 6.10 Topographical guidance of fibroblasts by microscale grooves. (From Douglas W. Hamilton, Carol Oakley, Nicolas A. F. Jaeger, and Donald M. Brunette, "Directional change produced by perpendicularly-oriented microgrooves is microtubule-dependent for fibroblasts and epithelium," *Cell Motil. Cytoskeleton* 66, 260–271, 2009. Adapted with permission from John Wiley and Sons.)

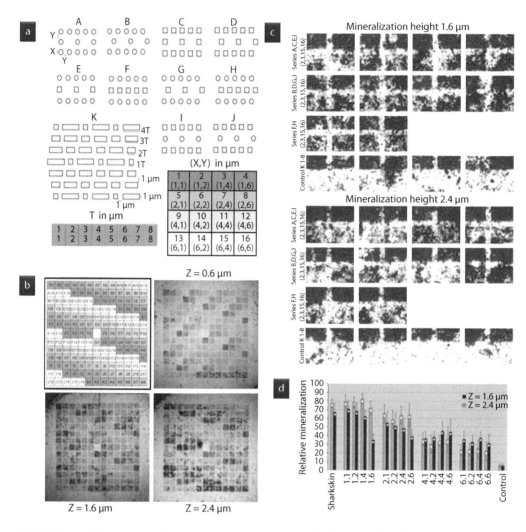

FIGURE 6.11 Microtopographical libraries to screen for cell differentiation. (From Jette Lovmand, Jeannette Justesen, Morten Foss, Rune Hoff Lauridsen, Michael Lovmand, Charlotte Modin, Flemming Besenbacher, Finn Skou Pedersen, and Mogens Duch, "The use of combinatorial topographical libraries for the screening of enhanced osteogenic expression and mineralization," *Biomaterials* 30, 2015–2022, 2009. Reprinted with permission from Elsevier.)

6.2.3 Muscle Cell Differentiation

Muscle cell cultures provide several advantageous features as cell biology testbeds: (a) their gene expression profiles are relatively robust and similar to those seen in vivo; (b) cell lines are available for differentiation studies from cellular to tissue-level functions—the cells can be induced to fuse (skeletal muscle cells) or form junctions (cardiac muscle cells); and (c) their electrophysiological addressability and mechanical output is invaluable for fundamental studies of the neuromuscular junction (NMJ), muscle waste, neuroprosthetic devices, and microactuators. Utilizing the full power of muscle cells in culture, however, requires a good control over muscle cell differentiation. Microengineering can be a powerful tool to achieve degrees of differentiation that cannot be achieved in a plain petri dish.

In 2004, Dennis Discher's group from the University of Pennsylvania tested the hypothesis that muscle cells sense both their molecular and mechanical microenvironment by culturing them on collagen strips attached to glass or polymer gels of varied elasticity (**Figure 6.12**). Although fusion into myotubes occurred independently of substrate flexibility, myosin/actin striations formed only on gels with stiffness typical of normal muscle (Young's modulus, E ~12 kPa). Cells did not form striations on glass and much softer or stiffer gels. Also, myotubes grown on top of a compliant bottom layer of glass-attached myotubes (but not softer fibroblasts) formed striations (**Figure 6.12a** and **b**), whereas the bottom cells only assembled stress fibers and vinculin-rich adhesions (**Figure 6.12c**). As time progressed, striation of the top (but not the bottom) myotubes increased (**Figure 6.12d** and **e**). Unlike sarcomere formation, adhesion strength increased monotonically versus substrate stiffness with strongest adhesion on glass.

To mimic neuromuscular synaptogenesis processes in vitro, the author's laboratory at the University of Washington in Seattle devised a microfluidic platform capable of sustaining long-term perfusion and focal stimulation of an array of single C2C12 myotubes (**Figure 6.13**). The device is assembled atop a pattern of lines of dried Matrigel separated by cell-repellent PEG-IPN (**Figure 6.13a**; see also **Figure 2.29** in Section 2.6.1.2). C2C12 myoblasts are then injected into the device, allowed to attach, and the nonattached cells are removed, leaving myoblast line patterns within the microfluidic device (**Figure 6.13b**). The cells are perfused in the direction orthogonal to the main channel (parallel to the lines) for approximately 1 week in low-serum conditions (so they do not divide) until they fuse into myotubes. At this point, clusters of acetylcholine receptors (AChRs) form spontaneously on the myotube surface and the myotubes are

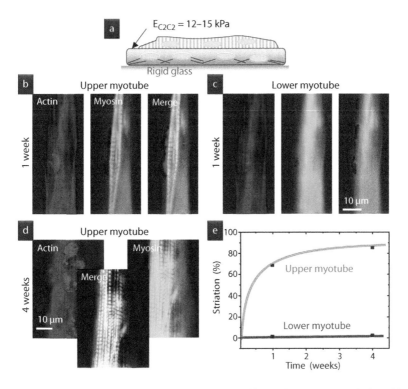

FIGURE 6.12 Micropatterned myotubes provide optimal stiffness for myotube striation. (From Adam J. Engler, Maureen A. Griffin, Shamik Sen, Carsten G. Bönnemann, H. Lee Sweeney, and Dennis E. Discher, "Myotubes differentiate optimally on substrates with tissue-like stiffness: Pathological implications for soft or stiff microenvironments," *J. Cell Biol.* 166, 877–887, 2004. Figure contributed by Dennis Discher.)

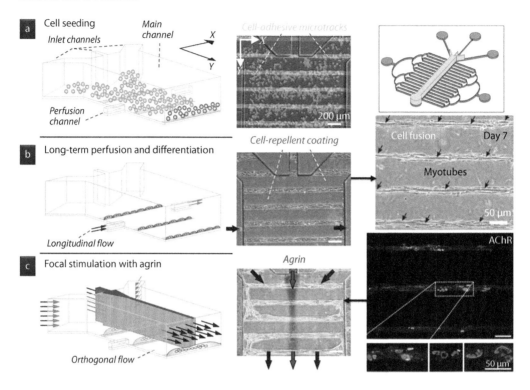

FIGURE 6.13 A microfluidic device to study muscle cell differentiation and neuromuscular synaptogenesis. (From Tourovskaia, A., Figueroa-Masot, X. and Folch, A., "Long-term microfluidic cultures of myotube microarrays for high-throughput focal stimulation," *Nat. Protoc.* 1, 1092, 2006. Figure contributed by Anna Tourovskaia.)

ready to be focally stimulated with agrin or other synaptogenic compounds (**Figure 6.13c**). The formation of these AChR clusters of convoluted morphologies—nicknamed "pretzels"—(similar to the "pretzels" found in vivo) is indicative of an advanced stage of differentiation.

6.3 Cell–Cell Communication

Cells send signals to other cells to command, alert, or coordinate a variety of cellular behaviors. Often, these signals are sent by direct membrane contact, such as through receptor activation or the formation of synaptic contacts or gap junctions, so there is great interest in methods that allow for placing pairs of single cells in direct contact, methods that modulate the physical distance between two cell populations, and methods that simulate the presence of another cell through a biomimetic microfluidic device or surface micropattern.

6.3.1 Control of Cell–Cell Contacts at Single-Cell Scale

Chris Chen's laboratory, then at Johns Hopkins University, used in 2002 surface "bow tie" micropatterns to control the amount of contact between two adhered endothelial cells, which in turn influences their cell division cycle (**Figure 6.14**). The substrates were prepared by microfluidic patterning of agarose (as a cell repellent, ~10 µm thick) on glass physisorbed with fibronectin, to which cells adhered normally (**Figure 6.14a**). The area available for cells to attach was restricted to a bow tie shape such that, if a pair of cells landed by chance on either side of the bow tie micropattern, they were able to communicate only through the spatial restriction; the

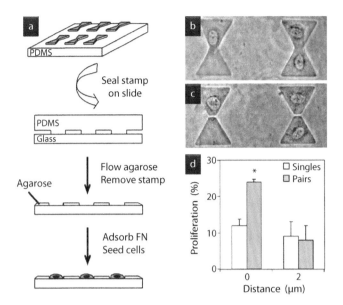

FIGURE 6.14 Cellular micropattern to engineer cell–cell contacts. (From Celeste M. Nelson and Christopher S. Chen, "Cell-cell signaling by direct contact increases cell proliferation via a PI3K-dependent signal," *FEBS Lett.* 514, 238–242, 2002. Adapted with permission from Elsevier. Figure contributed by Chris Chen.)

spatial restriction contained either no gap, a 2-μm gap, or a 5-μm gap (the gaps were filled with agarose, so cells would not spread over it). Intercellular adhesion molecules expressed in these cells, including connexin 43, N-cadherin, VE-cadherin, and β-catenin, were found to localize to cell–cell contacts within 4 hours after plating onto the bow tie micropatterns. Proliferation for pairs of cells on bow tie patterns without a gap was almost triple as proliferation for pairs of cells (or single cells) on bow tie patterns with a 2-μm gap (**Figure 6.14d**).

Note that surface micropatterns are limited to adherent cells. A direct cell trapping method is required to control the contact between nonadherent cells. A variety of other methods are available for trapping pairs of cells using microwells (at random, thus many wells have single occupancies or are occupied by two of the same cell type, see Section 5.3.2) and using hydrodynamic weirs or microtunnels, which trap exactly two cells (see Section 5.3.6).

6.3.2 Control of Cell–Cell Spacing Using Micromechanical Actuators

In cellular micropatterns, the cell–cell spacing is fixed by design and cannot be changed once the cells are seeded, which limits the range of questions that can be studied with patterning technologies. During development, wound healing, or tumor growth, for example, the cells rely on cell–cell communication and larger-scale tissue-level processes in which not only space but also *time* is of the essence. To control the time parameter in cellular micropatterning experiments, Sangeeta Bhatia's group, then at the University of California (San Diego), microfabricated interlocking combs whose fingers fit into each other (**Figure 6.15**). The combs, etched in silicon, were spin-coated with polystyrene, resulting in a surface comparable to tissue culture plastic. The slope of the tapered comb fingers results in a 20:1 mechanical transmission ratio; that is, sliding the parts 1.6 mm changes the gap between the fingers by only 80 μm. The combs can be manually "clicked" into a position in which the gap between the fingers is at one of several notched spacings (~80–400 μm). By seeding different cell types on the two different honeycombs, it is possible to study how the two cell types interact across the gap (which can be changed) via soluble factors at any time set by the user. The positioning resolution is extremely

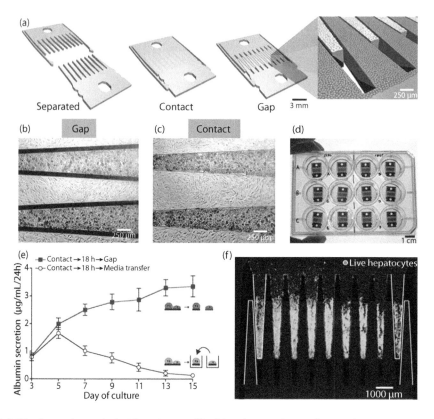

FIGURE 6.15 Dynamic control of hepatocyte–fibroblast interactions using a micromechanical actuator. (From Elliot E. Hui and Sangeeta N. Bhatia, "Micromechanical control of cell-cell interactions," *Proc. Natl. Acad. Sci. U.S.A.* 104, 5722–5726, 2007. Copyright (2007) National Academy of Sciences, USA. Figure contributed by Sangeeta Bhatia.)

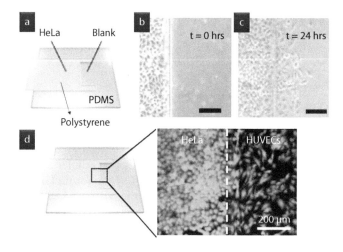

FIGURE 6.16 Control of cell migration using detachable substrates. (From Hirokazu Kaji, Takeshi Yokoi, Takeaki Kawashima, and Matsuhiko Nishizawa, "Controlled cocultures of HeLa cells and human umbilical vein endothelial cells on detachable substrates," *Lab Chip* 9, 427–432, 2009. Reproduced with permission from The Royal Society of Chemistry.)

reliable despite the fact that positioning is done manually, without the help of micropositioners or microscopy equipment. This work is important not only because it achieves dynamic control of cell–cell interactions (which could have been achieved with other technologies, in principle, such as traditional micropositioners) but also because (through clever use of microfabrication) it provides a user-friendly platform.

Unfortunately, Bhatia's interlocking combs are expensive to produce and they are opaque, so they are not ideal for cell imaging. Matsuhiko Nishizawa and coworkers at Tohoku University in Japan have produced a simple version of the combs in polystyrene (**Figure 6.16**). These blocks are then shape-fitted into matching PDMS molds in which cells have been preseeded. Cell migration from one substrate onto another can be straightforwardly observed.

6.3.3 Quorum Sensing in Bacteria

Bacteria, protozoa, and, in general, all microbes, are essential for the existence of all other life forms on Earth. It is thus important to perform studies of microbial responses as they compete in microhabitats for gradients of nutrients and other natural substances. **Quorum sensing** is used by several species of bacteria (e.g., *Vibrio fishcheri*, *Salmonella enterica*, *Pseuromonas aeruginosa*, and *Eschericia coli*) to coordinate gene expression according to the local population density. A team led by Bob Austin at Princeton University (New Jersey) has studied the behavior of GFP-expressing *E. coli* confined to microfabricated mazes (**Figure 6.17**). As shown in **Figure**

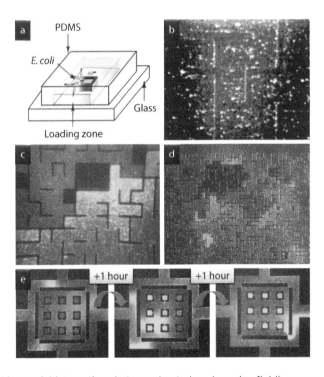

FIGURE 6.17 Probing social interactions between bacteria using microfluidic mazes. Because the bacteria express GFP, fluorescence intensity is a measure of cell density. The images in (b), (c), and (d) have been obtained at 600×, 400×, and 40× magnification, respectively. Note that there is no clustering at the open region at the top of (d), where there is no maze. The small squares in (e) are 250-µm-wide and have a 30-µm-wide opening. (From Sungsu Park, Peter M. Wolanin, Emil A. Yuzbashyan, Hai Lin, Nicholas C. Darnton, Jeffry B. Stock, Pascal Silberzan, and Robert Austin, "Influence of topology on bacterial social interaction," *Proc. Natl. Acad. Sci. U. S. A.* 100, 13910–13915, 2003. Copyright (2003) National Academy of Sciences, U. S. A. Figure contributed by Bob Austin.)

6.17e, the cells dynamically accumulate after approximately 2 hours in the "dead-end" parts of the maze. These results show that under nutrient-deprived conditions, *E. coli* search out each other in a collective manner, using the environmental topology of complex structures to create traveling waves of high cell density, the first step in quorum sensing. Through periods of stress, directed probing of confined structures could be vital for bacterial survival. Research on quorum sensing is important for producing anti-quorum sensing antibiotics. All current antibiotics aim to kill bacteria by inhibiting the synthesis of new bacteria (so natural selection rapidly selects for resistant strains), whereas anti-quorum sensing antibiotics would disrupt the bacteria's ability to communicate (and thus diminishing its ability to become pathogenic). Because anti-quorum sensing antibiotics do not kill the bacteria, there is no survival advantage for the resistant mutations (hence these are unlikely to occur).

6.3.4 Signal Transduction Studies Using Biomimetic Devices

In most cell biology experiments in vitro (i.e., based on cell cultures or explants), the response of cells to various ligands is studied by bathing the cells in a homogeneous ligand solution and seeding them on a homogeneous substrate. In addition to being a poor analogue of the physiological scenario in the organism (in which cells are stimulated locally) homogeneous stimulations can produce confounding information on how the signal is transduced from the membrane to the cytoplasm. By stimulating the cell in only part of its membrane, some modes of transmission of the signal can be revealed. (Particular types of cell–cell local signaling found in neural systems, such as those involved in synapse formation, synapse transmission, and neuron–glia interactions, are discussed later.)

As an example, here we take the case of how **epidermal growth factor** (EGF) signaling spreads within a cell after activation of its receptor (EGFR). In general, activation of the EGFRs by EGF is signaled to the nucleus through the Ras-Raf-MEK-ERK pathway, but it also intervenes in a variety of phenomena such as membrane trafficking, cell adhesion, and cytoskeletal organization. Does EGF signaling remain localized or is it propagated in response to local stimulation? Researchers at RIKEN in Japan used a simple, two-inlet and one-outlet microfluidic device (**Figure 6.18a**) to study how single COS cells respond to local EGF stimulation (using fluorescent indicators for Ras activation and tyrosine phosphorylation). One inlet carried plain cell culture medium and the other one carried medium containing fluorescent EGF, producing a heterogeneous laminar flow as the two inlets merged into one channel (**Figure 6.18b**). Stimulation was localized by positioning the EGF interface (visible in fluorescence) halfway onto a single COS cell (**Figure 6.18c**). In normal cells, both Ras activation and tyrosine phosphorylation signals were localized to the portion of the cell stimulated by EGF (**Figure 6.18e**)—a result that is

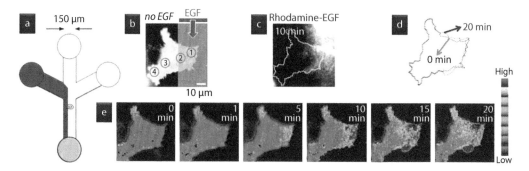

FIGURE 6.18 Localized activation of EGFR-dependent pathways using laminar flow streams. (From Asako Sawano, Shuichi Takayama, Michiyuki Matsuda, and Atsushi Miyawaki, "Lateral propagation of EGF signaling after local stimulation is dependent on receptor density," *Dev. Cell* 3, 245–257, 2002. Adapted with permission from Elsevier.)

consistent with the notion that EGFR-dependent pathways (such as those involved in cell motility and organogenesis) must be spatially restricted. However, in cells overexpressing EGFRs or when receptor/ligand endocytosis was blocked, those two signals spread over the entire cell even though EGF stimulation was localized—a result that is consistent with dysregulated cell motility and EGFR overexpression that are observed in tandem in carcinoma cells. Thus, ligand-independent propagation of EGF signaling may be the mechanism underlying the dysregulated cell motility that characterizes tumor invasion and metastasis.

This example illustrates how rapidly prototyped microfluidic systems can be used as inexpensive perfusion chambers for studies in which fast changes of the cell environment are required. Unlike in open-air perfusion chambers, in which the flow patterns can be erratic due to convection and evaporation at the air–liquid interface, in microchannels, the flow profile is parabolic and the concentration profile can be easily modeled, yielding more quantitative analysis of cell response as a function of analyte concentration changes.

Another example in which local signal transduction is manifested dramatically is in the **immunological synapse**. Cells of the immune system such as T cells cannot react to free antigen—they can only "see" it once it has been processed and presented to them via a molecule called major histocompatibility complex (MHC) present on the surface of antigen-presenting cells (APCs). The immunological synapse is the approximately 75 μm^2 contact that forms (through MHC+ antigen) between a lymphocyte (a type of white blood cell in the vertebrate immune system, including natural killer cells, T cells, and B cells) and an APC. Recent studies have shown that the microscale protein structure of the contact point is important in modulating T-cell activation. Hence, a team led by Lance Kam from Columbia University used microstamping to simulate the presence of proteins on the APC membrane in different spatial arrangements (colocalized versus segregated), as shown in **Figure 6.19**. Naïve CD4+ T cells were seeded on top of patterns of antibodies against CD3 and CD28 (surrounded by ICAM-1), and interleukin-2 (IL-2) secretion was measured by fluorescence microscopy (the cells were treated with an IL-2 capture reagent so IL-2 was detected on the cell surface). Naïve CD4+ T cells attached to the features of anti-CD3 antibodies, forming a stable synapse. When anti-CD28 was presented in the cell periphery, surrounding an anti-CD3 feature, this "segregated pattern" enhanced IL-2 secretion compared to having these signals combined at the center of the immunological synapse. Overall, these experiments show that costimulation geometries modulate IL-2 secretion

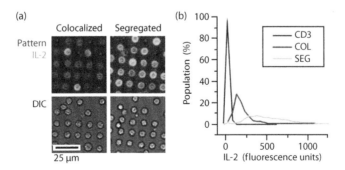

FIGURE 6.19 The immunological synapse. (a) CD4+ cells are seeded on surfaces that simulate the microscale organization of ligands presented by APC cells; patterns of anti-CD3 and anti-CD28 antibodies are shown in red, where IL-2 secretion is shown in green; the remainder of the surface was coated with ICAM-1 (omitted for clarity); (b) histogram of IL-2 secretion for a CD3-only pattern (CD3), the colocalized pattern (COL), and the segregated pattern (SEG). (From Keyue Shen, V. Kaye Thomas, Michael L. Dustin, and Lance C. Kam, "Micropatterning of costimulatory ligands enhances CD4+ T cell function," *Proc. Natl. Acad. Sci. U. S. A.* 105, 7791–7796, 2008. Copyright (2008) National Academy of Sciences, U. S. A.)

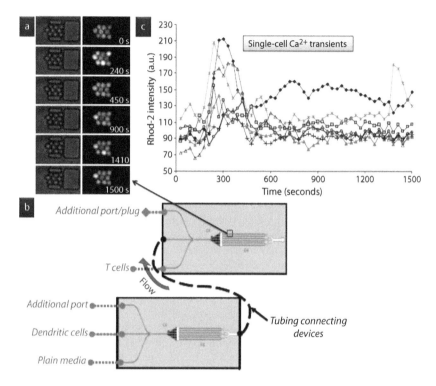

FIGURE 6.20 Microfluidic probing of T-cell signaling. (From Shannon Faley, Kevin Seale, Jacob Hughey, David K. Schaffer, Scott Van Compernolle, Brett McKinney, Franz Baudenbacher, Derya Unutmaz, and John P. Wikswo, "Microfluidic platform for real-time signaling analysis of multiple single T cells in parallel," *Lab Chip* 8, 1700–1712, 2008. Reproduced with permission from The Royal Society of Chemistry.)

by naïve CD4$^+$ T cells. By imaging the nuclear p65 subunit, it was determined that the ability of T cells to discriminate between the colocalized and the segregated patterns correlates with NF-κβ translocation. Also, *Akt* signaling might be important in discriminating between the two geometries because inhibition of the *Akt* pathway decreased IL-2 secretion on both geometries to levels that were higher than on CD3 alone and similar to each other.

A team led by John Wikswo at Vanderbilt University in Nashville, Tennessee has used 440-unit arrays of hydrodynamic "corrals" to trap groups of naïve CD4$^+$ T helper cells to study how they communicate with APCs (**Figure 6.20**). Either APCs (in particular dendritic cells, one of the main types of APCs) were delivered to the array to contact the T cells, or compounds usually secreted by APCs were injected into the device, or another array with dendritic cells was connected upstream (**Figure 6.20b**) to release chemical signals such as chemokines and cytokynes that, in the body, are autocrine and paracrine in nature. Dendritic signaling, which was detected by the downstream T cells, was visualized using Ca^{2+} imaging (**Figure 6.20a** and **c**).

6.4 Cell Migration

Cells need to move (or "migrate") during development, to heal a wound, to respond to a bacterial attack, to colonize a new site, and to grow a tumor. The force produced by a cell during cell migration, or **cellular traction**, can be measured using microfabricated sensors. The process by which cells migrate in response to external soluble signals is termed **chemotaxis**. Microfluidic gradient generators have become an indispensable tool for creating quantitative soluble gradients that elicit cell chemotaxis and directed growth in a reproducible way.

6.4.1 Cellular Traction

The mechanical interactions between cells and the surrounding ECM are a crucial aspect of tissue function. Hence, there has been a great interest in developing methods to directly measure the forces between the cell and the ECM. In 1997, Catherine Galbraith and Michael Sheetz at Duke University devised a micromachined silicon cantilever device to measure fibroblast traction forces (**Figure 6.21**). The device was formed of an array of 5904 pads, ranging in area from 4 to 25 μm², and resting at the end of a cantilever. The pads were flush with the cell culture surface, with only a 2-μm gap to allow for deflection of the pads in and out of the surface plane as the cell exerts traction forces. (The cells were never seen to penetrate the gap.) The deflection of each pad, and thus the locally exerted force, can be tracked by optical microscopy. The fronts of

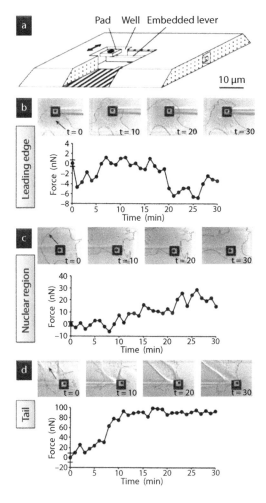

FIGURE 6.21 A micromachined cantilever sensor to measure cellular traction forces. (From Catherine G. Galbraith and Michael P. Sheetz, "A micromachined device provides a new bend on fibroblast traction forces," *Proc. Natl. Acad. Sci. U. S. A.* 94, 9114–9118, 1997. Copyright (1997) National Academy of Sciences, U. S. A.)

migrating fibroblasts produced intermittent rearward forces (~4 nN, see **Figure 6.21b**) and the tails produced larger forward-directed forces (~80 nN; see **Figure 6.21d**).

Unfortunately, the Sheetz force sensor is expensive to fabricate. Also, silicon is opaque, which precludes high-resolution phase-contrast microscopy of intracellular organelles. In 2001, Benjamin Geiger's group at the Weizmann Institute of Science in Israel patterned a wafer with photoresist markers (dots or lines), which they were able to transfer into and embed into an elastomeric cell culture surface (**Figure 6.22**). As the cell migrated across the transparent substrate, the deformation of the substrate could be measured as a set of displacements of the markers and translated to the force applied at individual focal adhesions in real time (the cells expressed GFP vinculin).

Embedded markers require complicated algorithms to compute traction forces because the displacement of each marker is linked to adjacent markers. Chris Chen's laboratory, then at Johns Hopkins University, proposed in 2003 a way to decouple the markers from each other: make them in the shape of PDMS needles or pillars, the bending of which can be predicted by classical elasticity theory (**Figure 6.23**). The average contractile force per post for a cell spreading larger than 900 μm^2 (4 × 4 posts) was measured to be on average approximately 20 nN after 20 hours, but half that amount (~10 nN) if the cell was constrained to 440 μm^2 (3 × 3 posts). This approach has also been used successfully to map forces during epithelial cell migration.

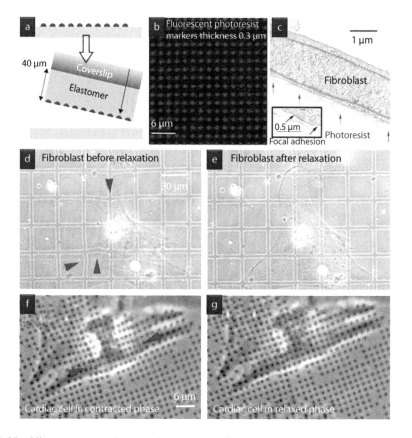

FIGURE 6.22 Micropatterned elastomeric substrates for measuring cellular forces. (From Nathalie Q. Balaban, Ulrich S. Schwarz, Daniel Riveline, Polina Goichberg, Gila Tzur, Ilana Sabanay, Diana Mahalu, Sam Safran, Alexander Bershadsky, Lia Addadi, and Benjamin Geiger, "Force and focal adhesion assembly: A close relationship studied using elastic micropatterned substrates," *Nat. Cell Biol.* 3, 466–472, 2001. Reprinted with permission from the Nature Publishing Group.)

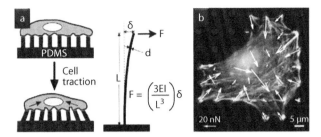

FIGURE 6.23 Measuring cell traction using PDMS microneedles. (From John L. Tan, Joe Tien, Dana M. Pirone, Darren S. Gray, Kiran Bhadriraju, and Christopher S. Chen, "Cells lying on a bed of microneedles: An approach to isolate mechanical force," *Proc. Natl. Acad. Sci. U. S. A.* 100, 1484–1489, 2003. (a) Figure contributed by Chris Chen; Copyright (2003) National Academy of Sciences, U. S. A; (b) Figure contributed by Nate Sniadecki.)

6.4.2 Chemotaxis

Chemotaxis is the phenomenon in which cells direct their movement through the action of certain chemicals. The first observation of chemotaxis is credited to the German microbiologist, Theodow W. Engelmann, who in 1881 observed that bacteria move toward the chloroplasts in algae—and he correctly hypothesized that the chemoattractant is the oxygen produced by the chloroplasts. In general, if a cell senses a difference in chemoattractant concentration across the length of its cell body, then it mounts a migration response, typically (but not always) toward the increasing chemoattractant concentration. (A parallel definition is used for chemorepellent substances.) Exceptions to this rule occur when the chemoattractant concentration saturates all the receptors on the cell surface (making the cell insensitive to chemoattractant spatial variations) and when the chemoattractant concentration field changes so fast that the signal transduction machinery cannot adapt to the changes fast enough. Because chemotaxis is essentially a phenomenon in which the cell needs to sense chemical gradients on a cellular scale with high temporal resolution, many groups have seen a great opportunity for using microfluidic gradient generators to control the gradients to which the cells are exposed. Nonmicrofluidic methods to create gradients, such as glass pipettes, the **Boyden chamber**, the **Zigmond chamber**, and the **Dunn chamber**, have been traditionally used by biologists (and are commercially available) but are slowly being displaced by the more quantitative microfluidic assays described in the next section.

6.4.2.1 Neutrophil Chemotaxis

Neutrophils are the body's first line of defense against bacterial infections and they are constantly roaming the blood, which they exit in response to bacterial or macrophage signals when there is an infection (Figure 6.24). They are a convenient source of cells for cell culture experiments, as they are straightforwardly extracted by pricking a human volunteer's finger. The various neutrophil chemoattractants, such as **interleukin-8** (**IL-8**) or **fMLP** (*N*-formyl-methionine-leucine-phenylalanine), a peptide chain produced by some bacteria, are commercially available. Hence, in addition to their intrinsic scientific interest, neutrophils have become a popular cell model for studying cell migration and, historically, they were the first cell type to be used in a microfluidic chemotaxis assay.

In 2002, a team led by Mehmet Toner at Harvard Medical School used the Dertinger gradient generator (see Figure 3.74 in Section 3.9.2) to expose neutrophils to various gradients of IL-8 (Figure 6.25). The use of the Dertinger generator allowed for exposing the neutrophils to complex (linear as well as nonlinear) gradients. As expected, the cells migrate toward increasing IL-8 concentrations in linear gradients and stop abruptly when they find a sudden "cliff" gradient (drop to zero concentration). However, they keep migrating without reversing direction if they find a "hill" (a maximum) in chemoattractant concentration, despite the fact that they are migrating toward decreasing concentrations. It is important to stress that the cells react quickly around zero concentrations (they do not overshoot the minimum concentration point), but they

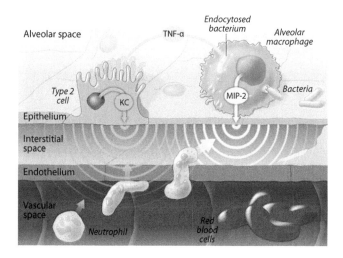

FIGURE 6.24 Neutrophil chemotaxis biology. (Figure contributed by Chuck Frevert.)

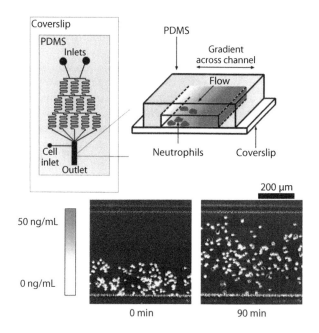

FIGURE 6.25 Neutrophil chemotaxis studied with the Dertinger gradient generator. (From Noo Li Jeon, Harihara Baskaran, Stephan K. W. Dertinger, George M. Whitesides, Livingston Van De Water, and Mehmet Toner, "Neutrophil chemotaxis in linear and complex gradients of interleukin-8 formed in a microfabricated device," *Nat. Biotechnol.* 20, 826–830, 2002. Reprinted with permission from the Nature Publishing Group. Figure contributed by Noo Li Jeon.)

are sluggish around the hill (they overshoot the maximum concentration point), likely because of a slow response time of the chemotaxis transduction machinery when the receptor occupancy is overloaded. Although these gradients are unlikely to exist in vivo, they help us to understand the importance of adaptation and temporal dynamics in chemotaxis.

The Dertinger device, as explained in Section 3.9.2, has several limitations, mainly that it exposes the cells to shear forces (potentially biasing the migrating direction of the cells). A collaborative team led by Chuck Frevert and the author at the University of Washington in Seattle has developed a microvalve-triggered gradient generator that is designed to simulate the radially symmetric, temporally evolving gradients that neutrophils experience in vivo (Figure 6.26). The cells are confined to a closed chamber which connects to another closed chamber through a microvalve (Figure 6.26a)—the opening of the microvalve and the evolution of the gradient can be followed by fluorescence microscopy (Figure 6.26c and e) and modeled by finite-element modeling; gradient formation is extremely reliable, so the only source of experimental variability is in the cells. Although migration occurs radially (Figure 6.26b and d), it is straightforward to make a coordinate transformation and plot all the migratory paths on the same axis for comparison (Figure 6.26f). Cells are observed to migrate up the IL-8 concentration and gradually "hesitate" as they approach the target (Figure 6.26g).

The microvalve-triggered gradient generator cannot, by design, sustain constant gradients, so it is difficult to extract from it any quantitative relationship between the gradients and the migration behavior. In addition, the enclosed nature of the microchambers reduces the viability of the neutrophils over time. Hence, the same University of Washington team used a "microjets" gradient generator (see Figure 3.81 in Section 3.9.3) to expose neutrophils to gradients in open cell culture conditions (Figure 6.27a). The cells are simply dropped on an open cell culture reservoir and the gradient is created by injecting cell culture medium (concentration $C = 0$) from one side and the chemoattractant ($C = C_0$) from the other side of the reservoir (Figure 6.27b). Desensitization experiments were performed by applying the same gradient,

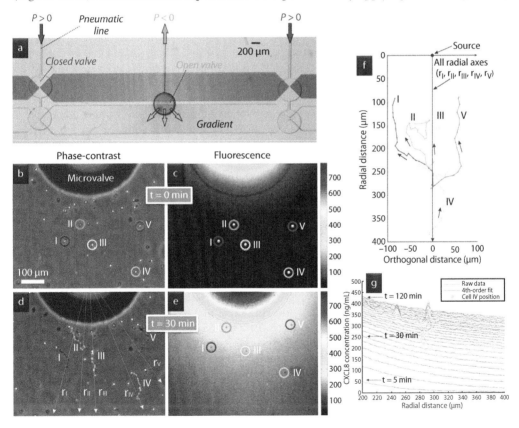

FIGURE 6.26 Control of chemotactic gradients using microvalves. (From Charles W. Frevert, Gregory Boggy, Thomas M. Keenan, and Albert Folch, "Measurement of cell migration in response to an evolving radial chemokine gradient triggered by a microvalve," *Lab Chip* 6, 849–856, 2006. Reproduced with permission from The Royal Society of Chemistry.)

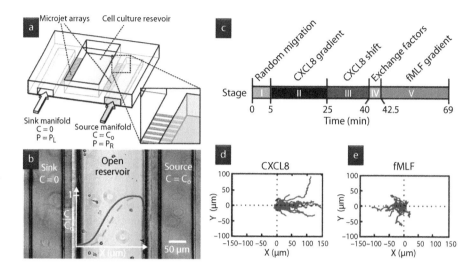

FIGURE 6.27 Chemotaxis studies using a microjet gradient generator. (From Thomas M. Keenan, Charles W. Frevert, Aileen Wu, Venus Wong, and Albert Folch, "A new method for studying gradient-induced neutrophil desensitization based on an open microfluidic chamber," *Lab Chip* 10, 116–122, 2010. Reproduced with permission from The Royal Society of Chemistry.)

then shifting it, then applying an inverse gradient of a different chemoattractant (Figure 6.27c). Distinct responses to the two chemoattractants were recorded (Figure 6.27d and e). The unexpected presence of "tethered cells" and the fact that many cells responded to the second CXCL8 or fMLF gradients but not the first CXCL8 gradient provides evidence that neutrophils have initial activation states, and long-term chemical sensitivities that differ dramatically. The population-based approaches and analyses provided by traditional methods are simply ineffective at parsing out how specific complements of chemotactic factor receptors and their relative activation states influence neutrophil migratory behavior.

These devices require, at the very least, some amount of microfluidics know-how (either for fabrication or for modeling, or both). Why not produce a microfluidic device that relies on purely diffusive processes (no flow) to produce the gradient? In 2007, a team from Cornell University led by Michael Shuler and Mingming Wu presented a three-channel gradient generator made in agarose (Figure 3.88 in Section 3.9.5.2); agarose is readily available in biological laboratories, it is inexpensive and safe to use, it molds well, and it is biocompatible (it has been used in the past for chemotaxis in what is called the "under-agarose" assay). In Shuler and Wu's device, the central channel contained the cells (similar to the microjets' central reservoir) and the lateral channels acted as sink and reservoir for the chemoattractant (similar to the microjets' sink and reservoir manifolds). HL-60 cells, a human promyelocytic leukemia cell line, were differentiated into neutrophil-like cells by culturing them in 1.3% DMS culture media for 4 to 7 days, and attached to fibronectin-coated glass. Less than 5 minutes after the addition of fMLP (250 nM, gradient = 0.27 nM/μm) into the source channel, the cells started to move (Figure 6.28). Hydrogel devices are extremely useful when conducting long-term experiments because nutrients and gases necessary for cell survival diffuse readily through the hydrogel.

None of the designs shown above are capable of delivering very sharp chemical gradients: the concentration field goes from zero to maximum over a distance that is much larger than the cell. How can a gradient be delivered in such a way that it is ensured that one end of the cell "sees" zero concentration and the other end sees the maximum concentration—if possible for all the cells? Mehmet Toner and colleagues at Harvard Medical School built a device that cleverly uses the cells themselves as barriers for the chemotaxis agent (Figure 6.29). The device consists of many channels, each of which has the width of a single neutrophil, so as the neutrophils

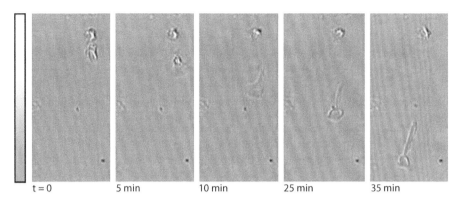

t = 0   5 min   10 min   25 min   35 min

FIGURE 6.28 Chemotaxis of neutrophils in an agarose gradient system. (From Shing-Yi Cheng, Steven Heilman, Max Wasserman, Shivaun Archer, Michael L. Shuler, and Mingming Wu, "A hydrogel-based microfluidic device for the studies of directed cell migration," *Lab Chip* 7, 763–769, 2007. Reproduced with permission from The Royal Society of Chemistry.)

start penetrating the channels, they completely occlude them and the concentration they see on the front end can be abruptly different than the concentration seen by the tail (**Figure 6.29b** through **d**). Actin filaments are seen to accumulate at the leading edge and at the channel walls (**Figure 6.29e**). (The device also features an innovative "pressure balancing" zone to guarantee that the inlet and outlet of the channels are at the same exact pressure yet they are fed the inlet and outlet concentrations; see **Figure 6.29a**.)

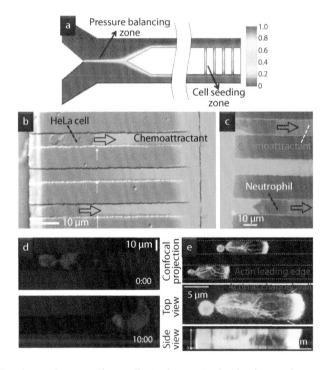

FIGURE 6.29 Ultrasharp chemotactic gradients in constrained microenvironments. (From Daniel Irimia, Guillaume Charras, Nitin Agrawal, Timothy Mitchison, and Mehmet Toner, "Polar stimulation and constrained cell migration in microfluidic channels," *Lab Chip* 7, 1783–1790, 2007. Reproduced with permission from The Royal Society of Chemistry.)

6.4.2.2 Cancer Cell Migration

Some cancer cells in primary tumors acquire the ability to penetrate and infiltrate surrounding normal tissues in the local area or in distant sites after becoming circulating tumor cells, forming a new tumor. Therefore, in the process by which cancer spreads to other organs, called **metastasis**, cell migration needs to be activated and plays an important role. As with neutrophils, cancer cells are also a convenient source of cells to study cell motility in general, because several cancer cell lines exist and the cells exhibit high motility rates in response to commercially available factors. Hence, cancer cells have historically been the second-most popular choice of cells (behind neutrophils) in BioMEMS chemotaxis studies.

In 1998, Jeffrey Segall's group at Albert Einstein's College of Medicine in New York used SAM hexadecanethiol micropatterns on a background of HEG-terminated alkanethiol to study the chemotaxis of rat mammary carcinoma cells, in particular, the relationship between adhesion and chemoattractant-stimulated lamellipod extension. When a cell was attracted into a nonadhesive region with a pipette delivering EGF, it still extended lamellipods over the HEG-derivatized areas but rapidly retracted them. The absence of cell–substrate contacts on the nonadhesive areas (as determined by immunostaining of talin, one of the proteins that links the cytoskeleton to integrins) indicated that lamellipod extension can occur without establishing focal adhesions but requires focal adhesions to stabilize.

In 2004, the laboratory of Noo-Li Jeon (who coinvented the Dertinger gradient generator in 2001, see Figure 3.74, and was the first author in the 2002 article on neutrophil chemotaxis using Dertinger's generator, see Figure 6.25), then at the University of California, Irvine, CA, again used the Dertinger gradient generator to probe the response of the breast cancer cell line MDA-MB-231 to various EGF gradients (Figure 6.30). Linear concentration gradients were not able to induce chemotaxis, whereas a nonlinear polynomial gradient induced a chemotactic movement.

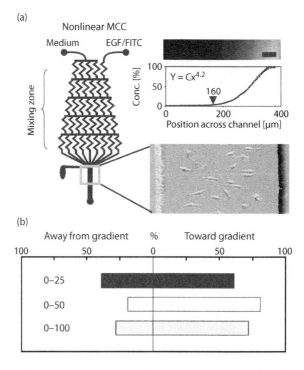

FIGURE 6.30 Chemotaxis of cancer cells in a microfluidic gradient. (From Shur-Jen Wang, Wajeeh Saadi, Francis Lin, Connie Minh-Canh Nguyen, and Noo Li Jeon, "Differential effects of EGF gradient profiles on MDA-MB-231 breast cancer cell chemotaxis," *Exp. Cell Res.* 300, 180–189, 2004. Reprinted with permission from Elsevier.)

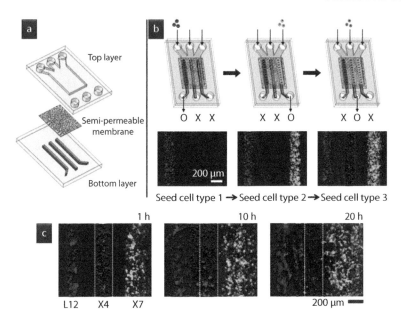

FIGURE 6.31 Chemotactic gradients created by CXCL12 source-sink cells. In panel (c), "L12" denotes the CXCL12-secreting cells (labeled with CellTracker red), "X4" the CXCR4-expressing cells (nuclear-labeled with Hoechst, in blue), and "X7" the CXCR7-expressing cells (labeled with CellTracker green). (From Yu-suke Torisawa, Bobak Mosadegh, Tommaso Bersano-Begey, Jessica M. Steele, Kathryn E. Luker, Gary D. Luker, and Shuichi Takayama, "Microfluidic platform for chemotaxis in gradients formed by CXCL12 source-sink cells," *Integr. Biol.* 2, 680–686, 2010. Reproduced with permission from The Royal Society of Chemistry.)

The chemokine CXCL12 has been proposed to promote metastasis in breast cancer and various other malignancies. CXCL12 is secreted by fibroblasts in primary human breast tumor. CXCL12 binds to receptor CXCR4 on breast cancer cells to promote chemotaxis, and also to receptor CXCR7 (expressed by stromal cells and subsets of breast cancer cells), but the role of CXCR7 in CXCL12-dependent chemotaxis is not well understood. Shuichi Takayama and colleagues at the University of Michigan have built a microfluidic system to test a hypothesis for the possible role of CXCR7-expressing cells (Figure 6.31). According to the "sink-source" hypothesis (based on a zebrafish observation), the CXCR7-expressing cells act as a sink to sequester the CXCL12 chemokine, and chemotaxis of CXCR4-expressing cells toward CXCL12-expressing cells (chemokine source) is critically dependent on the presence and location of the sink cells. Takayama and coworkers verified this hypothesis using their hydrodynamic trapping technology (see Section 2.6.2.6), whereby cells are trapped when a cell suspension is forced through a porous membrane (the cells do not fit through the pores of the membranes, so they get trapped on top of the membrane). The membrane supports cell culture and cell migration.

6.4.2.3 Bacterial Cell Migration

Bacterial chemotaxis is very different from chemotaxis of adherent cells and, consequently, the engineering challenges associated with investigating bacteria are also very different. Bacteria move in three-dimensional space in a biased random walk, constantly sampling their local microenvironment. They propel themselves with one or more flagella into straight trajectories called "runs" or erratic reorientation events called "tumbles," depending on which direction the molecular motors (that drive the flagella filaments) turn. To achieve gradient sensing, the cell simply modulates the fraction of time spent tumbling or running by doing a temporal comparison: the receptor occupancy in the past is compared to that in the more distant past. The swimming speed (~30 μm/s in *E. coli*) thus converts the spatial gradient into a temporal one and determines the steepness of the gradients that can be sensed.

The first bacterial chemotaxis assay on a chip is credited to Michael Manson's group from Texas A&M University, who in 2003 designed a simple T-mixer (with an extra central inlet for the bacteria) to measure *E. coli* motility under gradients of various chemoeffectors (**Figure 6.32**). As in any T-mixer (see Section 3.9.1), the chemoeffector gradient forms gradually down the channel as the streams mix by diffusion (**Figure 6.32a**). The 22 outlets allow for counting cells and obtaining transverse distribution of bacteria across the chamber. For example, it was found that wild-type *E. coli* responded to 3.2 nM gradients of L-aspartate (cells accumulate maximally in outlet 11, where the concentration is maximal), a concentration three orders of magnitude lower than the detection limit in the standard capillary assay (**Figure 6.32b**). The response to buffer (0 nM) is not a Gaussian distribution because (a) the initial distribution of cells is a 151-μm-wide stream (not a point), (b) the cells flow at different speeds (due to the parabolic flow profile, see Section 3.2.4), and (c) there is a distribution of swim speeds (given by *E. coli*'s different run/tumble biases). Surprisingly, L-leucine was sensed (by the receptor *Tar*) as an attractant at low concentrations and (by *Tsr*) as a repellent at higher concentrations (**Figure 6.32c**).

Obviously, the mixer part of the gradient generator does not need to be a T-mixer, which is limited to very simple monotonic gradients. Michael Manson's group has collaborated with Arul Jayaraman's group, also at Texas A&M, to incorporate a Dertinger mixer (see Section 3.9.2) into Manson's flow-based bacterial chemotaxis device (**Figure 6.33**). This improvement has allowed for the investigation of *E. coli* chemoattraction to the quorum-sensing molecule autoinducer-2, chemorepulsion by indole, as well as competing simultaneous gradients of both. (By using cells that express GFP or RFP, the cells could be counted by fluorescence instead of by collection into outlet channels.)

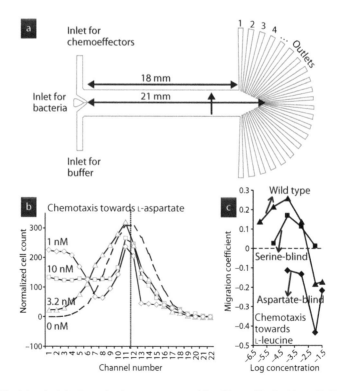

FIGURE 6.32 First bacterial chemotaxis assay on a chip. (From H. B. Mao, P. S. Cremer and M. D. Manson, "A sensitive, versatile microfluidic assay for bacterial chemotaxis," *Proc. Natl. Acad. Sci. U. S. A.* 100, 5449–5454, 2003. Copyright (2003) National Academy of Sciences, U. S. A.)

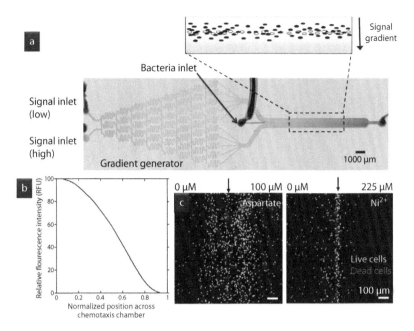

FIGURE 6.33 Bacterial chemotaxis in a Dertinger gradient generator. (From Derek L. Englert, Michael D. Manson, and Arul Jayaraman, "Flow-based microfluidic device for quantifying bacterial chemotaxis in stable, competing gradients," *Appl. Environ. Microbiol.* 75, 4557–4564, 2009. Figure contributed by Arul Jayaraman.)

One of the most powerful attributes of microfluidics is that it allows for *mimicking* the physicochemical properties of cellular microenvironments. Martin Polz and coworkers at MIT used microfluidic devices to simulate the tracks of dissolved organic matter, such as coming from snow particles or lysed algae, that slowly settle by gravity in the ocean, and accumulate feeding bacteria around them. These "bacterial hot spots," simulated in the microfluidic devices as plumes ejected from a hole (**Figure 6.34**), are able to attract the marine bacterium *Pseudoalteromonas haloplanktis* within tens of seconds, more than 10 times faster than *E. coli* (leading to twice the nutrient exposure). In other words, the rapid chemotactic response of *P. haloplanktis* substantially enhances its ability to exploit nutrient patches before they dissipate.

Roseanne Ford's group at the University of Virginia has used microfluidic devices (containing 200-µm-diameter tightly packed cylindrical posts) to simulate the percolating flow through dirt (**Figure 6.35**). The average flow velocity in the channel (5–20 m/day) was used to match that of groundwater flows. Gradients of α-methylaspartate were formed with a T-mixer upstream and cells counted within each pore (between the posts). The observed chemotactic response (**Figure 6.35b**) was larger than predicted, which could be due to (a) within-pore dynamics, or (b) hydrodynamic effects (e.g., trapping) confounding some of the bacteria–surface interactions.

Bacterial cells are swimmers; they do not attach. Hence, in bacterial chemotaxis experiments, flow is a dangerous proposition (unless it is one of the biological variables under investigation). The biggest problem is that, because of the parabolic flow profile, the cells are not all being carried at the same flow velocity; if the cells are moving quickly across the velocity profile field, the problem becomes even more complicated. A secondary problem is that, when there are concentration fields superimposed with flow, dispersion phenomena such as the butterfly effect (see **Figure 3.73**) will occur; cells that are at different z distances from the surface will experience different gradients.

Norman Stocker's group at MIT was the first to apply microfluidic design principles to reduce the exposure of bacteria to flow in bacterial chemotaxis experiments (**Figure 6.36**). A major advantage of "stopping flow" is that microscopic observation becomes straightforward, and

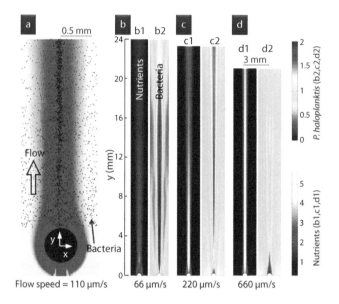

FIGURE 6.34 Chemotactic plumes of organic matter dissolved in the ocean simulated with a microfluidic device. (From Roman Stocker, Justin R. Seymour, Azadeh Samadani, Dana E. Hunt, and Martin F. Polz, "Rapid chemotactic response enables marine bacteria to exploit ephemeral microscale nutrient patches," *Proc. Natl. Acad. Sci. U. S. A.* 105, 4209–4214, 2008. Copyright (2008) National Academy of Sciences. Figure contributed by Roman Stocker.)

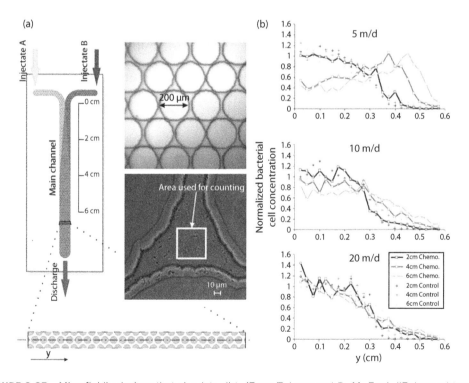

FIGURE 6.35 Microfluidic devices that simulate dirt. (From T. Long and R. M. Ford, "Enhanced transverse migration of bacteria by chemotaxis in a porous T-sensor," *Environ. Sci. Technol.* 43, 1546–1552, 2009. Figure contributed by Tao Long.)

high-throughput single-cell analysis becomes feasible. Their device consisted of a simple linear channel (in which the cells are loaded) with one narrow side branch (through which the chemoattractant α-methylaspartate and fluorescein as a tracer are loaded through a valve). The flow in the large cell-containing channel did not significantly penetrate the chemoattractant-containing channel, but created a transient gradient to which cells responded. Strong chemotaxis is visible in Figure 6.36g, where most cells swim up the gradient (high chemotactic velocity V_C), but not in Figure 6.36h (low V_C), likely due to receptor saturation. This type of microfluidically generated ephemeral gradients provide realistic microenvironmental signals that mimic the signals encountered by bacteria in their natural habitats and their responses to nutrient pulses.

Mingming Wu's group at Cornell University was able to fully suppress flow using a three-channel gradient generator in which the fluids are separated by semiporous agarose barriers (conceptually similar to Manson's device, in which the barriers were made of paper) by building the device in agarose (Figure 6.37). The bacteria were introduced in the central channel whereas the flanking channels served as sink and source to create a stable gradient without exposing the cells to flow (all channels are 400 μm wide and 160 μm deep). Single-cell chemotactic trajectories could be tracked for various concentration values of methyl-aspartate (the ligand used in this study). Both the ligand concentration ([L]) profile (as visualized by adding fluorescein) and the

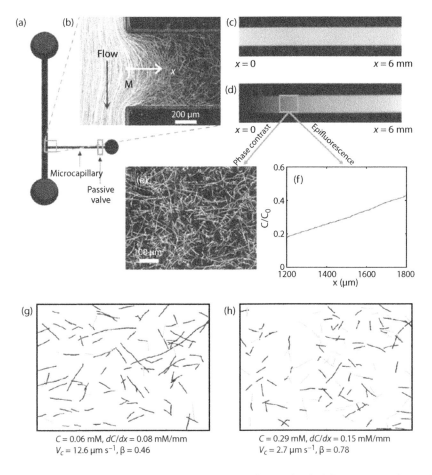

$C = 0.06$ mM, $dC/dx = 0.08$ mM/mm
$V_C = 12.6$ μm s^{-1}, β = 0.46

$C = 0.29$ mM, $dC/dx = 0.15$ mM/mm
$V_C = 2.7$ μm s^{-1}, β = 0.78

FIGURE 6.36 Tracking of single bacterial cells in flow-free microfluidic chemotactic gradients. (From Tanvir Ahmed and Roman Stocker, "Experimental verification of the behavioral foundation of bacterial transport parameters using microfluidics," *Biophys. J.* 95, 4481–4493, 2008. Reprinted with permission from Elsevier. Figure contributed by Roman Stocker.)

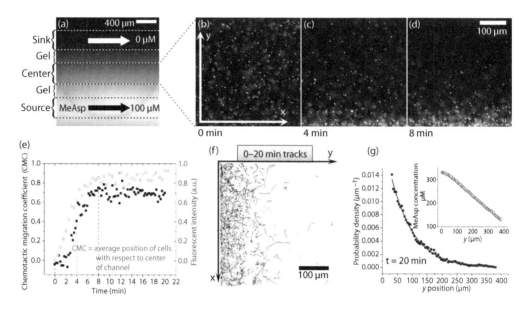

FIGURE 6.37 Logarithmic sensing of chemotactic gradients by bacteria. (From Yevgeniy V. Kalinin, Lili Jiang, Yuhai Tu, and Mingming Wu, "Logarithmic sensing in *Escherichia coli* bacterial chemotaxis," *Biophys. J.* 96, 2439–2448, 2009. Reprinted with permission from Elsevier.)

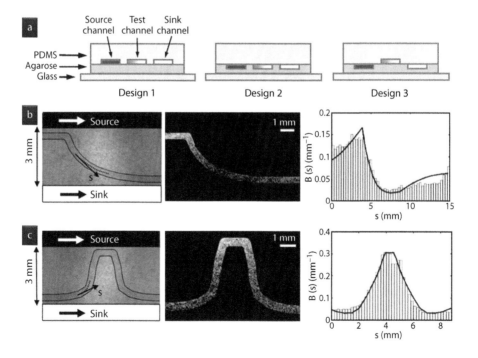

FIGURE 6.38 Agarose arbitrary gradient generator for bacterial chemotaxis. (From Tanvir Ahmed, Thomas S. Shimizu, and Roman Stocker, "Bacterial chemotaxis in linear and nonlinear steady microfluidic gradients," *Nano Lett.* 10, 3379–3385, 2010. Figure contributed by Roman Stocker.)

bacterial density distribution equilibrated in approximately 8 minutes. The mean chemotactic drift velocity of *E. coli* cells increased monotonically with $\Delta[L]/|L|$ or $\sim\Delta(\log[L])$—in other words, *E. coli* cells are able to sense the spatial gradient of the logarithmic ligand concentration.

The three-channel system does not allow for the production of gradients of arbitrary shapes. Roman Stocker's group at MIT has recently adapted Zare's gradient "spatial-derivative" generator (see **Figure 3.89**), also based on agarose, to bacterial chemotaxis (**Figure 6.38**). This gradient generator creates a uniform gradient in a hydrogel slab using one source and one sink channel and uses the gradient created in an arbitrarily-shaped test channel whose fluid is in equilibrium with the hydrogel slab. This way, the gradient in the test channel is a spatial derivative of the slab's gradient, and can be customized to approximate any continuous function. The measured distribution of bacteria (*B*(s), vertical bars in **Figure 6.38b** and **c**) closely followed the solution to the bacterial transport equation (solid line in **Figure 6.38b** and **c**).

Although the ability to probe arbitrary gradients is important, none of the devices above are very suited for studying one of the most critical variables in chemotaxis: *time*. The gradients above cannot be easily changed with time. Laurie Locascio's group at NIST has used microjets (see Section 3.9.3) to produce shear stress–free chemotactic gradients in circular chambers; the chamber typically contain three inlets: one for injecting the chemoeffector (e.g., glucose), one

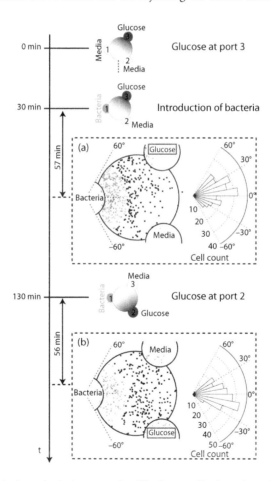

FIGURE 6.39 Bacterial chemotaxis in a purely diffusive gradient produced with microjets. (From Javier Atencia, Jayne Morrow, and Laurie E. Locascio, "The microfluidic palette: A diffusive gradient generator with spatio-temporal control," *Lab Chip* 9, 2707–2714, 2009. Reproduced with permission from The Royal Society of Chemistry.)

inlet for injecting the bacteria or medium, and one to inject medium. The delivered gradients are a function of the flow rates injected into the chamber as well as a function of which inlet is used for the chemoeffector (so by changing the chemoeffector inlet, the gradient can be rotated). These customizable gradients offer the possibility of mimicking the dynamics of bacterial adaptation to nutrient gradients much more realistically than previous devices (**Figure 6.39**).

6.5 BioMEMS for Cellular Neurobiology

A large proportion of the research efforts in BioMEMS devoted to basic biology have been focused on cellular neurobiology. The percentage of funding historically devoted to the neurosciences has undoubtedly helped fuel this trend. This is a lucky coincidence because, as we will see here, BioMEMS techniques are (or should be) particularly appealing to neuroscientists precisely because the static, homogeneous substrate known as the petri dish may be more limiting in neurobiology than in any other biological field (together, probably, with the field of developmental biology).

6.5.1 Axon Guidance

During development of the nervous system the response of growing axons to their environment is critical to the formation of the complex wiring pattern between neurons. Growth and guidance factors combined with extracellular matrices influence the speed and direction of axonal growth. Although much progress has been made in identifying the factors that influence axonal growth, as well as how axons respond to these factors individually, much less is known about how axons behave in response to the combined effects of multiple factors. Ideally, to fully understand how axonal growth is regulated it should be studied in vivo. The ability to express fluorescent protein markers in neurons has made such an approach possible. However, it is limited by a variety of factors. Imaging individual axons deep in the brain tissue is still restricted to layers less than 600 μm. Phototoxicity limits the time lengths of such observations. More importantly, the environment is difficult to characterize or control. As a complementary approach, many groups have developed in vitro (cell culture or explant) environments that potentially mimic some of the complexity found when studying the whole animal ("in vivo").

TRADITIONAL AXON GUIDANCE GRADIENT-GENERATION APPROACHES

IN **AXON GUIDANCE** STUDIES, the main advantages of using **in vitro** over **in vivo** experimentation are: (a) the ability to isolate specific cell functioning parameters without interference from more complex whole-organism responses; and (b) the accessibility for microscopic observation and manipulation (of both the cells and of their environment). Generally speaking, two in vitro setups are widely used for creating gradients; in both cases, the gradient never reaches equilibrium, cannot be quantified, and as a result it cannot be reproduced, yielding unnecessary experimental variability:

Explant cultures: In this system (based on a thin tissue slice), cells within the explant continue their development by growing long neurites beyond the edge of the explant. Soluble guidance factor gradients can be generated by embedding the explant culture in collagen gel within the proximity of another explant or transfected cells secreting the guidance molecule. Albeit convenient and robust, in the explant system, the amount of signaling molecule being released by the source is difficult to quantify, single axons are hard to track, and axon guidance from these cultures is confounded by factors secreted by the explant.

Dissociated cultures: In this system (whereby single cells have been separated mechanically or enzymatically from the source tissue), the cells can be stimulated by a concentration gradient of soluble guidance factors delivered from a manipulator-controlled micropipette placed at a 45-degree angle relative to the previous growth cone navigation direction. The growth cone turning and growth responses are measured by turning angle and elongation distance, respectively. Albeit simple, pipette-stimulation of dissociated cultures also has several crucial limitations: (1) Small geometric differences between each pulled micropipette make it extremely difficult to accurately predict the concentration of the signaling molecule delivered to the growth cone. (2) The concentration gradient produced by the micropipette changes over time, making it difficult to correlate different axon turning responses with specific gradients. (3) The extensive user involvement in choosing the neuron and positioning the pipette results in studies with low throughput, ultimately limiting the number of neurons that can be analyzed (poor statistical significance) and the number of combinations of signaling molecules that can be studied. (4) Simultaneous examination of more than one guidance factor is difficult because of the space taken by the micropipette setup on the microscope stage.

Cells have been exposed to diffusible gradients in cultures using biological gels (e.g., collagen, fibrin, or agarose), glass micropipettes, and a variety of static-chamber devices (e.g., the Boyden, Zigmond, and Dunn chambers). However, the chemical gradients generated by such "traditional" methods often evolve unpredictably or uncontrollably over space and time, greatly limiting the cell types and questions that can be studied, and can be difficult to characterize quantitatively.

Recently, a wide variety of microfluidic device designs have been presented for generating gradients that are predictable, reproducible, and easily quantified. A most successful one has been the Dertinger design (presented by Whitesides and colleagues, see Figure 3.74), which has been used to study the effects of soluble biomolecule gradients on neutrophil migration (see Figure 6.25 in Section 6.4.2), neural stem cell differentiation, breast cancer cell chemotaxis, and rat intestinal cell migration. The device has also been used to create substrate-bound biomolecule gradients to direct the growth of hippocampal neurons (see Section 6.5.1.3) and to study the intestinal cell cycle. The Dertinger design, however, has several important limitations with regard to neuronal cultures: (1) The device can only generate gradients under fluid flow, which induces shear and drag forces that may detach the cells; flow-induced changes in intracellular signaling and in cell shape/attachment (which led to neutrophil migrational bias for shear stresses >0.7 dyn/cm^2) have been documented; (2) the gradient evolves as the fluid flows downstream such that no two cells in the microchannel can be expected to experience the same concentration gradient; (3) downstream cells are exposed to higher concentrations of cell-secreted molecules (e.g., metabolites and growth factors) than upstream cells, which precludes true redundancy in single-cell data; and (4) the observation area is an enclosed microfluidic channel, which limits gas and nutrient exchange required for proper long-term cell viability/differentiation (e.g., neurons or stem cells that must be cultured for days before gradient application).

Myriad designs by many groups have been developed with various capabilities. Some designs do not suffer from the Dertinger design's limitations but suffer from others. For example, flow and gradient nonuniformity can be eliminated altogether with a three-microchannel system, with the cells in a center channel that is separated from the side channels (which act as sink and source) by a hydrogel wall (which provides transport of solutes by diffusion without convection); however, here the cells are still entrapped in a microfluidic channel and the hazy hydrogel walls hinder microscopy (see Figure 6.28). In other cases, it may be possible to present the gradients to the cells in substrate-bound form, typically done by creating it a priori with a microfluidic gradient generator or by microstamping the gradient of the protein of interest with a PDMS stamp (see Section 6.5.1.3).

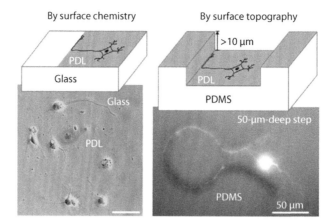

FIGURE 6.40 In vitro guidance of axons by surface chemistry and surface microtopography. (From Li, N. and Folch, A., "Integration of topographical and biochemical cues by axons during growth on microfabricated 3-D substrates," *Exp. Cell Res.* 311, 307, 2005. Figure contributed by Nianzhen Li.)

The tip of the axon, also termed growth cone due to its morphology, is a highly specialized cellular structure in charge of determining whether the substrate ahead is suitable for growth. In time-lapse movies, the growth cone can be seen to extend finger-like protrusions (termed **filopodia**) that seem to explore the substrate like blind "tentacles"; only if the "tentacle" adheres, the axon grows after it, extending more tentacles ahead. In vitro, if no soluble factors are used, axon growth can also be microengineered by either modulating the surface composition (because the axon follows the areas where the filopodium is adhered) or the surface topography (because a topographical change necessarily biases the exploratory range of the filopodia), as depicted in **Figure 6.40**. These studies not involving soluble gradients are technologically less challenging and logically were developed first, so they will be considered first here.

6.5.1.1 Axon Guidance by Biochemical Surface Micropatterns

As early as 1987, Friedrich Bonhoeffer and colleagues, then at the Max Planck Institute in Tübingen, Germany, created an ingenious device for immobilizing membrane fragments (**Figure 6.41**). The device contains microchannels molded in silicone rubber (in essence, PDMS) from a photolithographically etched master. Atop the PDMS microchannels, a membrane (a nucleopore filter with pore size 0.1 μm) was placed, such that when a suspension of cell membrane debris was flowed on top of the device and suction was applied through the microchannels, the cell membrane debris were hydrodynamically trapped on the top surface of the filter until it gets clogged (an idea that has been successfully recycled recently by Shuichi Takayama's laboratory to capture and pattern cells, see **Figure 2.42** in Section 2.6.2.6). After the filter gets clogged, and stripes of membrane are formed, the filter is moved on top of another matrix that has homogeneous porosity (not containing microchannels; not shown in **Figure 6.41**) and another cell extract is added on top of the filter; when suction is restarted, the membrane fragments from this new cell extract get immobilized in the areas between the old stripes, forming stripes of two different kinds. (Here, we note for the BioMEMS student that Bonhoeffer and his colleagues had just invented the first microfluidic microdevice in history—and the first PDMS device in history—6 years before George Whitesides published the first microstamping article!)

Friedrich Bonhoeffer's group devised this micropatterning technique so that they could test the hypothesis that specific molecules present in the membrane of cells from the optic tectum were able to guide the growth of chick retinal axons. They observed that temporal axons showed a preference for growth on membranes of the anterior tectum (their natural target area) over posterior tectum (**Figure 6.41b**), whereas nasal axons did not show a preference (**Figure 6.41c**).

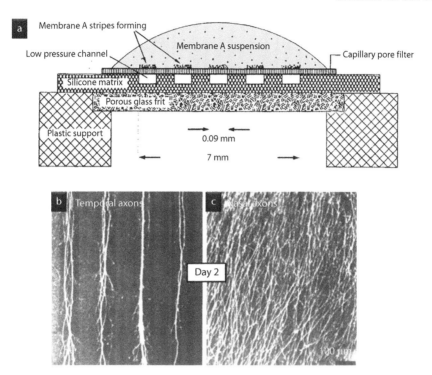

FIGURE 6.41 Hydrodynamical immobilization of membrane fragments: the biochemical stripe assay for axon guidance. (From Jochen Walter, Brigitte Kern-Veits, Julita Huf, Bernd Stolze, and Friedrich Bonhoeffer, "Recognition of position-specific properties of tectal cell membranes by retinal axons *in vitro*," *Development* 101, 685–696, 1987. Figure contributed by Friedrich Bonhoeffer.)

This setup, which became known as the "**stripe assay**," was used by neurobiologists for many years for the limited use of patterning axons, without realizing the technique had an immense potential for patterning other biological solutions, including cells, and building a rich variety of microfluidic architectures.

Uwe Drescher (who had trained with Bonhoeffer), also at the Max Planck Institute at Tübingen, Germany, was interested in elucidating why only temporal axons respond to repellent axon guidance cues of the caudal tectum in the stripe assay. He first changed the cell culture support surface (the filter), which was somewhat impractical for microscopy, and changed it to a solid support (plastic surface) coated with nitrocellulose. This coating procedure traces back to seminal work in 1987 by Vance Lemmon and Carl Lagenaur (University of Pittsburgh), who reasoned that "since nitrocellulose is a convenient substrate for rapid noncovalent attachment of proteins, we coated sterile Petri plates with nitrocellulose dissolved in methanol." Drescher's group also used a soluble chimeric ephrinA2 molecule in which the hydrophobic C terminus is replaced by the Fc part of a human immunoglobulin (ephrinA2-Fc), making it insensitive to cleavage by PI-PLC. Drescher observed that overexpression of ephrinA ligands on temporal axons abolished the sensitivity previously observed by Bonhoeffer's group (see Figure 6.41), whereas treatment with PI-PLC both removed ephrinA ligands from retinal axons and induced a striped outgrowth of formerly insensitive nasal axons (Figure 6.42). This work suggests that differential ligand expression on retinal ephrinA5 axons is a strong determinant of topographic targeting in the projection of retinotectal axons.

Philip Hockberger and colleagues, then at Bell Labs, pioneered in 1998 the use of surface chemistry to microengineer neuronal cell attachment and axon growth. In their case, the surface patterns were produced by photolithography, using aminosilane SAMs as the adhesive areas

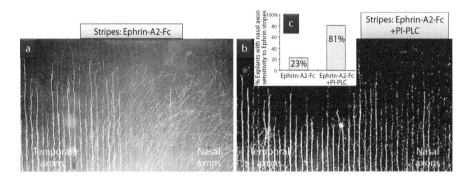

FIGURE 6.42 Biochemical modulation of the stripe assay. (From Martin R. Hornberger, Dieter Dütting, Thomas Ciossek, Tomoko Yamada, Claudia Handwerker, Susanne Lang, Franco Weth, Julita Huf, Ralf Weßel, Cairine Logan, Hideaki Tanaka, and Uwe Drescher, "Modulation of EphA receptor function by coexpressed ephrinA ligands on retinal ganglion cell axons," *Neuron* 22, 731–742, 1999. Reprinted with permission from Elsevier.)

and methyl-terminated silane SAMs as the nonadhesive areas (see **Figure 2.7** in Section 2.3.1). Although the mechanism by which axons grew preferentially on the **aminosilane** areas was not explored, this study demonstrated that microengineering axon growth with cultured cells was possible, effectively paving the way for others to explore more rational surface chemistries and even three-dimensional approaches. Aebischer and colleagues, for example, used a focused laser beam to create three-dimensional patterns of cell-adhesive laminin oligopeptide fragments; the laser produced a photochemical reaction that cross-linked the oligopeptides to selected locations of an agarose gel background. Even though the resolution did not allow for guiding single axons (imaging itself was very challenging, and removal of the unreacted oligopeptides was an issue), it was encouraging that the somas were seen to adhere preferentially to the oligopeptide-conjugated areas. It is foreseeable that a variation of this technique could be used to immobilize cell-surface molecules such as those found in glial cells to simulate the natural paths of axon guidance.

A collaborative team led by Gary Banker (Oregon Health & Science University at Portland, Oregon) and Harold Craighead (Cornell University at Ithaca, New York) have demonstrated a versatile protein micropatterning scheme for axon guidance experiments (**Figure 6.43**). The first step is the immobilization of **poly-L-lysine (PLL)** on glass, either over the whole substrate from solution or in a pattern (by microstamping). The second step is the microstamping of protein A, a protein that binds the Fc fragment of immunoglobulins. A chimeric protein (the extracellular domain of the guidance protein L1 recombinantly linked to the Fc fragment of IgG) is next applied from solution, which causes the L1-Fc chimera to bind to the protein A pattern. The L1-FC chimera micropattern on a PLL background selectively guides axon growth whereas the somas attach preferentially on PLL and the dendrites grow preferentially on PLL. This patterning scheme could be useful for inducing neuronal polarization.

Even simple micropatterns such as a border between two adhesion signals can be used not only to investigate guidance cues and signal transduction mechanisms acting at the nerve growth cone but can also be a very valuable tool to probe the intracellular mechanisms operating to change the direction of axon outgrowth. Paul Bridgman's laboratory at Washington University in St. Louis, Missouri, using PDMS blocks as masks to define the borders of stripes, has shown that growth cones (either from 13.5-day-old mouse embryo explants or from dissociated cells) integrate myosin II–dependent contraction for rapid, coordinated turning at borders of stripes of laminin plus poly-L-ornithine (PLO; on a background of PLO) in response to signals from laminin-activated integrin receptors (**Figure 6.44a** and the time-lapse series in **Figure 6.44d**). However, outgrowth continues across the borders when the neurons are pharmacologically treated with blebbistatin (which inhibits myosin II activity, see **Figure 6.44b** and

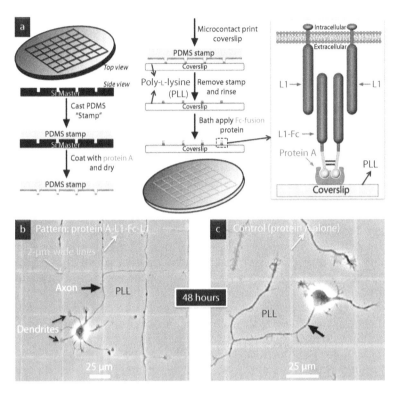

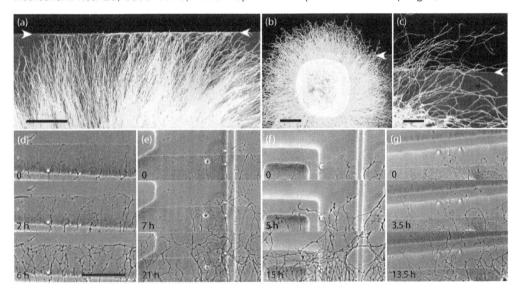

FIGURE 6.43 Microstamping of protein A for immobilizing axon guidance molecules. (From Anthony A. Oliva, Jr., Conrad D. James, Caroline E. Kingman, Harold G. Craighead, and Gary A. Banker, "Patterning axonal guidance molecules using a novel strategy for microcontact printing," *Neurochem. Res.* 28, 1639–1648, 2003. Reprinted with permission from Springer.)

FIGURE 6.44 Laminin stimulates axon outgrowth via growth cone myosin II activity. Scale bars: (a) 500 µm, (b) 250 µm, (c) 80 µm, and (d–g) 40 µm. (From Stephen G. Turney and Paul C. Bridgman, "Laminin stimulates and guides axonal outgrowth via growth cone myosin II activity," *Nat. Neurosci.* 8, 717, 2005. Reprinted with permission from the Nature Publishing Group.)

the time-lapse series in **Figure 6.44e**) or when neurons from a myosin IIB knockout mouse are used (**Figure 6.44c** and the time-lapse series in **Figure 6.44f**). Blebbistatin does not affect the outgrowth of myosin IIB knockout mouse axons (**Figure 6.44g**).

6.5.1.2 Axon Guidance by Microtopography

In 1997, a collaborative Scottish team led by Stephen Britland at the University of Glasgow and by Colin McCaig at the University of Aberdeen in Scotland (United Kingdom) performed the first study of the effects of microtopography on axon growth (**Figure 6.45**). They observed that the axons of *Xenopus* and rat hippocampal neurons could be guided by parallel quartz grooves as shallow as 14 nm (much shallower than the axon!) and as narrow as 1 μm.

If the objective is to guide axons, why not use microchannels directly so the axons are confined by four walls? For the same price, one could incorporate many connections, control perfusions, and produce gradients... In 2001, the laboratory of Hiroyuki Fujita at the University of Tokyo demonstrated basic axon growth in the first microfluidic neural guidance device, which was built by soft-lithographic techniques from a SU-8 master (**Figure 6.46**).

In 2003, Noo Li Jeon's group at the University of California designed a microfabricated "Campenot chamber" that allows for fluidically isolating the somas from the neurites in a neuronal culture (**Figure 6.47**). In traditional Campenot chambers, which are made of Teflon, grooves are scratched on collagen-coated plastic so that axons can grow under the Teflon barrier; unfortunately, the scratches are dimensionally undefined, the Teflon needs to be sealed with an unreliable vacuum grease (only ~30% of the devices work), and the Teflon is opaque (so it is not compatible with live cell imaging). Jeon's microfabricated Campenot chamber is transparent and consists of a set of parallel grooves that guide the growth of axons by topographical guidance from the somal chamber to the neuritic chamber (after ~4 days in culture). If a

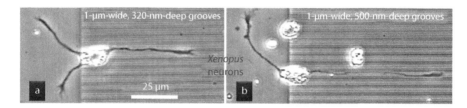

FIGURE 6.45 Axon guidance by microtopography. (From Ann M. Rajnicek, Stephen Britland, and Colin D. McCaig, "Contact guidance of CNS neurites on grooved quartz: Influence of groove dimensions, neuronal age and cell type," *J. Cell Sci.* 110, 2905–2913, 1997. Figure contributed by Stephen Britland.)

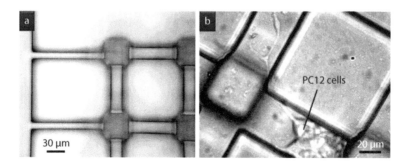

FIGURE 6.46 Microfluidic channels used as topographic guideposts. (From Laurent Griscom, Patrick Degenaar, Bruno LePioufle, Eichi Tamiya, and Hiroyuki Fujita, "Cell placement and neural guidance using a three-dimensional microfluidic array," *Jpn. J. Appl. Phys.* 40, 5485–5490, 2001. Reprinted with permission from the Japan Society of Applied Physics.)

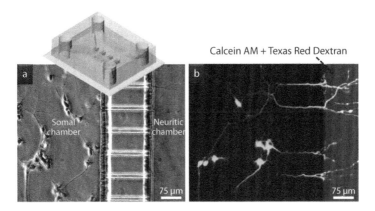

Calcein AM + Texas Red Dextran

FIGURE 6.47 A PDMS-based microfabricated Campenot chamber. (From Anne M. Taylor, Seog Woo Rhee, Christina H. Tu, David H. Cribbs, Carl W. Cotman, and Noo Li Jeon, "Microfluidic multi-compartment device for neuroscience research," *Langmuir* 19, 1551–1556, 2003. Figure contributed by Noo Li Jeon.)

solution is added to the neuritic chamber (and the right pressure balance is added to the inlets of the somal chamber), then there is only negligible flow from the somal chamber toward the neuritic chamber and diffusion can never overcome flow in the microchannels—thus, the neuritic chamber is effectively isolated from the somal chamber. (In practice, the balance of pressure can be difficult to perform because the inlet reservoir volumes are very small.) Thus, this device allows for applying drugs and insults to axons without disrupting the somas.

McCaig's group had studied the behavior of axons on shallow grooves (Figure 6.45), but how do axons react to much deeper topographies, say on the order of the size of their somas? The author's laboratory at the University of Washington in Seattle undertook a study of axons crossing microfabricated steps of various heights (ranging from 2.5 μm to 69 μm) and observed that the percentage of crossing axons decays monotonically as the step height increases (Figure 6.48). Axons of mouse cortical neurons were able to cross very shallow (2.5 μm deep) steps almost as

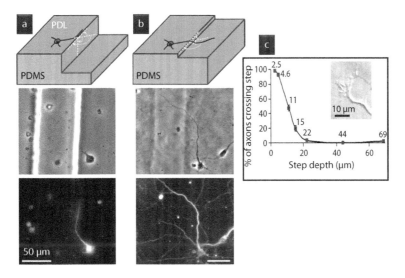

FIGURE 6.48 Growth cone sensing of microtopographical step height. (From Li, N. and Folch, A., "Integration of topographical and biochemical cues by axons during growth on microfabricated 3-D substrates," *Exp. Cell Res.* 311, 307, 2005. Figure contributed by Nianzhen Li.)

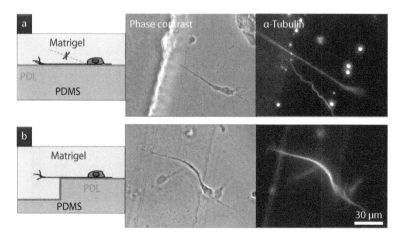

FIGURE 6.49 Integration of biochemical and microtopographical cues by the growth cone. (From Li, N. and Folch, A., "Integration of topographical and biochemical cues by axons during growth on microfabricated 3-D substrates," *Exp. Cell Res.* 311, 307, 2005. Figure contributed by Nianzhen Li.)

if the steps did not exist. On the other extreme, neurons always turned at steps deeper than 22 μm. In the middle, approximately 50% of axons turned when the step was about 11-μm-deep, possibly reflecting the size of the growth cone in culture (see **Figure 6.48c**, inset). The reason for this behavior is probably that the growth cone was "feeling" the substrate before deciding the growth direction. Indeed, for the 11 μm step, the incidence of turning was a strong function of the angle of approach: 100% of the axons that arrived at the step very tangentially (0–15 degrees) continued on the plateau whereas approximately 90% of the axons that arrived at the step orthogonally to it (76.90 degrees) ended up crossing it.

How about studying the reaction of the growth cone to topographical cues while conflicting biochemical cues are also present? The author's laboratory added Matrigel (a good neural growth substrate) to PDMS steps to observe the growth preferences of mouse cortical neurons (**Figure 6.49**). On flat **poly-D-lysine (PDL)**-coated PDMS substrates, the neurons preferred to grow on the PDL-coated substrates (**Figure 6.49a**), suggesting that neurons prefer PDL compared with Matrigel. However, on 22-μm-deep PDMS steps, the neurons that started growing on PDL-coated substrates preferred to grow into Matrigel (**Figure 6.49b**); if they had made any effort to "feel" the step, they would have turned, as they did for substrates that are not covered with Matrigel (see **Figure 6.48**). It is clear that these neurons have a preference to grow straight if the substrate is permissible, and that in doing so they integrate both biochemical and topographical cues.

6.5.1.3 Axon Guidance by Insoluble (Surface) Gradients

Studies in the last two decades have revealed that axons in the developing embryo are guided by extracellular signals to grow along specific paths toward their final synaptic targets. Axon guidance factors can be categorized according to their effect (attractive or repulsive) or according to their acting range (contact or long range), thus yielding four categories of axon guidance mechanisms. This is a simplification because different axons can respond to a given signal differently and even the same axon's response may change with time and space (e.g., with developmental clock, previous exposures to other signals, coincidence with other competing signals, etc.). The growth cone senses contact guidance factors through cell-adhesion receptors (integrins, **cadherins, NCAM, L1**, etc.) that enable it to move by modulating its adhesion to the ECM and to other cells. The growth cone also senses gradients of diffusible long-range factors through specialized receptors that are unique for each guidance factor and integrates all the stimuli to produce growth and motility. Yet reproducing these interactions in vitro has been difficult because neurons are not always are as responsive in culture as they are in the embryo and the concentrations/gradients used have to be guessed. The first studies used rudimentary gradient generators such as pipettes or the Boyden/Zigmond/Dunn chambers, which result in very low throughputs and poor reproducibility.

The first axon guidance study of neurons in a microfabricated gradient was performed by a collaborative team led by Venkatesh Murthy and George Whitesides at Harvard University (Figure 6.50). To simplify the experiment, the cells (rat hippocampal neurons) were exposed to an *immobilized* laminin gradient, which had been produced before seeding the cells. It is not clear whether such laminin gradients are present in vivo (in the hippocampus or elsewhere in the brain). Nevertheless, the work does raise an intriguing question: what is the appropriate mode of presentation of an axon guidance molecule (soluble or immobilized) in a cell culture experiment? In vivo, most axon guidance factors have motifs that make them bind to the ECM, at least partially, so they are not strictly "freely diffusing." It may be that a freely diffusing molecule is too mobile for the receptor to bind to it and transduce the signal to the intracellular signaling machinery. On the other hand, if the axon guidance factor is too bound by the surface (or the ECM), it may not signal properly to the cell.

Does the growth cone need to receive continuous feedback from the gradient to navigate, or can the signal be discontinuous as long as it gets integrated more often than the reaction time of the growth cone? In 2006, a German team led by Friedrich Bonhoeffer (Max Planck Institute in Tübingen) and Martin Bastmeyer (University of Karlsruhe, Germany) proposed to substitute the continuous gradients with discrete gradients, which are easier to microstamp as dots of varying spacing/dimensions (Figure 6.51). To form the densest part of the gradient, stripes as narrow as 0.3 µm separated by 0.3 µm were stamped. Growth cones of chick temporal retinal axons (which display a "stop reaction" in vivo at a particular location of ephrin gradients), but not nasal axons, were able to integrate discontinuous ephrin gradients and stop at a distinct zone in the gradient while still undergoing filopodial activity.

However, as we have seen, microstamping can be impractical (proteins must be dried and the stamp needs to be fabricated at high resolution), so Bonhoeffer and Bastmeyer repeated the experiment with a microfluidic gradient generator capable of patterning a "ladder" of proteins from solution (Figure 6.52). The biological findings were identical: temporal axons (but not nasal ones) were guided by discrete, immobilized ephrin gradients, and they were also stopped by certain concentration-dependent gradient values. This suggests a general validity of the

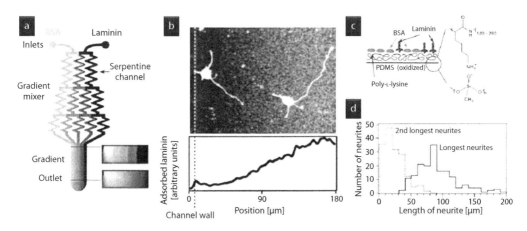

FIGURE 6.50 Axon growth dictated by absorbed laminin gradients. (From Stephan K. W. Dertinger, Xingyu Jiang, Zhiying Li, Venkatesh N. Murthy, and George M. Whitesides, "Gradients of substrate-bound laminin orient axonal specification of neurons," *Proc. Natl. Acad. Sci. U. S. A.* 99, 12542–12547, 2002. Copyright (2002) National Academy of Sciences, U. S. A. Figure contributed by George Whitesides.)

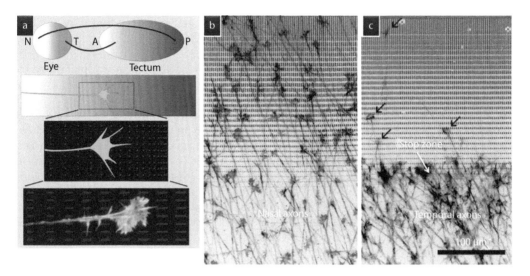

FIGURE 6.51 Micropatterned discrete ephrin gradients to simulate the optic tract development. (From Anne C. von Philipsborn, Susanne Lang, Jürgen Loeschinger, André Bernard, Christian David, Dirk Lehnert, Friedrich Bonhoeffer, and Martin Bastmeyer, "Growth cone navigation in substrate-bound ephrin gradients," *Development* 133, 2487–2495, 2006. Figure contributed by Friedrich Bonhoeffer and Martin Bastmeyer.)

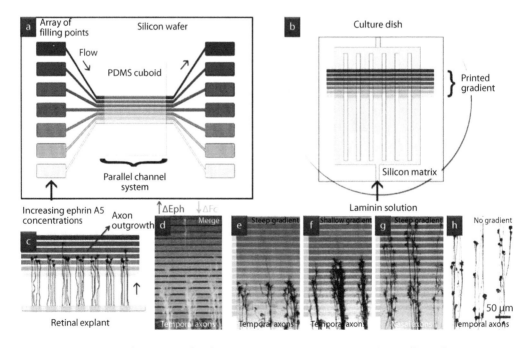

FIGURE 6.52 Axon guidance by microfluidically patterned discrete gradients. (From Susanne Lang, Anne C. von Philipsborn, André Bernard, Friedrich Bonhoeffer, and Martin Bastmeyer, "Growth cone response to ephrin gradients produced by microfluidic networks," *Anal. Bioanal. Chem.* 390, 809–816, 2008. Figure contributed by Friedrich Bonhoeffer and Martin Bastmeyer.)

micropatterned discrete gradient assay to probe axon guidance responses, at least for retinal axons.

FOR YOUR EYES ONLY

THE RETINA HAS A *FLAT* ANATOMY that (as brilliantly shown by modern fluorescence microscopy) is "painted" with *gradients* of axon guidance molecules. Therefore, retina gradients are ideally suited to mimicking by micropatterning and microfluidics techniques. Moreover, the retina, unlike the cortex and the hippocampus (the focus of most axon guidance research efforts to date), is a good source of highly homogeneous cell populations, the retinal ganglion cells (RGCs). RGCs respond to Netrin-1 gradients. During the development of the anterior visual pathway, the axon trajectories are simple, multiple relevant guidance molecules have already been identified (many tested with explants in vitro), and a common cause of blindness (optic nerve hypoplasia) is associated with defects in this process—which makes for great grant proposal opportunities. It is surprising that until very recently, Friedrich Bonhoeffer and his prolific academic offspring (Drescher, Bastmeyer, etc.) have been the only major group exploiting this rich mine with microfluidic tools. You might wonder how you didn't see this opportunity right before your eyes…

To further probe the role of bound protein gradients, a team led by Mu-ming Poo from the University of California at Berkeley has investigated the responses of hippocampal neurons to bound gradients of Netrin-1 and brain-derived neurotrophic factor (BDNF). Their novel hydrogel-based gradient generator is extremely easy to use and inexpensive, which should make it easy to disseminate to other biological laboratories. The gradient generator consists of an agarose stamp whose microchannels are filled with the protein solution of interest (**Figure 6.53a**). What

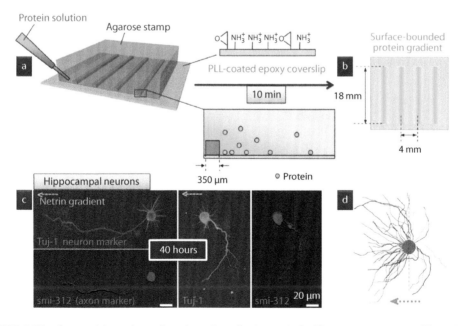

FIGURE 6.53 Axon guidance by surface-bound gradients created with an agarose stamp. (From Junyu Mai, Lee Fok, Hongfeng Gao, Xiang Zhang, and Mu-ming Poo, "Axon initiation and growth cone turning on bound protein gradients," *J. Neurosci.* 29, 7450–7458, 2009. Figure contributed by Mu-ming Poo.)

generates the gradient is the diffusion of protein through the agarose matrix; the protein gets immobilized on PLL by binding to the PLL-coated epoxy coverslip (Figure 6.53b). (Note: the protein could have been immobilized directly on the amine-binding epoxy groups, bypassing the PLL step.) This method is clever because, by contrast, methods that use PDMS devices (a) require fine control of the fluid flow rates to create gradients, and (b) agarose is very inexpensive and is molded in minutes. Poo and colleagues observed that the gradients of Netrin-1 and BDNF polarize the initiation and turning of axons in hippocampal neurons (Figure 6.53c and d). The response to BDNF can be attractive (at low-average density) or repulsive (at high-average density) depending on the basal level of cyclic adenosine monophosphate (cAMP) in the neuron.

6.5.1.4 Axon Guidance by Soluble Gradients

Although it is likely that, in vivo, secreted axon guidance factors are presented to neurons in some immobilized form, it is undeniable that the factors must *diffuse* to form the gradient; also, immobilized factors cannot be removed, so several groups have been designing microfluidic devices that allow for a dynamic control of the growing axon's microenvironment.

In 2005, Peter Asbeck's group from the University of California at San Diego designed a simple "alternate-choice" perfusion chamber for explants, whereby an explant is exposed to a laminar flow of two streams and the growing axons must ultimately decide for the attractive-factor stream (Figure 6.54). Neurites from spiral ganglion explants were shown to preferentially grow in medium containing neurotrophin-3 (NT-3) as opposed to medium without NT-3. The assay has a straightforward result but does not expose the axons to a well-defined gradient (it changes as the axon grows and as perfusion proceeds into the well).

Perhaps as a testimony to the technical difficulties involved in the experiment, the first article that exposed live neurons to a soluble gradient of an axon guidance factor (from the laboratory of Andre Levchenko at Johns Hopkins University) was not published until 2008. By using microvalves, the team was able to switch between two types of gradients ("N" or "\" shape). The cells were protected from shear stress because they were seeded in trenches

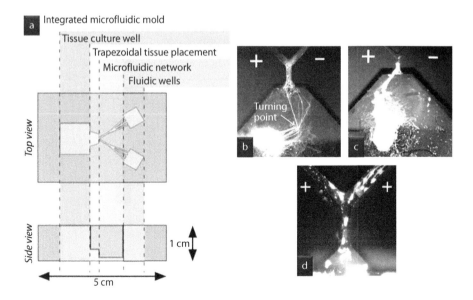

FIGURE 6.54 A microfluidic-choice device for studying axon guidance. (From John H. Wittig Jr., Allen F. Ryan, and Peter M. Asbeck, "A reusable microfluidic plate with alternate-choice architecture for assessing growth preference in tissue culture," *J. Neurosci. Methods* 144, 79–89, 2005. Reprinted with permission from Elsevier.)

(orthogonal to flow). Single (BDNF or laminin) as well as composite (BDNF plus laminin) gradients were studied. Each trench was subdivided into three zones for analysis, as the BDNF gradient was different in each of the zones. It was found that growth cones exhibit similar repulsive responses when an underlying laminin coating is uniform; on the other hand, when a linear "\"-shaped BDNF gradient was superimposed to a surface-bound "N"-shaped laminin gradient, the cells in different zones responded differently. In zone 1, growth cones were repelled by BDNF and extended up the laminin gradient; in zone 2, there was no preferred turning; and in zone 3, growth cones were attracted by BDNF and extended down the laminin gradient (Figure 6.55). This result suggests the possibility that axon guidance can be modulated simply by the relative abundance of a small number of factors in the microenvironment of the growth cone.

One of the best studied diffusible axon guidance factors are the netrins, a small family of proteins of approximately 600 amino acids that are expressed in various locations in the central nervous system (**CNS**), well known for their conserved role in attracting commissural axons to the midline. They also function in other attractive axon guidance pathways, such as the guidance of mammalian retinal axons and cortical efferent axons. Netrin-induced attraction is mediated by the DCC family of receptors. Netrins can also act as chemorepulsive signals; repulsion by netrins is important in directing axons that grow away from the midline, such as vertebrate trochlear motor axons. Although axonal turning by netrin was demonstrated in vitro by a team led by Marc Tessier Lavigne (then at the University of California, San Francisco) and

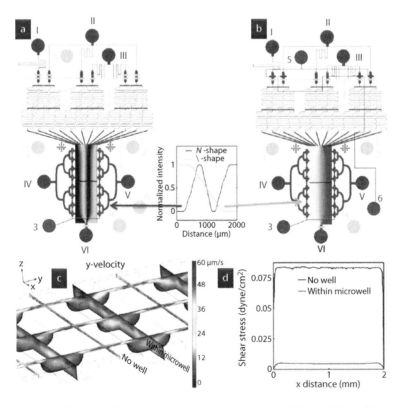

FIGURE 6.55 A low-shear gradient generator for probing axon guidance. (From C. Joanne Wang, Xiong Li, Benjamin Lin, Sangwoo Shim, Guo-li Ming, and Andre Levchenko, "A microfluidics-based turning assay reveals complex growth cone responses to integrated gradients of substrate-bound ECM molecules and diffusible guidance cues," *Lab Chip* 8, 227–237, 2008. Reproduced with permission from The Royal Society of Chemistry.)

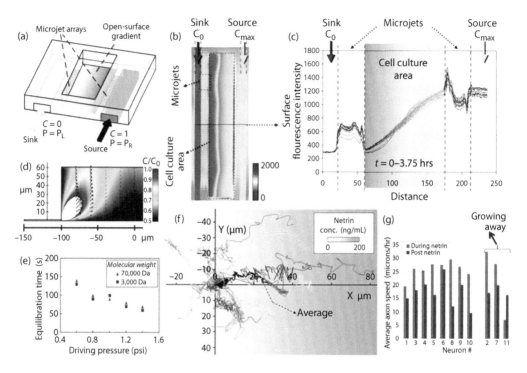

FIGURE 6.56 A "neuron-benign" microfluidic platform for detecting axon guidance responses of cortical neurons to netrin. (From Nirveek Bhattacharjee, Nianzhen Li, Thomas M. Keenan, and Albert Folch, "A neuron-benign microfluidic gradient generator for studying the response of mammalian neurons towards axon guidance factors," *Integr. Biol.* 2, 669–679, 2010. Reproduced with permission from The Royal Society of Chemistry. Figure contributed by Nirveek Bhattacharjee.)

Mu-ming Poo (then at the University of California, San Diego) using pipette-generated gradients in a now-legendary article in *Neuron* in 1997, they used frog neurons (their axons grow very fast), a nonmammalian cell type. Tessier Lavigne's group did show mammalian axon responses to netrin gradients, but using mammalian explants (which might be responding to other factors simultaneously). Hence, the exposure of slow-growing (isolated) mammalian neurons to a stable gradient for enough time (hours) to see a cell response is technically extremely difficult. The author's laboratory made use of a previously developed "cell-benign" gradient-generation technology, the "microjets" (see **Figure 3.81** in Section 3.9.3), to create gradients on open surfaces without exposing the cells to shear forces and without exposing the cells to the adverse conditions of microchannel confinement (**Figure 6.56a** through **e**). This technology allowed, for the first time, for visualizing axon turning responses toward netrin by isolated mouse cortical neurons (**Figure 6.56f**). Neuron growth slowed down immediately after netrin was retrieved, indicating that netrin also acts as a growth factor (**Figure 6.56g**).

6.5.1.5 Axon Guidance by Glial Cells

In 2006, Helen Buettner's group at Rutgers University in New Jersey cultured monolayers of Schwann cells (a type of glial cells) on top of stripes of laminin, which aligned and elongated the glial cells on top of the laminin stripes. Next, on top of the glial cells, they seeded rat spinal neurons and observed that their axons grew preferentially in the direction of the laminin stripes (**Figure 6.57**). Growth on unaligned substrates was not preferential. Thus, Schwann cell alignment can direct axon growth even in the absence of other axon guidance cues.

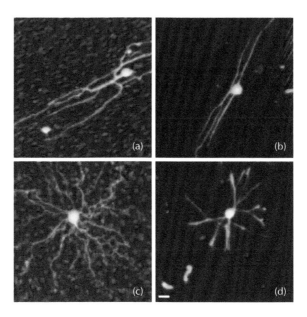

FIGURE 6.57 Directed neurite outgrowth on aligned Schwann cell monolayers. The images depict neurite outgrowth on (a) aligned Schwann cell monolayer, (b) micropatterned laminin, (c) unaligned Schwann cell monolayer, and (d) uniform laminin. Scale bar is 50 μm. (From Deanna M. Thompson and Helen M. Buettner, "Neurite outgrowth is directed by Schwann cell alignment in the absence of other guidance cues," *Ann. Biomed. Eng.* 34, 161–168, 2006. Reprinted with permission from Springer.)

6.5.2 Neuronal Polarization

In vivo, neurons grow their axons in prespecified directions because of the presence of gradient signals that "polarize" the neurons and determine which of the nascent processes become an axon. On traditional cell culture substrates, however, these directional signals are absent and the axon is determined at random. This limitation is a serious obstacle for the development of "neuronal networks on a chip," which aim to reproduce and study some of the characteristics of real networks on a reduced scale.

In 1998, a large collaborative group led by Carl Cotman at the University of California at Irvine and David Stenger at the Naval Research Laboratory was able to bias the polarity of embryonic hippocampal neurons in a specified direction given by a micropattern of aminosilane lines on a fluoroalkylsilane background (Figure 6.58). Each somata, adhered to a 25-μm-diameter aminosilane island, was "offered" four 5-μm-wide aminosilane paths along which the cell was able to extend processes; however, three of the aminosilane paths were broken in 10-μm-long segments, each separated by 10-μm-long fluoroalkylsilane spacings, whereas the remaining aminosilane path connected to the soma was continuous. Approximately 76% of the island-confined cells developed a process that was 100 μm or longer (along the continuous aminosilane line) which was identified as an axon, whereas the other processes were identified as dendrites. Identification was possible by immunostaining of the microtubule-associated proteins MAP2 (depleted in developing axons) and MAP5 (concentrated in axons undergoing rapid elongation) and of the neurofilament polypeptide NF150 (localized in axons late in postnatal development). Despite the fact that the surface patterning might be considered "outdated," this early work remains an outstanding example of micrometer-scale engineering and control of cell function.

Recently, a team led by Sarah Heilshorn at Stanford University and Luke Lee at the University of California (Berkeley) designed a microfluidic device that presents hippocampal neurons with biochemically different attachment sites and axonal guidance patterns, which direct neuronal polarization (Figure 6.59). To facilitate the filling of the intricate microfluidic networks with

351

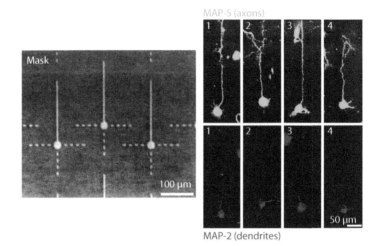

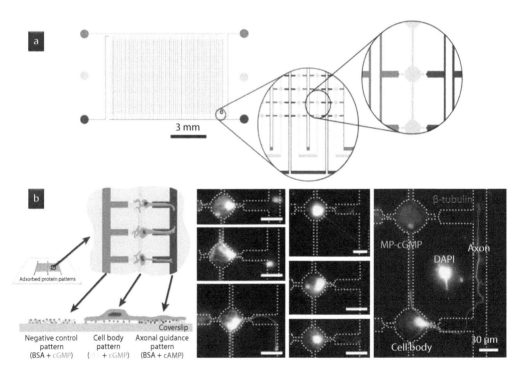

FIGURE 6.58 Determination of neuronal polarity using SAMs. (From David A. Stenger, James J. Hickman, Karen E. Bateman, Melissa S. Ravenscroft, Wu Ma, Joseph J. Pancrazio, Kara Shaffer, Anne E. Schaffner, David H. Cribbs, and Carl W. Cotman, "Microlithographic determination of axonal/dendritic polarity in cultured hippocampal neurons," *J. Neurosci. Methods* 82, 167–173, 1998. Figure contributed by David Stenger.)

FIGURE 6.59 Neuronal polarization directed by cAMP/cGMP micropatterns. (From J. Tanner Nevill, Alexander Mo, Branden J. Cord, Theo D. Palmer, Mu-ming Poo, Luke P. Lee, and Sarah C. Heilshorn, "Vacuum soft lithography to direct neuronal polarization," *Soft Matter* 7, 343–347, 2011. Reproduced with permission from The Royal Society of Chemistry.)

three different solutions, a "vacuum PDMS pumping" scheme (see Figure 3.62 in Section 3.8.4.2) was chosen. To induce polarization and axon pathfinding, they used a membrane-permeable, fluorescently tagged form of cAMP (MP-cAMP), a cytosolic second messenger implicated in axon guidance and neuronal polarization; MP-cAMP happens to bind or associate with macro-molecules such as bovine serum albumin (BSA), so it could be coabsorbed on the surface at the same time as BSA. Because cyclic guanosine monophosphate (cGMP) often plays antagonistic roles to cAMP, MP-cGMP was coadsorbed with BSA as a negative control to deter axon forma-tion. For the same reason, MP-cGMP was coadsorbed with PLL to discourage axon growth on the cell body patterns. It was necessary to locate the BSA/MP-cAMP and BSA/MP-cGMP patterns approximately 2 μm away from each PLL/MP-cGMP island to orient the neurites in the correct direction (even though it requires that the growth cone momentarily contacts the nonadhesive, plain glass substrate), similarly to Stenger's previous work (see Figure 6.58). With the orientation projections added to the design, approximately 60% of the cells attached to PLL/MP-cGMP sites grew axons along the intended BSA/MP-cAMP pattern, whereas approximately 31% grew them on the more adhesive PLL/MP-cGMP regions. Axons were rarely observed on the BSA/MP-cGMP pattern (~8%) and were never observed on plain glass.

6.5.3 Synaptogenesis

A paramount goal in neuroscience is to understand the cellular and molecular mechanisms of synapse formation ("synaptogenesis"). The nerve–muscle synapse or NMJ, because of its accessi-bility to experimental manipulation, has provided many insights into the molecular and cellular mechanisms of synaptogenesis.

NMJ SYNAPTOGENESIS

THE SEQUENCE OF MOLECULAR SIGNALS leading to NMJ formation is qualitatively well known. During development, axons grow along large distances before the tip contacts a muscle cell. Contact occurs at the same developmental stage when myoblasts are fusing to form myotubes. Three key molecules are secreted by the axon tip: **agrin**, **neuregulin**, and the neurotransmitter **acetylcholine** (ACh), which then interact with specific recep-tors in a small area of the muscle cell and induce different aspects of synapse formation. Synaptogenesis is initiated by the secretion of agrin (a heparan sulfate proteoglycan) by the nerve terminal, which is involved in the clustering and stabilization of AChRs at the future synapse. (Both the nerve terminal and the muscle cell produce agrin, but muscle agrin does not play a significant role in AChR clustering; therefore, we will henceforth refer to neural agrin only.) Not surprisingly, agrin-deficient mutant mice undergo defec-tive neuromuscular synaptogenesis. The muscle-specific tyrosine kinase (MuSK) has been identified as an essential component of the agrin receptor complex, if not agrin's receptor itself.

AChR clustering. Approximately two dozen proteins coaggregate postsynaptically at the NMJ. Rapsyn, in particular, is essential for the aggregation of AChRs. Embryo imag-ing experiments have shown that innervated AChR clusters in mice are highly dynamic, with the AChRs continuously being degraded, inserted, and recycled (receptor recycling is turned off when synaptic activity is blocked), with extrasynaptic AChRs contributing to the synaptic AChR pool.

The laminin pathway. Laminin-1, a major component of the muscle cell extracellular basal lamina, also induces AChR clustering, but through an alternative pathway, possi-bly associated with the well-known laminin-induced clustering of integrins. Denervation experiments in vivo have shown that there are components in the basal lamina that are sufficient to trigger AChR clustering on the muscle membrane, but possibly because agrin is by then already bound to the basal lamina. In addition, unless the muscle basal lamina is removed, AChRs remain focalized to the former NMJ site after denervation, and the nerve terminal continues to release ACh vesicles in the absence of the muscle. It seems that the first 130 amino acids from the N-terminus of agrin bind to laminin-1 in the basal lamina; it is therefore conceivable that the basal lamina's role in the maintenance of the NMJ is to provide a physicochemical support for agrin.

It was long held that agrin was the *cause* of AChR clustering (the "agrin hypothesis"), which initiated a decades-long controversy that has only been recently resolved. The observation that traditional myotube cultures form punctated clusters of AChRs on bath or focal application of agrin or an agrin fragment contributed to fuel the agrin hypothesis. For example, a team led by Bruce Wheeler, then at the University of Illinois at Urbana-Champaign, cultured **C2C12** myotubes on microstamped agrin micropatterns and observed that the myotubes formed AChRs clusters above some of the agrin lines (**Figure 6.60**). In this experiment, however, the myotubes are exposed to agrin for days, not briefly in time like in vivo, and the agrin is presented to them as an immobilized molecule. The author's group at the University of Washington in Seattle confirmed that if the focal agrin stimulus is delivered in soluble form using a microfluidic delivery device (**Figure 6.13**) at the right developmental time, i.e., right after their fusion into myotubes, then the myotubes also react forming small AChR clusters.

There is now firm evidence that AChR clustering precedes nerve contact (and thus, agrin secretion). The midline of the muscle forms clusters of AChRs even in the absence of innervation. Imaging of embryos (both in mouse and zebrafish) has revealed that large "protosynaptic" AChR aggregates of normal synaptic morphology (similar to a "pretzel") form on the muscle membrane before the arrival of the nerve. In myotube cultures, AChR clusters of complex morphologies (strikingly similar to those found in vivo) can, also form in the absence of agrin. What is, then, the role of agrin, given that it is not the cause of AChR clustering?

To find out, the author's group used a microfluidic device where single myotubes had been microengineered across the width of the device (see **Figure 6.13** in Section 6.2.3), making sure that the myotubes expressed AChR clusters before agrin application (myotubes cultured on laminin or Matrigel stripes work well). Next, agrin was focally delivered to the center of the device as a laminar stream (**Figure 6.61a**) and the evolution of AChR clusters was followed over

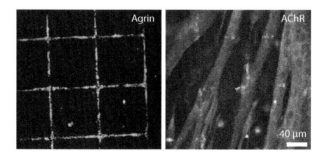

FIGURE 6.60 Local AChR clustering induced by surface-bound agrin micropatterns. (From Toby Cornish, Darren W. Branch, Bruce C. Wheeler, and James T. Campanelli, "Microcontact printing: A versatile technique for the study of synaptogenic molecules," *Mol. Cell. Neurosci.* 20, 140–153, 2002. Reprinted with permission from Elsevier. Figure contributed by Bruce Wheeler.)

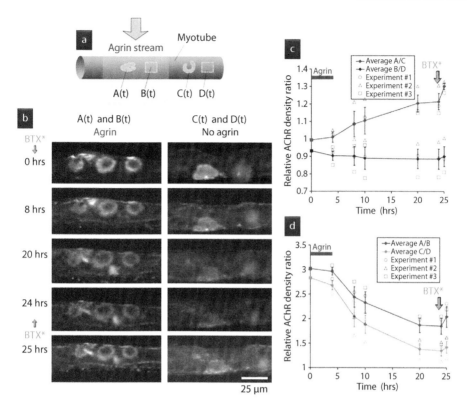

FIGURE 6.61 Localized dynamics of AChR clustering in response to microfluidic focal agrin stimulation. (From Tourovskaia, A., Li, N., and Folch, A., "Localized acetylcholine receptor clustering dynamics in response to microfluidic focal stimulation with agrin," *Biophys. J.* 95, 3009, 2008. Figure contributed by Anna Tourovskaia.)

time (see **Figure 6.61b**). It was found that preexisting ("pretzel") clusters that had been exposed to agrin became noticeably more resistant to degradation than those that were not exposed to agrin on the same myotube (**Figure 6.61c**). In support of the hypothesis that agrin's role is to stabilize existing clusters, in vivo experiments have shown that the agrin-predating protosynaptic clusters are stabilized if "found" by the nerve but dissolve if not innervated. An additional role of agrin seems to be to counteract the declustering action of ACh because synapses form in the absence of agrin provided that ACh is also absent. Another in vivo study has found that in denervated muscle fibers AChR degradation slows down when the muscle is transfected to express neural agrin.

Taher Saif's laboratory from the University of Illinois at Urbana-Champaign has recently challenged the established dogma that synaptic receptor clustering is only due to a series of biochemical processes: his group has shown that *forces* are also important, at least in the development of the neuromuscular system of **Drosophila**. Conveniently, the NMJs of anterior corner cell motoneurons in Drosophila are readily accessible and can be visualized by fluorescence microscopy in GFP-transgenic animals, with highly stereotyped innervation patterns. A microfabricated cantilever probe was used to measure the applied forces (**Figure 6.62b**). Neurotransmitter vesicle at the NMJs was shown to vanish after surgically severing the axon from the cell body (**Figure 6.62c**). The clustering was restored when tension was reapplied to the severed end of the axon and it was intensified in the NMJs of intact axons when these were stretched mechanically by pulling on the postsynaptic muscle.

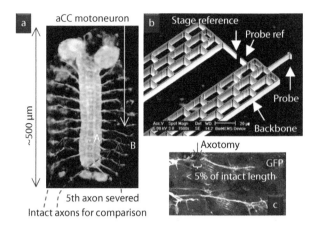

FIGURE 6.62 Neurotransmitter clustering induced by mechanical forces. (From Scott Siechen, Shengyuan Yang, Akira Chiba, and Taher Saif, "Mechanical tension contributes to clustering of neurotransmitter vesicles at presynaptic terminals," *Proc. Natl. Acad. Sci. U. S. A.* 106, 12611–12616, 2009.)

6.5.4 Emergent Properties of Neuronal Networks

One of the paramount goals of neuroscience is to understand how brain behavior emerges from the summed electrophysiological activity of each neuron. Unfortunately, no technology is within reach to record (or make sense of) all neuronal activity in a mammalian brain, so researchers are forced to study simpler systems, such as the intact nervous systems of slugs or worms, brain slices or ganglia, or neuronal cultures. Each system has its technical advantages, presents its own technical challenges, and has a horde of supporters and detractors that fight about their biological legitimacy with a vehemence that is probably unnecessary, because each system has a well-established trade-off between tractability and complexity.

6.5.4.1 Bottom-Up Approach: Neuronal Cultures

Dissociated neuronal cultures represent a "bottom-up" approach to try to understand the very basic properties of neuronal networks, with the advantage that cells can be accessed for recording by surface microelectrodes. However, the architecture of the networks lacks the complexity found in vivo and the cell's physiology is likely irreversibly damaged by the isolation and culture protocols. Early work in 1975 by Paul Letourneau, then at Stanford University (**Figure 2.25**), showed that dissociated neurons were able to grow along micropatterned proteins, which promised chips that would wire up a "brain on a chip." In 1988, Bruce Wheeler's group at the University of Illinois at Urbana-Champaign used MEAs to record from an *Aplysia* ganglion and, later, applied correlation algorithms to record from neuronal cultures. Jerry Pine's laboratory at Caltech has developed "neuro-cages" since the 1990s to facilitate trapping neurons on MEA arrays (**Figure 5.16**).

The extreme difficulty of correlating the information being recorded with a real neurophysiological process has tempered the early feeling that these cell culture systems could be used to understand the whole brain ("Brain on a Chip") and has changed the focus of the field to more practical goals. Steve Potter's group, then at Caltech, using rat cortical neurons randomly seeded on a MEA, claimed that the signals recorded by the MEA could be used to control an artificial animal or "Animat." Special software recognizes, in approximately 8 minutes, patterns of activity that arise spontaneously from the neuronal network and uses them to tell the Animat which way to move. The Animat sends a signal back to designated electrodes in the MEA when, for example, it collides with a wall—a mimicry of a sensory input into the "brain" on the MEA. Eventually, it is hoped that such cultures could be used to control "intelligent" robotic devices. However, the randomness of the connections that arise in the dish make it difficult to envision how in vitro "thought" processes could be reproducibly produced in a meaningful way. Clearly, the system has interesting applications in creating random patterns, as vividly illustrated by the art drawing in Figure 6.63d, which was drawn by a set of robotic arms under the command of a neuronal network on a MEA; a video camera recording the progress of the drawing sent signals back to some of the electrodes, which presumably altered in an unpredictable fashion the neural activity of the network and, thus, the drawing (dubbed "MEArt").

For some specific questions, small neuronal networks may be best suited. Peter Fromherz's group, then at the Max Planck Institute for Biochemistry in Martinsried (Germany), cultured pairs of snail neurons on linear protein patterns to precisely measure synaptic conductance (Figure 6.64). In classical measurements of the synaptic conductance, the conductance is estimated from voltage recordings at the cell's body assuming that each cell is isopotential, which neglects the contribution of the neurites. Fromherz's group used cable theory to derive an analytical expression that relates the synaptic conductance to voltage recordings at the cell bodies and to the neurite properties. To obtain the cable properties, they used a voltage-sensitive dye to

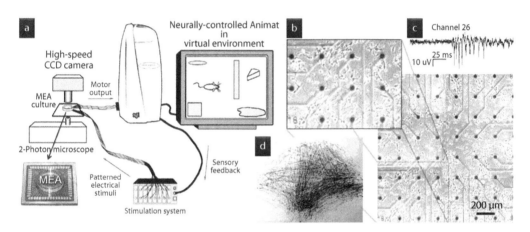

FIGURE 6.63 Neuronal network cultures on microelectrode arrays. (From Thomas B. DeMarse, Daniel A. Wagenaar, Axel W. Blau, and Steve M. Potter, "The neurally controlled Animat: Biological brains acting with simulated bodies," *Auton. Robots* 11, 305–310, 2001. Reprinted with permission from Springer. Figure contributed by Steve M. Potter.)

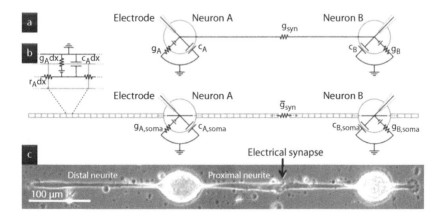

FIGURE 6.64 Neuron pair with electrical synapse. (From Astrid A. Prinz and Peter Fromherz, "Effect of neuritic cables on conductance estimates for remote electrical synapses," *J. Neurophysiol.* 89, 2215–2224, 2003. Figure contributed by Peter Fromherz.)

record spatiotemporal maps of signal propagation in the neurites and concluded that the non-isopotential model deviated from the synaptic conductance estimates of the isopotential theory by approximately 13%, which is not negligible.

Eshel Ben-Jacob's group from Tel-Aviv University in Israel has been able to induce collective modes of neuron firing in the activity of cultured (randomly organized) neural networks (**Figure 6.65**). The recordings are performed via a standard MEA and stimulation of selected neurons is performed by injecting microdroplets just above the desired locations. According to Ben-Jacob, previous attempts to trigger memories in neuronal networks in vitro failed because they focused on excitatory neurons, which resulted in randomly escalated activity that does not mimic what occurs when new information is learned. In this study, microdroplets of inhibitory antagonist (picrotoxin, an antagonist of **GABA**, the primary inhibitory neurotransmitter) were delivered locally instead. The chemical suppression of the inhibitory neuron created a pattern kicked off by a neighboring excitatory neuron that was now free to fire. Other neurons in the culture began to fire one by one as they received an electrical signal from one of their neighbors. This continued in the same pattern, which repeated for over a day. This new sequence of activity coexisted with the electrical pattern that was spontaneously generated when the neural culture was initially linked (correlation matrix in **Figure 6.65f**). A day later, they imprinted a third pattern starting at a different inhibitory synapse (correlation matrix in **Figure 6.65e**). Surprisingly, it was able to coexist with the other motifs, showing that chemical signaling might play a crucial role in memory and learning.

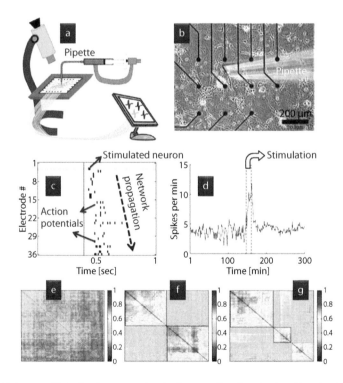

FIGURE 6.65 Using multielectrode arrays to study neural networks in vitro. In (c), each row is a binary bar code representation of the activity of a different neuron, and a bar is plotted each time the neuron fires an action potential. The time axis is divided into 10-ms bins. One injection is sufficient to initiate a single synchronized bursting event with a distinct pattern of neuronal firing. (From Itay Baruchi and Eshel Ben-Jacob, "Towards neuro-memory-chip: Imprinting multiple memories in cultured neural networks," *Phy. Rev. E* 75, 050901(R), 2007. Copyrighted by the American Physical Society. Figure contributed by Eshel Ben-Jacob.)

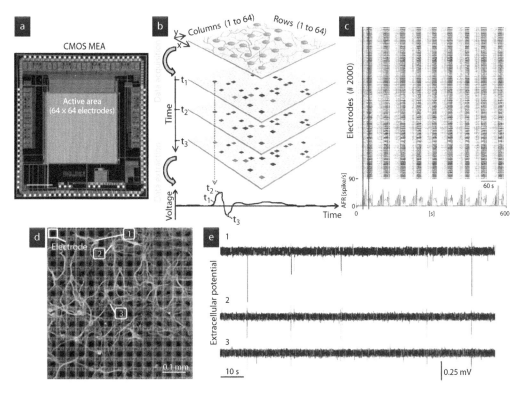

FIGURE 6.66 High spatiotemporal resolution electrophysiological recordings from in vitro neuronal networks using a pixel sensor array. (From Luca Berdondini, Kilian Imfeld, Alessandro Maccione, Mariateresa Tedesco, Simon Neukom, Milena Koudelka-Hep, and Sergio Martinoia, "Active pixel sensor array for high spatio-temporal resolution electrophysiological recordings from single cell to large scale neuronal networks," *Lab Chip* 9, 2644–2651, 2009. Reproduced with permission from The Royal Society of Chemistry.)

In 2009, a large collaboration headed by Luca Berdondini for the Italian Institute of Technology and Sergio Martinoia from the University of Genova used a CMOS pixel sensor array to record from cultured embryonic rat hippocampal neurons (**Figure 6.66**). The CMOS MEA has the advantage that each microelectrode has an integrated low-noise amplifier and the array has a real-time acquisition/processing board (for multiplexing the signals, so addressability by rows and columns does not require additional electronics and a temporal resolution of 8 μs/pixel on 64 selected pixels is possible). Because the cells can be imaged on the array, this tool could be very useful to study the interplay between architecture and function in small neuronal assemblies.

6.5.4.2 Brain Slices on a Chip

Neuroscientists have long been aware that dissociated neurons on a petri dish do not form very interesting networks, both because they have been damaged by the dissociation process and because they lack the microenvironmental cues to wire up the proper three-dimensional networks they form in the real brain. Brain slices, on the other hand, with thicknesses ranging from tens to hundreds of microns, can also be kept in a petri dish but contain semi-intact network architectures, so neuroscientists prefer them when they are trying to address network-related questions.

Peter Fromherz' group from the Max Planck Institute for Biochemistry in Martinsried (Germany) built a field-effect transistor (FET) array with 16,384 elements (each 7.8 μm in size, array size ~1 mm²) capable of recording (~2 kHz bandwidth) the electrical activity from the surface of 400-μm-thick cultured hippocampal slices (**Figure 6.67**). Reading fewer elements

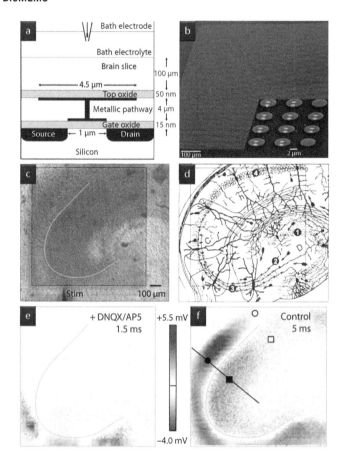

FIGURE 6.67 FET array for recording electrical field potentials from brain slices. (From M. Hutzler, A. Lambacher, B. Eversmann, M. Jenkner, R. Thewes, and P. Fromherz, "High-resolution multi-transistor array recording of electrical field potentials in cultured brain slices," *J. Neurophysiol.* 96, 1638, 2006. Figure contributed by Peter Fromherz.)

allowed for enhanced time resolution. Note that the FET design was improved from previous work in which the gate of the FET was exposed to electrolytes from the bath, thus causing migration of salt ions into the silicon lattice and the inevitable change in conductivity (see **Figure 5.46**); here, the gate is protected by an insulating TiO_2 layer and the cell transmits the voltage to the gate via capacitive coupling through a metallic pathway (**Figure 6.67a**). The spatial continuity of the records provided time-resolved images of evoked field potentials and allowed the detection of functional correlations over large distances. Slices were stimulated with a tungsten microelectrode positioned in the pyramidal layer of the CA3 area (black triangle in **Figure 6.67c**), the circuitry of which is shown in **Figure 6.67d**. Fast-propagating waves of presynaptic action potentials and patterns of excitatory postsynaptic potentials (not manifested in the presence of the inhibitory toxins DNQX/AP5) across and along the arch of cornu ammonis (drawn as a gray curved line in **Figure 6.67c, e,** and **f**) were demonstrated.

Although Fromherz' work addresses the challenge of recording from multiple points of a slice, another challenge remains: how can we *biochemically stimulate* multiple points or different areas of a slice? Justin Williams' laboratory at the University of Wisconsin has built microfluidic perfusion chambers that allow for delivering heterogeneous laminar flows onto slices (~530–700-μm-thick medullary brain slices from **P0–P4** neonatal rats), so that the slices are locally perfused in stripes of different biochemical environments (**Figure 6.68**). The design

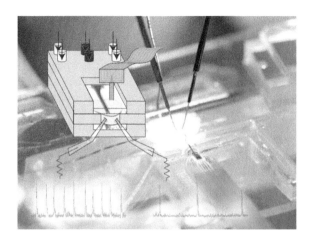

FIGURE 6.68   Brain slice in a microfluidic chip. (From A. J. Blake, T. M. Pearce, N. S. Rao, S. M. Johnson, and J. C. Williams, "Multilayer PDMS microfluidic chamber for controlling brain slice microenvironment," *Lab Chip* 7, 842–849, 2007. Reproduced with permission from The Royal Society of Chemistry.)

uniquely achieved independent control of fluids through multiple channels in two separate fluid chambers, one above and one below the slice. Simultaneous electrophysiological recordings from the edge of the slice while the biochemical environment was modulated were also possible.

6.5.4.3 *Caenorhabditis elegans* in a Chip

Many neuroscientists interested in studying neuronal networks will contend that even brain slices are a damaged system, and far too complex—it's better to start with lower organisms that happen to have a well-defined number of neurons with well-known connectivity, such as the sea slug *Aplysia* or the roundworm *Caenorhabditis elegans*, and that can also be manipulated genetically. In 2004, the laboratory of Cornelia Bargmann, then at the University of San Francisco in California (and since 2005 at the Rockefeller University in New York), pioneered the use of microfluidics for chemosensation research as well as the use of microfluidics in combination with *C. elegans*.

THE ELEGANT WORM

C. elegans IS A REMARKABLE, TRANSPARENT NEMATODE approximately 1 mm in length. It has become a model organism for molecular biologists, developmental biologists, and neurobiologists since South African biologist Sydney Brenner first started studying it in 1974 (which eventually awarded him the Nobel Prize in Physiology or Medicine in 2002). It remains viable after repeated cycles of freezing and thawing. In the wild, *C. elegans* feeds on bacteria. It is one of the simplest organisms with a nervous system, which consists of just 302 neurons whose location and synaptic connections are all known. The developmental fate of every single cell in the adult animal (1031 in male worms and 959 in hermaphrodites), which is largely invariant from individual to individual, has also been mapped out. It has not escaped the attention of researchers that a significant proportion of its nervous system is devoted to chemosensation: 32 chemosensory neurons (generally labeled with three letters: ASA, AWC, ASH, etc.), which penetrate the cuticle to expose their sensory cilia to the environment.

Using a simple oxygen gradient generator (one inlet with oxygen, one inlet with nitrogen, and a central PDMS chamber filled with the worms, all assembled on agar—a substrate onto which *C. elegans* crawls easily and bacteria thrive, so it is a good mimic of the worm's natural soil substrate), in 2004, the Bargmann group found that the worms avoid both high and low oxygen concentrations (**Figure 6.69**); however, they cluster where the concentration of oxygen is between 5% and 10%, which suggests a plausible mechanism for seeking the environments created by oxygen-consuming bacteria (on which the worms feed).

As it happens, *C. elegans* can also be manipulated very efficiently with microfluidics, as we saw in Chapter 5 (see **Figure 5.60**). Fatih Yanik's laboratory at MIT has developed microfluidic systems containing pneumatic microactuators that immobilize the worms in specific locations of the device in which they undergo femtosecond laser "surgery" (e.g., axotomy) or simply imaging (e.g., for screening), as shown in **Figure 6.70**. In a typical experiment, a neuron with a

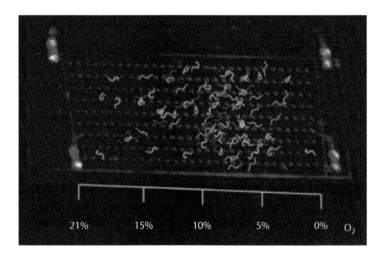

FIGURE 6.69 First use of *C. elegans* in microfluidics. The worms have been digitally added to convey their relative density in the device. (From J. M. Gray, D. S. Karow, H. Lu, A. J. Chang, J. S. Chang, R. E. Ellis, M. A. Marletta, and C. I. Bargmann, "Oxygen sensation and social feeding mediated by a *C. elegans* guanylate cyclase homologue," *Nature* 430, 317, 2004. Figure contributed by Cori Bargmann.)

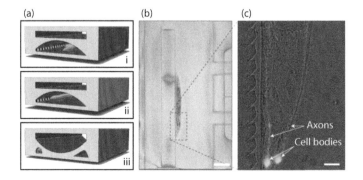

FIGURE 6.70 Recording and manipulation of *C. elegans* nervous system. Scale bar is 250 μm. In (c), the labeled neurons are GFP-labeled fluorescent posterior lateral mechanosensory neurons. (From Fei Zeng, Christopher B. Rohde, and Mehmet Fatih Yanik, "Sub-cellular precision on-chip small-animal immobilization, multi-photon imaging and femtosecond-laser manipulation," *Lab Chip* 8, 653–656, 2008. Reproduced with permission from The Royal Society of Chemistry.)

known network role is partially cut off from the network and the associated behavior is monitored from some time after the "surgery". Note that the device only needs to ensure the *overall viability* of the organism while the organism takes care of neuronal maintenance (e.g., recovery from surgery). Because the exact developmental lineage of the neuron being ablated is often known, it may be possible to infer from these experiments general rules on how to build small neuronal networks.

6.5.5 Olfaction

Using simple PDMS microfluidic mazes assembled on agar, in 2005, the Bargmann laboratory (Rockefeller University, New York) found out that, by modifying its olfactory preferences, *C. elegans* can learn to avoid pathogenic bacteria (the worms had previously learned the taste of this bacterial strain by eating it). The worms, placed in a central chamber approximately 1.5 cm away from the bacteria, were consistently repelled by the pathogenic bacteria strain and attracted by the nonpathogenic one (**Figure 6.71**).

Microfluidics can also be used to design traps in which the animal can be locally perfused and their neuronal responses imaged in real time with cellular resolution (**Figure 6.72**). The animal is introduced into a funnel-shaped microchannel so that, as it crawls forward, it gets stuck, sticking its nose out into a perfusion chamber. At this point, the animal is immobile and its chemosensory neurons can be functionally imaged by calcium imaging without anesthesia (see ASH neuron in **Figure 6.72b**) while they are perfused with various stimulants. **Figure 6.72c** shows the distinct patterns of ASH neuron response when comparing young and old (3 days older) worms, suggesting that the functionality of sensory neurons is altered with age.

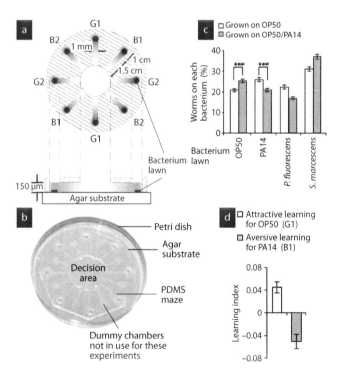

FIGURE 6.71 Microfluidic "olfactory mazes" reveal worm learning behaviors. (From Yun Zhang, Hang Lu, and Cornelia I. Bargmann, "Pathogenic bacteria induce aversive olfactory learning in *Caenorhabditis elegans*," *Nature* 438, 179–184, 2005. Adapted with permission from the Nature Publishing Group.)

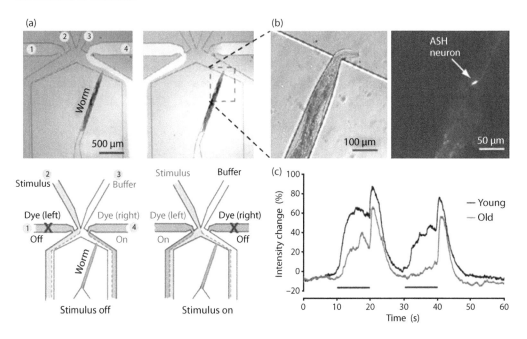

FIGURE 6.72 Trapping and local stimulation of *C. elegans* using microfluidics. (From Nikos Chronis, "Worm chips: Microtools for *C. Elegans* biology," *Lab Chip* 10, 432–437, 2010; see also Nikos Chronis, Manuel Zimmer, and Cornelia I. Bargmann, "Microfluidics for in vivo imaging of neuronal and behavioral activity in *Caenorhabditis elegans*," *Nat. Methods* 4, 727, 2007. Reproduced with permission from The Royal Society of Chemistry.)

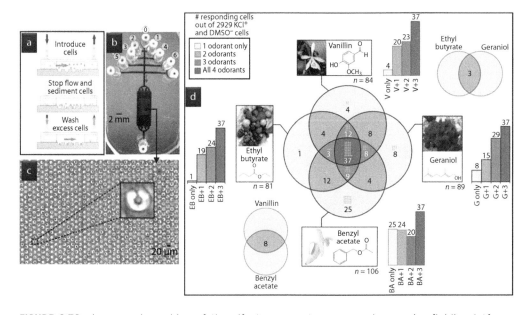

FIGURE 6.73 Large-scale probing of the olfactory receptor space using a microfluidic platform. (From Xavier A. Figueroa, Gregory A. Cooksey, Scott V. Votaw, Lisa F. Horowitz, and Albert Folch, "Large-scale investigation of the olfactory receptor space using a microfluidic microwell array," *Lab Chip* 10, 1120–1127, 2010. Reproduced with permission from The Royal Society of Chemistry.)

The author's group (University of Washington, Seattle) in collaboration with Lisa Horowitz from Linda Buck's laboratory at the Fred Hutchinson Cancer Research Center (Seattle) has used large PDMS microwell arrays to screen for odorant responses from dissociated mouse olfactory sensory neurons (OSNs). The microwell array was integrated into the floor of a simple nine-inlet microfluidic device (Figure 6.73a and b). Introducing a cell suspension into the chamber traps the OSNs into the microwells with high efficiency after the flow is restarted to remove excess cells (Figure 6.73c). The array allows for simultaneously screening more than 20,000 cells using calcium imaging, of which approximately 2900 were KCl-responsive (and assumed to be OSNs) by the end of the experiment. Therefore, the vast majority of the olfactory receptor space (mice express ~1000 olfactory receptors) are represented in the experiment. During the experiment, the cells also responded to one or more of four chemicals, chosen from a group of fruity pleasant smells (vanillin, berry, geraniol, and banana), and they did not respond to two control pulses of DMSO (the solvent in which the odorants are dissolved). The results confirmed the common finding that "broadly tuned" cells (those that respond to a variety of compounds) are more abundant, and "narrowly tuned" cells are the least abundant ones (except for those tuned to benzyl acetate; see the Venn diagram in Figure 6.73d).

6.5.6 Glial Biology

Signaling between glia and neurons is vital for the development and physiology of the CNS. Micropatterning and microfluidic techniques offer tools to accurately control the signaling between different cell types by controlling the spacing between microfabricated patterns. A team led by Marc Porter and Phil Haydon, both formerly of Iowa State University, coated glass slides with (cell-adhesive) PLL, microfluidically patterned them with (cell-repellent) agarose tracks, and plated astrocytes on top of them (Figure 6.74a). This team was able to settle a long-standing

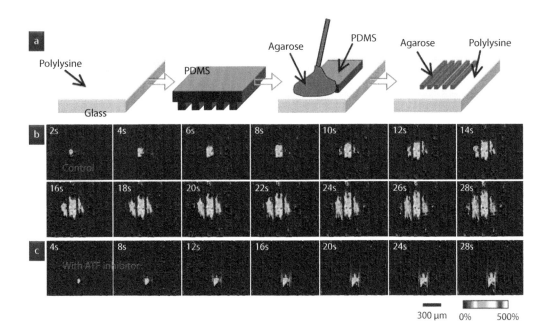

FIGURE 6.74 Transmission of calcium waves across microscale gaps between glial cells. (From Hajime Takano, Jai-Yoon Sul, Mary L. Mazzanti, Robert T. Doyle, Philip G. Haydon, and Marc D. Porter, "Micropatterned substrates: Approach to probing intercellular communication pathways," *Anal. Chem.* 74, 4640–4646, 2002. Figure contributed by Phil Haydon.)

question regarding the second messengers responsible for the propagation of calcium waves: are they released and diffuse to neighboring cells through the extracellular fluid, or do they diffuse intercellularly through the gap junctions? **Figure 6.74b** shows a sequence of images of astrocytes loaded with a calcium indicator immediately after mechanical stimulation, showing that the calcium wave is able to spread across extracellular gaps (i.e., gap junctions are not necessary to mediate diffusion of second messengers). Communication across the lanes does not happen in the presence of an ATP inhibitor (**Figure 6.74c**), which supports the existence of an extracellular communication pathway via a messenger like ATP.

6.6 Developmental Biology on a Chip

During development, the embryo (initially formed from just a few cells) is constantly changing the microenvironment of every cell by following a finely orchestrated gene expression program. Clearly, these time-changing microenvironments cannot be accurately reproduced in traditional petri dishes. On the other hand, microfluidic systems—with their ability to deliver fluids to particular locations on demand—offer an enormous potential, mostly untapped to this day, to investigate development in vitro. We have already seen examples in Chapter 5 of how microfluidics can be used to manipulate embryos for practical applications (e.g., fertilization studies). Here we review research efforts that use microfabrication tools to address basic developmental biology questions.

Rustem Ismagilov and colleagues at the University of Chicago have used spatial patterns of gene expression in *Drosophila* to study perturbations in biochemical networks. The perturbations can be compensated during development, which makes them an especially attractive case study for probing the spatiotemporal dynamics of biochemistry on a whole-cell scale. To that end, a Drosophila embryo was placed inside a microfluidic channel that had two inputs, one carrying a warm (27°C) solution and the other a cool (20°C) solution, which produced a nonphysiological temperature "step" that caused different parts of the embryo to develop at different rates. Some embryos were exposed to warm solution in the anterior part, and others

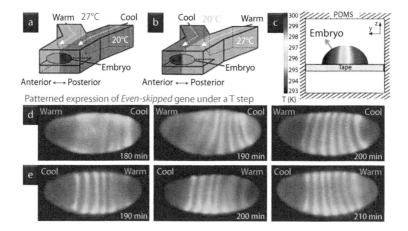

FIGURE 6.75 Patterns of gene expression in *Drosophila* embryos microfluidically stimulated with a temperature gradient. (From Elena M. Lucchetta, Ji Hwan Lee, Lydia A. Fu, Nipam H. Patel, and Rustem F. Ismagilov, "Dynamics of Drosophila embryonic patterning network perturbed in space and time using microfluidics," *Nature* 434, 1134–1138, 2005. Adapted with permission from the Nature Publishing Group. Figure contributed by Rustem Ismagilov.)

in the inverse orientation. Although initially the gene expression patterns in the anterior and posterior parts were rather different, after approximately 200 minutes, the patterns were very similar regardless of the environmental differences that the embryos had been exposed to (see Figure 6.75).

6.7 Yeast Biology

In 2005, Jeff Hasty's group at the University of California (San Diego) was able to solve a critical technical problem that had prevented researchers from studying yeast in microfluidic environments: yeast are not adherent and, as they divide, grow forming multilayers that are difficult to observe and perfuse. The Hasty group designed a cell culture chamber in the shape of a Tesla valve (Figure 6.76a), which has a loop region of high-resistivity slow flow (4 μm tall, just tall enough for one yeast cell) and a bypass region of low-resistivity fast flow (8 μm tall). In the loop region, the cells are always constrained to divide in the plane of focus (laterally, see Figure 6.76b) and the flow is so slow that they are not sheared away (see flow velocity modeling in Figure 6.76c). Pictures of the cells dividing are shown in Figure 6.76d and e. Segmentation algorithms could be applied to identify the newly divided cells (Figure 6.76f), so that the daughter, granddaughter, and great-granddaughter cells of each cell could be tracked to obtain senescence data.

The Tesla valve is not essential for achieving low flow and cell trapping. The essence of the design is that the chamber containing the cells must have a low ceiling (4 μm) to (a) produce a high resistivity to flow and (b) facilitate microscopic observation as the cells divide. Andre Levchenko's group at Johns Hopkins University has achieved the same goals with a different design (Figure 6.77) that also allows for producing gradients in the cell chambers. The Levchenko design features 5-μm-deep cell-containing chambers perpendicularly connected to two 25-μm-deep flow-through channels (which are each fed with different solutions, creating a gradient in the cell chambers); if the two flow-through channels are run at the same exact pressure, then the flow through the cell-containing chambers should equal zero. The team has used the design to investigate MAPK-mediated bimodal gene expression and phenotypic changes associated with the mating response of *Saccharomyces cerevisiae* under microfluidically generated pheromone gradients.

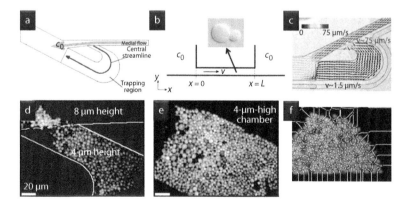

FIGURE 6.76 High-throughput studies of senescence with yeast in a microfluidic environment. (From Scott Cookson, Natalie Ostroff, Wyming Lee Pang, Dmitri Volfson, and Jeff Hasty, "Monitoring dynamics of single-cell gene expression over multiple cell cycles," *Mol. Syst. Biol.*, msb4100032, 2005. Reprinted with permission from the Nature Publishing Group.)

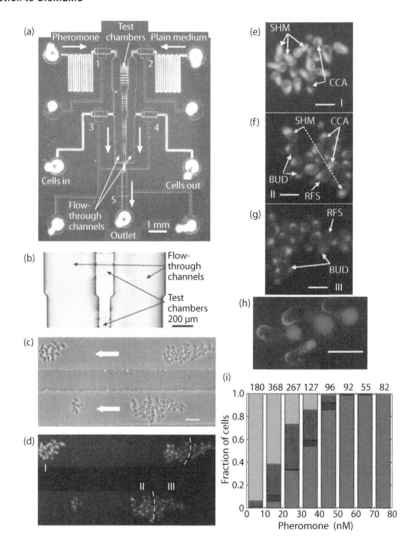

FIGURE 6.77 Adaptive gradient sensing in yeast in a microfluidic device. (From Saurabh Paliwal, Pablo A. Iglesias, Kyle Campbell, Zoe Hilioti, Alex Groisman, and Andre Levchenko, "MAPK-mediated bimodal gene expression and adaptive gradient sensing in yeast," *Nature* 446, 46–51, 2007. Reprinted with permission from Nature Publishing Group.)

In 2008, a collaborative team led by Tau-Mu Yi and Noo Li Jeon from the University of California (Irvine) used a very simple Y-mixer gradient generator to expose *S. cerevisiae* to several gradient slopes of the mating pheromone α-factor (**Figure 6.78**). The gradient became shallower downstream as the α-factor and the tracking dye (Dextran) had more time to diffuse (**Figure 6.78c**). Flow rates two times slower and five times faster than the flow rate used for this study (1 μL/min) did not produce a change in cell behavior. The observed cell morphologies, and specifically the robustness of projection formation, depended on the amount of α-factor the cells were exposed to (**Figure 6.78d**). The authors found that mutations that impair the downregulation of heterotrimeric G protein activation exhibited gradient-sensing defects that correlated with the supersensitivity of the mutants cells to α-factor.

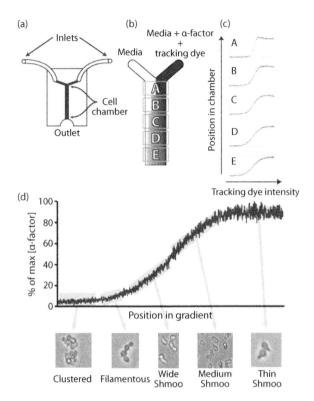

FIGURE 6.78 Mating pheromone gradient sensing in yeast. (From Travis I. Moore, Ching-Shan Chou, Qing Nie, Noo Li Jeon, and Tau-Mu Yi, "Robust spatial sensing of mating pheromone gradients by yeast cells," *PLoS ONE* 3, e3865, 2008.)

6.8 Plant Cell Biology

In 1987, practically at the dawn of BioMEMS, a group led by Harvey Hoch at Cornell University reported what has been, for more than two decades, the only BioMEMS article that deals with plants. This elegant article used microfabrication to elucidate the role of surface topography in leaves during fungal infection (Figure 6.79). The spores of the fungi need to find a pore on the surface of the leaf, and when they do so, they form an infection structure called an appressorium that blocks the pore and from which they enter the interior of the leaf (Figure 6.79a). How does the spore find the pore? Hoch and coworkers built various substrates, some containing a pore with concentric ridges (Figure 6.79b) and some containing parallel ridges of various heights (Figure 6.79c), to conclude that spore growth and appressoria formation is stimulated by a particular range of ridge heights (Figure 6.79c and d), which coincides with the ridges observed in natural leaves.

Recently, Rustem Ismagilov's laboratory at the University of Chicago has built a two-part microfluidic device that allows for inserting plant roots into a microchannel (Figure 6.80a). The microchannel features three inlets so that the roots can be locally perfused with a solution of choice using heterogeneous laminar flow (Figure 6.80b). The device was used to demonstrate the local exposure of live roots of *Arabidopsis thaliana* (a model organism in plant biology and genetics, the first plant genome to be sequenced, and well-suited for light/fluorescence microscopy because the seedling and the roots are translucent); local exposure to auxin resulted in local GFP expression in the root and local epidermal hair growth (Figure 6.80b through d).

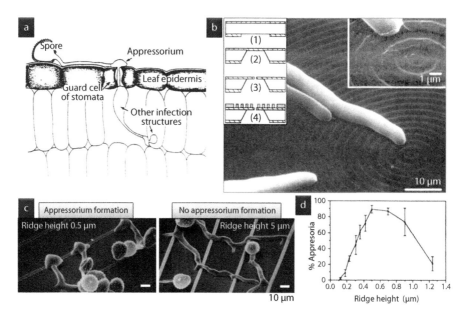

FIGURE 6.79 The role of leaf surface microtopography in fungal infection. (From Harvey C. Hoch, Richard C. Staples, Brian Whitehead, Jerry Comeau, and Edward D. Wolf, "Signaling for growth orientation and cell differentiation by surface topography in uromyces," *Science* 235, 1659–1662, 1987. Figure contributed by Harvey Hoch.)

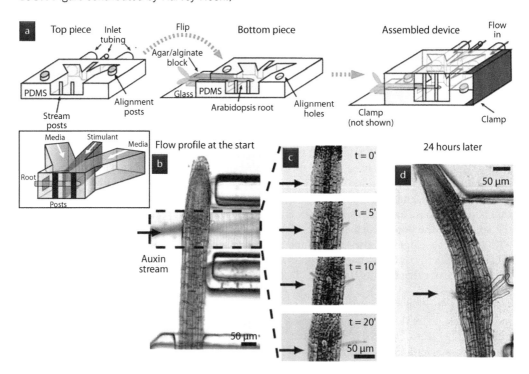

FIGURE 6.80 Root in a channel: plant on a chip. (From Matthias Meier, Elena M. Lucchetta, and Rustem F. Ismagilov, "Chemical stimulation of the *Arabidopsis thaliana* root using multi-laminar flow on a microfluidic chip," *Lab Chip* 10, 2147–2153, 2010. Reproduced with permission from The Royal Society of Chemistry.)

6.9 Microfluidics for Studying Cellular Dynamics

Time is a crucial parameter in most cellular processes. Rapid changes are ever-present during development, reproduction, cell migration, and healing, to mention a few key phenomena that are central to life. During these processes, cells change through activation of genetic programs by extracellular signals and cells, in turn, change their environment by secretion of more extracellular signals. Microfluidic systems, in combination with recent advances in imaging and biochemical readouts, provide an ideal platform for the quantitative stimulation of cells under precise spatiotemporal control conditions. We have already seen a variety of microfluidic cell culture systems for general cell maintenance in Section 5.4. Here we will cover systems that allow for a great spatiotemporal control of the cellular microenvironment (while obtaining some particular readout).

Owe Orwar's group from the Chalmers University of Technology in Goteborg (Sweden) has found a very practical way of modulating solution composition around a single cell very quickly and at high throughput (Figure 6.81). The cell is held at the tip of a probe and held stationary with respect to a stage, which holds a microfluidic device and is scanned with respect to the cell. The microfluidic device simply produces an outflow of 16 parallel laminar flows into a large open reservoir area (accessible to the pipette); the mixing of the solutions away from the outlets is irrelevant to the experiment. The experimenter decides which solutions and which mixtures are introduced into which wells, so that as the cell passes in front of the outlets in sequence, it "sees" a "chemical waveform." By varying the trajectory of the probe with respect to the stage (varying the stage speed or the stage–probe distance), it was possible to create very complex waveforms even with simple input concentration profiles (e.g., using only two different concentrations, or a binary input). The probe could also be used to obtain electrical (patch clamp) measurements from the cell while agonist or antagonist substances to ion channel receptors were applied. (A similarly conceived laminar flow switching system that features a piece of tubing that ends in an open reservoir has been used in ion channel research for a long time, but it requires much larger flow rates.)

Rustem Ismagilov's group at the University of Chicago has interfaced a droplet generator with Teflon tubing and a piece of PDMS (termed a "chemistrode") that allows for a brief exit of the droplets from the system (Figure 6.82)—for example, to expose cells to the droplets' contents. It is crucial that the outflow rate be very well balanced with the droplet production rate, especially

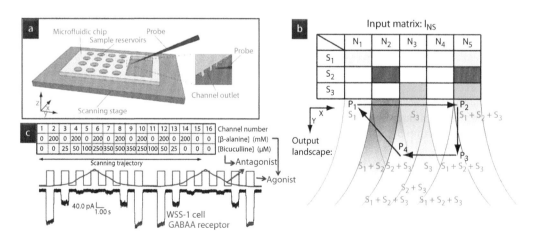

FIGURE 6.81 A chemical waveform synthesizer. (From Jessica Olofsson, Helen Bridle, Jon Sinclair, Daniel Granfeldt, Eskil Sahlin, and Owe Orwar, "A chemical waveform synthesizer," *Proc. Natl. Acad. Sci. U. S. A.* 102, 8097–8102, 2005. Copyright (2005) National Academy of Sciences, U. S. A. Figure contributed by Owe Orwar.)

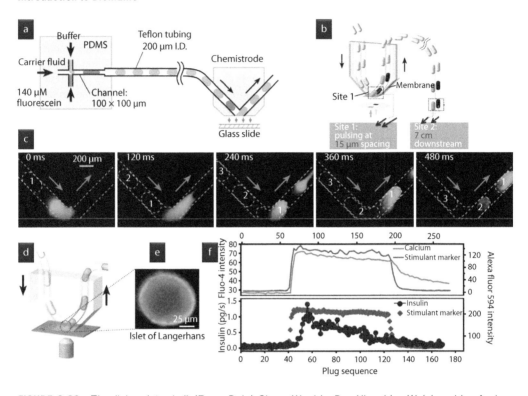

FIGURE 6.82 The "chemistrode." (From Delai Chen, Wenbin Du, Ying Liu, Weishan Liu, Andrey Kuznetsov, Felipe E. Mendez, Louis H. Philipson, and Rustem F. Ismagilov, "The chemistrode: A droplet-based microfluidic device for stimulation and recording with high temporal, spatial, and chemical resolution," *Proc. Natl. Acad. Sci. U. S. A.* 105, 16843–16848, 2008. Copyright (2008) National Academy of Sciences, U. S. A.)

when there are two lines operating in parallel (otherwise, one line could be capturing some of the other line's flow). The chemistrode allows for exposing cells to pulses of chemicals as short as a few tens of milliseconds (more than an order of magnitude faster than Orwar's waveform synthesizer) and spreading in space approximately more than 200 μm (similarly to Owar's waveform synthesizer, although it could potentially be improved). The researchers used the chemistrode to stimulate a mouse islet of Langerhans on a glass-bottomed dish and recorded insulin secretion every 1.5 seconds and an increase in intracellular calcium by fluorescence microscopy. An advantage of the chemistrode over Orwar's waveform synthesizer is that it allows for sampling the chemical microenvironment of the cell (the plugs can be analyzed offline); however, Orwar's patch clamp measurements would be difficult with the chemistrode.

6.10 Summary

In sum, BioMEMS constitutes a set of alternative techniques that are very appealing for basic cell biology studies because they potentially allow for *a more precise spatiotemporal design* of the signals (artificially) delivered to cells than traditional cell culture techniques. BioMEMS enables biologists to design heterogeneous *substrates* (containing topographical features or micropatterns of biomolecules) and heterogeneous *microfluidic environments*, even with some temporal control. In most cases, this increased precision is aimed at creating cell cultures of *increased physiological relevance*. For example, microfluidic systems can be used to mimic the release of biochemicals by other cells, in effect, substituting the presence of the signaling cells by a fluidic stream that presents the soluble signal in a highly localized and quantitative manner (i.e., duration, concentration, and position are known and can be changed). Although micropipette stimulation can be seen as a form of "microfluidic" delivery, it does not feature the throughput, precision, and amenability to modeling of "real" microfluidics.

Further Reading

Ahmed, T., T. S. Shimizu, and R. Stocker. "Microfluidics for bacterial chemotaxis," *Integrated Biology* **2**, 604–629 (2010).

Choi, C. K., M. T. Breckenridge, and C. S. Chen. "Engineered materials and the cellular microenvironment: a strengthening interface between cell biology and bioengineering," *Trends in Cell Biology* **20**, 705–714 (2010).

Desai, T., and S. Bhatia (editors). "Therapeutic micro/nanotechnology," vol. III, *BioMEMS and Biomedical Nanotechnology Series*. Springer (2006).

Irimia, D. "Microfluidic technologies for temporal perturbations of chemotaxis," *Annual Review of Biomedical Engineering* **12**, 259–284 (2010).

Kaji, H., G. Camci-Unal, R. Langer, and A. Khademhosseini. "Engineering systems for the generation of patterned co-cultures for controlling cell–cell interactions," *Biochimica et Biophysica Acta* **1810**, 239–250 (2011).

Keenan, T. M., and A. Folch. "Biomolecular gradients in cell culture systems," *Lab on a Chip* **8**, 34–57 (2008).

Kim, D.-H., P. K. Wong, J. Park, A. Levchenko, and Y. Sun. "Microengineered platforms for cell mechanobiology," *Annual Review of Biomedical Engineering* **11**, 203–233 (2009).

Raghavan, S., and C. S. Chen. "Micropatterned environments in cell biology," *Advanced Materials* **16**, 1303–1313 (2004).

Taylor, A. M., and N. L. Jeon. "Micro-scale and microfluidic devices for neurobiology," *Current Opinion in Neurobiology* **20**, 640–647 (2010).

Velve-Casquillas, G., M. Le Berre, M. Piel, and P. T. Tran. "Microfluidic tools for cell biological research," *Nano Today* **5**, 28–47 (2010).

Wang, J., L. Ren, L. Li, W. Liu, J. Zhou, W. Yu, D. Tong, and S. Chen. "Microfluidics: a new cosset for neurobiology," *Lab on a Chip*, **9**, 644–652 (2009).

Weibel, D. B., W. R. DiLuzio, and G. M. Whitesides. "Microfabrication meets microbiology," *Nature Reviews Microbiology* **5**, 209–218 (2007).

7

Tissue Microengineering

TISSUE ENGINEERING IS A SUBFIELD OF BIOENGINEERING that seeks to apply the principles of engineering and life sciences toward the development of biological substitutes that restore, maintain, or improve tissue function or a whole organ. The subdiscipline of tissue microengineering has the same goals as tissue engineering but uses microfabrication approaches and tools to achieve those goals.

A quick glance at a real tissue under the microscope is enough to convince us of the daunting challenge of the task of "developing biological substitutes" for them (Figure 7.1). Some cellular structures, like the hair cells of the inner ear (Figure 7.1a), are precisely arranged and mechanically coupled to membranes such that their acoustic sensitivity and performance is a function of their position on the membrane. Obviously most tissues are not even single-layered like the hair cell layer—they derive their function from their packing in three dimensions, like the flat cells of the eye lens, which are layered in an arrangement that minimizes the diffraction of light as it travels perpendicular to their plane (Figure 7.1c). Even if we succeed in engineering such complex cellular microstructures, we will need to address the challenge of their vascularization in three dimensions, which may well be a major microengineering problem on its own if one looks at the vasculature of an organ like the kidney (Figure 7.1b). Finally, these challenges seem dwarfed when one looks at the structure of multicellular tissues, such as a simple neural network such as the retina (Figure 7.1d), whose function is the result of the interplay of its cellular components in that particular arrangement: it would not function in any other spatial permutation of its components. How can we ever hope to develop a biological substitute of such tissues? Clearly, we will need a bit of help from nature—using the fact that cells have programs to build these structures—to achieve this; engineering scaffolds alone will not do.

7.1 Microscaffolding

In its attempt to better replicate in vivo tissue functions, tissue engineering has driven the development of methods for producing three-dimensional (3-D) scaffolds, textured cell culture substrates, and 3-D cell patterns. These **microscaffolds** have served to build tissue constructs for skin grafts, cardiac patches, and the investigation of vasculature growth.

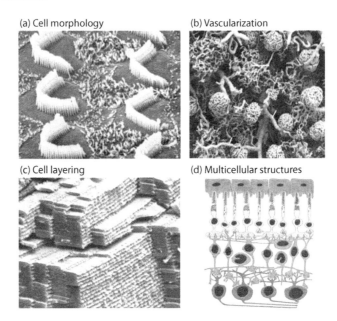

(a) Cell morphology   (b) Vascularization

(c) Cell layering   (d) Multicellular structures

FIGURE 7.1 Challenges of tissue engineering. Scanning electron micrographs of (a) outer hair cells, (b) kidney vasculature, and (c) eye lens cells. (From Richard G. Kessel and Randy H. Kardon, *Tissues and Organs: A Text Atlas of Scanning Electron Microscopy*, W. H. Freeman and Co., 1979; (d) is a cross-sectional drawing of the retina is adapted from Bruce Alberts et al., *Molecular Biology of the Cell*, 4th ed., Garland Science/Taylor & Francis LLC, New York, 2002. Reproduced with permission from Garland Science/Taylor & Francis LLC.)

7.1.1 Cellular Micropatterns in 3-D Microscaffolds

In the late 1990s, Patrick Aebischer and coworkers at the Lausanne University promoted neurite outgrowth in three dimensions by seeding neurons onto agarose gels that had been derivatized with laminin oligopeptides (**Figure 7.2**). The oligopeptides were first conjugated to a photoreactive cross-linker (such as benzophenone maleimide) and selectively immobilized within a transparent agarose gel using a laser beam. The nonimmobilized peptides were removed by thorough rinsing and electroelution. Neurites were observed to grow along the 3-D paths only. Such patterned gels could potentially have a clinical application in nerve regeneration, although first other important concerns would need to be addressed, such as immunorejection of the implant and interference of agarose with neurotransmitter release at synapses.

In the span of just over a decade, rapid advances in microscopy and biomaterials science have brought technological solutions to some of the challenges sketched earlier. A team from the University of Texas at Austin led by Christin Schmidt and Jason Shear has recently demonstrated a system for patterning neurons in hydrogels on the basis of multiphoton lithography (see Section 1.4.5), which represents a substantial improvement over Aebischer's system. Like Aebischer's system, it is based on the laser immobilization of cell-attachment peptides, but it uses (a) multiphoton excitation to photo-cross-link 3-D bovine serum albumin (BSA) structures (because BSA is biotinylated, the microstructures can be decorated with biotinylated peptides via an avidin linkage); and (b) hyaluronic acid, a highly transparent and biocompatible material that plays an essential role in the extracellular matrix (ECM) of the nervous system, as the hydrogel (**Figure 7.3**). Importantly, hyaluronic acid is not cell adhesive to mature cell

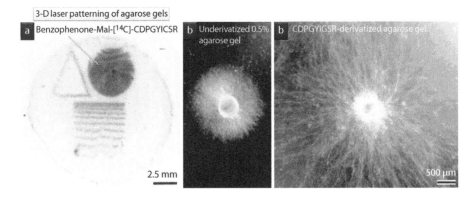

FIGURE 7.2 The 3-D-patterned laser immobilization of cell-attachment peptides in hydrogels. (From M. Borkenhagen, J.-F. Clémence, H. Sigrist, and P. Aebischer, "Three-dimensional extracellular matrix engineering in the nervous system," *J. Biomed. Mater. Res.*, 40, 392–400, 1998. Reprinted with permission from John Wiley and Sons.)

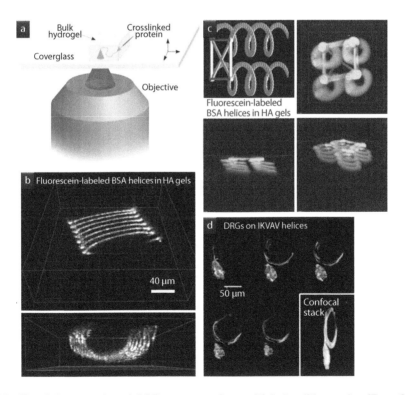

FIGURE 7.3 The 3-D patterning of DRG neurons using multiphoton lithography. (From Stephanie K. Seidlits, Christine E. Schmidt, and Jason B. Shear, "High-resolution patterning of hydrogels in three dimensions using direct-write photofabrication for cell guidance," *Adv. Funct. Mater.*, 19, 3543–3551, 2009. Reprinted with permission from Wiley.)

types. A path of IKVAV peptides (the cell-attachment peptides of laminin) was "built" first by photo-cross-linking a path of biotinylated BSA, then by exposing the gel to avidin, and finally by exposing the gel to biotinylated IKVAV peptides, which attached selectively to the photo-cross-linked BSA path. It was demonstrated that DRG neurons and hippocampal neural progenitor cells selectively grew and migrated on the IKVAV 3-D paths (**Figure 7.3d**).

Laser writing is not widely accessible to biology laboratories and only works with very specific photochemical groups (which may require specialized chemical synthesis not available in most biological laboratories); so many groups have pursued different ways of building 3-D microscaffolds for tissue engineering. In 1998, Linda Griffith's group at the Massachusetts Institute of Technology (MIT) printed biodegradable polymer scaffolds using a solid free-form printing process. The microscaffolds were polymerized layer by layer using biodegradable polylactic acid [modified with **poly(ethylene-oxide)**–poly(propylene-oxide) copolymers] and cell-adhesive carbohydrate ligands (specific for the hepatocyte asialoglycoprotein receptor) to promote hepatocyte attachment. The technique has the potential for creating low-resolution scaffolds in which each point can be tailored at will as adhesive or nonadhesive to cells.

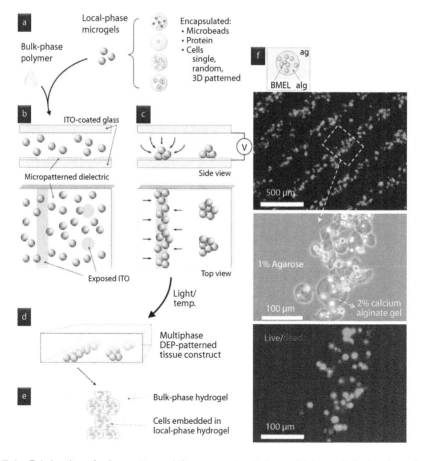

FIGURE 7.4 Fabrication of micropatterned tissue constructs by multiphase dielectrophoresis. (From Dirk R. Albrecht, Gregory H. Underhill, Avital Mendelson, and Sangeeta N. Bhatia, "Multiphase electropatterning of cells and biomaterials," *Lab Chip*, 7, 702–709, 2007. Reproduced with permission from The Royal Society of Chemistry.)

A major driving force in tissue microengineering has been the goal of trying to produce multicellular tissue constructs, which benefit from a richer range of cell–cell heterotypic interactions. To produce such constructs, Sangeeta Bhatia's laboratory at MIT first encapsulated bipotential mouse embryonic liver cells in alginate microspheres (<100-µm-diameter) using a standard needle-based extrusion method, then the microspheres were patterned by dielectrophoresis into lines within low-melting-point (gelling) agarose gels so that their final position became "frozen" in place when the temperature of the gel was lowered (Figure 7.4). In its present form, the method still suffers from some cell death (see Figure 7.4f), but other gel formulations and protocols will likely improve cell viability.

In an effort to broaden the range of 3-D cellular microstructures that can be produced, Sunghoon Kwon's group in South Korea's Inter-University Semiconductor Research Center have used guided self-assembly in microchannels, a sister technique of stopped-flow lithography (see Section 1.6.4 and Figure 1.26), to assemble polymeric microstructures containing cells (Figure 7.5). The polymer is **poly(ethylene glycol) diacrylate (PEG-DA)**, with 5 wt.% of photoinitiator, to synthesize the polymer with a focused laser within the microchannel. The method is generally applicable to many cell types and to arbitrary polymeric shapes, although the long-term biocompatibility of the polymer and microchannel environment is a concern.

Instead of using a microfluidic device to guide the assembly of multicellular units, it may be more practical to use nature's forces to guide their assembly. Ali Khademhosseini's group at MIT has used another UV-cross-linkable polymer (PEG-methacrylate polymer, or PEGmA) that forms hydrogels when exposed to light to entrap cells into microgels (Figure 7.6). In a second step, the microgel units could be assembled in lock-and-key multicellular microstructures by self-assembly (hydrophobic forces from surrounding mineral oil held together the assemblies of microgels). Importantly, the first step requires neither a microchannel nor specialized projection equipment.

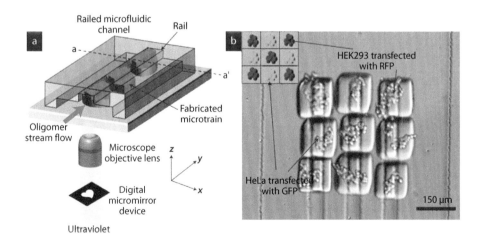

FIGURE 7.5 Microfluidic guided self-assembly and packaging of cells. (From Su Eun Chung, Wook Park, Sunghwan Shin, Seung Ah Lee, and Sunghoon Kwon, "Guided and fluidic self-assembly of microstructures using railed microfluidic channels," *Nat. Mater.*, 7, 581–587, 2008. Reprinted with permission from the Nature Publishing Group. Figure contributed by Sunghoon Kwon.)

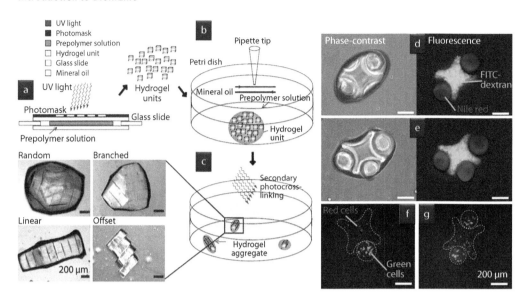

FIGURE 7.6 Fabrication of cell-laden microgels. (From Yanan Du, Edward Lo, Shamsher Ali, and Ali Khademhosseini, "Directed assembly of cell-laden microgels for fabrication of 3D tissue constructs," *Proc. Natl. Acad. Sci. U. S. A.*, 105, 9522–9527, 2008. Copyright (2008) National Academy of Sciences, U. S. A.)

7.1.2 Skin Microengineering

On a cellular level, the skin is not flat—it contains deep invaginations (termed **rete ridges**) that play a crucial role in maintaining the skin's mechanical integrity and flexibility. In 2000, a group led by Jeff Morgan, then at the Shriners Burns Hospital in Boston, fabricated micromolded collagen analogs of the basal lamina (**Figure 7.7a** and **b**) to test the levels of differentiation of keratinocytes compared with keratinocytes cultured on flat collagen layers. Their results showed differentiation and stratification of the keratinocytes cultured in deep ridges (**Figure 7.7c**), suggesting that the ridges provide a differentiation cue for keratinocytes in the normal process of skin growth. Because, until now, implanted artificial skin has not contained rete

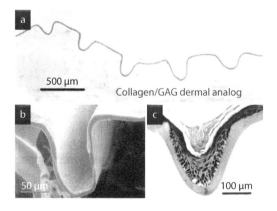

FIGURE 7.7 Microfabricated rete ridges. (From George D. Pins, Mehmet Toner, and Jeffrey R. Morgan, "Microfabrication of an analog of the basal lamina: Biocompatible membranes with complex topographies," *FASEB J.*, 14, 593–602, 2000. Figure contributed by Mehmet Toner.)

ridges (resulting in skin that is extremely sensitive to shear and a hairless "shiny" appearance that deeply concerns the patient), this work offers hope for improved skin and hair replacement therapies.

7.1.3 Vasculature on a Chip

For tissue-engineered organs to function properly, researchers must devise methods to design or grow capillary networks that will allow for the delivery of blood to the organs. A team led by Donald Ingber at Children's Hospital in Boston has shown that microvascular endothelial cells cultured on 10-μm lines of adhesive fibronectin (surrounded by nonadhesive PEG-thiol SAM) form tubelike structures (Figure 7.8). The confocal sectioned image of the cells clearly shows a hollow lumen inside. The same cells cultured on 30-μm lines, on the other hand, spread as ribbons and do not form a hollow lumen.

A large collaborative team led by Joseph Vacanti (MIT), Robert Langer (MIT), and Yadong Wang (Georgia Tech) were able to produce endothelialized capillary networks within molded microchannels made of **poly(glycerol sebacate) (PGS)**. PGS is a transparent biodegradable elastomer that replicates following a procedure very similar to that of PDMS, although with lower resolution (Figure 7.9). To produce an enclosed PGS device, the molded PGS side containing grooves was capped with a half-cured PGS flat cap, and the assembled device was cured overnight (a similar procedure works for bonding PDMS). To enhance cell adhesion, the devices were modified with the pentapeptide Glycine Serine Rarginine Dasparatic acid (GRGDS) (although endothelial cells adhered to PGS too). The capillary networks were endothelialized to confluence and were perfused up to 4 weeks at physiological flow rates without leakage, thus opening the way for organlike devices that contain their own, predesigned vasculature.

For certain applications, it may be more physiological to directly create the starting matrix in an ECM protein such as **collagen**. Abraham Stroock's group at Cornell University has produced microfluidic devices that are micromolded in collagen (Figure 7.10a), which are then seeded with human umbilical vein endothelial cells (**HUVECs**). The cells were found to endothelialize the device (Figure 7.10b), that is, they completely covered the collagen walls and formed an impermeable lumen that could be perfused without leakage.

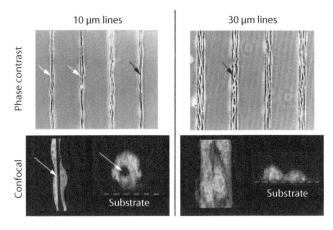

FIGURE 7.8 Microengineered endothelial cell tubules. (From Laura E. Dike, Christopher S. Chen, Milan Mrksich, Joe Tien, George M. Whitesides, and Donald E. Ingber, "Geometric control of switching between growth, apoptosis, and differentiation during angiogenesis using micropatterned substrates," *In Vitro Cell. Dev. Biol. Anim.*, 35, 441–448, 1999. Reprinted with permission from Springer. Figure contributed by George Whitesides.)

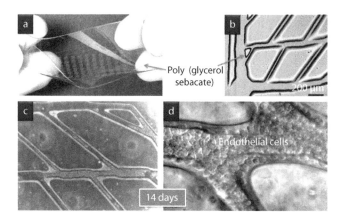

FIGURE 7.9 Endothelial cell growth in micromolded biodegradable capillary networks. (From Christina Fidkowski, Mohammad R. Kaazempur-Mofrad, Jeffrey Borenstein, Joseph P. Vacanti, Robert Langer, and Yadong Wang, "Endothelialized microvasculature based on a biodegradable elastomer," *Tissue Eng.*, 11, 302–309, 2005. Figure contributed by J.P. Vacanti.)

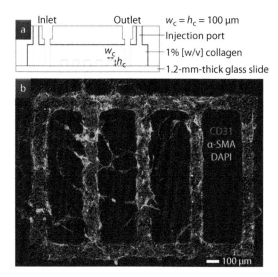

FIGURE 7.10 Endothelialization of micromolded collagen capillaries. (From Valerie L. Cross, Ying Zheng, Nak Won Choi, Scott S. Verbridge, Bryan A. Sutermaster, Lawrence J. Bonassar, Claudia Fischbach, and Abraham D. Stroock, "Dense type I collagen matrices that support cellular remodeling and microfabrication for studies of tumor angiogenesis and vasculogenesis in vitro," *Biomaterials*, 31, 8596–8607, 2010. Figure contributed by Ying Zheng.)

7.1.4 Muscle Cells

Biologists have known for a long time that muscle cells derive their great strength by joining forces, literally: they *fuse*, only in the direction of force generation, to form myotubes according to a genetic program that happens at a precise time during development. For this cell fusion to be effective, the cells prealign and elongate along features present on the ECM. When such cues are not absent, as on a traditional Petri dish, myoblasts form myotubes in random orientations.

Myoblasts can be oriented to fuse into single myotubes when seeded on the narrow (~35–50 μm wide) tracks of ECM surrounded by nonadhesive material (say, PEG; see, for example, Figure 6.13). However, such isolated myotubes are not useful for tissue engineering because they do not constitute usable tissue, in the sense that it cannot be eventually implanted. Tissue engineers have attempted to recreate substrates with aligned ECM and microtopographies that induce myoblast alignment and fusion in the same direction over the whole substrate. The real tissue engineering challenge is to produce functional cardiac tissue that can be implanted into a human for heart repair (after an infarct, etc.). The myotubes should be (a) electrically active (i.e., display spontaneous beating); (b) removable from the substrate; and (c) implantable into the host. Andre Levchenko's group at Johns Hopkins University has been able to produce wafer-sized biomimetic nanoscale topographies that resemble the ECM underlying rat heart myocardium (Figure 7.11). Nanotopographies molded in the polymer PEG diacrylate allowed for creating large functional patches of myocardium that transmitted action potentials preferentially in the direction of the grooves.

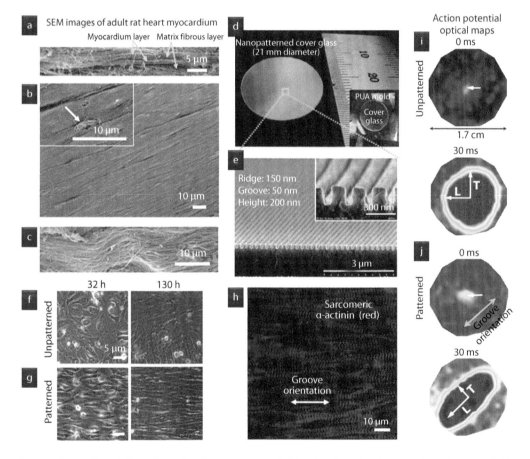

FIGURE 7.11 Regulation of cardiac function through biomimetic microtopography. (From Deok-Ho Kim, Elizabeth A. Lipke, Pilnam Kim, Raymond Cheong, Susan Thompson, Michael Delannoy, Kahp-Yang Suh, Leslie Tung, and Andre Levchenko, "Nanoscale cues regulate the structure and function of macroscopic cardiac tissue constructs," *Proc. Natl. Acad. Sci. U. S. A.* 107, 565–570, 2010. Figure contributed by Andre Levchenko.)

7.2 Micropatterned Cocultures

Large organisms owe their ability to develop complex behaviors to the cooperation of multiple cell types, each with a specialized function. It is of no surprise, then, that many cell types function poorly when separated from their neighbors, which may secrete signals that regulate expression of key genes involved in growth, proliferation, motility, and differentiation, among other processes. There is a reason why the body is made of multiple components, and separating it into pieces is always a source of problems.

Certain cell types are more sensitive than others at being separated from their neighbors—probably because certain tissues rely on the 3-D natural structure to perform their function more than others. Hepatocytes (liver cells), which in the liver are surrounded not by one but by many other cell types (endothelial cells, fibroblasts, and Kupffer cells), are notorious for losing all their liver-specific functions (e.g., albumin secretion, urea synthesis, p450 detoxification, etc.) within a couple of days of culture—unless they are cocultured with a second cell type (such as fibroblasts). Neurons, which in the brain are surrounded by glial cells and whose long axons and dendrites are severely damaged by the isolation process that reduces them temporarily to a spherical cell, are extremely sensitive in culture and do not survive well unless kept in the presence of a "feeder layer" of **glial cells**.

Mehmet Toner's group at Harvard Medical School is credited for pioneering the concept of micropatterned cocultures. The hope of the approach, depicted in **Figure 7.12**, is that by reproducing some key elements of the cellular architecture found in vivo, one is restoring the signaling present in vivo, thus enhancing the function observed with respect to the control random cocultures. This group used microfabrication to optimize the liver-specific functions in hepatocyte–fibroblast cocultures.

7.2.1 Liver Cells

Mehmet Toner and colleagues found interesting trends in liver-specific functions measured from these micropatterned cocultures (see also Section 2.6.2.1 and **Figure 2.34**). To optimize the amount of contact between the two cell types, they designed patterns with various hepatocyte/fibroblast contact that had the same available surface for both cell types, as shown in **Figure 7.13**. (Fibroblasts that landed in between hepatocytes were neglected.) In these micropatterned cocultures, unlike in a random coculture configuration, the seeding density can be specified independently of the length of contact between the two cell populations. For example, in the three patterns of **Figure 7.13**, the ratio of area available to fibroblasts versus area avail-

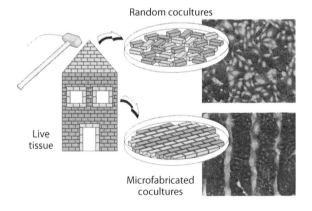

FIGURE 7.12 Random and micropatterned cocultures.

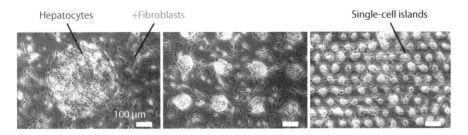

FIGURE 7.13 Micropatterned hepatocyte/fibroblast cocultures of various sizes. (From S.N. Bhatia, U.J. Balis, M.L. Yarmush, and M. Toner, "Probing heterotypic cell interactions: Hepatocyte function in microfabricated co-cultures," *J. Biomater. Sci. Polym. Ed.*, 9, 1137, 1998. Figure contributed by Sangeeta Bhatia.)

able to hepatocytes is 4:1 in all three patterns—but the amount of contact between the two cell populations increases as the size of the islands decreases.

At least two liver-specific functions (albumin production and urea synthesis) were seen to improve with decreasing island size (see graph in Figure 7.14), and intracellular albumin staining revealed that the signal provided by the fibroblasts is of short range (see Figure 7.14). It was also found that albumin production and urea synthesis, which were known to decay with time in random cocultures, are sustained for lengthy periods of time (~2–6 weeks) in certain microfabricated cocultures featuring the same cell–cell ratio as the random coculture. In addition, the response of hepatic function to changes in fibroblast number was distinct from that attributed to increased contact between hepatocytes and fibroblasts, suggesting that fibroblast number plays a role in the modulation of hepatic function through homotypic fibroblast interactions. This work suggests the possibility of building bioartificial livers on the basis of highly efficient micropatterned liver cell cocultures.

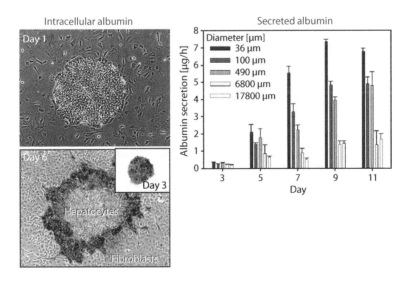

FIGURE 7.14 Enhanced liver function in micropatterned hepatocyte/fibroblast co-cultures. (From S.N. Bhatia, U.J. Balis, M.L. Yarmush, and M. Toner, "Probing heterotypic cell interactions: Hepatocyte function in microfabricated co-cultures," *J. Biomater. Sci. Polym. Ed.*, 9, 1137, 1998. Figure contributed by Sangeeta Bhatia.)

7.2.2 Lung Cells

Perhaps microfluidic systems are most needed to model those organs in which two vascularizations meet each other—such as in the blood–brain barrier, the lung, the liver, and so on—because traditional two-dimensional cell culture systems cannot mimic mass transport in the third dimension (usually a porous membrane or cell fenestrations). In this area, one of the most imaginative work has come from the laboratory of Shuichi Takayama at the University of Michigan, who has developed a microfluidic cellular model of the lung (Figure 7.15). Primary human small airway epithelial cells (**SAECs**) are seeded on a porous polyester membrane containing 400-nm pores (which mimics the in vivo basement membrane and limits transport of solutes by diffusion). The size of the microchannels (300 and 100 μm in width and height, respectively) was chosen to recreate the dimensions of distal conducting airways and respiratory bronchioles. Once the SAECs form a confluent monolayer (in ~6 days), their apical surface is exposed to an air–liquid interface (i.e., they are exposed to air), as shown in Figure 7.15c. This system allows for studying how pathologic fluid mechanical stresses (e.g., the propagation and rupture of liquid plugs; Figure 7.15d and e) can induce injury of SAECs.

Building on Takayama's model, Don Ingber's group at Harvard University has presented a "lung-on-a-chip" design, whereby the porous membrane and the whole epithelial cell layer can be stretched to mimic the cyclic mechanical strain "seen" by lung cells (Figure 7.16). Stretching is applied via two side vacuum channels (Figure 7.16a). In vivo, inhalation due to diaphragm contraction results in distension of the alveoli (see Figure 7.16b); in this study, "inhalation" also results in the stretching of the alveolar–capillary interface. This bioinspired microdevice shows that cyclic mechanical strain accentuates toxic and inflammatory organ-level lung responses to silica nanoparticles. Furthermore, mechanical strain enhances the uptake of nanoparticles by epithelial and endothelial cells and stimulates their transport into the underlying microvascular channel, an effect that is also seen in mouse lung in nanoparticle breathing experiments.

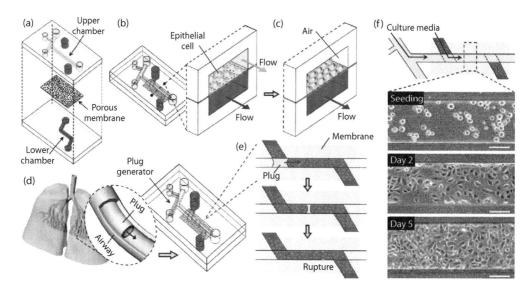

FIGURE 7.15 A microfluidic model of lung. Scale bars: 150 μm. (From Dongeun Huh, Hideki Fujioka, Yi-Chung Tung, Nobuyuki Futai, Robert Paine III, James B. Grotberg, and Shuichi Takayama, "Acoustically detectable cellular-level lung injury induced by fluid mechanical stresses in microfluidic airway systems," *Proc. Natl. Acad. Sci. U. S. A.* 104, 18886–18891, 2007. Copyright (2007) National Academy of Sciences, U. S. A.)

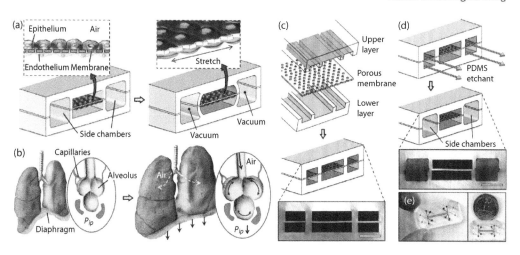

FIGURE 7.16 Organ-level lung functions reconstituted on a chip. (From Dongeun Huh, Benjamin D. Matthews, Akiko Mammoto, Martín Montoya-Zavala, Hong Yuan Hsin, and Donald E. Ingber, "Reconstituting organ-level lung functions on a chip," *Science*, 328, 1662, 2010. Figure contributed by Don Ingber.)

7.3 Stem Cell Engineering

BELIEVING IN THE POTENTIAL

STEM CELLS ARE CELLS that have the ability to proliferate while maintaining their undifferentiated state and, at the same time, the capacity to differentiate into specialized cell types. Embryonic stem (ES) cell lines are cultures of cells derived from the inner cell mass of blastocysts (blue cells in **Figure 7.17**). ES cell lines are pluripotent, that is, they can differentiate into nearly all cells. In vivo, ES cells can develop into more than 200 cell types of the adult body when given sufficient and necessary stimulation for a specific cell type. During development, ES cells cause all derivatives of the three primary germ layers (they do not contribute to the extra-embryonic membranes or the placenta): ectoderm (which gives rise to the nervous system and skin), endoderm (composed of the entire gut tube and the lungs), and mesoderm (which gives rise to muscle, bone, blood, and basically everything else that connects the endoderm to the ectoderm). Almost all research to date has been carried out using mouse or human ES cells (**hESCs**). Typically, mouse ES cells are grown on a layer of gelatin and require the presence of leukemia inhibitory factor to maintain an undifferentiated state, whereas hESCs are grown on a feeder layer of mouse embryonic fibroblasts and require the presence of basic fibroblast growth factor (bFGF or FGF-2) to stay undifferentiated. Their combined abilities of unlimited expansion and pluripotency confer ES cells an invaluable source for tissue replacement and regenerative medicine as well as after injury or disease. Despite this potential, some religious groups fiercely oppose and successfully lobby against the use of hESCs. Because the basic argument of these groups—that human embryos are sacrosanct and thus should never be used as a source of cells—is fundamentally flawed (it is not like the embryos are "saved" for some other purpose better than research or saving lives—they are placed in a garbage bag), progress has been slowed down considerably (mostly in the United States).

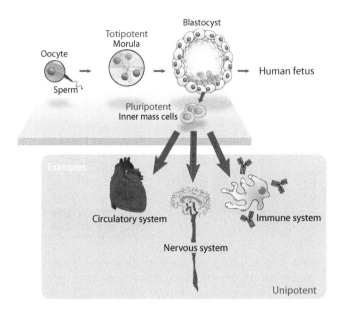

FIGURE 7.17 Pluripotency of ES cells. (Figure contributed by Michael D. Jones, MSc.)

Without optimal culture conditions or genetic manipulation, ES stem cells differentiate rapidly. Hence, there have been great research efforts (many microfluidic-based) into optimizing the culture and growth conditions of ES stem cell lines. It is also possible to form **embryoid bodies**, or aggregates of cells from ES cell lines, that although they are largely disorganized compared with a real embryo, they recapitulate some of the normal events of embryonic development (beating heart muscle cells and neurons appear commonly in embryoid bodies). We have seen how Shuichi Takayama's group was able to immobilize ES cells and form embryoid bodies in particular locations of microchannels using a hydrodynamic trapping technique (see **Figure 2.42** in Section 2.6.2.6).

Conventional techniques for forming embryoid bodies do not allow for a precise control over the size of the initial ES cell aggregates. Robert Langer's group at MIT cocultured hESCs on mouse embryonic fibroblast feeder layers inside 200-μm-diameter PDMS microwells to produce hESC aggregates of controlled size (**Figure 7.18**). Within the microwells, the hESCs maintained their undifferentiated state (as confirmed by Oct-4 and ALP immunostaining) and displayed a similar viability but superior homogeneity in aggregate size compared with flat substrates (± 8300 mm^2 within microwells versus $\pm 46{,}000$ mm^2 on flat surfaces).

A newer version of this process using PEG microwells instead of PDMS microwells has further improved the homogeneity in size and shape of the embryoid bodies achieved, without the need for a fibroblast feeder layer (**Figure 7.19**).

A team led by Maish Yarmush at the Massachusetts General Hospital in Boston performed an in-depth study of mouse ES cell aggregate formation in PDMS microwells. The microwells were produced by applying a PDMS stencil that contained holes ranging from 100 to 500 μm in diameter (**Figure 7.20a**). They found that germ layer differentiation (after 20 days of induction of differentiation), assessed by gene and protein expression assays as well as biochemical functions, depends on the initial size of the ES cell aggregate. The smallest (100 μm) aggregates showed an increased expression of ectodermal markers compared with the largest (500 μm) aggregates. On the other hand, the 500-μm aggregates showed an increased expression of mesodermal and endodermal markers compared with 100 μm aggregates. Hence, the initial conditions on which the embryoid bodies are formed can be critical for the extent of differentiation.

Several groups have already started developing microfluidic cell culture laboratories (see Section 5.4), and it can hardly be a coincidence that most of the devices are being applied to

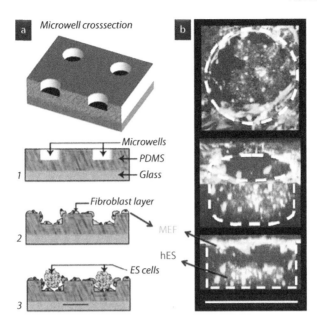

FIGURE 7.18 Formation of embryoid bodies in PDMS microwells. Scale bar is 200 μm. (From Ali Khademhosseini, Lino Ferreira, James Blumling III, Judy Yeh, Jeffrey M. Karp, Junji Fukuda, and Robert Langer, "Co-culture of human embryonic stem cells with murine embryonic fibroblasts on microwell-patterned substrates," *Biomaterials*, 27, 5968–5977, 2006. Reprinted with permission from Elsevier. Figure contributed by Ali Khademhosseini.)

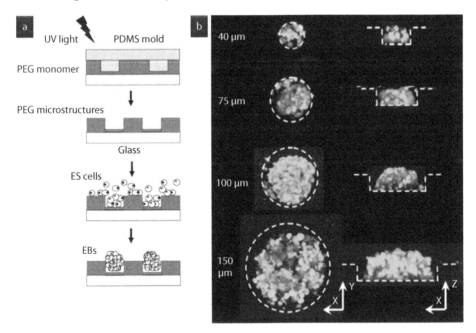

FIGURE 7.19 Formation of embryoid bodies in PEG microwells. (From Jeffrey M. Karp, Judy Yeh, George Eng, Junji Fukuda, James Blumling, Kahp-Yang Suh, Jianjun Cheng, Alborz Mahdavi, Jeffrey Borenstein, Robert Langer, and Ali Khademhosseini, "Controlling size, shape and homogeneity of embryoid bodies using poly(ethylene glycol) microwells," *Lab Chip*, 7, 786–794, 2007. Reproduced with permission from The Royal Society of Chemistry.)

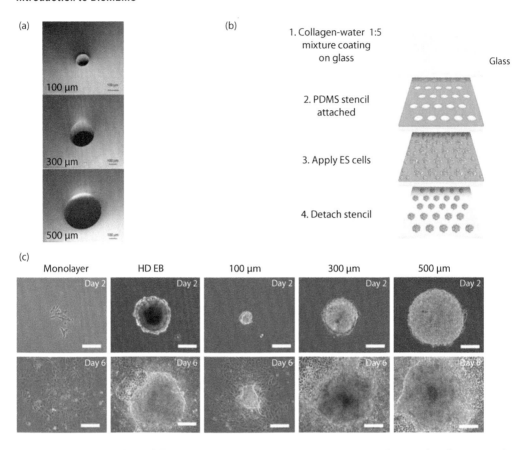

FIGURE 7.20 Formation of homogeneous embryoid bodies using PDMS stencils. Scale bar is 200 µm. (From Jaesung Park, Cheul H. Cho, Natesh Parashurama, Yawen Li, François Berthiaume, Mehmet Toner, Arno W. Tilles, and Martin L. Yarmush, "Microfabrication-based modulation of embryonic stem cell differentiation," *Lab Chip*, 7, 1018–1028, 2007. Reproduced with permission from The Royal Society of Chemistry.)

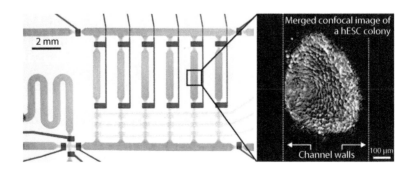

FIGURE 7.21 Microfluidic stem cell culture. (From Ken-ichiro Kamei, Shuling Guo, Zeta Tak For Yu, Hiroko Takahashi, Eric Gschweng, Carol Suh, Xiaopu Wang, Jinghua Tang, Jami McLaughlin, Owen N. Witte, Ki-Bum Lee, and Hsian-Rong Tseng, "An integrated microfluidic culture device for quantitative analysis of human embryonic stem cells," *Lab Chip*, 9, 555–563, 2009. Reproduced with permission from The Royal Society of Chemistry. Figure contributed by Hsian-Rong Tseng.)

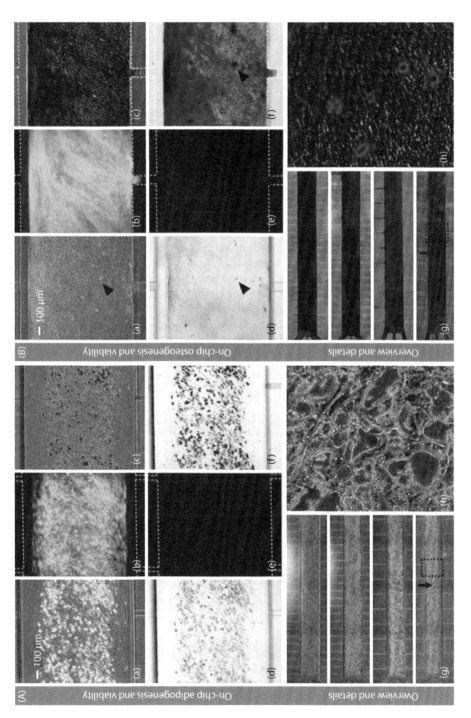

FIGURE 7.22 Control of fat and bone cell differentiation in a microfluidic device. (From Ellen Tenstad, Anna Tourovskaia, Albert Folch, Ola Myklebostd, and Edith Rian, "Extensive adipogenic and osteogenic differentiation of patterned human mesenchymal stem cells in a microfluidic device," *Lab Chip*, 10, 1401–1409, 2010. Reproduced with permission from The Royal Society of Chemistry.)

control stem cell culture differentiation (see Figures 5.35 and 5.36 in Section 5.4). Hsian-Rong Tseng's laboratory at the University of California (Los Angeles) has also demonstrated a fully automated, valved microfluidic device capable of culturing and manipulating hESC lines (Figure 7.21). Culturing hESCs is challenging because they have to be passaged in clusters and cocultured in the presence of growth-arrested mouse embryonic fibroblast (mEF) feeder layers, and their pluripotency must be confirmed on-chip over the culture period (using sequences of immunoassays for several pluripotency markers).

Often, the goal is to control and to direct the fate of ES cells to obtain a particular cell type (i.e., for therapeutics). A collaborative team led by Ellen Tenstad from Vestfold University College in Tønsberg (Norway) has reported the long-term (3-week) differentiation process of human mesenchymal stem cells into either osteoblasts (bone cells) or adipocytes (fat cells) in a microfluidic device (Figure 7.22). The success of the differentiation processes is demonstrated by the presence of cells that stain positive for molecular markers of osteogenesis and adipogenesis.

7.4 Morphogenesis

Tissue morphogenesis is the biological process by which an organ develops its shape. Soluble molecules that can diffuse and carry signals that direct tissue morphogenesis are called **morphogens**. Mina Bissell's group at the Lawrence Berkeley Laboratory has applied a micropatterning approach to study a particular type of morphogenesis, "branching morphogenesis," responsible for the growth of many organs with treelike structures from a preexisting epithelium; to initiate a branch, cells have to deform from their positions within the tubule and invade the surrounding tissue. The group produced micropatterns of mouse mammary epithelial tubules in collagen gel sandwich cultures (Figure 7.23). They then used a computer model to predict the

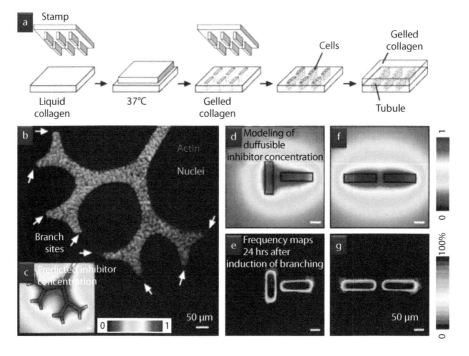

FIGURE 7.23 Tissue morphogenesis determined by microscale design. (From Celeste M. Nelson, Martijn M. VanDuijn, Jamie L. Inman, Daniel A. Fletcher, and Mina J. Bissell, "Tissue geometry determines sites of mammary branching morphogenesis in organotypic cultures," *Science*, 314, 298–300, 2006. Figure contributed by Celeste Nelson and Mina Bissell.)

extent of branching and confirmed that mouse mammary epithelial tubules initiate branching at sites with a local minimum in concentration of transforming growth factor-β, an autocrine inhibitory morphogen. As seen in the cell micropatterns of Figure 7.23b, e, and g, branching is most prominent at the edges, where the diffusible inhibitor concentration secreted by the cells is lowest (compared with Figure 7.23c, d, and f, respectively). The invasion of epithelial cells into

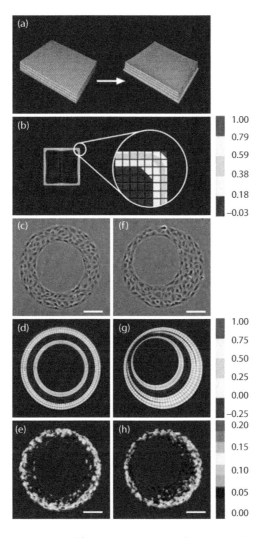

FIGURE 7.24 Correlation between proliferation and mechanical stress in microfabricated multicellular islands. Scale bars are 100 μm. (a) A contracting monolayer modeled in finite element model (FEM). (b) FEM calculations of relative maximum principal tractional stress exerted by cells in a small square island. Cell proliferation on a symmetric cell annulus shown in (c) phase contrast, (d) FEM results, (e) and colorimetric stacked image. In colorimetric stacked images, a pixel value of 0.20 indicates that 20% of cells at that location proliferated. Cell proliferation on an asymmetric cell annulus shown in (f) phase contrast, (g) FEM results, and (h) colorimetric stacked image. (From Celeste M. Nelson, Ronald P. Jean, John L. Tan, Wendy F. Liu, Nathan J. Sniadecki, Alexander A. Spector, and Christopher S. Chen, "Emergent patterns of growth controlled by multicellular form and mechanics," *Proc. Natl. Acad. Sci. U. S. A.* 102, 11594–11599, 2005. Copyright (2005) National Academy of Sciences, U. S. A.)

surrounding tissues bears great resemblance with (and thus has relevance for the control of) metastasis.

Chris Chen's laboratory at the University of Pennsylvania has investigated how a different kind of signal affects morphogenesis: mechanical stress. Morphogenesis is known to be driven by the activation of specific genes that activate growth patterns. Chen's laboratory has conclusively demonstrated that the resulting tissue form is not only a consequence but also an active regulator of tissue growth by feeding back to regulate patterns of proliferation. In micropatterns of sheets of cells, regions of high tractional stress in the sheet corresponded with regions of high proliferation (Figure 7.24). The stress was predicted with a finite-element model of multicellular mechanics and measured with a PDMS **microneedle** array (see Figure 6.23).

7.5 Summary

Microfabrication technologies, in the form of micropatterned substrates, microfluidic devices, and devices for the encapsulation of cells in hydrogels on a microscale, offer an invaluable tool to study the formation of microscale tissue assemblies and to produce tissues for therapeutic applications. We are now starting to grasp, in the horizon, the possibility of a simple organ that has fully grown in vitro with seeded cells surrounded by a scaffold of hydrogels and/or biodegradable polymers and a vasculature network that has been designed de novo using micromolding techniques.

Further Reading

Ashton, R.S., A.J. Keung, J. Peltier, and D.V. Schaffer. "Progress and prospects for stem cell engineering," *Annual Review of Chemical and Biomolecular Engineering* 2, 479–502 (2011).

Berthiaume, F., T.J. Maguire, and M.L. Yarmush. "Tissue engineering and regenerative medicine: history, progress and challenges," *Annual Review of Chemical and Biomolecular Engineering* 2, 403–30 (2011).

Coutinho, D., P. Costa, N. Neves, M.E. Gomes, and R.L. Reis. "Micro- and nanotechnology in tissue engineering," *Tissue Engineering* Part 1, 3–29 (2011).

Desai, T., and S. Bhatia (editors). "Therapeutic micro/nanotechnology," in *BioMEMS and Biomedical Nanotechnology Series* (Vol. III), Springer (2006).

Drury, J.L., and D.J. Mooney. "Hydrogels for tissue engineering: scaffold design variables and applications," *Biomaterials* 24, 4337–4351 (2003).

Gupta, K., D.-H. Kim, D. Ellison, C. Smith, and A. Levchenko. "Using lab-on-a-chip technologies for stem cell biology," *Stem Cell Biology and Regenerative Medicine* 5, 483–498 (2011).

Khademhosseini, A., R. Langer, J. Borenstein, and J.P. Vacanti. "Microscale technologies for tissue engineering and biology," *Proceedings of the National Academy of Sciences of the United States of America* 103, 2480–2487 (2006).

Lanza R., R. Langer, and J. Vacanti (editors). *Principles of Tissue Engineering* (3rd ed.), Academic Press (2007).

Palsson, B., and S. Bhatia. *Tissue Engineering*, Prentice Hall (2003).

Shan, J., K.R. Stevens, K. Trehan, G.H. Underhill, A.A. Chen, and S.N. Bhatia. "Hepatic tissue engineering," in *Molecular Pathology of Liver Diseases*, S.P.S. Monga (ed.), Springer Science, 321–342 (2011).

Toh, Y.-C., K. Blagovic, and J. Voldman. "Advancing stem cell research with microtechnologies: opportunities and challenges," *Integrative Biology* 2, 305–325 (2010).

8

Implantable Microdevices

ENGINEERS AND DOCTORS FORESAW VERY EARLY the need for miniaturizing devices that penetrate the human body, recognizing that any amount of penetration would be an invasion of a foreign body. These penetrating devices would range from "passive implants," such as dental implants, to "active implants," such as microelectrodes. As a result, the field of BioMEMS produced early on a plethora of implantable microdevices using traditional silicon micromachining. Not all the envisioned devices were as useful as doctors would have liked them to be because, as it turned out, the materials that were used in their development were not biocompatible for long-term human use. It would require a few more decades of research to fine-tune or redesign most of these processes.

Note that, for obvious reasons of space, we do not cover here the vast body of literature on drug delivery, a field that has often resorted to clever microfabrication and microfluidics tricks to produce nanoparticles that encapsulate drugs and that are then implanted into the body; although it can be argued that the nanoparticles are, in fact, implantable nanodevices—and we agree that they have extremely interesting biophysical properties and an enormous therapeutic potential—we do not have space to cover them here.

The reader should note that, this being the final chapter, it tends to contain less details on processing and microfabrication (referring to earlier chapters whenever possible) and tries to focus on the final goal of these studies: the use of these devices inside the human body (skin included). Paradoxically, for several decades, the field of implantable microdevices struggled to produce biomedically relevant data because it was technologically behind in its vision. This is no longer the case, and the field is now (as we speak) blooming like a desert after a rainfall. Here is a short chapter with very exciting developments.

8.1 Dental Implants

Donald Brunette, from the University of British Columbia in Vancouver (Canada), envisioned early in the 1980s the importance of exploiting **contact guidance**, the tendency of cells to align to edges such as those present on micromachined groove substrates (see Section 6.2.2). Brunette's laboratory was housed in the Department of Oral Biology, so he was deeply interested in controlling the migration and penetration into dental implants of epithelial cells (which form a seal with the implants) and gingival fibroblasts; this critical process determines the success of the implantation procedure in the long term. Translation of the early in vitro studies into animal studies took more than a decade, but eventually it was possible to produce implants containing microtopographies in the form of titanium-coated micromolded tapered pits (**Figure 8.1a** through c). It was shown that these topographies can stimulate connective tissue and bone

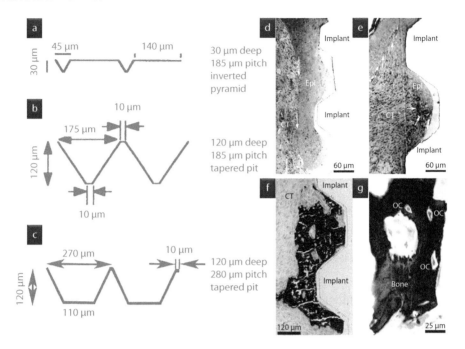

FIGURE 8.1 Dental implants with microfabricated topographies. (a–c) Titanium-coated replica-molded epoxy substrates used in the study. (d–g) Toluidine blue staining of sectioned implants showing (d) epithelial perimigration, (e) inhibition of epithelial migration by connective tissue attachment, (f) bone-like tissue formation 8 weeks after placement, with (g) osteocyte-like cells evident close to the base of the tapered pit. Epi, epithelium; CT, connective tissue; OC = osteocyte. (From Douglas W. Hamilton, Babak Chehroudi, and Donald M. Brunette, "Comparative response of epithelial cells and osteoblasts to microfabricated tapered pit topographies *in vitro* and *in vivo*," *Biomaterials*, 28, 2281–2293, 2007. Reprinted with permission from Elsevier.)

attachment to these implants and prevent migration of epithelial cells (**Figure 8.1d** through **g**), overall promoting long-term implant survival.

8.2 Implantable Microelectrodes

Interfacing electrical sensors with the body's electroactive nervous system is a highly attractive proposition. It is attractive because, in principle—if nothing went wrong—they should allow for recording and commanding bioelectrical signals with electronic equipment such as amplifiers, computers, and so on. The challenge, however, is by definition insurmountable: where are the sensors to be placed, and how many? The body has its own network architecture, with (to be conservative) billions of interconnections, so we have to give up on recording from *every* potentially interesting node of the network. Hence, scientists have focused on trying to learn about the nervous system by recording from only a few nodes. Let us consider first two examples of widely spread, successful technologies that use nonmicrofabricated electrical sensors: **electroencephalograms (EEGs)** and cochlear implants for the deaf.

EEGs are based on recording signals from outside the scalp using contact electrodes (about a few dozens of them) that are wired to a recording terminal. EEGs are of paramount importance to diagnose, predict, and prevent epileptic seizures, to facilitate therapies and recovery of delicate neurologic patients (e.g., patients in sleep, under anesthesia, or in a coma), and to enable research on naturally behaving animals. In some cases, however, those signals are not sufficient, and then size matters a lot. One issue is portability: if it is a small animal such as a mouse or a child, we cannot really ask them to stay still and tethered to dozens of wires (requiring constant supervision), so wearable, wireless EEG devices have been developed. The most fundamentally limiting issue, however, is the amount of information that can be extracted with scalp-based recordings and the amount of intervention that can be done with them. Sometimes, intracranial recordings (i.e., through a hole in the skull) can provide precious additional information, such as in severe epilepsy for providing a warning and time for therapeutic intervention before a seizure starts. Importantly, implanted electrodes (unlike scalp-based ones) allow for electrical stimulation. The clinical standard of intracranial recordings is based on implanting into the dura a set of four platinum wires (80 μm in diameter, spaced 5 mm apart) directly tethered to an EEG rack. Here is where microfabrication can help, and it can help a lot of people. (Approximately 3% of the U.S. population is expected to develop epilepsy by the age of 75 years; ~20%–30% of epilepsy patients cannot be effectively treated with current medical or surgical techniques—this constitutes approximately 7000 patients a year in the U.S. alone.)

WISE LESSONS

IN THE 1980S, A GROUP at the University of Michigan's Ann Arbor campus led by Kensall Wise and Khalil Najafi started microfabricating silicon-based neural probes, and soon after, Richard Normann started another large group at the University of Utah. Despite the sizes of these centers, progress has been slow by many measures. A title of a 2009 article, "An implantable 64-channel wireless microsystem for single-unit neural recording," reminds of a 1984 abstract submitted by the same group: "A multichannel probe for intracortical single-unit recording," implying that 25 years of progress by the best engineers in the field is summarized with little more than the addition of wireless capability. The problem is that these engineers decided to take on one of the most difficult challenges in the field of BioMEMS: back then, technology was not quite ripe to address the formidable challenges associated with implanting a microfabricated device and obtaining meaningful biomedical information from the body in real time. On the way, they also produced some very clever designs of other products (such as thermal detectors and inertial sensors) and 12 start-ups, but that may not be their biggest legacy yet. Their struggle has taught the rest of the neural engineering field many valuable lessons, among them the fact that they persevered in their goals is what makes them stand as pioneers.

The major driving need for fabricating microelectrodes has come from neuroscientists who demanded the integration of more electrodes per unit volume to record from more cells simultaneously. Several groups jumped at this opportunity with great enthusiasm in the 1980s after the advent of silicon micromachining techniques that made possible the fabrication of the first microfabricated multichannel probes. However, in the 1980s, the field of multisite intracortical recordings presented three major challenges that, together, seemed almost intractable. First, the probes necessarily caused damage during insertion—which was a compelling reason to microfabricate them and in particular to microfabricate them on thin substrates. Second, if the probe had to have many electrodes, the idea of moving them relative to one another (the design concept behind the "tetrode" where four wires can be independently lowered) had to be abandoned, thereby losing some functionality and ability to optimize the electrode-to-cell contact. Last but not least, the materials (silicon and metals) might cause a foreign-body reaction in the form of electrode encapsulation (massive ECM deposition preventing neuron-electrode contact), inflammation, and infections. These challenges were taken up first and mostly by two big groups

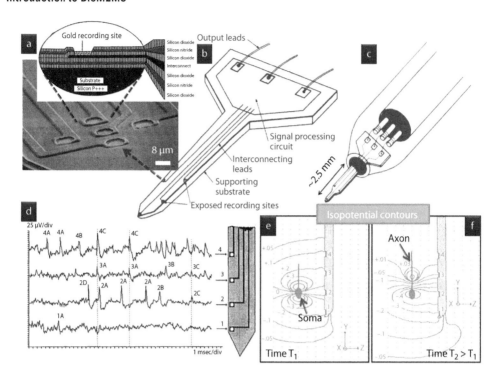

FIGURE 8.2 Micromachined electrodes for intracortical recordings. (a–c: from S. L. Bement, K. D. Wise, D. J. Anderson, K. Najafi, and K. L. Drake, "Solid-state electrodes for multichannel multiplexed intracortical neuronal recording" *IEEE Trans. Biomed. Eng.*, 33, 230, 1986.) (d–f: from K. L. Drake, K. D. Wise, J. Farraye, D. J. Anderson, and S. L. Bement, "Performance of planar multisite microprobes in recording extracellular single-unit intracortical activity," *IEEE Trans. Biomed. Eng.*, 35, 719, 1988. Figure contributed by Khalil Najafi and Kensall Wise.)

in the United States, one at the University of Michigan (Ann Arbor campus, led by Kensall Wise and Khalil Najafi) and one at the University of Utah (led by Richard Normann, who in 1995 founded the Center for Neural Interfaces). The research that came about from these biomedical challenges resulted in the development of the "Michigan probes" at the University of Michigan (**Figure 8.2**) and the **"Utah electrode array" (UEA)** at the University of Utah (see the following sections). The exact microfabrication details can be found in the literature references provided in this chapter and, in any case, are only interesting from a historical perspective; they are conceptually similar to the methods described in the microfabrication of cantilevers in Section 1.4.2 (**Figure 1.11**).

8.2.1 The Michigan Probes

The so-called **Michigan probes** were developed in the 1980s by a team led by Kensall Wise and Khalil Najafi and consisted of metal microelectrodes, half-exposed by insulator layers (**Figure 8.2a**) and patterned on silicon shanks (**Figure 8.2b**), which were etched by micromachining techniques and mounted on glass pipettes before operation (**Figure 8.2c**). Some of the probes contained active signal-processing circuits such as amplifiers. These first multisite simultaneous recordings showed that activity between adjacent sites can be correlated by comparing temporal coherence (**Figure 8.2d**). Using biophysical models, the measured signals were described in terms of axon potential propagation and soma depolarization (**Figure 8.2e** and **f**).

The materials science, electrical engineering, and biomaterials challenges associated with the Michigan probes as outlined earlier have largely been resolved, at least for short-term small animal studies, and are now routinely fabricated in various multishank, multielectrode-per-shank formats from a single silicon wafer incorporating amplifiers (see Figure 8.3a through c); the integration of amplifiers greatly reduces the pickup noise, the movement artifacts, and the mass of headgear (a crucial improvement for small-animal research and long-term, implantable devices). A fruitful collaboration between Kensall Wise's group (University of Michigan at Ann Arbor) and György Buzsáki's laboratory at Rutgers University has yielded abundant single-neuron data on freely behaving rodents (Figure 8.3d) and increased our understanding on how networks of neurons produce complex behavior.

Most notably, the Buzsáki group has addressed and resolved the critical problem of finding the optimal locations of each electrode with respect to each neuron by using many electrodes and triangulation algorithms that deduct the relative locations of the neurons as a function of the detected signal decays: when a spike is detected in, say, electrode 3, then the triangulation algorithm "knows" that nearby electrode 4 should detect the same cell with a similar amplitude; if the amplitude is lower, it means the cell is further away from electrode 4 than it is from electrode 3. As the signal decay function is known and the calculation is repeated for every pair of electrodes, it also becomes possible to infer the three-dimensional location and connectivity of cells around the electrodes (Figure 8.4).

Achieving very low impedance at the electrode–tissue interface is a paramount requirement for obtaining a high-quality signal. The electrode material influences the impedance directly (since its first contact with the cell) as well as indirectly (by triggering cellular responses, such

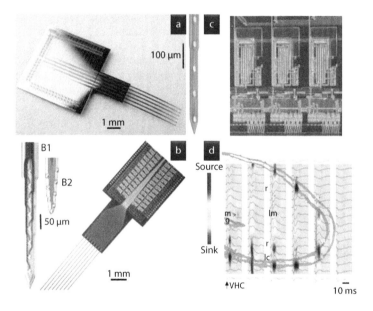

FIGURE 8.3 Multiunit recording using the Michigan probes. (a) Michigan probe with six shanks (16 sites/shank, 100-μm vertical spacing, 300-μm intervals between shanks). (b) Probe with eight shanks (eight sites/shank); the sites can be fabricated in either linear (B1) or staggered (B2) configuration. (c) On-chip buffering circuitry showing 3 of the 64 amplifiers and associated circuits. (d) Single evoked responses in response to ventral hippocampal commissure (VHC) stimulation (vertical arrow) recorded at 96 sites simultaneously (using the probes shown in panel a) allows for online calibration of recording site positions in a freely moving rat. The color maps are made by current source density analysis using the second spatial derivative of the local field potentials, assuming that the resistivity of the extracellular medium is homogeneous and isotropic. (From Jozsef Csicsvari, Darrell A. Henze, Brian Jamieson, Kenneth D. Harris, Anton Sirota, Péter Barthó, Kensall D. Wise, and György Buzsáki, "Massively parallel recording of unit and local field potentials with silicon-based electrodes," *J. Neurophysiol.*, 90, 1314–1323, 2003. Figure contributed by Ken Wise.)

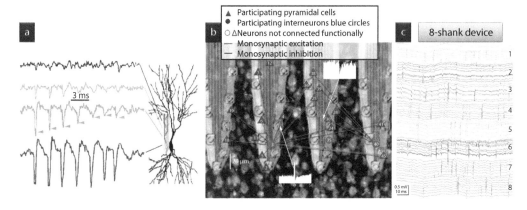

FIGURE 8.4 Large-scale recordings and network analysis using the Michigan probes. (a) Arrowheads show the progressive attenuation of spike backpropagation recorded at dendritic recording sites during a complex spike burst. (b) Functional topography in the somatosensory cortex of the rat. Recording sites are spaced 20 μm vertically; the shanks were 200 μm apart but have been digitally placed closer for illustration purposes. Interneurons (e.g., cells 3 and 40) are activated by many pyramidal cells, and an interneuron inhibits several local and distant pyramidal cells. The relative positions of the neurons were determined by calculating the "center of mass" of spike amplitude recorded from multiple sites. Cross correlograms between an interneuron-pyramidal cell pair (35–25) and a reciprocally connected pair (3–4) are shown in white. (c) A short epoch of raw recording using an eight-shank device, illustrating both field and unit activity (1–5 kHz). The presence of spikes on several sites of the same shank (color-coded) and lack of the same spikes on dissimilar shanks show that electrodes placed ≥200 μm record laterally from different cell populations. (From György Buzsáki, "Large-scale recording of neuronal ensembles," *Nature Neurosci.,* 7, 446–451, 2004. Figure reprinted with permission from Nature Publishing Group.)

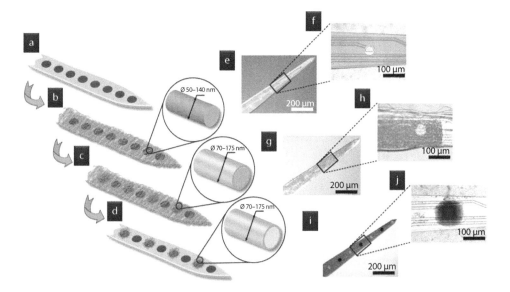

FIGURE 8.5 Conductive polymer nanotubes on neural microelectrodes. (From Mohammad Reza Abidian, Kip A. Ludwig, Timothy C. Marzullo, David C. Martin, and Daryl R. Kipke, "Interfacing conducting polymer nanotubes with the central nervous system: Chronic neural recording using poly(3,4-ethylenedioxythiophene) nanotubes," *Adv. Mater.,* 21, 3764–3770, 2009. Reprinted with permission from John Wiley and Sons. Figure contributed by Daryl Kipke.)

as protein deposition). The tissue conducts currents in the form of ions, and the electrode conducts current in the form of electrons, so the current needs to be transduced either through capacitive currents or through faradaic currents from redox reactions at the electrode surface. A fundamental problem in electrode technology has been the few options available: gold, tungsten, platinum, and iridium—which have a well-defined chemistry but are difficult to "improve." Daryl Kipke's group from the University of Michigan (Ann Arbor) has taken a radical turn and has reengineered the surface of the electrodes with nanotubes made of conducting polymers, such as poly(pyrrole) and poly(thiophene) (Figure 8.5). They demonstrated that these nanotube-based electrodes significantly enhance the quality of the recording signals, showing that the signal at day 49 after surgery is virtually the same as immediately after surgery, and its impedance has a flat spectrum over more than four orders of magnitude in frequency (whereas the traditional metal electrodes decay linearly).

8.2.2 The Utah Electrode Array

The UEA, a 10×10 array of penetrating silicon needle electrodes (1.5 mm tall, 400 μm spacing), is the brainchild of Richard Normann's group at the University of Utah. It is also available in a configuration in which the needles have a range of heights, from 0.5 to 1.5 mm. (The Michigan group has produced an array similar to the UEA with integrated amplifiers.) Perhaps the biggest success of the UEA was a study performed by John Donoghue (who studied with Kensall Wise) at Brown University in 2006, which showed that the UEA allowed a tetraplegic patient to communicate via a brain-computer interface for 9 months (Figure 8.6).

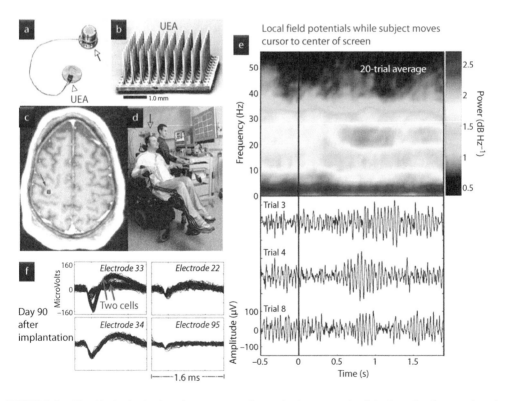

FIGURE 8.6 The Utah electrode microarray used as a brain–computer interface for the paralyzed. (From Leigh R. Hochberg, Mijail D. Serruya, Gerhard M. Friehs, Jon A. Mukand, Maryam Saleh, Abraham H. Caplan, Almut Branner, David Chen, Richard D. Penn, and John P. Donoghue, "Neuronal ensemble control of prosthetic devices by a human with tetraplegia," *Nature*, 442, 164–171, 2006. Reprinted with permission of the Nature Publishing Group.)

8.2.3 Microfabricated Cochlear Implants

Kensall Wise's group has worked on the development of implantable electrodes for improved cochlear implants (**Figure 8.7**). From the point of view of microfabrication, the cochlea is a challenging site because of its spiral anatomical architecture (**Figure 8.7a**), but on the other hand, from the point of view of bioaccessibility of the signal, the cochlea is an engineer's dream: the hair cells are on the surface, and even as the hair cells start dying in adult individuals (the cause for hearing loss as we age), the nerve terminals stay intact (so the nerve can be stimulated through a thin layer of skin without need for penetrating electrodes). The Wise group has designed a linear array of extracellular electrodes patterned on the outside of an inflatable helix-shaped plastic structure (**Figure 8.7b** through **d**). Upon application of air pressure, the plastic support chamber causes the electrodes to straighten up, facilitating insertion and electrode–tissue contact (**Figure 8.7e** through **g**). Improved electrode-tissue contact is beneficial for reducing current density at the electrode surface, which otherwise can produce harmful biproducts through electrochemical processes. The ability to fold and unfold should prove essential to reach the apical tip of the cochlea, where the cells that sense low-frequency sounds reside and should be less damaging to the cochlea than present implants, which bend by friction against the walls and reach only half-way into the cochlea (indeed, cochlear implants are known to cause irreversible hearing loss).

It has not been determined yet whether these microfabricated, actuatable cochlear implants featuring many electrodes will bring additional benefits to the deaf who presently enjoy the

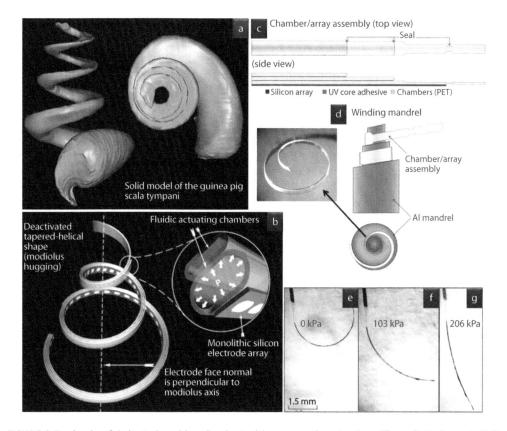

FIGURE 8.7 A microfabricated cochlear implant with pneumatic actuation. (From B. Y. Arcand, P. T. Bhatti, N. V. Butala, J. Wang, C. R. Friedrich, and K. D. Wise, "Active positioning device for a perimodiolar cochlear electrode array," *Microsyst. Technol.*, 10, 478–483, 2004. Figure contributed by Ken Wise.)

benefits of traditional (24-electrode) implants. With current technology, a person can typically recover hearing to the point of being able to carry on conversations routinely. However, traditional implants are not capable of transducing music (music sounds horrible to people with cochlear implants) or fine differences in intonation. The hope is that implants with many electrodes will be able to transduce high-fidelity sounds.

8.2.4 Microfabricated Electrocorticography Arrays

Electrocorticography (ECoG) is based on the extracellular recording of electrical activity using electrodes placed on the surface of the brain, just above the pia membrane after the surgical removal of the dura layer and the skull. In humans, ECoG is most often performed with approximately 1-cm-diameter electrodes to detect the onset of epileptic seizures before epilepsy surgery. These patients are tethered to long wires and electronic racks during data acquisition, and the size of the electrodes does not allow for fine mapping of cortical activity. David Rector's group from Washington State University in Pullman (Washington) has developed a microfabricated ECoG array made as a sandwich of Kapton and patterned SU-8, which leaves an 8 × 8 grid of 150-μm-diameter electrodes (the "ham" of the sandwich) exposed at selected locations and spaced 750 μm (see **Figure 8.8a** through **c**). Recordings of a rat's whisker barrels (the area of the rat somatosensory cortex that receives inputs from the whiskers) yield excellent stability during more than 8 hours, which allowed for producing spatial amplitude maps of evoked responses

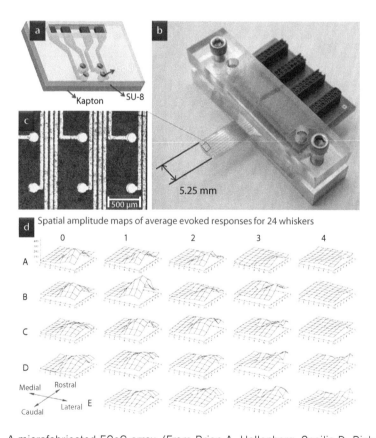

FIGURE 8.8 A microfabricated ECoG array. (From Brian A. Hollenberg, Cecilia D. Richards, Robert Richards, David F. Bahr, and David M. Rector, "A MEMS fabricated flexible electrode array for recording surface field potentials," *J. Neurosci. Methods*, 153, 147–153, 2006. Reprinted with permission from Elsevier.)

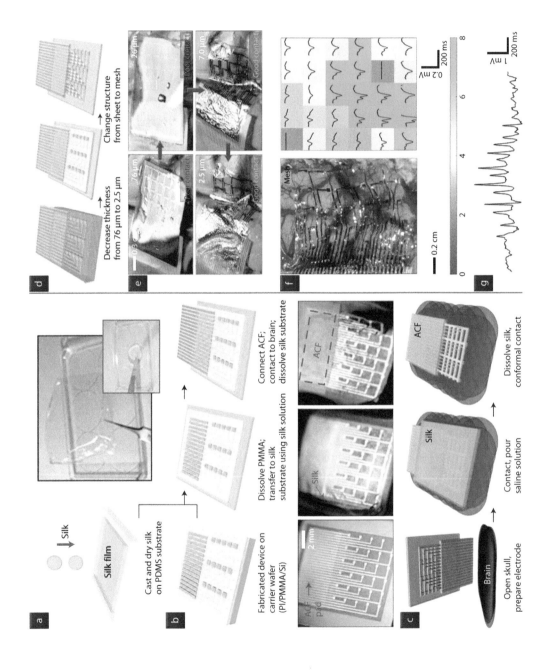

for 24 whiskers (Figure 8.8d). The maps in Figure 8.8d depict the surface electrical activity of the rat's cortex (for a particular whisker region) when that whisker is twitched. Although this device, containing Kapton, gold, and SU-8, would not be approved for human implantation, similar microfabricated ECoG arrays made in appropriate materials approved by the Food and Drug Administration (FDA) should find exciting clinical uses very soon and allow for high-resolution mapping of cortical activity in humans and higher primates.

A fundamental problem with microfabricated circuits is that they are produced on planar surfaces, yet the human brain is highly curvilinear, containing deep "crevasses" (named *sulci*). John Rogers' laboratory at the University of Illinois (Urbana-Champaign) has invented a powerful strategy for adapting electronic circuits to almost arbitrary contours on the basis of a dissolvable, resorbable silk fibroin substrate and applied it in producing functional ECoG recordings of the visual cortex of a cat (Figure 8.9). The arrays consisted of 30 measurement (gold) electrodes in a 6 × 5 configuration, each with dimensions of 500 × 500 μm and spaced by 2 mm. The silk fibroin film was obtained by spinning and drying for 12 hours at room temperature (Figure 8.9a). The electrode array was first fabricated by traditional photolithography on a silicon wafer coated with poly(methyl methacrylate) (PMMA) (acting as a sacrificial layer) and polyimide (PI, acting as a support mesh for the electrodes); after dissolving the PMMA in acetone, the PI-supported electrodes were transferred onto the silk fibroin film (Figure 8.9b). When the electrodes were mounted on tissue and the silk was allowed to dissolve and resorb, capillary forces initiated a process of spontaneous, conformal wrapping at the biotic–abiotic interface (Figure 8.9c and d). Thick (76–26 μm) silk films resulted in poor conformal contacts, whereas thin (7–2.5 μm) silk films resulted in good contacts, as measured on realistic brain models (Figure 8.9e). Excellent recordings of the visual cortex of cats were obtained using 2.5-μm-thick mesh electrodes (Figure 8.9f). This technology opens the way for long-term integration of a variety of electronic devices (including optoelectronics) in neural systems.

8.2.5 Microelectrodes for Visual Prostheses

The restoration of vision to the blind by artificial electrical stimulation has been an intensive area of research for some time now. Unfortunately, contrary to the cochlea and the auditory nerves, here the engineer faces a fundamental biological challenge: the optic nerve and the

---

FIGURE 8.9   An ECoG microarray supported on a dissolvable silk protein matrix. (a) Schematic depiction of casting and drying of silk fibroin solution on a sacrificial PDMS substrate. (b) Schematics and images of the fabrication of the electrode arrays and transfer printing of electrodes onto silk and connecting them to a cable. (c) Schematic depiction of clinical use of a device in an ultrathin mesh geometry with a dissolvable silk support. (d) Schematics and (e) pictures illustrating how the thickness of the electrode array contributes to conformal contact on a brain model. (g) Image of an electrode array on a cat brain (left) and the average evoked response from each electrode (right) with the color showing the ratio of the root mean square (RMS) amplitude of each average electrode response in the 200-ms window (plotted) immediately after the presentation of the visual stimulus to the RMS amplitude of the average 1.5-s window (not shown) immediately preceding the stimulus presentation for a 2.5-μm mesh electrode array. The stimulus presentation occurs at the left edge of the plotted window. The color bar at the bottom of panel f indicates the RMS amplitude ratios. (g) Representative voltage data from a single electrode in a 2.5-μm mesh electrode array showing a sleep spindle. (From Dae-Hyeong Kim, Jonathan Viventi, Jason J. Amsden, Jianliang Xiao, Leif Vigeland, Yun-Soung Kim, Justin A. Blanco, Bruce Panilaitis, Eric S. Frechette, Diego Contreras, David L. Kaplan, Fiorenzo G. Omenetto, Yonggang Huang, Keh-Chih Hwang, Mitchell R. Zakin, Brian Litt, and John A. Rogers, "Dissolvable films of silk fibroin for ultrathin conformal biointegrated electronics," *Nat. Mater.*, 9, 511–517, 2010. Reprinted with permission from the Nature Publishing Group.)

visual cortex of a blind person degrade with time if unused. Recent reports in which the cortex of blind volunteers has been directly stimulated with penetrating electrodes are not very encouraging: those people only see "phosphenes," or transient light spots, in locations that we will never be able to reproduce again. There was a hope at some point that the UEA could be implanted to stimulate the visual cortex directly, providing 100 inputs—but what are 100 inputs compared with the millions that we need to produce an image? A better strategy is the "retinal approach" (for people with an intact optic nerve, who were not born blind and lose vision because of a disease such as retinitis pigmentosa). Here, microfabrication will help in producing microelectronic chips that can efficiently collect enough light, convert it into charge, and deliver the charge to a layer of intact cells (ideally, ganglion cells that are directly connected to the brain). Several groups are actively working on this incredibly challenging problem, and although there have not been breakthroughs to date, we foresee exciting developments in the coming years. (At least one of these "artificial silicon retinas," developed by a company, is already in clinical trials.)

8.2.6 A Microelectronic Contact Lens

Contact lenses are passive optical devices for people that need optical corrections. Babak Parviz's laboratory at the University of Washington (Seattle) has pioneered the idea that contact lenses do not need to be passive—there is plenty of room for sensors, actuators, and displays that could be connected, in principle, to biomedical devices and games via wireless links. The technological challenges are enormous, in terms of both fabrication (some of the microfabrication processes are not compatible with the polymeric substrate) and biocompatibility (e.g., some materials used for making laser diodes and batteries are highly toxic). Nevertheless, Parviz and colleagues have produced contact lenses that contain an LED and an antenna (so the LED can be wirelessly powered via radiofrequency [RF]). The antenna coil is micropatterned onto the polymeric substrate of poly(ethylene terephthalate) using traditional liftoff. Miniature planar LEDs were soldered onto the metal patterns, making use of capillary forces to align the components in place (See Section 1.9 on fabrication method based on self-assembly). Finally, the planar sheet of poly(ethylene terephthalate) was pressed in a heated mold to produce the contact lens shape (**Figure 8.10**).

8.2.7 Flexible, Thin-Film Microelectronic Circuits for Monitoring Clinical Parameters

We have already seen the dissolvable ECoG microelectrodes of John Rogers' laboratory and the microelectronic contact lens of Babak Parviz's laboratory. Both devices adopted novel technological solutions for integrating microelectronic components onto thin, flexible (polymeric,

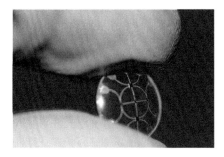

FIGURE 8.10 A contact lens with active microelectronic circuitry. (From B. Parviz, "For your eye only," *IEEE Spectrum*, 46, 9, 36–41, 2009. Figure contributed by Babak Parviz.)

even organic) substrates, departing from the rigid, "traditional" silicon or glass substrates that posed so many problems for implantation. What these laboratories have brought to light is a new dimension of hope: not only the old problems of biocompatibility have been solved, but, as it turns out, the flexibility of thin substrates can be used to great advantage to obtain a better contact because surface tension overcomes the bending of the substrate.

A collaborative group led by John Rogers and by Brian Litt (University of Pennsylvania) has exploited this principle to produce a variation of the dissolvable ECoG microarray for mapping the cardiac electrophysiology in a beating heart (Figure 8.11). The approximately 48-μm-thick device consists of $18 \times 16 = 288$ gold microelectrodes (250×250 μm, the tissue measurement points) on a 25-μm-thick Kapton (PI) substrate, with each microelectrode connected to an associated amplifier and multiplexer (Figure 8.11a); the associated multilayer electronics are transfer-printed onto plastic from a silicon wafer and are encapsulated (using SU-8, PI, SiO_2, and Si_3N_4 as insulators) so that they are protected from the tissue and surrounding biofluids. As shown in Figure 8.11b, the device adhered to the curvilinear surface of the (porcine) heart by surface tension alone, even at rapid (77 beats/minute) pacing. The color map in Figure 8.11b shows a plot of the relative time of depolarization from paced activation close to the LAD coronary artery (the heart was sometimes artificially paced with an external electrode a few cm from the device, see white arrow in Figure 8.11b). The mapping of normal and paced cardiac wave front propagation was demonstrated.

Recently, the Rogers laboratory has extended the thin-film cardiac patch concept to the skin, converting it into an "electronic tattoo." The main difference with the cardiac patch is the substrate, which here is an approximately 30-μm-thick, gas-permeable elastomeric polyester sheet; the total thickness of the device is only 37 μm, and as a result it can conformally adhere to the skin through van der Waals interactions alone (from the point of microelectronics, the device is simpler to fabricate because the requirements for encapsulation are not as strict and because the device is not immersed in biofluids). The device incorporates miniature temperature sensors, strain gauges, LED, and electrocardiogram or electromyograph sensor as well as solar cells and

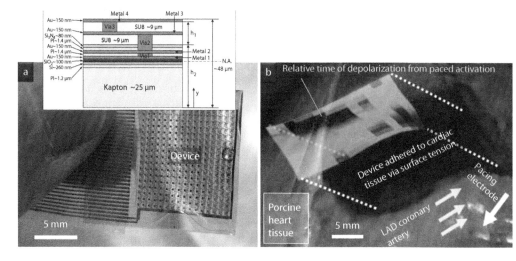

FIGURE 8.11 A cardiac patch for measuring cardiac electrophysiology. The inset in panel a shows a cross-section of the device. (From Jonathan Viventi, Dae-Hyeong Kim, Joshua D. Moss, Yun-Soung Kim, Justin A. Blanco, Nicholas Annetta, Andrew Hicks, Jianliang Xiao, Younggang Huang, David J. Callans, John A. Rogers, and Brian Litt, "A conformal, bio-interfaced class of silicon electronics for mapping cardiac electrophysiology," *Sci. Transl. Med.*, 2, 24ra22, 2010. Figure contributed by John Rogers.)

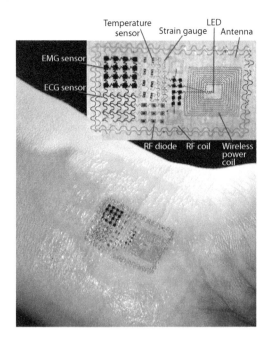

FIGURE 8.12 Thin-film electronics for measuring skin temperature and bioelectricity. (From Dae-Hyeong Kim, Nanshu Lu, Rui Ma, Yun-Soung Kim, Rak-Hwan Kim, Shuodao Wang, Jian Wu, Sang Min Won, Hu Tao, Ahmad Islam, Ki Jun Yu, Tae-il Kim, Raeed Chowdhury, Ming Ying, Lizhi Xu, Ming Li, Hyun-Joong Chung, Hohyun Keum, Martin McCormick, Ping Liu, Yong-Wei Zhang, Fiorenzo G. Omenetto, Yonggang Huang, Todd Coleman, and John A. Rogers, "Epidermal electronics," *Science*, 333, 838–843, 2011. Figure contributed by John Rogers.)

wireless coils for power supply. The transcutaneous monitoring of electrical activity from heart, brain, and muscle was demonstrated (**Figure 8.12**).

8.3 Delivery of Soluble Signals into the Body

As the Michigan probes and the UEA grew increasingly sophisticated with the integration of electronics, it became more and more obvious that they lacked the most fundamental mode of bio-communication: the ability to deliver fluids. Our cells do not understand the language of electrons, only that of ions and molecules dissolved in fluids. A new generation of implantable microfluidic devices is now taking advantage of this concept.

8.3.1 Microneedles

Microfabricated needles ("**microneedles**") that insert themselves only into the skin layer have precious advantages with respect to traditional medical needles: (1) they reduce tissue damage; (2) as a result, they reduce insertion pain and eliminate bleeding (so are more likely to be viewed as acceptable by people who have phobia of needles or have concerns, e.g., a newborn's parents, peoples who are just coming into contact with modern medicine, etc.); (3) they can be integrated with microfluidic devices or storage microchambers (if they are hollow); (4) they can be multiplexed, that is, multiple applications at once or at multiple spots very close to each other is possible, so that large doses can be administered as the sum of many small doses; (5) they can be incorporated in a user-friendly "patch" format for self-administration, which allows nonexperts to vaccinate themselves, for example, people in remote villages and the elderly; and (6) in its polymeric format, they do not constitute a sharp waste biohazard so they can be disposed of by, for example, incineration.

The microneedle field has a very interesting history and has produced some of the most beautiful MEMS images. The first microneedles were reported in micromachining conferences by Albert Pisano (University of California at Berkeley) in 1993 and Ken Wise (University of Michigan at Ann Arbor) in 1994 and were made entirely in single-crystal silicon on the same plane as the wafer. For 10 years, the field improved the fluidic performance of these "in-plane microneedles," introducing other materials such as polysilicon and silicon nitride to produce better channels (Bruno Frazier from Georgia Tech even produced metallized microneedles by electroplating in 2003), but no relevant biological data were produced with them. A big concern was that they were brittle, and if they stayed inside the patient, there would be no easy way to remove them (except for costly surgery, which would defeat their very existence in the first place). An example of an in-plane polysilicon microneedle is shown in Figure 8.13a. The field made a big turn in 1998, when a team led by Mark R. Prausnitz and Mark G. Allen at Georgia Institute of Technology was able to produce short microneedles that stuck out of the plane of the wafer (Figure 8.13b). These "out-of-plane microneedles" were much less brittle and could

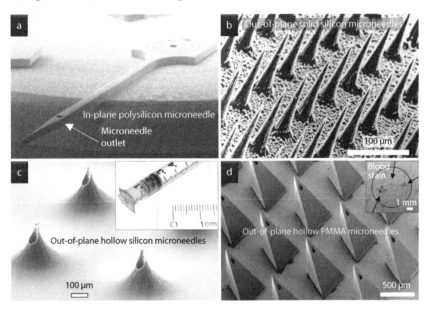

FIGURE 8.13 In-plane versus out-of-plane microneedles. (a) Polysilicon microneedle fabricated with planar micromachining. (From Jeffrey D. Zahn, Ajay Deshmukh, Albert P. Pisano, and Dorian Liepmann, "Continuous on-chip micropumping for microneedle enhanced drug delivery," *Biomed. Microdevices*, 6, 183–190, 2004. Reprinted with permission from Springer.) (b) First array of out-of-plane microneedles for transdermal use. (From Sebastien Henry, Devin V. McAllister, Mark G. Allen, and Mark R. Prausnitz, "Microfabricated microneedles: A novel approach to transdermal drug delivery," *J. Pharm. Sci.*, 87, 922, 1998. Image contributed by Mark Allen.) (c) Microarray of out-of-plane hollow silicon microneedles (inset: chip mounted at the end of the syringe). (From Raja K. Sivamani, Boris Stoeber, Gabriel C. Wu, Hongbo Zhai, Dorian Liepmann, and Howard Maibach, "Clinical microneedle injection of methyl nicotinate: Stratum corneum penetration," *Skin Res. Technol.*, 11, 152–156, 2005. Reprinted with permission from John Wiley and Sons.) (d) Array of hollow, out-of-plane PMMA microneedles. The blood stains shown in the inset (produced with a three-microneedle array) demonstrate the full penetration of the stratum corneum layer of dead cells through the skin of a volunteer. (From Sang Jun Moon and Seung S. Lee, "A novel fabrication method of a microneedle array using inclined deep x-ray exposure," *J. Micromech. Microeng.*, 15, 903–911, 2005. Figure contributed by Sang Jun Moon.)

be easily arrayed in large two-dimensional arrays. The group reported an increase in the permeability of skin on microneedle penetration by three to four orders of magnitude; however, it was observed that occasionally the tip of a few microneedles would break, which raised a biocompatibility concern.

Prausnitz' first silicon microneedles were essentially simple solid sharp tips that were so small (and short) that did not cause pain (they never reached the nerve layers) but were also limited in what they could be loaded with: they could only deliver drugs in dry form. One year later, the same group reported the first hollow microneedles (made in metal and featuring an aperture), which allowed for delivering arbitrary solutions through the back of the wafer without complicated interfaces. **Figure 8.13c** shows an example of out-of-plane, 200-μm-tall hollow microneedles fabricated in silicon by Dorian Liepmann's group from the University of California at Berkeley. The 40-μm-diameter hole at the center of the tips is a through hole, so the tips can be fed from the opposite side of the wafer from a regular syringe (see inset in **Figure 8.13c**) to inject compounds past the stratum corneum, the 10- to 15-μm-thick outer layer of the skin. However, these hollow microneedles tended to get clogged during insertion by the tissue, which changed the fluidic resistance of the array (because they were all connected to the same inlet). In 2003, Göran Stemme's group from the Royal Institute of Technology in Stockholm, Sweden (in June), and van der Berg's group from the University of Twente in the Netherlands (in December) independently reported different ways of producing out-of-plane silicon microneedles with side openings (which prevented the clogging problem). What else could one ask for? There was still the risk of breakage because silicon's biocompatibility was not optimal. Hence, Seung Lee's group from KAIST in Daejeon (Korea) fabricated polymeric tips in PMMA (a very biocompatible polymer, used in many implants and contact lenses) using inclined deep X-ray

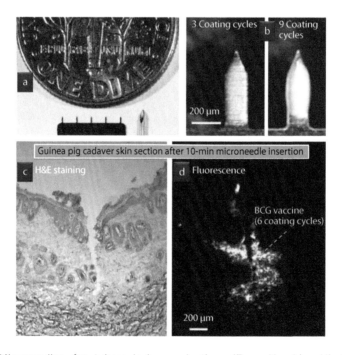

FIGURE 8.14 Microneedles for tuberculosis vaccination. (From Yasuhiro Hiraishi, Subhadra Nandakumar, Seong-O Choi, Jeong Woo Lee, Yeu-Chun Kim, James E. Posey, Suraj B. Sable, and Mark R. Prausnitz, "Bacillus Calmette-Guérin vaccination using a microneedle patch," *Vaccine*, 29, 2626, 2011. Reprinted with permission of Elsevier.)

exposure (Figure 8.13d). As shown in the inset, these microneedles are deep enough to penetrate through the stratum corneum, the top skin layer of keratinized dead cells, and reach the blood vessels for potential blood extraction (shallower microneedles for diffusive drug delivery are also possible).

For some applications, it may be important to develop very simple, dirt-cheap devices. Mark Prausnitz's laboratory has recently fabricated planar solid stainless-steel microneedles (700 μm long, 170 × 55 μm at the base and tapered at the tip with a 5-μm radius of curvature) by laser-cutting stainless steel sheets. These solid microneedles, meant only for delivery of transdermal drugs that can be dried on the tips (see next section), do not suffer from the silicon brittleness and the biocompatibility concerns of older microneedles. Figure 8.14a shows an array of five microneedles next to a 26-gauge hypodermic needle and a U.S. dime coin for size comparison. The microneedles are used to deliver the bacillus Calmette–Guérin (BCG)—the only licensed vaccine for human use against tuberculosis—in dry form. The BCG-coated microneedles (Figure 8.14b) produce minimal tissue damage (Figure 8.14c) and efficiently deliver the (fluorescently tagged) BCG vaccine intradermally (Figure 8.14d). The microneedle-delivered vaccine was able to induce an immune response in both the lungs and the spleen of guinea pigs comparable with intradermal vaccines delivered by traditional hypodermic needles, suggesting a promising potential for future human use. This group has also been able to develop a process for fabricating microneedles in biodegradable polymers (polylactic acid, polyglycolic acid, and their copolymers), which addresses safety concerns of previous microneedle designs made of metal or silicon.

8.3.2 Microfluidic Drug Delivery to the Eye

Drug delivery to the eye is used to treat retinal diseases such as glaucoma, age-related macular degeneration, diabetic retinopathy, and retinitis pigmentosa, but it can be traumatic to the patient, especially when administered via surgery or intraocular injection. Ellis Meng's group at the University of Southern California (Los Angeles) has reported a refillable drug delivery PDMS microfluidic device that is implanted under the conjunctiva (Figure 8.15a) and that delivers the contents of its reservoir into the eye when it is manually pressurized (e.g., with a cotton swab) beyond the cracking pressure of the check valve (Figure 8.15b and c). The reservoir features posts to prevent stiction of the roof against the floor when it is collapsed by pressure application. The contents of the reservoir are refilled with a sharp needle that punctures the PDMS, leaving a hole that self-seals.

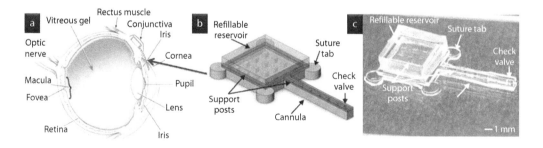

FIGURE 8.15 A microfluidic drug delivery device for ocular diseases. (From Ronalee Lo, Po-Ying Li, Saloomeh Saati, Rajat Agrawal, Mark S. Humayun, and Ellis Meng, "A refillable microfabricated drug delivery device for treatment of ocular diseases," *Lab Chip*, 8, 1027–1030, 2008. Reproduced with permission from The Royal Society of Chemistry.)

8.4 Microtools for Surgery

Since the beginning of surgical practice, surgeons have been limited by the size of their tools. This constraint has stimulated the development of technologies that enable the manipulation of tissue in real time, such as robotics, computer tomography, optical systems, and miniaturized tools. Here microfabrication is playing an increasingly important role.

8.4.1 Micromachined Surgical Tools

We saw in Chapter 1 that with state-of-the-art metal and ceramic layer-by-layer fabrication, it is now possible to fabricate incredibly complex three-dimensional microstructures, some even containing microgears (see **Figure 1.12**). This technology has been applied (and is being tested) to produce better surgery microtools, ranging from tiny forceps, microneedle arrays, and tissue debriders (**Figure 8.16**). Only 10 years ago, these would have belonged to a science fiction book.

8.4.2 Catheters and Surgical Gloves Equipped with Microsensors

Conventional balloon catheters are essentially built from electronically and optically inactive materials (e.g., polyurethane or silicone); therefore, they cannot be programmed to respond to physiological data in real time. John Rogers' laboratory at the University of Illinois (Urbana-Champaign) has been able to integrate electronics on the flexible, curvilinear substrates of balloon catheters (**Figure 8.17**) and of surgical gloves, hence rendering them "multifunctional." An engineering marvel (the article is full of delightful sensor designs), this multifunctional device provides the ability to measure electrical, tactile, optical, temperature, and flow properties at the balloon or glove–tissue interface in real time. Simultaneous, multiple electrocardiogram or

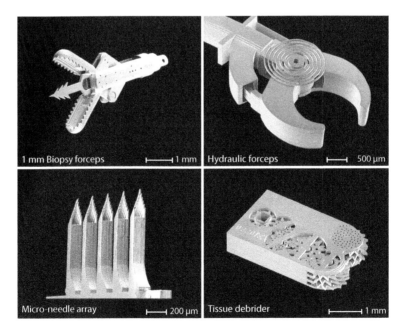

1 mm Biopsy forceps ⊢———⊣ 1 mm

Hydraulic forceps ⊢———⊣ 500 µm

Micro-needle array ⊢———⊣ 200 µm

Tissue debrider ⊢———⊣ 1 mm

FIGURE 8.16 Microfabricated surgical tools. The structures are made of ceramics. Images courtesy of Microfabrica, Inc. (Note: the figure is not meant to endorse Microfabrica's technology over other competing technologies.)

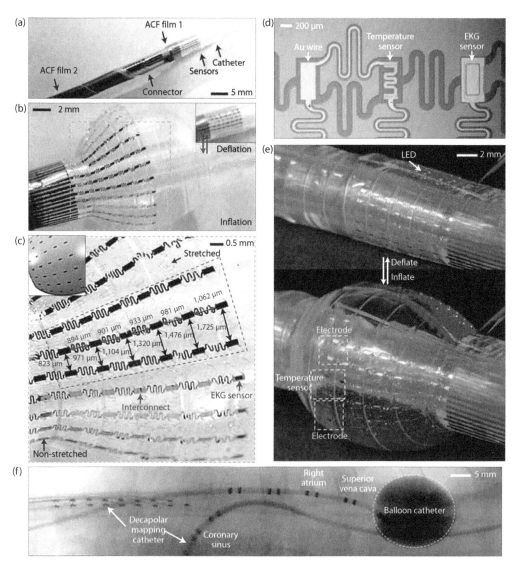

FIGURE 8.17 Multifunctional balloon catheters. (a) Interconnected passive network mesh integrated on a balloon catheter (deflated). (b) Balloon of a catheter inflated by approximately 130% relative to its deflated state (inset). (c) Magnified view of noncoplanar serpentine interconnects on the balloon in its inflated state (same area as defined by the green dotted line in panel b). (d) Magnified image showing a temperature sensor, the gold lines used to apply bias voltages, and the electrodes for simultaneous electrogram mapping (electrocardiogram). (e) Multifunctional balloon catheter in deflated and inflated states; the image shows arrays of temperature sensors (anterior), microscale light-emitting diodes (posterior), and tactile sensors (facing downward). (f) X-ray angiography image of an instrumented balloon catheter deployed in the heart (right atrium) of a pig for in vivo recording of electrophysiology near the superior vena cava. (From Dae-Hyeong Kim, Nanshu Lu, Roozbeh Ghaffari, Yun-Soung Kim, Stephen P. Lee, Lizhi Xu, Jian Wu, Rak-Hwan Kim, Jizhou Song, Zhuangjian Liu, Jonathan Viventi, Bassel de Graff, Brian Elolampi, Moussa Mansour, Marvin J. Slepian, Sukwon Hwang, Joshua D. Moss, Sang-Min Won, Younggang Huang, Brian Litt, and John A. Rogers, "Materials for multifunctional balloon catheters with capabilities in cardiac electrophysiological mapping and ablation therapy," *Nat. Mater.*, 10, 316, 2011. Reproduced with permission from the Nature Publishing Group.)

temperature recordings and mechanical contact during ablation were demonstrated on beating rabbit and rat hearts.

Similarly, the company CardioMEMS has developed a wireless, battery-less, RF-activated pressure sensor the size of a paper clip that is implanted into the pulmonary artery and that has demonstrated a 30% lower risk of ending up in the hospital for heart failure. The device, which is now approved by the FDA, is activated with a wand-like device as it is positioned over the implant area. The pressure sensor transmits real-time data to an external unit, which then communicates this information to the patient's physician (who can effectively monitor the patient remotely without requiring the patient to come to the hospital, thus reducing the cost of health-care and increasing the quality of life substantially).

8.4.3 "Gecko" Surgical Tape

A large collaborative team led by Jeffrey Karp at Harvard Medical School has developed a bio-degradable polymeric tape that sticks to tissue because its surface is covered with nanopillars (**Figure 8.18**)—a strategy inspired on the feet of Geckos. The adhesion strength of nanopatterned **poly(glycerol sebacate acrylate) (PGSA)** was nearly twofold greater than the adhesion strength of flat unpatterned polymer. The surface of the pillars is coated with a sugar-based "glue" to close a surgical incision, and because it is made of poly(glycerol sebacate acrylate) (PGSA), it completely degrades over time. PGSA is an elastomer that can be made more or less stretchable (e.g., for a heart patch) and can be loaded with drugs that help tissue heal (i.e., "active band-aids").

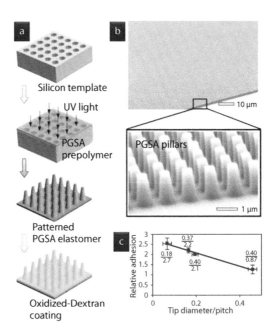

FIGURE 8.18 Biodegradable "gecko-inspired" adhesive surface. (From A. Mahdavi, L. Ferreira, C. Sundback, J. W. Nichol, E. P. Chan, D. J. D. Carter, C. J. Bettinger, S. Patanavanich, L. Chignozha, E. Ben-Joseph, A. Galakatos, H. Pryor, I. Pomerantseva, P. T. Masiakos, W. Faquin, A. Zumbueh, S. Hong, J. Borenstein, J. Vacanti, R. Langer, and J. M. Karp, "A biodegradable and biocompatible gecko-inspired tissue adhesive," *Proc. Natl. Acad. Sci. U. S. A.*, 105, 2307–2312, 2008. Copyright (2008) National Academy of Sciences, U. S. A.)

8.5 Insect Research

The idea that microsystems, due to their small size and low weight, can be applied to monitor and control the behavior of individual insects has recently attracted great interest from several research groups. Nigel Franks, a behavioral ecologist at the University of Bristol in England, glued small RF identification tags to the backs of individual ants to track, register, and analyze what the ants were doing using a computer. (Note that the FDA has approved—not without controversy—a human-implantable RF identification tag for subcutaneous use, the VeriChip, to help track patients in acute care.) In principle, an insect could carry optical and chemical sensors and be remotely directed to inaccessible or hazardous locations (e.g., chemical spill sites, stealthy reconnaissance, etc.).

The large flying moth **Manduca sexta** is a widely used model because big loads (~1 g, about half their body mass) do not significantly affect its flight patterns. In 2008, David Erickson's group at Cornell University (New York) implanted a microfluidic device into the thorax of *M. sexta* pupae to deliver reversibly paralyzing agents (such as L-glutamic acid, L-aspartic acid, or γ-aminobutyric acid, which are also excitatory neurotransmitters at insect skeletal neuromuscular junctions but at high concentrations they are venoms produced by many spiders and wasps). The device was implanted at the pupal stage (**Figure 8.19a** and **b**) because the wound heals better—and presumably is integrated more harmonically with the physiology of the animal—at this developmental stage than at the adult stage. Once the device reached the adult stage

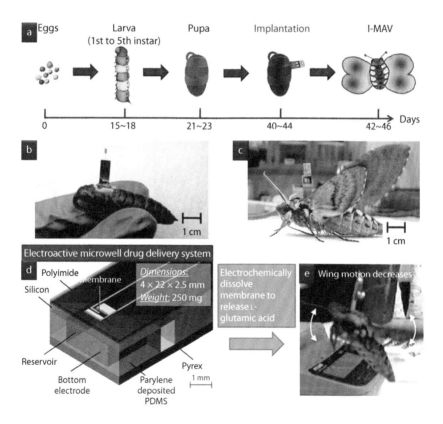

FIGURE 8.19 Microfluidic control of insect metabolism. (From Aram J. Chung and David Erickson, "Engineering insect flight metabolics using immature stage implanted microfluidics," *Lab Chip*, 9, 669–676, 2009. Reproduced with permission from The Royal Society of Chemistry.)

(Figure 8.19c), its respiratory CO_2 output was monitored in a chamber where it was allowed to flap its wings (i.e., "fly" without moving). The neurotoxins (L-glutamic acid, L-aspartate acid [LLA], or γ-aminobutyric acid) were loaded in a reservoir capped by a gold membrane that could be dissolved electrochemically on command by application of a small voltage, which caused the release of the chemicals into the thorax of the insect and the (reversible) wing paralysis within 90 seconds (Figure 8.19d and e).

Erickson's remarkable achievement, however, has the obvious shortcoming that the moth is not free to go, so it is not practical for field applications. A team led by Tom Daniel at the University of Washington (Seattle) and a team led by Joel Voldman at the Massachusetts Institute of Technology (Cambridge) have accomplished what only a few years ago was the subject of science fiction talk: to be able to remote control a moth in free flight (Figure 8.20). The clever design of the electrodes, which clamp circumferentially around the nerve cord in a spoke-like manner, plays an important role in the success of the design (Figure 8.20a and b). Implantation, as with Erickson's system, is performed at the pupal stage (2 days before eclosion) to minimize rejection. Stimulation of the adult elicits abdominal motions (presumably by activating motoneurons or interganglionic neurons) that are used to bias the flight path. There is a design trade-off that occurs when more electrode tips are introduced to provide for more stimulation points: the spokes cannot penetrate as deep into the tissue. (Four tips did not provide enough stimulation points, and eight tips were too small, so a compromise design was reached with six electrodes.)

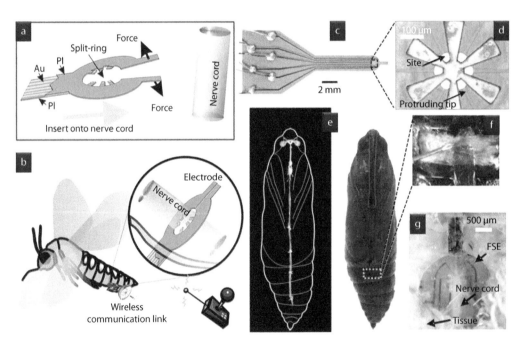

FIGURE 8.20 Moth flight control with a wireless neural microstimulation system. (From Wei Mong Tsang, Alice L. Stone, Zane N. Aldworth, John G. Hildebrand, Tom L. Daniel, Akintunde Ibitayo Akinwande, and Joel Voldman, "Flexible split-ring electrode for insect flight biasing using multisite neural stimulation," *IEEE Trans. Biomed. Eng.*, 57, 1757, 2010. Figure contributed by Joel Voldman.)

8.6 Summary

Compared with other microsystems, implantable microsystems have always faced the most difficult challenges in the BioMEMS field. The sample from which the microsystems are extracting a measurement (the individual) has a virtually unknown substrate, and repetition of the measurement is often unacceptably painful or too unreliable because of anatomical and physiological imprecision. A daunting amount of paperwork is usually required to make sure all the ethical guidelines are being followed before implantation. As a result, experiments have typically been sketchy or have taken years to produce conclusive results: the field has been progressing very slowly. Yet, this is the part of the field that the public is most excited about, so scientists are persevering: Who does not dream of a deep-brain microstimulator that one day might be the routine cure for Parkinson's disease, just as pacemakers are routinely implanted to treat heart arrhythmias nowadays? Or a wireless micro-ECoG electrode grid that cures infant epilepsy? Or a pill the size of a grain of rice that reports to your doctor your internal constants, including temperature and pH, while it travels through your intestines and you are at work?

I am certain that in a few years, microfabrication technology will be ubiquitous in almost every medical device because there are vast benefits, with little incremental cost per device, derived from miniaturization. We are only limited by our own imagination.

Further Reading

Ainslie, Kristy M., and Desai, Tejal A. "Microfabricated implants for applications in therapeutic delivery, tissue engineering, and biosensing," *Lab on a Chip* **8**, 1864–1878 (2008).

Cheung, K. C. "Implantable microscale neural interfaces," *Biomedical Microdevices*, **9**, 923–938 (2007).

Desai, Tejal and Bhatia, Sangeeta (editors). "Therapeutic micro/nanotechnology," in *BioMEMS and Biomedical Nanotechnology Series* (Vol. III), Springer (2006).

Kipke, D. R., W. Shain, G. Buzsaki, E. Fetz, J. M. Henderson, J. F. Hetke, and G. Schalk. "Advanced neurotechnologies for chronic neural interfaces: new horizons and clinical opportunities," *Journal of Neuroscience* **28**, 11830–11838 (2008).

Vandervoort, J., and A. Ludwig. "Micro-needles for transdermal drug delivery: a minireview," *Frontiers in Bioscience* **13**, 1711–1715 (2008).

Wang, Wanjun, and Soper, Steven A. (editors). *Bio-MEMS: Technologies and Applications*, CRC Press (2007).

Wise, K. "Integrated sensors, MEMS and microsystems: reflections on a fantastic voyage," *Sensors and Actuators. A, Physical* **36**, 39–50 (2007).

Zhou, David, and Greenbaum, Elias (editors). *Implantable Neural Prosthesis 1: Devices and Applications*," Springer (2009).

Zhou, David, and Greenbaum, Elias (editors). *Implantable Neural Prosthesis 2: Techniques and Engineering Approaches*, Springer (2009).

Appendix. Teaching Materials

A.1 Suggested Exercises for Chapter 1

Exercise A.1.1. Briefly describe the basic elements of and differences between photolithography, soft lithography, and micromachining.

Exercise A.1.2. Describe how to make a trench in silicon with (a) pyramidal walls and (b) vertical walls. (If you know more than one way, please describe them.)

Exercise A.1.3. Describe the conceptual differences between photolithography, micro-stamping, and microfluidic patterning, putting an emphasis on their relative advantages and disadvantages.

Exercise A.1.4. Describe five important properties of PDMS in the context of soft lithography. Briefly explain whether each property is advantageous or disadvantageous for building microdevices (note that it can be both, depending on the context of its use), and why you think so.

Exercise A.1.5. Choose from the literature (e.g., Figure A.1a) a picture of a metal pattern on a flat surface and describe at least four processes, two photolithographic and two soft lithographic, that could have been used to produce it. Is it possible to use the same technique to produce the reverse pattern, and if so, how would it be done? Be as specific as possible about the chemistry employed at each step and use step-by-step cross-sectional schematics to illustrate your explanations.

Exercise A.1.6. Choose from the literature (e.g., Figure A.1b) a picture of a structure that has been micromachined on Si(100) and describe at least two processes, one photolithographic and one soft lithographic, that could have been used to produce it. Be as specific as possible about the chemistry employed at each step and use step-by-step cross-sectional schematics to illustrate your explanations.

Exercise A.1.7. Imagine that you are prototyping a device and need to write the patterns shown in Figure A.2 in 40-μm-thick SU-8 photoresist on a silicon wafer (required exposure energy of 160 mJ/cm^2).

You have the following tools available:

(a) A variable aperture flash-and-repeat laser writer—this tool has a laser light source and a rectangular aperture with dimensions (length and width) that can be varied from 2 μm to 1.5 mm in 0.5-μm increments. The aperture can rotate and the sample stage can translate. Each flash exposes a rectangular feature at an exposure energy of 40 mJ/cm^2. The exposure region corresponds to the aperture size at a location that is determined by the stage translation. The machine is capable of 3000 flashes/hour. The stage translation time is negligible compared with the exposure time.

(b) A raster-scan laser writer. This tool has a 3-μm spot size laser light source that is raster-scanned over the entire region of the sample to be patterned and the light is either allowed to reach the substrate or deflected for pixels that are not to be written. The tool writes 30 mm^2/min at a scan rate of 1×, corresponding to an exposure energy of 40 mJ/cm^2. Scan rates of 2× and 4× are also available.

(c) A laser-illuminated digital mirror device (DMD). The DMD has an array of 1024 × 768 square mirrors that are 13 μm × 13 μm. The DMD is illuminated with the laser and the light is reflected through a lens with a demagnification factor of 5:1 onto the sample surface. The optical power density that reaches the sample surface is 4 mW/

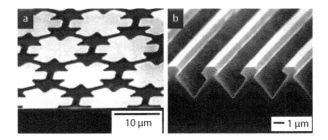

FIGURE A.1 SEMs of microstructures. (a) The structures are made of a 200-nm-thick layer of silver on SiO2. (b) The structures are fabricated using shadow evaporation and anisotropic etching of Si(100). (From Xia, Y. and G. M. Whitesides. "Soft lithography," *Angew. Chem. Int. Ed.* 37, 550–575, 1998. Reproduced with permission from John Wiley and Sons.)

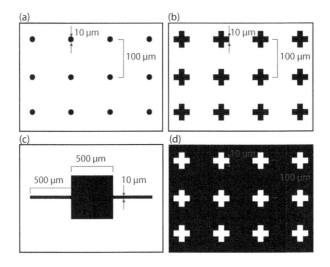

FIGURE A.2 Photolithography patterns. The dark region is to be exposed and the white region will be developed away.

cm². There is a precision sample *x–y* translation stage for multiple exposure and pattern stitching. The translation time is negligible compared with the exposure time.

(d) A high-resolution laser printer for printing photomasks and a contact mask aligner. The laser printer has a resolution of 60,000 dots per inch (DPI), an ink spot diameter of 5 μm, and prints at a rate of 1 page/min. The contact aligner has a 5 mW/cm² mercury arc lamp.

Calculate how long it will take to write one sample of each pattern using each of the tools. For (a), (b), and (d), imagine that you are writing a 1.5 cm × 1.5 cm area filled with an array of the features shown at the given pitch. For (c), imagine that you are writing one copy of the pattern.

For each pattern, calculate how long it will take to write 1000 samples with each of the tools.

Comment on any features in the patterns that will not be properly resolved with each of the given methods.

Exercise A.1.8. If a wafer is spun with the photoresist SU-8 (specifically, SU8 2035) at 2000 rpm for 30 seconds, what will be the film thickness? What if it is spun for 2 minutes at 2000 rpm? (Please check the data sheet available at http://www.microchem.com/Prod-SU82000.htm). What speed would you have to spin SU-8 2050 to achieve the same thickness?

Exercise A.1.9. A photolithography machine with known dose of UV light (6.5 mJ/cm² per second) is used to expose a silicon wafer that has been coated with 75-μm-thick SU-8 photoresist. The SU-8 data sheet suggests that the exposure energy should be 150 to 215 mJ/cm² for a thickness of 45 to 80 μm of SU-8 (please check the data sheet from www.microchem.com). How many seconds should the wafer be exposed to properly cross-link 75-μm-thick SU-8?

(a) 20–26 seconds

(b) 28–34 seconds

(c) 37–43 seconds

A.2 Suggested Exercises for Chapter 2

Exercise A.2.1. Describe three different methods for creating protein micropatterns. After you have described all three methods step by step, describe four different applications of protein patterning. For each application, choose which method you think would be the best of the three and explain why you have chosen that one (and not the other two).

Exercise A.2.2. Describe (using cross-sectional schematics) two alternative microfabrication processes that yield the cellular microstructures seen in Figure A.3. The inset shows a large-scale view to give an idea of the pattern. Assume that the substrate is polystyrene for one of the processes and glass for the other alternative.

Exercise A.2.3. Draw, in cross-sectional schematics, two different step-by-step microfabrication processes (one using only photolithography, the other using soft lithography) for producing the cellular microarray in Figure A.4 on a flat surface. Please be specific when explaining how the materials/molecules are immobilized on the surface—that is, is it by physisorption or by chemisorption, and if the latter, make sure that the chemical reaction that you are proposing is not vague or ambiguous.

Exercise A.2.4. A suspension of cells (fluorescently labeled blue, in serum-free medium) is added onto a surface that contains a micropattern of many islands just like the fluorescence micrograph shown in Figure A.5a. The X, Y, and Z letters denote surface coatings or grafts (i.e., they can be physisorbed or chemisorbed; you will have to figure it out in this exercise).

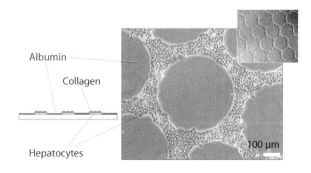

FIGURE A.3 Cellular micropattern.

Appendix

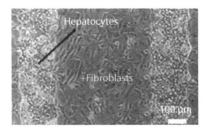

FIGURE A.4 Micropatterned coculture.

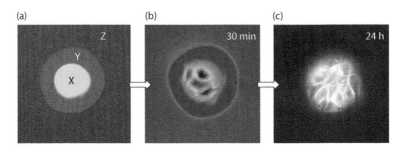

FIGURE A.5 Micropatterning cellular adhesiveness.

After 30 minutes, the unattached cells are removed, and the cell culture surface appears like the micrograph on Figure A.5b, with four blue cells attached on the central "X" region. After 24 hours, the cells have proliferated and have been able to attach and migrate onto the "Y" region (but not onto the "Z" region!), with the result that the island looks approximately like that in Figure A.5c.

Provide one name of surface coating/graft and cell type for which this patterning scheme would work and that would be consistent with *all three* images; please *briefly* explain why you chose it:

> Cell type = ...Why?
>
> X = ... Why?
>
> Y = ... Why?
>
> Z = ... Why?

Exercise A.2.5. The picture in Figure A.6 shows two pairs of cells attached to a layer of ECM protein (on a substrate of your choice) and separated by a chemically grafted layer containing

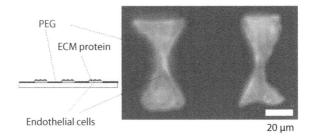

FIGURE A.6 PEG micropattern.

422

PEG groups. Consider three different ways of producing the cellular micropattern: (a) using photolithography, (b) using microcontact printing, and (c) using microfluidic patterning. In all cases, the techniques must be used to define the ECM and PEG micropatterns (not to directly deposit the cells). Explain the biological basis of why the cells prefer to attach to the ECM protein micropattern and never spread beyond, into the PEG areas.

Exercise A.2.6 (Design Challenge). "A microfabricated clonogenic assay." The goal of this experiment is to measure the replication rates of approximately 100,000 single cells. You are part of a large research team, but your job is the most critical one. You are asked to design a microarray of single cells on a substrate that will allow you to visualize the cells dividing in real time using a microscope. Optimization of the time-lapse microscope is not your job, so you work under the assumption that the cells, under the microscope, will be living and dividing happily. Another person in your research team will take care of counting the cells from the images you acquire, so you don't need to worry about that either. Your job is only to design a substrate on which you will be seeding a suspension of adherent cells that a technician will hand to you. (You can choose a cell type if you want.)

Remember that as the cells divide, the "colony" will grow in size—after the first division, you won't have single cells, you will have pairs of single cells, and so on. You are only going to acquire data for the first 10 divisions. You have one design constraint: the colonies cannot overlap for the duration (10 divisions) of the experiment. You finally decide that you will try two different microfabrication techniques (no robotics to position anything, please). Describe them in detail, drawing the cross-section schematics for clarity if you think you need to.

Obviously, you have to make a combination of design choices on the type of substrate, substrate coatings, and microfabrication technique. Challenge yourself by choosing two combinations that work *and* correctly explaining why you made those choices. Instead of using vague, obvious statements such as "we will use a cell-adhesive protein," specify which cell-adhesive protein. Similarly, do not say "albumin will be immobilized"—instead, specify the physical principles or chemical reactions (or both) you used to immobilize it, and instead of indicating "we will micropattern collagen," specify which micropatterning technique you used.

Exercise A.2.7. To promote immobilization of proteins on a Si-OH modified surface, you would like to use an epoxy functional silane coating. You have your choice of a (3-glycidyloxypropyl)trimethoxysilane and (3-glycidoxypropyl)methyldiethoxysilane. How many reactive groups does each silane have for proteins? For the surface? For other silanes? Which do you think will be most effective at making a stable and cross-linked surface coating?

Exercise A.2.8. Approximate the following molecular species of interest for surface coating by simple geometric shapes and calculate the maximum packing density (molecules/cm^2), assuming a minimum intermolecular spacing of 0.5 Å

(a) Alkanethiol—rod, 10 Å length, 1 Å diameter

(b) Bovine serum albumen (BSA)—ellipsoid, $140 \times 40 \times 40$ Å^3

You are coating the surfaces of a closed microwell that is $100 \times 100 \times 50$ μm^3 with the species above. The inlet and outlet dimensions of the chamber are 50×50 μm^2. The coating protocol is to load a solution of the desired species diluted with buffer into the chamber, allow it to sit for 1 hour to coat, and then flush the chamber with buffer. Assume that the molecules will pack at the density from part 1 of the problem, only one monolayer will form on the surface, and the rest of the molecules will stay in solution. The concentration of molecule in the solution should not change by more than 1% during the coating process. Calculate the minimum concentrations necessary.

Exercise A.2.9. A femtosecond titanium-sapphire laser, with a pulse frequency of 76 MHz and pulse duration of 150 fs, tuned to 740 nm was used for fabrication of albumin microstructures.

Appendix

The laser beam was focused through a 100× oil objective with a numerical aperture of 1.3 and an 80% transmittance. Calculate the peak intensity at the focal point of the laser corresponding to an average laser power of 30 mW. Assume a Gaussian laser beam profile, diffraction-limited optics, and a constant power during laser pulses.

Hint: Use the following equation to calculate the peak intensity, I_0.

$$I_0 = 2\, P_0 / \omega_0^2 \pi$$

Here P_0 is the peak power and ω_0 is the beam waist. The average laser power can be used to calculate P_0.

A.3 Suggested Exercises for Chapter 3

Exercise A.3.1. Describe five properties of PDMS that have been important in its success in microfluidics.

Exercise A.3.2. Sketch three different designs of microfluidic gradient generators.

Exercise A.3.3. Describe three different PDMS-to-PDMS or PDMS-to-glass bonding techniques. Specify their chemistry principles as accurately as possible. State their advantages and disadvantages.

Exercise A.3.4. Briefly describe two different ways of fabricating microchannels in (a) silicon, (b) glass, (c) PDMS, (d) PMMA, and (e) paper.

Exercise A.3.5. Describe two different platforms for creating and shuttling droplets within microfluidic channels. Explain their advantages and disadvantages.

Exercise A.3.6. Describe three PDMS microvalve designs and point to their comparative strengths and limitations.

Exercise A.3.7. Compare three micropump designs of your choice.

Exercise A.3.8. Describe three flow-based and three nonflow microfluidic gradient generators and compare their advantages and shortcomings with respect to each other.

Exercise A.3.9. The Good–Girifalco–Fowkes equation describes the relationship between θ, the contact angle, and γ_{sv} and γ_{lv}, the surface free energy values of the solid-vapor and liquid-vapor interfaces, respectively:

$$\cos\theta = -1 + 2(\gamma_{sv}/\gamma_{lv})^{0.5}$$

Find γ_{lv} given that $\theta = 70$ degrees and $\gamma_{sv} = 22.1$ ergs/cm^2 for methylene iodide on PDMS.

Exercise A.3.10. Consider a closed microfluidic cell-culture chamber of dimensions 31 mm × 1 mm × 120 μm ($l \times w \times h$). Neutrophils are attached to the floor of this chamber, and cell-culture medium is flowed through the chamber at 0.20 mL/min. The density and viscosity values of the cell-culture medium can be assumed to be the same as water.

(a) Calculate the Reynolds number and comment on the flow conditions inside the chamber.

(b) Calculate the maximum linear velocity (v_x) experienced by the cells. For convenience, assume that the velocity changes only as a function of the chamber height. Also, assume that the cells are of a uniform height. Justify the selected height of cells and any additional assumptions made.

Hint: Use equations describing pressure-driven flow through a narrow rectangular channel.

Exercise A.3.11. The dielectrophoretic force on a dielectric sphere in a dielectric medium given in Equation 3.34 can be written as

$$F = 2\pi r^3 \varepsilon_m \, \text{Re}\{K\} \nabla E_{rms}^2$$

where K is the complex Clausius–Mossotti function given by

$$K = \frac{\varepsilon_p^* - \varepsilon_m^*}{\varepsilon_p^* + 2\varepsilon_m^*}$$

where ε_m^* and ε_p^* are the complex dielectric constants of the medium and the particle, respectively. The complex dielectric constants can be expressed as $\varepsilon^* = \varepsilon - i\dfrac{\sigma}{\omega}$

(a) List two parameters that can be used in the design or operation of a device to increase the DEP force for a target cell of a given size in a given medium.

(b) Write a computer script to calculate and plot the real part of the normalized Clausius–Mossotti function as a function of frequency from DC to 1 GHz for the following cases:

  (i) $\varepsilon = \sigma$, $\varepsilon_m = 2\varepsilon_p$
  (ii) $\varepsilon = \sigma$, $2\varepsilon_m = \varepsilon_p$
  (iii) $\varepsilon = 2\sigma$, $\varepsilon_m = 2\varepsilon_p$
  (iv) $2\varepsilon = \sigma$, $\varepsilon_m = 2\varepsilon_p$

Exercise A.3.12. (a) For a microchannel with dimensions $w = 100$ μm, $h = 20$ μm, $l = 2$ mm, calculate the microfluidic resistance. (b) What would be the resistance of four identical channels with these dimensions in parallel?

Exercise A.3.13. You would like to design a binary branching tree to distribute flow evenly across a large area. You would like to maintain equivalent flow velocity at each level of the tree while the width of channels is larger by a factor of 2 relative to the next level. By what factor would you scale the length of the channels relative to the previous level?

Exercise A.3.14. There are two rectangular PDMS microchannels which are fabricated by soft lithography. The first microchannel is 3.5 mm in length, 50 μm in width, and 75 μm in height. The second microchannel is 4 mm in length and 60 μm in height. What should the width of the second microchannel be to have an equal fluidic resistance as the first microchannel?

Exercise A.3.15. Using Equation 3.14, calculate the radius of a circular-cross-section microchannel that has a length of 2 mm and a resistance of 7.0×10^{11} N·s·m^{-1} when filled with water ($\eta = 8.90 \times 10^{-4}$ Pa·s).

A.4 Suggested Exercises for Chapter 4

Exercise A.4.1. Enumerate three advantages and three disadvantages of Affymetrix's oligonucleotide microarrays. Repeat the exercise with Pat Brown's DNA microarrays.

Exercise A.4.2. Sketch three different designs of PCR chips.

Exercise A.4.3. Explain the basis of the ELISA-based microfluidic pregnancy test.

Exercise A.4.4. Assume a solid crystal-silicon cantilever (length = 100 μm, thickness = 1 μm, width = 10 μm) vibrating in vacuum. What is the expected frequency shift when an average-sized (dry) bacterium is deposited near the tip?

Exercise A.4.5. Calculate and compare the resonance frequency of the following two cantilevers A and B. Cantilever A has dimensions of 3.4 mm × 1.6 mm × 1 mm (length × width × thickness) and cantilever B has dimensions of 112 μm × 2.94 μm × 650 nm (length × width × thickness). Which cantilever would be better suited for making sensitive measurements and why?

Exercise A.4.6. Cantilevers are great tools for analyzing biomolecular binding forces. The force can be calculated easily by knowing the parameters of the cantilever material and the tip displacement. A silicon cantilever is used to detect the binding force between one type of biomolecule and a specific substrate with known parameters (width, 5 μm; thickness, 0.5 μm; length, 50 μm). If the displacement of the tip is 0.25 μm, the integrated binding force of all the biomolecules on the tip is _____.

Exercise A.4.7. Given a cantilever of spring constant k, resonance frequency ω_0, and dimensions width w, length L and thickness t, determine how k and ω_0 would change if dimensions w and L were both doubled, keeping t constant.

A.5 Suggested Exercises for Chapter 5

Exercise A.5.1. Describe at least three designs of microfluidic flow cytometers. Rank their relative performance qualitatively, in terms of their ability to focus flow, the throughput that can be achieved with them, and the ease of fabrication.

Exercise A.5.2. Describe the main advantages that microfluidic cell cultures confer over conventional cell cultures (e.g., in a petri dish). What are the main limitations of microfluidic cultures over conventional ones?

Exercise A.5.3. State the advantages and limitations of extracellular recordings (obtained with multielectrode arrays, MEAs) and compare to advantages and limitations of intracellular recordings (obtainable with current patch clamp chips). It is important that you demonstrate a knowledge of the microfabrication procedures (in particular, the dimensions and materials involved) and a critical discussion on the cell-electrode interface.

Exercise A.5.4. Describe at least three designs of microfluidic patch clamp chips. Explain their advantages and disadvantages in terms of ease of fabrication (i.e., aperture resolution, integration of microfluidics, etc.), fabrication throughput, and expected quality of giga-seal.

Exercise A.5.5. Hydrodynamic focusing has been widely used for cell sorting and focal stimulation of cells in microfluidic chips. Suppose a hydrodynamic focusing chip with a center flow of BSA and sheath flows of water flowing along the y direction under a constant flow rate. At the convergence point, the center flow is 100 μm wide. How much wider is the center flow due to diffusive broadening of BSA approximately 1 second after the convergence of the center and sheath flows? (Assume that the butterfly effect due to Taylor dispersion is negligible.)

Exercise A.5.6. Using a five-loop spiral microchannel $W = 100$ μm wide and $H = 50$ μm high, Ian Papautsky's group separated 7.3 μm beads from 1.9 μm beads at Dean number, $De = (\rho V D_h / \mu)(D_h/2R)^{1/2} = \mathrm{Re}\,(D_h/2R)^{1/2} = 0.47$, where ρ is the density of the fluid medium, V is the average fluid velocity, D_h is the microchannel's hydraulic diameter ($D_h \equiv 2HW/(H+W)$), μ is the fluid viscosity, and R is the radius of curvature of the path of the channel.

What would the Dean number be if D_h were doubled and the radius of curvature, R, were halved?

A.6 Suggested Exercises for Chapter 6

Exercise A.6.1. Describe how cell biology research benefits from microfabrication technology. Bring up at least three advantages.

Exercise A.6.2. Explain three different schemes for measuring cellular traction on a microscale.

Exercise A.6.3. Describe the designs of five different microfluidic gradient generators used in chemotaxis research. Point to their salient features (advantages and disadvantages in the context of chemotaxis).

Exercise A.6.4. We are going to use an array of PDMS posts to determine the forces exerted by an epithelial cell (e.g., see Figure 6.23). The micro-post array is made up of cylinders 2 μm in diameter and 6 μm in height. They are spaced 4 μm apart (center-to-center). The Young's modulus of PDMS is assumed to be $E = 2$ MPa. The formula for a beam that deflects an amount δ under a force F is:

$$F = \left(\frac{3EI}{L^3} \right) \delta \qquad (A.1)$$

where E is the Young's modulus of the beam material, I is the area moment of inertia, and L is the beam length. For a solid cylindrical beam of diameter D (load applied at the end),

$$I = \frac{1}{64} \pi D^4 \qquad (A.2)$$

The deflections of all the posts that lie under the cell under examination are measured (in micrometers) and collected (for 10 posts) in the table below:

Post 1	Post 2	Post 3	Post 4	Post 5	Post 6	Post 7	Post 8	Post 9	Post 10
0.2	0.12	0.04	0.35	0.09	0.26	0.11	0.14	0.18	0.28

(a) Calculate the force exerted by the cell at the post positions given above.

(b) Given the distribution of the forces that you calculated, which posts do you think are likely to be toward the cell periphery? Why?

(c) Instead of cylindrical posts, if the posts were square with the side measuring 2 μm, then for the same deflection as given in the table above, what would the forces exerted by the cell be? For a solid square-cross-section beam of width a (load applied at the end),

$$I = \frac{1}{12} \pi a^4 \qquad (A.3)$$

Exercise A.6.5. When measuring cell traction forces using a micropost substrate, what is the minimum force that can be measured using PDMS cylindrical microposts (Young's modulus of PDMS $E = 2$ MPa) that are 3 μm in diameter and 10 μm long with a microscope that can accurately image post deflections as small as 1 μm.

Exercise A.6.6. We are designing microfabricated cylindrical posts to be used as cell traction force probes but we are not yet sure which materials and dimensions to use. How much would the stiffness change if both the Young's modulus and the length of the post were doubled?

Exercise A.6.7. Myotubes can be patterned on laminin or Matrigel tracks (25 μm in width and separated by 200 μm) inside microchannels, as shown in Figure 6.13. Moreover, acetylcholine receptors (AChRs) get clustered, as "pretzels" in mature and differentiated myotubes (Figure 6.13c). Let us assume that an average AChR pretzel can be approximated by an ellipse with a major axis of 20 µm and a minor axis of 10 µm. We would like to design a three-input channel (as in Figure 6.13a) to focally stimulate these AChR pretzels with agrin (molecular weight = 90 kDa). Ideally, we would like the width of the middle stream to be less than or equal to the size of the average pretzel. Let us assume that the height and width of the main microchannel is 200 µm and 2 mm, respectively.

(a) Assuming the length of the three-input channels and the pressure at the inlets to be the same, design the ratio of the resistances of the input channels (the two flanking channels and the central channel), so that the width of the central stream in the main channel is 60 µm. In reality, will the width of the central stream be exactly the same as the calculated value? Please reason your answer.

(b) Now that we have found out the ratio of the resistances, what would be the heights and widths of the inlet channels (assuming the height of the flanking channels is 200 µm)?

(c) Assuming a parabolic flow profile and a flow rate of 20 µL/s, calculate the approximate width of the central stream of agrin (which you designed to be 60 µm wide) (i) 1 mm downstream, (ii) 1 cm downstream, and (iii) 5 cm downstream. Assume only diffusive broadening. Are there other effects that contribute to broadening of the agrin stream?

(d) Let's assume that you can allow the agrin stream to broaden by 20 µm on either side. How many myotube tracks can you focally stimulate, if the flow rate is 20 µL/s?

(e) Let's assume that you have two syringe pumps, so that you can control the flow of the central channels and the side channels separately. What flow rates will you use so that the width of the central stream remains between 20 and 100 µm for the entire length of the channel (5 cm)?

(f) Can you estimate an approximate value of fluidic shear stress at the bottom of the channel, where the myotubes are patterned (assuming parabolic flow) for the flow rates you used in the earlier parts? Are these values acceptable for cell culture?

Exercise A.6.7. Calculate the maximum shear stress experienced by endothelial cells inside a closed microfluidic cell-culture chamber of dimensions 21 mm × 150 µm × 14 µm (length × width × height). Endothelial cells are attached to the fibronectin-coated floor of this chamber, and cell culture medium is flowed through the chamber at 0.10 mL/min. Assume laminar flow conditions inside the chamber. For convenience, also assume that the velocity changes only as a function of the chamber height, and the cells are of a uniform height. Justify the selected height of cells and any additional assumptions made. The density and viscosity values of the culture medium can be assumed to be same as water.

Hint: Use equations describing pressure-driven flow through a narrow rectangular channel.

A.7 Suggested Exercises for Chapter 7

Exercise A.7.1. Elaborate (in four to five sentences/lines) on the importance of micropatterned cocultures of hepatocytes+fibroblasts. Why do people use them? What applications/subfields are they important for?

Exercise A.7.2. Describe three different methods for micropatterning gel-encapsulating hydrogels.

Exercise A.7.3. Encapsulating cells in a certain type of hydrogel is a common way to generate 3-D culture platforms. A microfluidic laboratory is trying to reproduce this technique by mixing C2C12 myoblasts with Matrigel and infuse the mixture into an open microtrench. We will use confocal microscopy to observe a 1 mm × 1 mm field of view with the ideal cell number of 500 in each z plane (depth of focus, ~10 µm). What is the concentration of the cell suspension in the Matrigel we should use initially? The dimension of the trench is 5 mm in length, 500 µm in width, and 200 µm in height.

A.8 Suggested Exercises for Chapter 8

Exercise A.8.1. Discuss (a) what affects the signal/noise ratio and (b) biocompatibility risks in (i) traditional EEG recordings, (ii) ECoG recordings, and (iii) recordings using microfabricated intracortical multielectrodes produced by silicon micromachining. Note that you need to state which material the electrode is made of (if different materials have been used in the literature, you may choose one and stick to that one for the purposes of discussion).

Exercise A.8.2. Name three different technologies for recording electrical signals from the brain at different length scales. Describe, for each technology: (a) the dimensions and materials used to make the electrodes, (b) the location/depth of recording, (c) the spatial resolution of the recording (order of magnitude), (d) the typical duration (order of magnitude) of a recording, and (e) the biocompatibility risks.

Exercise A.8.3. Research the literature to find published microfabrication processes for the three main types of microneedles (one example each): in-plane microneedles, out-of-plane solid microneedles, and out-of-plane hollow microneedles.

Exercise A.8.4. Discuss challenges and possible solutions in the design of a microfabricated cochlear implant with improved cell-electrode impedance and greater number of electrodes compared with the Nucleus 24, one of the most prevailing and popular cochlear implants.

Exercise A.8.5 (Design Challenge). Suppose you work for a consulting firm that assesses the feasibility of engineering ideas before they actually go into research and development. A major manufacturer of biomedical implants commissions you to assess whether it would be possible to develop an implant for blind people that would directly stimulate the brain with the output of a camera (rather than with retinal implants—the manufacturer has already tried that approach and is not happy with it). Think of the assessment as the first phase of the design, in which at the beginning you are still open to any design and you have to narrow down the options of, for example, materials of the device and biological principles of stimulation. The manufacturer is not looking for a final design—only an assessment of (a) the engineering challenges and (b) the fundamental obstacles, which the manufacturer needs in order to determine the cost (of fabrication, implantation procedure, etc.) and the targeted market (e.g., it may not work for all blind people or it may only allow for partial vision).

Exercise A.8.6 (Design Challenge). Propose the design of a retinal implant and discuss the various challenges ahead, for example, signal processing, biocompatibility, power consumption, surgery, field of vision, and dynamic range.

A.9 Tips for a Good BioMEMS Exam

- There is no substitute for serious study. Students often make the mistake of inventing things in exams. For example, when asked to come up with the microfabrication process that leads to a given device, they might invent processes that have never been shown in class. Unless the purpose of the exercise is to show your ability to invent new devices, it is a very risky strategy in an exam: most of them, I find, have serious flaws. The processes taught in class were optimized after considerable trial and error and are only the tip of the iceberg of what very smart researchers tried out. If you are an incoming student to the BioMEMS discipline, the way to succeed in designing microfabrication processes is by studying (i.e., memorizing *and* understanding) lots of them. Bottom line: if you invent in an exam or in a homework assignment, your teacher will be able to tell.

❋ Just answer the question(s). No student ever got credit for volunteering more information than what was being asked. On the contrary, if you add something that is not asked but is very wrong, it might influence negatively on your grades. For the same reason, do not write questions for the teacher or grader ("What protein will work better?")—you are the one being examined.

❋ Some students use the concepts "chemisorption" and "physisorption" too liberally. These are processes that only apply to molecules on surfaces. Therefore, it is incorrect to say that "cells attach by chemisorption" (or by physisorption).

❋ In photolithography, photoresist does not get "etched" (unless in the very particular case in which it is attacked by a plasma). It gets *developed* or *dissolved.*

❋ In exams, my students have repeatedly invented a type of photoresist that, much to my regret, does not exist: one that can be spun and UV-patterned like most photoresists, and after seeding cells on the photoresist pattern, it can be magically dissolved without affecting cells. The closest that anyone has ever come to this goal is shown in **Figure 1.6**, in which Stefan Diez and coworkers were able to pattern proteins with a photoresist whose water solubility changes with temperature. However, although the process is compatible with proteins, there is no evidence that the polymer that is crucial for the temperature phase transition, poly-NIPAM, is biocompatible enough as to be compatible with cells. I am looking forward to the day when my students are right!

A.10 Microfluidics Outreach and Education

Fluids offer endless opportunities to introduce scientific concepts to kids in many areas that are related to their everyday experiences, ranging from physics (swimming, rain formation) and chemistry (cooking, explosions) to biology (bodily fluids, marine life) and art (color mixing). In our experience with children in science fairs, microfluidics is equal to "fluids with a technology flavor" because it requires some component of fabrication or (if the device has already been made for them) of observation of a new phenomenon, generally with a microscope (which looks "cool"). Hands-on education (where students get to construct some device, even if rudimentary) is most effective because the children actively think and learn while they have fun. Two microfluidic technologies have emerged as ideal for the classroom: paper and *Jell-O* (**Figure A.7**). They are extremely low-cost, very safe, and simple to use and construct (~5–10 minutes per device). If a CO_2 lase is available, more sophisticated microfluidic devices can be produced by laser cutting of plastic laminates in a few minutes (see Section 1.4.4), which allows for teaching complex phenomena such as droplet formation and chaotic mixing (e.g., for advanced undergradute laboratories).

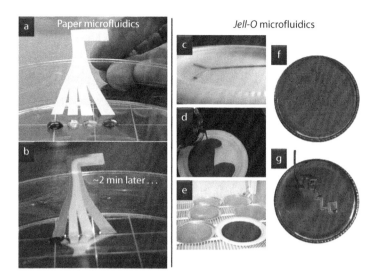

FIGURE A.7 Microfluidics outreach. (a and b) The paper channels are cut with scissors. Within seconds of contacting the droplets, fast laminar flow is established within the pores of the paper. Figure contributed by the author. (c–g) For the *Jell-O* process, a negative mold is prepared using foam plates, wooden coffee stirrers, and double-sided tape. A *Jell-O*/gelatin liquid mixture is then poured onto the mold and left to cure in a 4°C refrigerator. The solidified chips are peeled off and placed on aluminum pans for experiments at room temperature. (From Yang, C. W. T., E. Ouellet, and E. T. Lagally, "Using inexpensive *Jell-O* chips for hands-on microfluidics education," *Anal. Chem.*, 82, 5408–5414, 2010. Figure contributed by Eric Lagally.)

A.11 A Quick (and Personal) Overview of BioMEMS History

Writing this textbook has been a lot of fun. Now that I'm done, I can indulge myself in one last amusement: pick those articles that would be essential for a 5-minute summary (or 1-hour lecture) of the field. This is an exercise I have already been through for all the articles that are featured in this textbook (and many excellent ones that had to be left out due to space constraints, unfortunately), so it's not a very mind-struggling exercise at this point. I have organized them in a table below, chronologically, stating the year and the month (if known) of the manuscript acceptance or publication date (whichever was earlier), the merit that deserves them to be featured in the table, the citation, and a representative visual by which they can be easily remembered.

This table should not be regarded as a "BioMEMS Hall of Fame" or "Best 60 BioMEMS articles." It is rather a very personal look at BioMEMS history, and even so, with a clear bias for technical breakthroughs (rather than the scientific breakthroughs that the technology has enabled). I'm saying this before my many dear colleagues are offended because their work is not included in here. Great laboratories that do not waste time developing gadgets and instead advance science with well-established microtechnology are not even represented in this table but have my highest esteem. (If you look for thoroughness, please refer to the textbook.) The table could have been made longer and longer—everyone is "first" at something, as long as you include the whole title of the manuscript. It is with this spirit that I included, somewhat humorously, some of my own minor (but still cherished) work in this table—it is indeed a very personal glance at history. In any case, the table (or a modified version of it) should be useful to the BioMEMS educator to provide an end-of-the-year overview of the field.

I have decided to include this table because, despite the subjective lens with which it has been created, it does give a powerful instant view of BioMEMS history. Note that, as stated in the text, Carter should be credited as the first author of a BioMEMS article, despite earlier microfluidics publications. I hope the difference is now clear to the reader: to qualify as a BioMEMS work, it must contain a microtechnology contribution, as Carter did; however, it is possible to build a microfluidic system with a glass tube, a sheet of paper, or two glass plates and a spacer, as those clever pre-BioMEMS microfluidic pioneers did (Philpot in 1939 and Hannig in 1950). The most striking feature of the table is, perhaps, that the early history of this engineering field is dominated by cell biologists and studies of fundamental cell biology—not by biochemists seeking high-throughput assays as the future (our present) might have led us to believe. Should one conclude that cell biologists are collectively more open to embrace new technologies? If so, it announces a great future demand for cell-based microfluidics. Another salient feature is that many of these articles have not been properly acknowledged (judging by their number of citations), whereas articles that followed in their wake that were published in higher-visibility journals are still being cited. A third interesting feature of the table is that it puts into perspective the advances of the various subfields of BioMEMS with respect to each other. For example, the first microfluidic pattern of a biomolecule (to perform a fairly simple immunoassay) was not reported until 1997 (by Hans Biebuyck's group in *Science*), but by then, microfluidic chips had been reported for axon guidance (1987, Bonhoeffer's group), PCR (1993, Watson), sperm research (1993, Wilding), capillary electrophoresis (1993, Manz), blood cell deformability (1989, Sato; 1995, Barnes), and cell manipulation (1997, Harrison), so in retrospect, it seems like Biebuyck's device *should* have been developed a few years earlier—perhaps, if that excellent student in such and such institution had not left academia for an industry job, or funding had not run out . . . The table, then, highlights one of the cruxes of scientific discovery: if we could know when the time is right to develop an invention, we would be much more efficient scientists. Finally, the table is wonderful for relating to our dear BioMEMS field on a personal level: this is when you were born, this is when you entered kindergarten, this is when you got your first job, and here is when the Human Genome was completed.

Anyways, I hope you enjoy the table over and over as much as I do!

Year[a]	Merit	Reference	Image
1939, July	First microfluidic device based on laminar flow principles. First continuous-flow electrophoresis device.	J.St.L. Philpot, *Trans. Far. Soc.* **35**, 38 (1940)	
1950, March	First article on microfluidic device (electrophoresis confined to a sheet of filter paper).	W. Grassman and K. Hannig, *Naturwissenschaften* **37**, 397 (1950)	

(continued)

Year[a]	Merit	Reference	Image
1967, Oct	First BioMEMS article that uses photolithography. First cellular micropattern (shadow-evaporated Pd islands for cell attachment).	S.B. Carter, *Exp. Cell Res.* **48**, 189 (1967)	
1972, February	First cells on microelectrode arrays.	C.A. Thomas, Jr., P.A. Springer, G.E. Loeb, Y. Berwald-Netter, and L.M. Okun, *Exp. Cell Res.* **74**, 61 (1972)	
1975, Jan	First cellular micropattern with cells attached on a protein template. First micropattern of isolated neurons.	P.C. Letourneau, *Dev. Biol.* **44**, 92 (1975)	
1982, June	First microfluidic device in space.	McDonnell Douglas Co. (http://www .thermoelectric .com/2010/ archives/library/ Electrophoresis in Space 1985.PDF)	
1984, September	First cellular micropattern made exclusively with biomolecules (ECM protein).	J.A. Hammarback, S.L. Palm, L.T. Furcht, and P.C. Letourneau, *J. Neurosci. Res.* **13**, 213 (1985)	

(*continued*)

Year[a]	Merit	Reference	Image
1985	Commercialization of *ClearBlue*, the first microfluidic device containing paper. First microfluidic product to market.	Unipath, Inc.	
1985, September	First observation that cell behavior can be modulated by changing cell shape.	Charles O'Neill, Peter Jordan, and Grenham Ireland, *Cell* **44**, 489 (1986)	
1985, November	First observation that cells (fibroblasts) orient on micromachined substrates.	D.M. Brunette, *Exp. Cell Res.* **164**, 11 (1986)	
1987, February	First application of microfabrication to study plant biology.	H.C. Hoch, R.C. Staples, B. Whitehead, J. Comeau, and E.D. Wolf, *Science* **235**, 1659 (1987)	
1987, August	First PDMS microdevice. First microfluidic microdevice. First microfluidic pattern of biomolecules. First use of porous membranes for hydrodynamic immobilization.	J. Walter, B. Kern-Veits, J. Huf, B. Stolze, and F. Bonhoeffer, *Development* **101**, 685 (1987)	

(continued)

Year[a]	Merit	Reference	Image
1988, February	First micropattern of self-assembled monolayer (SAM). First cellular micropattern (neurons) on SAMs.	D. Kleinfeld, K.H. Kahler, and P.E. Hockberger, *J. Neurosci.* **8**, 4098 (1988)	
1989	First microfluidic device containing live cells. First microfluidic device for studies of blood rheology.	Y. Kikuchi, H. Ohki, T. Kaneko, and K. Sato, *Biorheology* **26**, 1055 (1989)	
1991, January	First oligonucleotide chip.	S.P.A. Fodor, J.L. Read, M.C. Pirrung, L. Stryer, A.T. Lu, and D. Solas, *Science* **251**, 767 (1991)	
1991, March	First recording of neurons using a field-effect transistor.	P. Fromherz, A. Offenhäusser, T. Vetter, and Jürgen Weis, *Science* **252**, 1290 (1991)	
1991, March	First PEG-thiol SAMs.	K.L. Prime and G.M. Whitesides, *Science* **252**, 1164 (1991)	

(*continued*)

Year[a]	Merit	Reference	Image
1993, June	First PCR chip.	M.A. Northrup, M.T. Ching, R.M. White, and R.T. Watson, *Transducers '93*, Yokohama, Japan (June 7–10, 1993), p. 924	
1993, June	First microfluidic device containing sperm. First biochemical treatment of cells on chip.	L.J. Kricka, O. Nozaki, S. Heyner, W.T. Garside, and P. Wilding, *Clin. Chem.* **39**, 1944–1947 (1993)	
1993, June	First capillary electrophoresis on a chip. First report of microfluidic valving (electrokinetic valve).	D.J. Harrison, K. Fluri, K. Seiler, Z. Fan, C.S. Effenhauser, and A. Manz, *Science* **261**, 895	
1993, July	First soft lithography article. First article on microstamping (alkanethiol printing using PDMS stamps).	A. Kumar and G.M Whitesides, *Appl. Phys. Lett.* **63**, 2002 (1993)	
1993, October	First article on PEG micropatterning (demonstrated protein patterns).	G.P. Lopez, H.A. Biebuyck, R. Harter, A. Kumar, and G.M. Whitesides, *JACS* **115**, 10774 (1993)	

(*continued*)

Year[a]	Merit	Reference	Image
1994, March	First use of microstamping of PEG-thiol SAMs to control cell shape and cell behavior.	R. Singhvi, A. Kumar, G.P. Lopez, G.N. Stephanopoulos, D.I.C. Wang, G.M. Whitesides, and D.E. Ingber, *Science* **264**, 696 (1994)	
1995, January	First microscale cell trapping and sorting (principle used: dielectrophoretic force).	F.F. Becker, X.-B. Wang, Y. Huang, R. Pethig, J. Vykoukal, and P.R.C. Gascoyne, *PNAS* **92**, 860 (1995)	
1995, March	First microfluidic assay of blood cell deformability.	M.C. Tracey, R.S. Greenaway, A. Das, P.H. Kaye, and A.J. Barnes, *IEEE Trans. Biomed. Eng.* **42**, 751 (1995)	
1995, July	First microfluidic molding of polymers ("micromolding in capillaries" or MIMIC).	E. Kim, Y. Xia, and G.M. Whitesides, *Nature* **376**, 581 (1995)	
1996, October	First microfluidic chip for electrospray mass spectrometry (no integrated emitter).	Q. Xue, F. Foret, Y.M. Dunayevskiy, P.M. Zavracky, N.E. McGruer, and B.L. Karger, *Anal. Chem.* **69**, 426 (1997)	

(*continued*)

Year[a]	Merit	Reference	Image
1997, January	First biochemical manipulation of cells using electrokinetic flow.	P.C.H. Li and J. Harrison, *Anal. Chem.* **69**, 1564 (1997)	
1997, March	First microfluidic pattern of purified proteins.	E. Delamarche, A. Bernard, H. Schmid, B. Michel, and H. Biebuyck, *Science* **276**, 779 (1997)	
1997, May	First report of hydrodynamic focusing in microchannels.	S.C. Jacobson and J.M. Ramsey, *Anal. Chem.* **69**, 3212 (1997)	
1997, August	First microstamped pattern of proteins.	D.W. Branch, J.M. Corey, J.A. Weyhenmeyer, G.J. Brewer, and B.C. Wheeler, *Med. Biol. Eng. Comput.* **36**, 135 (1998)	

(*continued*)

Year[a]	Merit	Reference	Image
1998, January	First optically addressable random bead array.	K.L. Michael, L.C. Taylor, S.L. Schultz, and D.R. Walt, *Anal. Chem.* **70**, 1242 (1998)	
1998, January	First out-of-plane (solid) microneedles for transdermal drug delivery.	S. Henry, D.V. McAllister, M.G. Allen, M.R. Prausnitz, *J. Pharm. Sci.* **87**, 922 (1998)	
1998, February	First micropatterned coculture.	S.N. Bhatia, U.J. Balis, M.L. Yarmush, and M. Toner, *J. Biomater. Sci. Polym. Ed.* **9**, 1137 (1998)	
1998, March	First continuous-flow PCR chip.	M.U. Kopp, A.J. de Mello, and A. Manz, *Science* **280**, 1046 (1998)	
1998, April	First micropattern of agarose as cell repellent for cellular micropatterning.	A. Folch and M. Toner, *Biotechnol. Prog.* **14**, 388 (1998)	

(continued)

Year[a]	Merit	Reference	Image
1998, September	Invention of PDMS bonding by oxygen plasma.	D.C. Duffy, J.C. McDonald, O.J.A. Schueller, and G.M. Whitesides, *Anal. Chem.* **70**, 4974 (1998)	
1999, February	First use of SU-8 for soft lithography.	A. Folch, A. Ayon, O. Hurtado, M.A. Schmidt, and M. Toner, *J. Biomech. Eng.* **121**, 29 (1999)	
1999, March	First demonstration of heterogeneous laminar flows for selective treatment of cells in microchannels.	S. Takayama, J.C. McDonald, E. Ostuni, M.N. Liang, P.J.A. Kenis, R.F. Ismagilov, and G.M. Whitesides, *PNAS* **96**, 5545 (1999)	
1999, October	First hollow out-of-plane microneedles.	D.V. McAllister, S. Kaushik, P.N. Patel, J.L. Mayberry, M.G. Allen, and M.R. Prausnitz, *Proc. of the 1st Joint BMES/EMBS Conf.*, vol. 2, p. 836 (1999)	
1999, November	First passive micromixer (uses chaotic advection to homogenize two dissimilar fluids), the "serpentine micromixer".	R.H. Liu, M.A. Stremler, K.V. Sharp, M.G. Olsen, J.G. Santiago, R.J. Adrian, H. Aref, and D.J. Beebe, *JMEMS* **9**, 190 (2000)	

(continued)

Year[a]	Merit	Reference	Image
1999, November	First active micromixer (uses pulsatile flow to shear flow in orthogonal channels), the "Dahleh micromixer".	M. Volpert, C.D. Meinhart, I. Mezic, and M. Dahleh, *Proc. MEMS ASME IMECE*, Nashville, TN (1999), p. 483	
2000, February	First experimental observation of diffusive broadening in microchannels ("butterfly effect").	R.F. Ismagilov, A.D. Stroock, P.J.A. Kenis, G. Whitesides, and H.A. Stone, *Appl. Phys. Lett.* **76**, 2376 (2000)	
2000, February	First PDMS microvalves (Quake microvalves); first PDMS micropump.	M.A. Unger, H.-P. Chou, T. Thorsen, A. Scherer, and S.R. Quake, *Science* **288**, 113 (2000)	
2000, February	First smart-polymer microvalve.	D.J. Beebe, J.S. Moore, J.M. Bauer, Q. Yu, R.H. Liu, C. Devadoss, and B.-H. Jo, *Nature* **404**, 588 (2000)	
2000, April	First PDMS "doormat" microvalves.	K. Hosokawa and R. Maeda, *J. Micromech. Microen.* **10**, 415 (2000)	
2000, March	First use of PDMS stencils for micropatterning cell cultures.	A. Folch, B.-H. Jo, O. Hurtado, D.J. Beebe, and M. Toner, *J. Biomed. Mater. Res.* **52**, 346 (2000)	

(*continued*)

Appendix

Year[a]	Merit	Reference	Image
2001, April	First microdroplets in microchannels.	See T. Thorsen, R.W. Roberts, F.H. Arnold, and S.R. Quake, *Phys. Rev. Lett.* **86**, 4163 (2001)	
2001, July	First PDMS emitter for electrospray mass spectrometry.	J.-S. Kim and D.R. Knapp, *J. Chromatogr. A* **924**, 137 (2001)	
2001, December	First microfluidic gradient generator.	S.K.W. Dertinger, D.T. Chiu, N.L. Jeon, and G.M. Whitesides, *Anal. Chem.* **73**, 1240 (2001)	
2001, December	First passive micromixer that induces chaotic advection with only surface grooves.	A.D. Stroock, S.K.W. Dertinger, A. Ajdari, I. Mezic, H.A. Stone, and G.M. Whitesides, *Science* **295**, 647 (2002)	
2002, January	First PDMS patch clamp chip (oocyte recordings).	K.G. Klemic, J.F. Klemic, M.A. Reed, and F.J. Sigworth, *Biosens. Bioelectron.* **17**, 597 (2002)	
2002, February	First GΩ seal and whole-cell recordings on mammalian cells with a patch clamp (glass) chip.	N. Fertig, R.H. Blick, and J.C. Behrends, *Biophys. J.* **82**, 3056 (2002)	

(*continued*)

Year[a]	Merit	Reference	Image
2002, September	First microfluidic multiplexer.	T. Thorsen, S.J. Maerkl, and S.R. Quake, *Science* **298**, 580 (2002)	
2003, October	First microfluidic device made of a hydrogel.	M.D. Tang, A.P. Golden, and J. Tien, *JACS* **125**, 1988 (2003)	
2004, April	First combinatorial micromixer.	C. Neils, Z. Tyree, B. Finlayson, and A. Folch, *Lab Chip* **4**, 342 (2004)	
2004, June	First use of *C. elegans* in microfluidics.	J.M. Gray, D.S. Karow, H. Lu, A.J. Chang, J.S. Chang, R.E. Ellis, M.A. Marletta, and C.I. Bargmann, *Nature* **430**, 317 (2004)	
2004, June	Invention of acoustophoresis, a separation technique based on ultrasound standing waves.	F. Petersson, A. Nilsson, C. Holm, H. Jönsson, and T. Laurell, *Analyst* **129**, 938–943 (2004)	

(continued)

Year[a]	Merit	Reference	Image
2004, July	First biodegradable microfluidic device.	K.R. King, C.C.J. Wang, M.R. Kaazempur-Mofrad, J.P. Vacanti, and J.T. Borenstein, *Adv. Mater.* **16**, 2007 (2004)	
2005, July	First report of cells in PDMS microwell arrays.	J. Rettig and A. Folch, *Anal. Chem.* **77**, 5628 (2005)	
2005, November	First microfluidic device containing patterned (laser-cut) paper.	J. Diao, L. Young, S. Kim, E.A. Fogarty, S.M. Heilman, P. Zhou, M.L. Shuler, M. Wu, and M.P. DeLisa, *Lab Chip* **6**, 381 (2006)	
2006, March	First microfluidic gradient generator that incorporates a hydrogel.	H. Wu, B. Huang, and R.N. Zare, *J. Am. Chem. Soc.* **128**, 4194 (2006)	

(*continued*)

Year[a]	Merit	Reference	Image
2006, July	First microfluidic gradients on open surfaces.	T.M. Keenan, C.-H. Hsu, and A. Folch, *Appl. Phys. Lett.* **89**, 114103 (2006)	
2007, March	First microfluidic gradient generator entirely made in hydrogel (agarose).	S.-Y. Cheng, S. Heilman, M. Wasserman, S. Archer, M.L. Shuler, and M. Wu, *Lab Chip* **7**, 763 (2007)	
2007, May	First tissue (brain) slice on a chip.	A.J. Blake, T.M. Pearce, N.S. Rao, S.M. Johnson, and J.C. Williams, *Lab Chip* **7**, 842 (2007)	
2007, June	Lateral patch clamp chips made of glass yield high GΩ seals and high microfluidic integration capabilities.	W.-L. Ong, K.-C. Tang, A. Agarwal, R. Nagarajan, L.-W. Luo and L. Yobas, *Lab Chip* **7**, 1357 (2007)	

(*continued*)

Year[a]	Merit	Reference	Image
2008, February	First microfluidic device made entirely with (photolithographically patterned) paper.	A.W. Martinez, S.T. Phillips, E. Carrilho, S.W. Thomas III, H. Sindi, and G.M. Whitesides, *Anal. Chem.* **80**, 3699 (2008)	
2010, April	First "in-cell" extracellular electrodes (providing long-term, intracellular-quality recordings).	A. Hai, J. Shappir, and M.E. Spira, *J. Neurophysiol.* **104**, 559 (2010)	

[a] Month indicates acceptance date, if known, or publication date.

A.12 Glossary of Terms—The First BioMEMS Dictionary (to Be Used Instead of *Wikipedia*)

Here, you will find a compilation of all the technical terms that appear in the text. This glossary of terms is included here for two reasons. First, to provide some additional rigor to the text, which I purposely wrote in a rather informal style (evading the engineering and physics style that is full of definitions). Second, I made this effort after I noticed that the definitions of many of these "BioMEMS-related terms" in *Wikipedia* had some imprecisions, or were just not written in a way that is amenable for study—and I know that today students use *Wikipedia* "a lot." (Conversely, if you notice that my definition is very similar to the one found in *Wikipedia*, it's because I agree with the *Wikipedia* definition and I didn't see a reason to reinvent the wheel.)

A

absolute viscosity: *See* dynamic viscosity.

acoustic streaming: Steady current in a fluid driven by the absorption of high-amplitude acoustic oscillations.

acoustophoresis: Separation of particles suspended in a fluid (e.g., cells and beads) by means of acoustic fields, usually forming a standing wave within a microchannel. The technique was invented by Thomas Laurell from Lund University (Sweden) in 2004.

adsorption: Adhesion of a chemical species (atoms, ions, or molecules) onto a surface.

alginate: Anionic polysaccharide from the cell walls of seaweed (such as the giant kelp) and bacteria. The commercial varieties of alginate are used to form hydrogels that are widely used in biotechnology (e.g., cell encapsulation and microfluidics) and the food/culinary industries (e.g., for thickening drinks and ice cream making).

alkanesilanes, alkylsilanes: In organic chemistry, alkane (or, equivalently, the alkyl prefix) refers to a chemical compound that consists only of carbon and hydrogen linked together exclusively by single bonds (i.e., the C and H are said to be "saturated"). A silane is the $-Si-R_3$ terminal group ($-SiCl_3$, for example). Thus, alkanesilanes are molecules that are terminated in the $-Si-R_3$ group and that have a fully saturated hydrocarbon tail. Example: $CH_3-CH_2-CH_2-CH_2-SiCl_3$, n-butyl-trichlorosilane. The "n" stands for "normal," the name of the configuration when the functional group (in this case a silane) hangs from the end of the chain.

alkanethiols, alkylthiols: As with alkanesilanes (see above), the "alkane" or "alkyl" portion of the name means that the compound contains a chain which consists only of carbon and hydrogen linked together exclusively by single bonds (i.e., the C and H are said to be "saturated"). A thiol is the $-SH$ terminal group or sulfhydryl group. Therefore, alkanethiols are molecules that are terminated in the $-SH$ group and that have a fully saturated hydrocarbon tail. Example: $CH_3-CH_2-CH_2-SH$, n-propanethiol.

aminosilane: In organic chemistry, any organosilane (such as an alkanesilane) that contains or terminates in an amine group. Example: $NH_2-CH_2-CH_2-CH_2-SiCl_3$, amino-propyl-trichlorosilane.

anisotropic etch: A chemical etch process that proceeds in a preferential direction. The etch direction may be biased as a result of an applied voltage (e.g., during a reactive ion etch) or as a result of a crystalline anisotropy of the substrate. The anisotropicity of an etch is a quality that depends on the substrate as well as on the etchant; changing the etchant may result in a different etch that proceeds isotropically.

anisotropic sieving: A strategy for separating DNA and proteins based on nanofluidic gaps (which act as the "sieve") and the simultaneous presence of flow (which helps separate the molecules) and electrical fields (which induce anisotropy, i.e., attract the molecules preferentially in one direction).

***Arabidopsis thaliana*:** A small flowering plant that is used as a model organism in plant biology and genetics, well-suited for light and fluorescence microscopy because the seedling and the roots are translucent. Its genome was the first plant genome to be sequenced. *Arabidopsis thaliana* roots can grow inside PDMS microchannels.

axon guidance: Process that occurs during neural development by which neurons send out axons to reach the correct target.

B

batch microfabrication: A process for microfabricating many units, devices or wafers in parallel, which reduces the final cost per device.

bead microarray: A microarray of beads, either self-assembled into microwells (e.g., for use as a biomolecular binding assay) or tightly packed in a hexagonal array (e.g., for use as a metal evaporation mask).

benzophenone: Photoinitiator that is widely used as the photoreactive functional group in UV-photoreactive cross-linkers.

biodegradable polymers: Polymers that gradually break down into their monomer, nontoxic components in the presence of water, usually by hydrolysis, such that (if implanted) they can be excreted by natural processes after breakdown.

BioMEMS: Abbreviation that stands for Biomedical MicroElectroMechanical Systems. By generalization, the term also includes micro-optical and microfluidic devices for biomedical applications, even though they are neither electrical nor purely mechanical.

Boltzmann distribution: Probability measure for the distribution of the states of a system. The Boltzmann distribution for the fractional number of particles N_i/N occupying a set of states i at energy E_i is:

$$\frac{N_i}{N} = \frac{g_i e^{-E_i/(k_B T)}}{\sum_i g_i e^{-E_i/(k_B T)}}$$

where k_B is the Boltzmann constant, g_i is the number of levels having energy E_i, and T is the temperature of the system. When applied to the distribution of velocities of gas particles, it is known as the Maxwell-Boltzmann distribution.

bovine serum albumin (BSA): Albumin is the most abundant protein in blood plasma. Bovine serum albumin is obtained from the serum of domestic cows for a large variety of biotechnology and biochemical applications such as ELISA and immunocytochemistry assays. In BioMEMS, it is mostly used to block background protein adsorption.

Boyden chamber: Passive device invented in 1962 by Stephen Boyden of the Australian National University (Canberra, Australia) to study chemotaxis. The device consists of a top chamber separated from a bottom chamber by a porous membrane (with a pore size smaller than the cells but that still allows cell migration through the pores). The test chemoattractant is dispensed into the bottom chamber and the cells are seeded on the top chamber. After a certain period of time, the experiment is stopped and the amount of cells that have transmigrated to the bottom side of the membrane is counted to provide a measurement of the relative efficacy of the chemoattractant.

BSA: *See* bovine serum albumin.

bulk micromachining: Micromachining of a wafer by additive or subtractive processes that affect the majority of its thickness; *see* micromachining.

Butterfly effect: Diffusive broadening of the concentration profile in a T- or Y-mixer of rectangular cross-section under pressure-driven flow. This effect is a special manifestation of the Taylor dispersion.

C

C. elegans: See Caenorhabditis elegans.

C2C12 cells: C2C12 is a mouse myoblast cell line capable of differentiation into myotubes.

cadherins: Class of transmembrane proteins that play a role in cell–cell adhesion. They are named after their calcium-dependent adhesion properties.

Caenorhabditis elegans: A transparent 1-mm-long roundworm that lives in soil and is used extensively as a model organism in molecular and developmental biology.

capillarity pump: Pump that relies on capillary action to move the fluid forward.

capillary action: Ability of a fluid to spontaneously wet or wick into a small conduit (a "capillary") such as a narrow tube or a porous material such as paper or sand, even against gravity.

capillary burst microvalve: A microvalve that consists of a microchannel constriction that is too hydrophobic for the fluid flow to overcome unless extra pressure is used to pass (or "burst") the constriction.

capillary electrophoresis: Technique for separating ionic species in an electrolyte-filled glass capillary based on the size-to-charge ratio of the analytes; an electric field causes the migration of the analytes (which are detected near the outlet) by electro-osmotic flow. See electro-osmotic flow.

CE: *See* capillary electrophoresis.

cellular traction: Force produced by a cell during cell migration.

centrifugal microvalves: Microvalves that can be opened or closed by the action of centrifugal force; centrifugal microvalves require spinning platforms that are generically termed "Lab-CD."

check microvalve: Type of microvalve whose mechanical design allows fluid to flow in only one direction (i.e., they close with reverse flow).

chemisorption: Adhesion of a molecule onto a surface via the formation of a chemical bond

chemotaxis: Cellular motion that is directed by chemicals or chemical gradients present in the extracellular environment.

CHO cells: Abbreviation that stands for Chinese hamster ovary cell, a cell line derived in 1957 by Theodore T. Puck from the ovary of a Chinese hamster (chosen because of its very low chromosome number for a mammal, 2n = 22, which yields stabler gene expression upon transfection). CHO cells are often used as a mammalian host for the in vitro production of recombinant proteins.

CMOS: Abbreviation that stands for Complementary Metal-Oxide-Semiconductor, a technology for building integrated circuits.

CNS: Abbreviation that stands for Central Nervous System.

Coandă effect: The tendency of a fluid jet to be attracted to a nearby surface. Named after its discoverer, Romanian inventor and aerodynamics pioneer, Henri Marie Coandă (1886–1972).

collagen: A family of connective tissue proteins (with more than two dozen members); one of the major components of the extracellular matrix.

contact angle: The angle at which a liquid-gas interface meets a solid surface.

contact guidance: Tendency of cells to align to topographical features such as edges and grooves.

continuous-flow electrophoresis: *See* free-flow electrophoresis.

cross-linker: A molecule with two reactive functional groups, which can then be used to link two molecules with functional groups that react selectively with those of the cross-linker.

D

Dahleh micromixer: Microfluidic mixer that homogenizes two input solutions by quickly alternating their injection into a microchannel, which results in Taylor dispersion effectively helping the mixing process; it was invented by Mohammed Dahleh (1961–2000) and Igor Mezic at the University of California (Santa Barbara) in 1999.

Darcy's law: Equation that describes fluid flow through a porous medium. It was derived experimentally in 1856 by French engineer Henry Darcy using columns of sand. It can also be derived from the Navier-Stokes equation.

Debye layer: Double layer of ions and counter-ions that form on a surface when the surface is placed in contact with a liquid.

Debye length: In the context of liquids contacting a surface, it is the characteristic thickness of the Debye layer. In general, it expresses the scale over which charges screen out other charges.

denaturation (of a protein): Loss of quaternary, tertiary, or secondary structure of a protein. Denaturation may occur due to heat, by application of forces or a solvent, and changes in the ionic environment.

Dertinger gradient generator: Microfluidic gradient generator invented by George Whitesides and Stephan Dertinger in 2001. The gradient generator is based on splitting and recombining the inlet streams a large number of times until a smooth gradient is created.

diaphragm micropump: Microfluidic pump whose main pumping mechanism is based on the deflection of a diaphragm.

diazirines: Class of photoreactive organic molecules that are used in cross-linkers for labeling nucleic acids and proteins.

DiCarlo cell counter: An ultra-high-throughput microfluidic cytometer invented in 2010 by young professor Dino DiCarlo from the University of California at Berkeley.

dielectrophoresis: The motion of particles (charged or uncharged) as a result of polarization induced by nonuniform electric fields; the induced force is a function of the polarizability of the particle and that of the surrounding medium, the particle shape and size, and the frequency and magnitude of the applied electric field. Dielectrophoresis was discovered by Herbert Pohl in 1951.

dielectrophoretic traps: Devices that allow for confining cells or particles using the principle of dielectrophoresis.

diffusion coefficient: The proportionality constant between the molar flux (due to molecular diffusion) and the concentration gradient (that causes the flux).

diffusion immunoassay: Immunoassay performed in solution, relying entirely on observing the diffusion of the antibodies as they bind to the analytes (as opposed to surface-bound or heterogeneous-phase immunoassays such as ELISA); the diffusion immunoassay was invented by Paul Yager's laboratory at the University of Washington (Seattle) in 2001. *See* heterogeneous-phase immunoassay, homogeneous-phase immunoassay, ELISA.

diffusion length: In considering Fick's second law of diffusion in one dimension (x), its solution is the complementary error function $C(x,t)$ that has a term $2\sqrt{Dt}$, called diffusion length, which provides a measure of how far the initial concentration has diffused in the x direction in the amount of time t.

diffusivity: *See* diffusion coefficient

digital microfluidics: Manipulation of droplets by electrowetting using electronic automated platforms.

digital micromirror device: Projection device invented by Texas Instruments in 1987 that consists of an array of hundreds of thousands of aluminum-coated silicon micromirrors (each 16 µm in size) that can be individually deflected at megahertz rates.

digital PCR: Strategy for performing PCR based on subdividing the sample into a large number of small, nanoliter-scale (or smaller) volumes; by virtue of its size, in each small volume there are only one or a few DNA molecules at most, so amplification always leads to a fluorescent signal that is quantized—hence the name "digital."

dilution generator: A microfluidic device that creates a number of dilutions (titrations) from a given input solution and a diluent (usually water).

direct laser writing: Microfabrication technique that is based on multiphoton polymerization of an adequate photoresist by means of a laser, allowing for the fabrication of arbitrary 3-D structures with submicron resolution.

dissociated culture: Culture of dissociated cells, that is, of tissue that has been enzymatically or mechanically separated into single cells.

DMD: *See* digital micromirror device.

DNA chip: *See* oligonucleotide chip.

DNA microarray: A square matrix of ~100-µm-diameter dried spots of cDNA, usually deposited on glass by means of metal pins; the microarray can detect the presence of complementary DNA strands (fluorescently labeled) from a sample. The technology was invented in 1995 by Patrick Brown at Stanford University.

DNA prism: A microfluidic device containing a hexagonal post lattice that acts as an asymmetrical sieve for DNA diffusion (i.e., DNA molecules of different length diffuse at different speed in different directions) when electrical fields of different magnitude are applied along the axes of symmetry of the lattice. The DNA prism was invented in 2002 by a team led by Edward Cox, Robert Austin, and James Sturm at Princeton University.

droplet microfluidics: A set of microfluidic techniques that allow for creating, manipulating, and analyzing miniature droplets.

Drosophila: Genus of small flies commonly referred to as "fruit flies" or "vinegar flies" that contains more than 1500 species. One *Drosophila* species, *Drosophila melanogaster*, has been widely used in genetics and developmental biology research.

dry etch: In microfabrication processing, an etch that uses gaseous chemistry (as opposed to an etch that uses liquid chemistry; *see* wet etch).

Dunn chamber: Passive device used for traditional chemotaxis assays, which consists of concentric rings etched in a glass slide and separated by a shallow ridge or weir. (The Dunn chamber operates on the same principles as the Zigmond chamber but has a radial topology with respect to the Zigmond chamber; *see* Zigmond chamber.) The rings are first covered with a glass coverslip before loading any solutions. The outer ring chamber is loaded (usually by capillarity filling) with the cells and the inner ring chamber is loaded with the chemoattractant of interest. The chemoattractant reaches the cells by diffusion across the weir.

duroplastic polymer: *See* thermoset polymer.

dynamic viscosity: The tangential force per unit area required to slide one plane with respect to another a unit distance apart at unit velocity. In cgs units, the dynamic viscosity of water at room temperature is 1 centipoise (cP), where $1 \text{ cP} = 10^{-2} \text{ P} \equiv 10^{-2} \text{ dyne s/cm}^2$.

E

E. coli: *See Escherichia coli.*

EBL, e-beam lithography: *See* electron beam lithography.

ECM: *See* extracellular matrix.

ECoG: *See* electrocorticography.

EEG: *See* electroencephalography, electroencephalogram.

elastomer: *See* elastomeric polymer.

elastomeric optics: A class of optical devices and, by extension, a family of optical techniques that use PDMS elements to manipulate the light path.

elastomeric polymer: Amorphous polymer, usually with a structure and properties similar to that of rubber, which can be described as having a comparatively high viscoelasticity, with low Young's modulus and high yield strain.

electrical double layer: *See* Debye layer.

electrocorticography: Electrophysiological technique for recording directly from the surface of the brain cortex with metal electrodes (after the cranium has been opened and the dura layer has been lifted).

electroencephalogram: EEG recording.

electroencephalography: Technique for recording electrical brain activity from the scalp surface.

electrokinetics: Field of study of the interaction of fluids (including fluids containing charged particles) with electric fields. Electrokinetics, also known as electrohydrodynamics, covers the following fluid transport mechanisms: electrophoresis, dielectrophoresis, electro-osmosis, and electrorotation.

electrokinetic valve: Microfluidic valve that opens or closes by the action of a change in flow driven by a voltage switch.

electron beam lithography (EBL, e-beam lithography): Nanoscale patterning technique that consists in illuminating an electron-beam resist with the electron beam of a scanning electron microscope and in subsequently developing the exposed areas away with a developer solution.

electro-osmosis, electro-osmotic flow: Motion of fluid near the walls of a conduit when an electric field is applied to the fluid.

electro-osmotic pump: Pump that drives fluid by producing electro-osmotic flow.

electrophoresis: The movement of charged molecules in a liquid under the influence of an applied electric field.

electrospray mass spectrometry: Mass spectrometry technique based on spraying the sample in liquid form into an aerosol, which forms charged ions as the aerosol droplets dry out during their flight in vacuum; the charged ions are accelerated by an electric field and sorted spatially according to their mass-to-charge ratio.

electrowetting: The modulation of the wettability of a surface (i.e., the contact angle) by means of an electric field.

ELISA: Abbreviation that stands for Enzyme-Linked Immunosorbent Assay, a method for measuring the presence of a given biomolecule (the antigen, usually a protein) in a sample; first, the sample is adsorbed onto a solid support and an antibody to the antigen is added to recognize the presence of the antigen; finally, a secondary antibody (which recognizes the first antibody) that is linked to an enzyme is added to initiate a reporting reaction (usually colorimetric, which provides a quantitative measure of the presence of the antigen).

elution: In chemistry, process of extracting one material from another by washing with a solvent.

embryoid bodies: Cellular aggregates derived from embryonic stem cells; with embryoid bodies, researchers attempt to recreate the series of differentiation events that occur in the embryo, although differentiation occurs in a much more disorganized manner in embryoid bodies than in the embryo.

EOF: *See* electro-osmotic flow.

***Escherichia coli*:** A rod-shaped bacterium that is commonly found in the lower intestine of warm-blooded organisms; it is the most widely studied prokaryotic model organism for microbiology and genetics research and it is universally used in biotechnology to produce recombinant proteins.

evaporation pump: Pump driven by the relative change in size of two droplets as they evaporate at different rates.

explant culture: Culture of a piece of nondissociated tissue, usually in the form of a slice or a natural clusters/layer of cells that can be easily excised (e.g., neuronal ganglia, a piece of skin, or a retina).

extracellular matrix: A mesh of fibrous proteins and glycosaminoglycans that forms the extracellular part of animal tissue, providing structural support to animal cells and a number of other signaling functions. It plays a central role in tissue elasticity, wound healing, tumor development, blood clotting, and inflammation, among others.

F

fabric microfluidics: A subspecialty of microfluidics that designs, builds, and studies the flow properties of microdevices made with fabric.

FET: *See* field-effect transistor.

fibronectin: A ~440 kDa glycoprotein that is a major component of the extracellular matrix; during cell attachment, transmembrane integrin receptors recognize short, specific peptide sequences present in fibronectin.

field-effect transistor: A device that uses electrical field applied to a terminal (called the "gate") to control the flow of charge (usually electrons) between two other terminals (called the "source" and the "drain").

filopodia: The Latin word used to designate the small "feet-like" protrusions formed by the membrane of cells during motility. Plural of "filopodium."

FISH: Abbreviation for Fluorescent In Situ Hybridization. FISH is a DNA-labeling technique that detects and localizes the presence/absence of DNA sequences that have a high degree of similarity with that of the fluorescent probe added to the sample.

flow cytometer: Device for counting cells (or other particles) based on suspending the particles in a fluid stream and "shooting" the particles past a detector at high speed.

fluoroalkylsilane: An organosilane molecule (a molecule containing an alkyl tail and a silane group) in which some or all of the hydrogens in the alkane chain have been replaced by fluorine atoms, for example, CF_3-$(CF_2)_5$-Si-$(OCH_3)_3$.

fMLP: Abbreviation that stands for *N*-formyl-methionine-leucine-phenylalanine, a peptide chain produced by some bacteria and a neutrophil chemoattractant; it is commercially available.

focal adhesions: Dynamic protein complexes (containing more than 100 different proteins that are constantly assembled and disassembled in motile cells) through which the cell cytoskeleton mechanically connects to the extracellular matrix.

free-flow electrophoresis: Mode of capillary electrophoresis in which the separation is performed under continuous flow and applying the electrical field orthogonally to the flow.

G

GABA: Abbreviation that stands for γ-aminobutyric acid, the main inhibitory neurotransmitter of the mammalian nervous system. Other GABAergic mechanisms exist in various peripheral tissues (e.g., intestine, stomach, pancreas, uterus, ovary, testis, kidney, lung, and liver). GABA has also been shown to regulate the growth of embryonic and neural stem cells.

gas permeation micropump: Microfluidic pump that makes use of the high permeability of PDMS to gases in order to pump fluids.

GFP: *See* green fluorescent protein.

giga-seal: Abbreviation for "gigaohm seal," the seal between the cell membrane and the recording pipette that is necessary for obtaining a good patch clamp electrophysiology measurement; *see* patch clamp technique.

glial cells: Nonneuronal cells of the central and peripheral nervous system that maintain homeostasis, form the myelin sheath around the neuronal axons, modulate neurotransmission, and support and protect the neurons.

glutaraldehyde: A common cross-linker with two reactive aldehyde (amine-reactive) functional groups; it is highly used in protein biochemistry for nonspecific cross-linking of proteins (amine groups are very abundant in proteins).

gradient generators: Microfluidic devices that allow for generating gradients of soluble compounds, usually for presenting the gradients to cells or for transferring the gradients onto a surface.

gravity pumps: Pumps that are powered by the weight of the column of fluid in an inlet reservoir; the name is given from the fact that gravity is the only force that powers the flow.

grayscale photolithography: Set of photolithography techniques that allow for patterning features containing multiple heights or even graded heights in one illumination step.

green fluorescent protein: A naturally occurring jellyfish protein (in *Aequorea victoria*) that exhibits bright green fluorescence (peak at 509 nm wavelength) when exposed to blue light; it is frequently used as a marker in molecular and cell biology.

H

Hagen–Poiseuille equation (or Hagen–Poiseuille law): Mathematical expression that gives the pressure drop in a fluid (assumed incompressible and flowing without turbulence) flowing in a cylindrical pipe (or microchannel) of constant cross-section; in particular, it states that the pressure required to push a fluid to a certain flow rate in such a pipe is inversely proportional to the fourth power of its radius and proportional to its length, to the dynamic viscosity of the fluid, and to the flow rate.

haptotaxis: Cell motion directed by a gradient of substrate-bound molecules that act as chemoattractants; the term was coined by S. B. Carter in 1965, after he observed the behavior on mouse fibroblasts migrating over palladium gradients created by shadow-evaporation.

HEG: *See* hexa(ethylene glycol).

HEK-293 cells: Abbreviation that denotes the name of a human cell line, a Human Embryonic Kidney Cell Line; sometimes referred to simply as "HEK cells." The cell line was established by Frank Graham from an aborted fetus in his experiment number 293 in Alex Van der Eb's laboratory in Leiden (Holland) in the late 1970s. Being very easy to culture and transfect, HEK-293 cells are particularly useful as protein expression systems.

hESC: Abbreviation that stands for Human Embryonic Stem Cell.

heterobifunctional cross-linker: Cross-linker with two different reactive functional groups; *see* cross-linker.

heterogeneous-phase immunoassay: Immunoassay that relies on binding the antibodies (or the antigen) on a surface; a well-known example is the assay named ELISA. *See* immunoassay, surface-bound immunoassay, ELISA.

hexa(ethylene glycol): An extremely short version of the polymer poly(ethylene glycol) (PEG), a well-known cell adhesion repellent, with similar cell-adhesion properties when linked to alkanethiol self-assembled monolayers; *see* PEG, alkanethiol, self-assembled monolayer.

H-filter: A microfluidic filter that consists of two inlets and two outlets, connected by a straight microchannel in which solutes from one of the two inlets are diluted and collected into the opposing outlet by virtue of laminar stream separation.

high-throughput: Expression that denotes high parallelization in the production of results; in biotechnology and biology, "high-throughput screening" refers to the identification of a drug or pathway after conducting a large number of tests.

HL-60 cells: Abbreviation that stands for human promyelocytic leukemia cell line; in BioMEMS, HL-60 cells have been used extensively as a precursor cell type for obtaining neutrophils (otherwise, neutrophils have to be obtained for each experiment from a human donor).

homobifunctional cross-linker: Cross-linker with the same reactive functional group at both ends; *see* cross-linker.

homogeneizer: A class of microfluidic device that mixes two or more input solutions until the concentration at the outlet is spatially homogeneous.

homogeneous-phase immunoassay: Immunoassay in which the antibodies that recognize the antigen are free-floating in solution (instead of attached to a solid support), that is, the assay is carried out entirely in solution; *see* immunoassay.

hot embossing: Micromolding technique that consists of pressing the mold against a thermoplastic polymer substrate heated above the glass transition temperature of the polymer, resulting in the imprinting of the mold features into the polymer surface; for this reason, hot embossing is also known as "nanoimprint lithography." The first academic report was published in *Science* in 1996 by Stephen Chou's laboratory from Princeton University.

HUVEC: Abbreviation that stands for Human Umbilical Vein Endothelial Cell.

hydration forces: Forces associated with the formation of complexes of ions and water (which may be itself highly dissociated in the form of protons H^+, hydroxyl ions OH^-, or hydronium ions H_3O^+); hydration forces are common around biomolecules.

hydraulic diameter: In any particular fluid mechanics formula that deals with noncircular ducts, the equivalent diameter that allows for using the (simpler) circular duct expression. It has the value of $D_H = 4A/P$, where A is the cross-sectional area of the duct and P is the wetted perimeter of the cross-section. For example, a fully wetted rectangular microchannel of width W and height H has the same fluid resistance as a circular microchannel of diameter $D_H = 4\,HW/2(H + W) = HW/2(H + W)$.

hydrodynamic trap: The confinement of particles in a fluid using forces generated by means of fluid flow.

hydrogel: A network of highly water-absorbent polymer chains; a hydrogel such as collagen gel or Matrigel may contain up to 99.9% water.

hydrophilic surface: A surface that is attracted to water, such that a droplet of water deposited on that surface spreads into a thin film.

hydrophilicity: The physical property of a molecule (or, in general, a body, such as a surface) that is attracted to a mass of water.

hydrophobic force: In bodies that have an uneven distribution of polar groups and nonpolar groups, the net repulsion that the nonpolar groups exert in contact with water in contrast with the strong (attractive) associations between the polar groups and water.

hydrophobic surface: A surface that is repelled by water, such that a droplet of water deposited on that surface beads up into a round sphere.

hydrophobicity: The physical property of a molecule (or, in general, a body, such as a surface) that is repelled by a mass of water.

I

IL-8: *See* interleukin-8.

immunoassay: A biochemical test that measures the concentration of a biomolecule by the binding of an antibody to its corresponding antigen (which is not necessarily the biomolecule of interest directly).

immunoblotting: *See* Western blot.

immunological synapse: Interface between an antigen-presenting cell and a lymphocyte.

in vitro: Latin expression ("within glass") that refers to studies conducted outside of the organism on a petri dish, a glass slide, etc.

in vivo: Latin expression ("within the living") that refers to studies conducted inside a living organism.

injection molding: Molding technique for high-speed production of plastic parts that consists of injecting a molten (thermoset or thermoplastic) polymer at high pressure into a hard mold followed by a rapid cooldown.

integrins: A family of heterodimeric transmembrane proteins that on the cytoplasmic side of the membrane are linked to the cytoskeleton and on the extracellular side of the membrane recognize specific peptide sequences (such as RGD) present in the extracellular matrix; *see* RGD, ECM.

interleukin-8: A chemokine (of the CXC subfamily) produced by macrophages (and other cell types, e.g., epithelial cells).

isoelectric focusing: A technique for separating proteins with different isoelectric points.

isotropic etch: Chemical etch that proceeds in all directions at the same rate.

K

keratinocyte: Predominant cell type of the epidermis (outer layer of the human skin).

kinematic viscosity: The ratio of the absolute viscosity (also called "dynamic viscosity") to the density of the fluid; *see* dynamic viscosity.

L

L1: A neuronal cell adhesion transmembrane protein that is involved in axon guidance and cell migration; it is also known as L1CAM.

Lab-CD: A spinning microfluidic platform invented by Gamera Bioscience in 1998 that makes use of the centrifugal force to modulate capillary forces, which act to fill microchannels and to open/close burst microvalves.

laminar flow: Flow characterized by the absence of turbulence; in general, the fluid flows in parallel streamlines, without forming eddies.

laminins: A family of proteins of the basal lamina, the layer of extracellular layer secreted by epithelial cells.

laser cutting: In microfabrication, the melting or ablation (or both) of material by means of a focused laser beam.

laser deposition: In microfabrication, the deposition of material by means of a focused laser beam.

L-aspartate: The salt of L-aspartic acid, one of the 20 basic amino acids, the building blocks of proteins. It was first synthesized in 1827 from asparagine (another amino acid), which in turn was isolated from asparagus juice in 1806.

lateral patch-clamp chip: A patch clamp chip design in which the aperture that produces the gigaohm seal with the cell is produced as the opening of a small side channel from the cell-containing chamber (in most patch-clamp chip designs, the aperture is on the floor of the chamber).

lift-off: In micromachining, a process for patterning a given material that consists of using a sacrificial layer which is removed ("lifted off") at the end of the process.

M

maleimide: In organic chemistry, a functional group that specifically reacts with thiol groups; hence, it is the basis of thiol-reactive cross-linkers (e.g., for linking molecules to proteins at the thiol group found on cysteines).

Manduca sexta: A large moth (also known as tobacco hornworm) commonly used in neurobiology for its easily accessible nervous system and short life cycle.

Matrigel: Trade name for a basement membrane extract mostly composed of laminin and collagen. Like collagen gel, it is liquid at 4°C and gels when warmed at 37°C.

mass spectrometry: Any analytical technique that measures the mass-to-charge ratio of charged particles; a mass spectrometer usually consists of an ion source (which converts the sample into ions), a mass analyzer (which sorts the ions by their masses by applying electromagnetic fields), and a detector (which counts the abundance of each ion).

MEAs: *See* microelectrode array.

MEMS: Abbreviation that stands for MicroElectroMechanical Systems.

mESC: Abbreviation that stands for Mouse Embryonic Stem Cell.

metastasis: The spread of a disease, most often used in the context of cancer, from one organ to another.

Michigan probes: Implantable multielectrode arrays micropatterned on thin silicon shanks. They were developed by the University of Michigan for neural (intracortical) recording applications in the 1980s.

micro-bioreactor: Microfluidic device capable of performing the functions of a traditional incubator while allowing optical access to the cells and biochemical sampling of the extracellular milieu.

microchannel resistance: Ratio between the applied pressure (P) and the volumetric flow rate (Q), $R = P/Q$. Note the similarity between this expression and Ohm's law, $R = V/I$, when one thinks of pressure as a "voltage" and flow rate as a "current."

microcontact printing: *See* microstamping.

microelectrode: General word to describe an electrode that has been fabricated with some miniaturization technology. Electrophysiologists use the term to refer to thin metal wires whose tips have been etch-sharpened. Engineers generally use the term to mean "micropatterned metal electrode," a thin flat element that is fabricated by photolithography and lift-off.

microelectrode array: Array of micropatterned electrodes; in BioMEMS, microelectrode arrays (MEAs) have been mostly used for electrophysiological recordings of cultured cells. The electrodes are usually made of metal or indium-tin-oxide (a transparent semiconductor), on a glass surface, to facilitate microscopic inspection of the cells.

microfluidic immunoassay: Immunoassay implemented in a microfluidic format; *see* immunoassay.

microfluidic patterning: Class of micropatterning techniques that rely on the use of microfluidic devices to produce patterns.

microfluidic photomask: A photomask whose features are fluids rather than solid patterns.

microfluidic resistor: A microfluidic component with a flow resistance that can be varied by the user.

microfluidics: Field that studies and exploits the behavior of fluids confined to small volumes, such as microchannels, droplets, jets, and thin water films.

microjets: Contraction of "microfluidic jets," a gradient-generation device that consists of an open reservoir that is fed by two sets of high-resistance microchannels (the "microjets") on both sides of the reservoir at different concentrations, which quickly creates a steady-state gradient. The device was invented by the author's laboratory in 2006.

micromachining: A set of etching, deposition, and patterning techniques collectively used to produce three-dimensional (or planar) microstructures.

micromixer: Microfluidic device that accepts two or more inputs and that fully or partially homogenizes the mixture before it exits through the outlet.

micromolding: A family of techniques that allows for molding microstructures.

micromolding in capillaries: Microfluidic technique that consists of the spontaneous wetting of small PDMS microchannels by an organic prepolymer solution by capillary action and the final curing of the polymer in the microchannel once the filling is complete. The technique was invented by George Whitesides' laboratory in 1995.

microneedle: A microfabricated needle; the term applies to both hollow and solid microneedles, which are built on planar wafers and are usually no longer than a few mm.

micropump: A microfabricated/microfluidic pump.

microscaffold: In tissue engineering, a temporary structure made of a biocompatible or biodegradable material such as a polymer, natural protein, or a hydrogel, that supports the growth of tissue and vasculature and is eventually biodegraded by natural processes such as hydrolysis or cell-secreted enzymes.

microstamping: Technique that consists of contact-transferring a material of interest from a microfabricated PDMS stamp onto a surface only on the areas contacted by the stamp. Initially named "microcontact printing," it was the first soft lithographic technique invented by George Whitesides' laboratory in 1993. *See* soft lithography, PDMS.

microstreaming: Acoustic streaming effect that occurs near a gas boundary or when a boundary is vibrating in a still medium, most notably around an ultrasound-vibrating air bubble. *See* acoustic streaming.

microvalve: A microfabricated valve.

MIMIC: *See* micromolding in capillaries.

morphogen: A chemical substance that forms a concentration gradient from its cellular source to trigger and govern a particular pattern of tissue development.

multi-photon lithography: *See* direct laser writing.

multiplexer: In microfluidics, as in microelectronics, a multiplexer is a device that selects one of several inputs and forwards the selected input into a single line.

N

nanoimprint lithography nanoimprinting: *See* hot embossing.

Navier–Stokes equation: Nonlinear partial differential equation that describes the motion of fluids; it results from applying Newton's second law ($F = d(mv)/dt$) to fluid motion. A solution of the Navier-Stokes equation is a description of the velocity of the fluid at a given point in space and time and is called a velocity flow field.

NCAM: Abbreviation that stands for Neural Cell Adhesion Molecule, a glycoprotein involved in cell–cell adhesion and neurite outgrowth, among other roles. It is expressed on the surface of neurons, glia, skeletal muscle, and natural killer cells.

negative dielectrophoresis: Dielectrophoresis scenario in which the solvent is of higher polarizability than the particle, resulting in the movement of the particle toward the electrical field minima; *see* dielectrophoresis.

negative photoresist: Photoresist that, if exposed to the correct light, becomes insoluble in the developer solution; *see* photoresist, developer.

neuroblastoma cells: Cells obtained from a neuroblastoma, a common extracranial childhood cancer that can grow in the abdomen, the chest, the spinal cord, bones, or bone marrow.

neurotrophins: Family of proteins that induce the development, survival, and function of neurons.

Newtonian fluid: A fluid whose stress versus strain curve is linear and passes through the origin (the proportionality constant is the viscosity); all Newtonian fluids are incompressible, uniform, and viscous, like water. Air, slurries, foams, and blood are examples of non-Newtonian fluids.

***N*-formyl-methionine-leucine-phenylalanine:** *See* fMLP.

NHS ester: *See* succinimidyl ester.

no-slip condition: Assumption of zero-velocity (no flow) next to a solid body (e.g., at the walls of a microchannel, or the surface of a ball) that is commonly used as a boundary condition to solve the Navier-Stokes equation.

O

oligonucleotide chip: A chip or a device that incorporates an array of microscopic spots of surface-immobilized short nucleic acid sequences (the oligonucleotides). Oligonucleotide chips are popularly known also as "DNA chips" (a poorly chosen name, because the nucleic acid sequences are not full DNA sequences).

oligopeptide: A molecule that consists of between 2 and 20 amino acids.

open microfluidics: A class of microfluidic devices, usually cell culture devices, in which the main cell-containing chamber or microchannel has no roof; the main advantages of using these microfluidic devices over closed-architecture ones are that traditional cell culture protocols (optimized for petri dishes in incubators) can be used without modification, and that the cells and reagents can be dispensed/exchanged with user-friendly procedures such as pipetting. The first open microfluidic system was invented by the author's laboratory in 2004 using PDMS exclusion molding.

Ormocer™: Abbreviation that stands for organically modified ceramics, a dental implant polymer that is used as a negative photoresist in BioMEMS.

P

P0, P1, P2, etc.: Abbreviation that refers to "postnatal day 0," "postnatal day 1," etc.

paper electrophoresis: The technique of performing electrophoretic separations in a paper sheet carrier, rather than in a gel or in a free-flowing solution.

paper microfluidics: The study and manipulation of fluids flowing in paper.

parabolic flow profile: Flow profile characterized by a parabolic distribution of flow velocities across the cross-section of the pipe, typically achieved under pressure-driven flows in pipes of circular or rectangular cross-sections.

particle imaging velocimetry: Imaging technique for visualizing the 3-D flow velocity distribution in microchannels based on tracing large numbers of small particles within the flow.

particle lithography: Technique that uses the spontaneous assembly of monolayers of beads as the first step to generate nanofabricated patterns.

patch clamp technique: Electrophysiology technique based on clamping and sealing a glass pipette against the cell membrane that allows for recording from single or multiple ion channels.

patch-clamp chip: Microdevice that contains a microfabricated aperture, which acts as a patch clamp pipette; a live cell is brought into proximity with the aperture for the establishment of a high-quality seal before electrophysiological recording.

PCR chip: Microfluidic chip designed to perform a polymerase chain reaction (PCR), which generates millions of copies of a given DNA sequence from a single piece or a few copies of DNA.

PDL: *See* poly-D-lysine.

PDMS: *See* poly(dimethyl siloxane).

PDMS peristaltic micropump: A peristaltic pump in which the fluid-carrying microchannel and the microvalves used to pump the fluid are built in PDMS; *see* peristaltic pump, PDMS.

PEG: *See* poly(ethylene glycol).

PEG thiol: In organic chemistry, PEG is the polymer poly(ethylene glycol), and thiol is the group –SH. Although thiol-functionalized PEG can be found, the expression "PEG thiol" is most commonly found in the BioMEMS literature referring to the short (six-carbon-long) PEG chains linked to an alkanethiol self-assembled monolayer developed by the Whitesides laboratory to repel cell adhesion; *see* alkanethiol, self-assembled monolayer.

PEG-DA: *See* poly(ethylene glycol) diacrylate.

PEG-IPN: *See* PEG-interpenetrating polymer networks.

PEO: *See* poly(ethylene-oxide).

peristaltic pump: Pump based on the successive "pinching" of a tube or microchannel at three or more locations (imitating peristalsis—the radially symmetric, wave-like propagation of muscle contraction in the digestive tract).

PGS: *See* poly(glycerol sebacate).

PGSA: *See* poly(glycerol sebacate acrylate).

photoinitiator: A chemical compound that decomposes into free radicals when exposed to light; photoinitiators are commonly used for starting polymerization reactions in combination with a light trigger.

photolithography: Light-projection process that forms the basis for modern microfabrication and is based on the patterning of a layer of photoresist by means of light; the exposure chemically alters the photoresist by modifying its solubility in a certain developer solution; *see* photoresist.

photomask: In photolithography, the template containing the pattern (in the form of opaque features on a transparent support) which is projected (usually with UV light) onto the unexposed photoresist.

photoreactive cross-linker: A cross-linker that becomes reactive upon exposure to light (of a particular wavelength); *see* cross-linker.

photoresist: Photosensitive polymer used as the basis for photolithography; photoresist is usually dissolved in a solvent and spun as a thin layer on the surface of a wafer before exposure to UV light (through a photomask); next, the photoresist is exposed to a developer solution, which leaves a photoresist pattern on the wafer; *see* photolithography, developer.

physisorption: Adsorption of a chemical species onto a surface by physical forces (i.e., not mediating a chemical bond); *see* adsorption.

piezoelectric micropump: Microfabricated pump that incorporates as its active element a piezoelectric microactuator.

pinched-flow fractionation: Technique for separating and sorting particles according to their size based on forcing the particles against a wall (using a secondary "pinching" flow); as they flow close to the wall, the bigger particles move faster and the smaller particles move slower because the parabolic flow profile dies off at the wall, so as they exit the microchannel into a large chamber, the particles come out sorted by size. The technique was invented by Minoru Seki and colleagues from Osaka Prefecture University in 2004.

pinch-valves: Valves that use mechanical force to reduce the diameter of a section of the tubing through which the fluid is flowing. In 2000, Stephen Quake and coworkers devised a way to "pinch" a PDMS microchannel using a second air-filled microchannel (called "control channel") on top of the fluid-carrying one and separated by a thin (~30 µm) PDMS gap; these microfabricated PDMS pinch-valves are popularly known as "Quake valves"; see Quabe microvalve.

PIV: *See* particle imaging velocimetry.

PLGA: *See* poly(DL-lactic-co-glycolide).

PLL: *See* poly-L-lysine.

plug flow profile: Flow profile characterized by a constant flow speed across the whole cross-section of the pipe, typically achieved in electro-osmotic flow; *see* electro-osmotic flow.

poly(dimethyl siloxane): Transparent elastomer that is widely used in BioMEMS and soft lithography for the molding of microdevices.

poly(DL-lactic-co-glycolide): Biodegradable thermoplastic polymer that is approved by the FDA for therapeutics.

poly(ethylene glycol): A biocompatible polymer used as an active ingredient or as an additive by the food and pharmaceutical industries (e.g., in laxatives, skin creams, eye drops, and ointments); in BioMEMS and tissue engineering, PEG is used to block protein adsorption on surfaces.

poly(ethylene glycol) diacrylate: A photopolymeryzable version of PEG.

poly(ethylene oxide): Synonymous of poly(ethylene glycol). *See* PEG.

poly(glycerol sebacate acrylate): Transparent biodegradable elastomer.

poly(glycerol sebacate): Transparent biodegradable elastomer.

poly-D-lysine: A common tissue culture coating for enhancing cell and protein adhesion. Poly-D-lysine and poly-L-lysine are enantiomers (they only differ in the sterical assembly of the atoms in the molecule); poly-D-lysine is convenient because, unlike poly-L-lysine, it cannot be degraded by cellular proteases.

poly-L-lysine: A small natural homopolymer of the amino acid L-lysine that is produced by bacterial fermentation and that is used as a common tissue culture coating for enhancing cell and protein adhesion.

poly(N-isopropylacrylamide): A hydrogel polymer that undergoes a reversible phase transition (becoming insoluble) when heated to more than 32°C.

poly-NIPAM: See poly(N-isopropylacrylamide).

positive dielectrophoresis: Dielectrophoresis scenario in which the solvent is of lower polarizability than the particle, resulting in the movement of the particle toward the electrical field maxima; *see* dielectrophoresis.

positive photoresist: Photoresist that, if exposed to the correct light, dissolves in the developer solution; *see* photoresist, developer.

protein crystallization chips: Microfluidic chips used to produce large numbers of microscale protein crystals in a parallel array of various conditions; the chips, thus help optimize which condition produces the best crystal and are an improvement in screening speed of several orders of magnitude over the serial trial-and-error procedure that has been used traditionally.

protein immunoblot: *See* Western blot.

Q

Quake microvalve: First design of microvalve fabricated integrally in PDMS; the microvalve was developed by Stephen Quake's laboratory, then at Caltech, in 2000; see pinch-valves.

quorum sensing: A system of communication signals with which certain species such as bacteria and some social insects are able to adapt to their environment based on their population density by controlling gene expression, where to nest, etc.

R

Re: *See* Reynolds number.

reactive ion etch: A chemical etch that is conducted in the presence of a highly ionized plasma of a chemically reactive gas (as opposed to, for example, a plasma of a noble gas such as argon, which can also etch a surface through the momentum with which accelerated argon ions impinge on the surface but attack all surface species nonselectively); for example, in MEMS processing, SF_6 is commonly used to produce a plasma of F^- ions, which etches Si but does not attack photoresist.

resonance frequency: Frequency at which a mechanical or electrical system that is being driven to oscillation reaches its maximum oscillation amplitude.

rete ridges: Thickenings in the human epidermis that extend downward between dermal papillae, the interdigitations of the dermis into the epidermis.

Reynolds number: Dimensionless number that gives a measure of the relative importance of inertial versus viscous forces in fluid flows.

RGD: Sequence of three amino acids (R, arginine; G, glycine; D, aspartic acid) that is present in many extracellular matrix proteins; when integrin receptors on the membranes of adherent cells recognize the RGD sequence, the cells attach and spread on the surface that is covered with RGD. This fundamental mechanism was discovered by Erkki Ruoslahti and Michael D. Pierschbacher in 1986. Other similar sequences responsible for cell attachment and survival have been discovered since then.

RIE: *See* reactive ion etch.

S

S. cerevisiae: *See Saccharomyces cerevisiae*.

Saccharomyces cerevisiae: A species of yeast, one of the most widely studied eukaryotic model organisms in molecular and cell biology.

SAEC: Abbreviation that stands for Small Airway Epithelial Cell, a human cell type.

SAM: *See* Self-Assembled Monolayer.

SAW: *See* Surface Acoustic Wave.

scanning probe lithography: In nanotechnology, the use of scanning probe microscopes for fabricating nanostructures and microstructures on a surface.

Scanning probe microscope: A class of non-light-based microscope that uses a sharp tip in proximity with a surface of interest to obtain a topographic image of the surface.

Schwann cells: Main glial cells of the peripheral nervous system; they are involved in nerve impulse conduction along axons, nerve development and regeneration, and ECM secretion, among other key processes.

self-assembled monolayer: Organized layer or film of single-molecule thickness in which one end of the molecule has a strong affinity for the substrate; the "tails" of the molecules, usually alkyl chains, attract each other via van der Waals interactions to form a tightly packed monolayer so that, after completion of the monolayer, the substrate is no longer accessible—the substrate has been effectively "replaced" by the functional group of the self-assembled monolayer.

self-assembly: Spontaneous creation of an ordered multi-component structure from a set of smaller (nanoscale, microscale, or mesoscale) components that does not require the formation of covalent bonds (usually involving electrostatic, capillary, van der Waals forces, or hydrogen bonds).

shadow evaporation: In microfabrication, a technique to produce microstructures based on casting the shadow of a previously patterned microstructure (a line, a post, etc.); as a material is evaporated from a source (usually a metal), it deposits over the patterned microstructure(s) but not behind it, thus creating a shadow which may have a shape or a thickness that is not accessible with traditional photolithography.

shear stress: A measure of the transverse force (per unit area perpendicular to the force) acting on a deformable body. For a Newtonian fluid, the shear stress is proportional to the velocity gradient and the proportionality constant is the dynamic viscosity of the fluid.

single-stroke PDMS peristaltic micropumps: Micropumps fabricated in PDMS that produce peristaltic pumping (i.e., reproduce the undulatory movements of the digestive tract or "peristalsis") with pulses from a single control line; traditional peristaltic micropumps achieve peristalsis with three or more microvalves actuated in sequence (requiring three control lines).

skin graft: A skin transplant; "grafting" is the name of the surgical procedure that involves transplanting tissue from one site of the body to another, or from another person, without transplanting the blood vasculature (the graft becomes vascularized after transplantation).

SlipChip: A microfluidic multichamber strategy for moving and mixing fluids that, paradoxically, does not require the use of microvalves; invented by Rustem Ismagilov's laboratory at the University of Chicago, it is based on sliding two silanized glass plates with a lubricating layer of fluorocarbon that facilitates the relative motion of the two plates; etched wells on either plate can be selectively filled with a special pipettor and mix with each other when they are overlapped.

smart polymer: A polymer material that can undergo a reversible phase transition (such as a change in volume or solubility) on application of a stimulus (such as a change in pH, temperature, or light); as the state is reversed on removal of the stimulus, these polymers can be used as a sensor for that particular stimulus.

soft lithography: Family of microfabrication techniques that have in common the use of a micromolded piece of the elastomer PDMS; *see* PDMS.

solid-binding peptides: A class of synthetic polypeptides that selectively bind to solid supports such as particular metals, semiconductors, and plastics.

SPM: *See* scanning probe microscope.

stencil patterning: Method for depositing materials on surfaces that consists of temporarily blocking the surface with a removable stencil; the material may be shadow-evaporated, or if the stencil seals well against the surface (i.e., a PDMS stencil), the material may be deposited from the liquid phase.

Stokes–Einstein relation: Equation derived by Einstein in 1905 that expresses the diffusion coefficient D of spherical particles diffusing through liquid with low Reynolds number (Brownian motion). It expresses that D is proportional to temperature and inversely proportional to the viscosity of the fluid and to the radius of the particle.

stopped-flow lithography: Modality of microfluidic patterning developed by Patrick Doyle's laboratory (MIT) in which a photocurable polymer is introduced in the microchannel and a pattern of light is projected into the microchannel to locally cure polymer microstructures inside the microchannel; during exposure, the flow of polymer is stopped temporarily, hence the name of the technique.

stripe assay: Biological assay for axon guidance developed by Friedrich Bonhoeffer's laboratory in 1987 that utilized a PDMS microfluidic device to hydrodynamically immobilize stripes of cell membrane fragments onto a porous filter; retinal temporal axons were shown to be guided by the stripes whereas nasal axons did not obey the pattern. Bonhoeffer's device is the first micromolded PDMS device in history.

SU-8: A negative photoresist (chemically speaking, a photosensitive epoxy which, when exposed, becomes a hard polyester), developed by IBM in the late 1990s, that can be patterned with high aspect ratios and develops with high selectivity, hence it has become the preferred photoresist of choice for microfluidics applications.

succinimidyl ester: Amine-reactive group commonly used in heterobifunctional cross-linkers; usually seen abbreviated as "NHS ester" in the literature.

surface acoustic wave: In MEMS, surface acoustic wave (SAW) devices are devices that contain interdigitated transducers that convert electrical signals into acoustic waves by means of piezoelectric materials.

surface-bound immunoassay: *See* heterogeneous-phase immunoassay.

surface energy: The excess energy at the surface of a material compared with the bulk; equivalently, a measure of the loss in total molecular cohesive energy (from the breaking of bonds) that occurs when a surface is created. It is measured in Joules per square meter.

surface engineering: Subdiscipline of materials science that involves manipulating and designing the properties and the structure of the surface of materials to obtain better materials.

surface micromachining: Micromachining of a wafer by additive or subtractive processes that affect only its surface layer; *see* micromachining.

surface tension: A property of the surface of a liquid (i.e., a liquid–liquid, a liquid–gas, or a liquid–solid interface) that allows it to resist external forces on a millimeter or submillimeter scale. It is measured in units of force per unit length (e.g., N/m, dyne/cm). The generalization of this concept to both liquids and solids and all length scales is termed surface energy (*see* surface energy). Surface tension governs the formation of bubbles, droplets and puddles and explains why certain water bugs are able to stride on water. Surface tension has been exploited in microfluidics to build micropumps and microvalves.

surface tension-driven micropump: Micropump design that takes advantage of surface tension effects to propel the fluid forward.

syringe pump: A type of laboratory and clinical-use pump that consists of a syringe the plunger of which is driven with an electrical (usually programmable) motor.

T

T- or Y-mixer: A two-inlet, one-outlet microfluidic mixer of planar architecture. The inlets are usually designed to converge in a symmetrical "T" or "Y" configuration, hence its name. It is arguably the simplest microfluidic design for mixing fluids on a microscale that can be conceived; it relies solely on diffusion across the flat interface between the two fluids to achieve mixing; therefore, it is also a highly inefficient design.

Taylor dispersion: Smearing of the concentration distribution (of a given solute) in a flow pipe or duct by effect of a nonuniform velocity flow profile. It was first described by Sir G. I. Taylor.

Tesla mixer: A type of microfluidic mixer that incorporates a Tesla valve to induce the formation of vortices, which enhance mixing; *see* Tesla valve.

Tesla valve: A valve that preferentially allows flow in one direction without moving parts. Tesla valves do not close in the reverse-flow configuration; they simply display a higher flow resistance.

thermoplastic polymer: A polymer that can be molded into shapes above its glass transition temperature.

thermoset polymer: A polymer that cures irreversibly into a hard matrix, either through heat, a chemical reaction, or irradiation.

tubeless microfluidics: A class of microfluidic devices that does not require tubing for the introduction of fluids into the inlets; in these devices, the inlets are an integral part of the device in the form of arrays of reservoirs that facilitate dispensing of fluids in parallel (e.g., with a multipipettor, if the array matches the spacing of a 96-well plate).

U

UEA: *See* Utah electrode array.

ultrasound-based micropump: Microfabricated pump that exploits the use of ultrasound to propel fluids.

Utah electrode array: A 10×10 array of microfabricated, individually addressable, high-aspect-ratio silicon needle electrodes (1.5 mm tall, 400 μm spacing) for intracortical implantation, developed at the University of Utah's Center for Neural Interfaces.

V

van der Waals force (between two bodies): The sum of all dipole–dipole interactions (attractive and repulsive) between the two given bodies under consideration, for example, the van der Waals force between two molecules or the van der Waals force between a molecule and a surface.

W

weir: In hydraulics, a type of dam consisting of a barrier that allows water to overflow over the barrier. In microfluidics, the weir is usually built inverted (i.e., hanging from a roof) so as to act as a particle trap.

Western blot: Technique used for the separation of specific proteins in a given extract or homogenate of tissue or cells. Separation is performed by gel electrophoresis according to the length of the polypeptide chain of the protein (if the protein has been denatured) or according to the 3-D structure of the protein (if the protein is in its native conformation). Once separated, the proteins are transferred to a PVDF or nitrocellulose membrane in which they are detected by antibodies to the protein of interest.

wet etch: In microfabrication processing, an etch that uses liquid chemistry (as opposed to an etch that uses gaseous chemistry; *see* dry etch).

X

Xenopus: Genus that includes 19 species of sub-Saharan aquatic frogs, of which *Xenopus laevis* is a highly used model organism in developmental biology research. *Xenopus* oocytes are convenient protein expression systems and are commonly used in ion channel research.

Y

YIGSR: Sequence of five amino acids (Y, tyrosine; I, isoleucine; G, glycine; S, serine; R, arginine) that is present in laminin; when integrin receptors on the membranes of adherent cells recognize the YIGSR sequence, the cells attach and spread on the surface that is covered with YIGSR.

Young equation: Equation derived by Thomas Young (1773–1829) that relates the surface tensions between the three phases surrounding a droplet on a surface (gas, liquid, and solid phase) and predicts the contact angle if the three surface tensions are known. Young is also known for the wave theory of light (including Young's double-slit experiment that demonstrates wave interference), Young's modulus of elasticity, the Young–Laplace equation that describes the capillary pressure difference across the interface between two static fluids, and other contributions to various other fields, including medicine, Egyptology, and music.

Young's modulus: Measure of the stiffness of an elastic material that is calculated by dividing the tensile stress (F/A_0) by the tensile strain ($\Delta L/L_0$) in the elastic portion of the stress–strain curve (i.e., when the curve is still linear, for small strain)

$$E \equiv \frac{F/A_o}{\Delta L/L_o}$$

where F is the force applied to the object, A_0 is the original cross-sectional area through which F is applied, ΔL is the amount by which the length of the object changes, and L_0 is the original length of the object.

Z

zeta potential: In colloidal science (which deals with suspensions of particles in fluids), it is the potential difference between the dispersion medium and the stationary layer of fluid attached to the dispersed particle. It is related to the quantification of Debye length in the electrical double layer (e.g., on microchannel walls in capillary electrophoresis on a chip) because the zeta potential coincides with the electric potential in the electrical double layer at the location of the "slipping plane" (a conventionally introduced plane that separates mobile fluid from fluid that remains attached to the surface). Hence, we can use as a definition of zeta potential the potential at the slipping plane.

Zigmond chamber: Passive device used for traditional chemotaxis assays, which consists of two chambers or reservoirs etched in glass and separated by a shallow ridge or weir. The chambers are first covered with a glass coverslip before loading any solutions. One of the chambers is loaded (usually by capillarity filling) with the cells and the other chamber is loaded with the chemoattractant of interest. The chemoattractant reaches the cells by diffusion across the weir.

Author Index

Author Index

Author Index

Author Index

Ionov, L., 9
Ireland, Grenham, 66, 304, 305, 434
Irimia, D., 49, 75, 153, 179, 266, 278, 327
Islam, Ahmad, 408
Ismagilov, R.F., 143, 160, 161, 177, 201, 235, 236, 239,
        268, 269, 304, 366, 369, 370, 371, 372, 440,
        441, 462
Ivester, Robin H.C., 155

J

Jacobson, S.C., 146, 248, 438
Jaeger, Nicolas A.F., 311
Jaenisch, Rudolf, 264
James, Conrad D., 341
Jamieson, Brian, 399
Jan, Lily, 265, 291
Javier, D.J., 16, 57
Jayaraman, Arul, 278, 331
Jean, John A., 113
Jean, Ronald P., 393
Jeng-Hao Pai, 268
Jenkner, M., 360
Jensen, Erik C., 155
Jeon, N.L., 324, 328, 343, 367, 369, 442
Jeong Won Park, 190
Jeong Woo Lee, 410
Jeong Yun Kim, 173
Jeong, O.C., 171
Jeonggi Seo, 291
Jernigen, Jeremy D., 64
Ji Hwan Lee, 366
Ji Yoon Kang, 159, 180
Ji Young Park, 153
Jian Wu, 408, 413
Jianbin Wang, 136
Jiang, Lili, 334
Jianhua Qin, 144
Jianjun Cheng, 389
Jianliang Xiao, 405, 407
Jianping Fu, 218
Jianzhong Xi, 134, 136
Jin-Woo Choi, 198
Jinbo Wu, 131
Jing-Tang Yang, 198, 200
Jinghua Tang, 390
Jinjie Shi, 122, 201
Jinpian Diao, 187
Jintae Kim, 148
Jizhou Song, 413
Jo, B.-H., 82, 441
Johnson, Matthew B., 64
Johnson, S.M., 361, 445
Johnston, I.D., 174
Jones, Michael D., 387
Jong Hwan Sung, 280
Jong Il Ju, 153
Jong Wook Hong, 230, 235
Jong-Myeon Park, 148
Jongyoon Han, 217, 218
Jönsson, H., 121, 443

Jordan, Peter, 305, 434
Jovic, A., 186
Jr-Lung Lin, 204
Ju Yeoul Baek, 153, 173
Juan Santiago, 140
Judy Yeh, 260, 388, 389
Julita Huf, 339, 340
Juluri, Bala Krishna, 122, 201
Jun Keun Chang, 195, 196
Juncker, D., 168, 185
Jung Kyung Kim, 196
Junha Park, 196
Junji Fukuda, 388, 389
Junyu Mai, 347
Justesen, Jeannette, 312

K

Kaazempur-Mofrad, M.R., 126, 381, 444
Kae Sato, 170
Kaehr, B., 16, 17, 57
Kahler, K.H., 53, 435
Kahp-Yang Suh, 383, 389
Kai-Yang Tung, 200
Kaigala, Govind V., 155
Kaji, H., 91, 260, 316
Kalinin, Yevgeniy V., 334
Kam, Lance C., 319
Kamei, Ken-ichiro, 390
Kamholz, A.E., 221
Kamm, Roger D., 191
Kandere-Grzybowska, Kristiana, 310
Kaneko, T., 435
Kang Xiao, 131
Kangsun Lee, 180
Kaplan, David L., 405
Kardon, Randy H., 376
Karger, B.L., 242, 437
Karlsson, J.O.M., 49, 75
Karlsson, Jens, 75
Karow, D.S., 362, 443
Karp, J.M., 260, 388, 389, 414
Kaushik, S., 440
Kaye, P.H., 254, 437
Kazuo Hosokawa, 150, 151, 170
Keay, Joel C., 64
Keenan, T.M., 183, 325, 326, 350, 445
Keh-Chih Hwang, 405
Kellogg, Gregory, 147
Kenis, P.J.A., 160, 177, 201, 304, 440, 441
Ker-Jer Huang, 200
Kern-Veits, B., 339, 434
Kessel, Richard G., 376
Keyue Shen, 319
Khademhosseini, A., 260, 379, 380, 388, 389
Khine, Michelle, 124
Ki Hwa Lee, 173
Ki Jun Yu, 408
Ki Wan Bong, 132
Ki-Bum Lee, 390
Kikuchi, Y., 435

Author Index

Author Index

Author Index

Author Index

Subject Index

Page numbers followed by *f* and *t* denote figures and tables.

Subject Index

Printed and bound by CPI Group (UK) Ltd, Croydon, CR0 4YY

01/11/2024

01782604-0011